Benchmark Papers in Acoustics

Series Editor: R. Bruce Lindsay
Brown University

PUBLISHED VOLUMES AND VOLUMES IN PREPARATION

UNDERWATER SOUND / *Vernon M. Albers*
ACOUSTICS: Historical and Philosophical Development / *R. Bruce Lindsay*
SPEECH SYNTHESIS / *James L. Flanagan and Lawrence R. Rabiner*
PHYSICAL ACOUSTICS / *R. Bruce Lindsay*
MUSICAL ACOUSTICS, PART I: Violin Family Components / *Carleen M. Hutchins*
MUSICAL ACOUSTICS, PART II: Violin Family Functions / *Carleen M. Hutchins*
ULTRASONIC BIOPHYSICS / *Floyd Dunn and William D. O'Brien, Jr.*
ARCHITECTURAL ACOUSTICS / *Thomas D. Northwood*
MUSICAL ACOUSTICS: Piano and Wind Instruments / *Earle L. Kent*
PSYCHOLOGICAL ACOUSTICS / *Arnold M. Small, Jr.*
ACOUSTICAL INSTRUMENTATION / *Benjamin B. Bauer*
SOUND RECORDING / *Benjamin B. Bauer*
LIGHT AND SOUND INTERACTION / *Osman K. Mawardi*
VIBRATION PROBLEMS / *Arturs Kalnins and Clive L. Dym*
NOISE AND NOISE CONTROL / *Malcolm J. Crocker*

Benchmark Papers
in Acoustics / 7

A BENCHMARK® Books Series

ULTRASONIC BIOPHYSICS

Edited by

FLOYD DUNN
University of Illinois at Urbana-Champaign

and WILLIAM D. O'BRIEN, JR.
Bureau of Radiological Health
Food and Drug Administration
Rockville, Maryland

STROUDSBURG, PENNSYLVANIA

Copyright © 1976 by **Dowden, Hutchinson & Ross, Inc.**
Benchmark Papers in Acoustics, Volume 7
Library of Congress Catalog Card Number: 75–30715
ISBN: 0-87933-206-9

All rights reserved. No part of this book covered by the copyrights hereon may be reproduced or transmitted in any form or by any means—graphic, electronic, or mechanical, including photocopying, recording, taping, or information storage and retrieval systems—without written permission of the publisher.

78 77 76 1 2 3 4 5
Manufactured in the United States of America.

LIBRARY OF CONGRESS CATALOGING IN PUBLICATION DATA
Main entry under title:

Ultrasonic biophysics.

(Benchmark papers in acoustics ; 7)
Includes bibliographical references and indexes.
1. Ultrasonic waves--Physiological effect--Addresses, essays, lectures. 2. Biological physics --Addresses, essays, lectures. I. Dunn, Floyd, 1924-   II. O'Brien, William D., Jr., 1942-
QP82.2.U37U47   574.1'9145   75-30715
ISBN: 0-87933-206-9

Exclusive Distributor: **Halsted Press**
A Division of John Wiley & Sons, Inc.
ISBN: 0 470–22691–9

ACKNOWLEDGMENTS AND PERMISSIONS

ACKNOWLEDGMENTS

AMERICAN ASSOCIATION FOR THE ADVANCEMENT OF SCIENCE—*Science*
   Production of Reversible Changes in the Central Nervous System by Ultrasound
   Ultrasonic Lesions in the Mammalian Central Nervous System

THE INTERNATIONAL FEDERATION FOR MEDICAL ELECTRONICS—*Proceedings of the 3rd International Conference on Medical Electronics*
   Characteristics of Intracellular Motion Induced by Ultrasound

THE INTERNATIONAL FEDERATION FOR MEDICAL ELECTRONICS and ILIFFE BOOKS LTD.—*Medical Electronics, Proceedings of the 2nd International Conference on Medical Electronics*
   Ultrasonically Induced Motions in Single Plant Cells

PERMISSIONS

The following papers have been reprinted or translated with the permission of the authors and copyright holders.

ACADEMIA, PUBLISHING HOUSE OF THE CZECHOSLOVAK ACADEMY OF SCIENCES—*Folia Biologica (Prague)*
   Direct and Indirect Effect of Ultrasound on Bone Marrow Cell Suspensions

ACADEMIC PRESS, INC.—*Experimental Cell Research*
   Morphological Changes Induced in the Frog Semitendinosus Muscle Fiber by Localized Ultrasound

THE ACOUSTICAL SOCIETY OF AMERICA—*The Journal of the Acoustical Society of America*
   Absorption of Sound Arising from the Presence of Intact Cells in Blood
   Acoustic Properties of Hemoglobin Solutions
   An Analysis of Lesion Development in the Brain and in Plastics by High-Intensity Focused Ultrasound at Low-Megahertz Frequencies
   Cavitation Microstreaming
   Deformation and Motion Produced in Isolated Living Cells by Localized Ultrasonic Vibration
   Degradation of DNA in High-Intensity Focused Ultrasonic Fields at 1 MHz
   Determination of the Acoustic Properties of Blood and Its Components
   Effects of Intense Noncavitating Ultrasound on Selected Enzymes
   Mechanism of Absorption of Ultrasound in Liver Tissue
   Mechanisms for Nonthermal Effects of Sound
   Physical and Chemical Aspects of Ultrasonic Disruption of Cells

Acknowledgments and Permissions

    Physical Factors Involved in Ultrasonically Induced Changes in Living Systems: I. Identification of Non-Temperature Effects
    Physical Factors Involved in Ultrasonically Induced Changes in Living Systems: II. Amplitude Duration Relations and the Effect of Hydrostatic Pressure for Nerve Tissue
    Production of Lesions in the Central Nervous System with Focused Ultrasound: A Study of Dosage Factors
    Quantitative Relationships Between Ultrasonic Cavitation and Effects upon Amoebae at 1 MHz
    The Role of Heat in the Production of Ultrasonic Focal Lesions
    Shear as a Mechanism for Sonically Induced Biological Effects
    Tabular Data of the Velocity and Absorption of High-Frequency Sound in Mammalian Tissues
    Temperature and Amplitude Dependence of Acoustic Absorption in Tissue
    Threshold Ultrasonic Dosages for Structural Changes in the Mammalian Brain
    Ultrasonic Irradiation of the Central Nervous System at High Sound Levels
    The Velocity of Sound Through Tissues and the Acoustic Impedance of Tissues

AMERICAN ASSOCIATION FOR THE ADVANCEMENT OF SCIENCE—*Science*
    Cell Disruption by Ultrasound
    Hemolysis near an Ultrasonically Pulsating Gas Bubble
    Hemolysis near a Transversely Oscillating Wire

AMERICAN CHEMICAL SOCIETY—*Journal of Physical Chemistry*
    Ultrasonic Absorption Mechanisms in Aqueous Solutions of Bovine Hemoglobin
    Ultrasonic Investigation of the Conformal Changes of Bovine Serum Albumin in Aqueous Solution
    Ultrasonic Studies of Proton Transfers in Solutions of Poly(lysine) and Poly(ornithine). Implications for the Kinetics of the Helix-Coil Transition of Polypeptides and for the Ultrasonic Absorption of Proteins

AMERICAN CONGRESS OF REHABILITATION MEDICINE—*Archives of Physical Medicine and Rehabilitation*
    Changes of Potentials and Temperature Gradients in Membranes Caused by Ultrasound

AMERICAN INSTITUTE OF PHYSICS—*The Journal of Chemical Physics*
    Ultrasonic Absorption in Aqueous Solutions of Dextran

BRITISH JOURNAL OF RADIOLOGY—*The British Journal of Radiology*
    A Study of the Production of Haemorrhagic Injury and Paraplegia in Rat Spinal Cord by Pulsed Ultrasound of Low Megahertz Frequencies in the Context of the Safety for Clinical Usage

ELSEVIER SCIENTIFIC PUBLISHING COMPANY—*Biochimica et Biophysica Acta*
    Ultrasonic Absorption of Aqueous Hemoglobin Solutions

S. HITZEL VERLAG
    Physikalische Zeitschrift
        On the Absorption of Ultrasound in Human Tissues and Their Dependence upon Frequency
    Akustiche Beihefte
        Investigation of Vibratory Cavitation in Liquids

INSTITUTE OF ELECTRICAL AND ELECTRONICS ENGINEERS, INC.—*IRE Transactions on Medical Electronics*
    Ultrasonically-Induced Movements in Cells and Cell Models

INSTITUTE OF PHYSICS, LONDON—*Physics in Medicine and Biology*
    Ultrasonic Absorption and Reflection by Lung Tissue

Acknowledgments and Permissions

MASSON & CIE, PARIS—*Journal de Physiologie*
  Nerve Excitation Due to High-Frequency Ultrasound

PERGAMON PRESS LTD.
  Biophysics
    Absorption of Ultrasound in Biological Media
  Ultrasound in Medicine and Biology
    The Production of Blood Cell Stasis and Endothelial Damage in the Blood Vessels of Chick Embryos Treated with Ultrasound in a Stationary Wave Field

SOCIETY FOR EXPERIMENTAL BIOLOGY AND MEDICINE—*Proceedings of the Society for Experimental Biology and Medicine*
  Effects of Ultrasonic Vibrations on Nerve Tissues

SPRINGER-VERLAG, BERLIN, HEIDELBERG, NEW YORK—*Die Naturwissenschaften*
  Measurement of Ultrasonic Absorption in Animal Tissue and Its Dependence on Frequency
  UHF Acoustic Interaction with Biological Media
  Ultrasonic Absorption Measurements in Human Skull Bone and Their Dependence upon Frequency

TAYLOR & FRANCIS LTD.—*Philosophical Magazine*
  The Physical and Biological Effects of High-Frequency Sound Waves of Great Intensity

JOHN WILEY & SONS, INC.—*Biopolymers*
  Ultrasonic Absorption and Relaxation Spectra in Aqueous Bovine Hemoglobin
  The Ultrasonic Degradation of Biological Macromolecules Under Conditions of Stable Cavitation: I. Theory, Methods and Application to Deoxyribonucleic Acid

THE WILLIAMS & WILKINS COMPANY—*American Journal of Physical Medicine*
  Physical Mechanisms of the Action of Intense Ultrasound on Tissue

THE WISTAR PRESS—*Journal of Cellular and Comparative Physiology*
  Effects of High Intensity Sound on Electrical Conduction in Muscle

SERIES EDITOR'S PREFACE

The "Benchmark Papers in Acoustics" constitute a series of volumes that make available to the reader in carefully organized form important papers in all branches of acoustics. The literature of acoustics is vast in extent and much of it, particularly the earlier part, is inaccessible to the average acoustical scientist and engineer. These volumes aim to provide a practical introduction to this literature, since each volume offers an expert's selection of the seminal papers in a given branch of the subject, that is, those papers which have significantly influenced the development of that branch in a certain direction and introduced concepts and methods that possess basic ultility in modern acoustics as a whole. Each volume provides a convenient and economical summary of results as well as a foundation for further study for both the person familiar with the field and the person who wishes to become acquainted with it.

Each volume has been organized and edited by authorities in the area to which it pertains. In each volume there is provided an editorial introduction summarizing the technical significance of the field being covered. Each article is accompanied by editorial commentary, with necessary explanatory notes, and an adequate index is provided for ready reference. Articles in languages other than English are either translated or abstracted in English. It is the hope of the publisher and editor that these volumes will constitute a working library of the most important technical literature in acoustics of value to students and research workers.

The present volume, *Ultrasonic Biophysics*, has been edited by Floyd Dunn of the Bioacoustics Research Laboratory of the University of Illinois at Urbana and William D. O'Brien, Jr., of the Bureau of Radiological Health, Rockville, Maryland. Through its 53 well-chosen articles it provides a thorough overview of the applications of ultrasonic radiation to biological media and systems, including tissues and organs. During the past 25 years this field has become of increasing importance and promises to provide significant advances in our knowledge of living systems as well as in medical diagnosis and therapy. The papers included cover the major seminal literature in the field and are accompanied by valuable editorial commentary.

R. BRUCE LINDSAY

To the legacy of William J. Fry

PREFACE

Ultrasonic Biophysics involves those portions of physics, biology, engineering, and chemistry which contribute to an understanding of the interactions of high-frequency acoustic waves and living systems. Because of its diverse and specialized pedagogy, it is a postgraduate and postdoctoral undertaking, and only a few research centers have emerged to deal specifically with its problems. As a result, no texts have appeared and the literature remains scattered in a great many periodicals, some of which are available in only the most complete university libraries. With the ever-increasing medical interest in the clinical usefulness of ultrasound, and, as a consequence, the opportunity for ever-larger fractions of the population to be exposed to ultrasound, questions of risk and hazard become preeminent and will be studied by specialists who enter the field at advanced stages in their academic careers. R. Bruce Lindsay's suggestion that we prepare this monograph was considered by us a unique opportunity to collect the important works in ultrasonic biophysics into this most convenient of forms. We believe this collection of papers will serve as an introduction for the novice and will provide a substantive background for which current publications complete the educational formalism.

In selecting papers for inclusion in this book, we sought and received opinions from several of our colleagues, and for this we are grateful to E. L. Carstensen, University of Rochester; W. L. Nyborg, University of Vermont; and F. J. Fry, Indiana University; and to W. T. Coakley, University College Cardiff; P. D. Edmonds, Institute of Electrical and Electronics Engineers; C. R. Hill, Institute of Cancer Research, Sutton, Surrey; L. W. Kessler, Sonoscan; J. B. Pond, Kingston Polytechnic Institute; K. J. W. Taylor, Royal Marsden Hospital, Sutton, Surrey; and A. R. Williams, University of Manchester.

FLOYD DUNN
WILLIAM D. O'BRIEN, JR.

CONTENTS

Acknowledgments and Permissions	v
Series Editor's Preface	ix
Preface	xi
Contents by Author	xix
Introduction	1

PART I: ULTRASONIC PROPERTIES OF BIOLOGICAL MEDIA

Editors' Comments on Papers 1 Through 18   8

1   POHLMAN, R: On the Absorption of Ultrasound in Human Tissues and Their Dependence upon Frequency   14
Translated from *Physik. Z.*, **40**(5), 159–161 (1939)

2   HUETER, T. F.: Measurement of Ultrasonic Absorption in Animal Tissues and Its Dependence on Frequency   19
Translated from *Naturwiss.*, **35**(9), 285–287 (1948)

3   LUDWIG, G. D.: The Velocity of Sound Through Tissues and the Acoustic Impedance of Tissues   23
J. Acoust. Soc. Amer., **22**(6), 862–866 (1950)

4   HUETER, T. F.: Ultrasonic Absorption Measurements in Human Skull Bone and Their Dependence upon Frequency   28
Translated from *Naturwiss.*, **39**(1), 21–22 (1952)

5   GOLDMAN, D. E., and T. F. HUETER: Tabular Data of the Velocity and Absorption of High-Frequency Sound in Mammalian Tissues   31
J. Acoust. Soc. Amer., **28**(1), 35–37 (1956)

6   CARSTENSEN, E. L., K. LI, and H. P. SCHWAN: Determination of the Acoustic Properties of Blood and Its Components   34
J. Acoust. Soc. Amer., **25**(2), 286–289 (1953)

7   CARSTENSEN, E. L., and H. P. SCHWAN: Absorption of Sound Arising from the Presence of Intact Cells in Blood   38
J. Acoust. Soc. Amer., **31**(2), 185–189 (1959)

Contents

8   CARSTENSEN, E. L., and H. P. SCHWAN: Acoustic Properties of Hemoglobin Solutions   43
J. Acoust. Soc. Amer., **31**(3), 305–311 (1959)

9   EDMONDS, P. D., T. J. BAULD III, J. F. DYRO, and M. HUSSEY: Ultrasonic Absorption of Aqueous Hemoglobin Solutions   50
Biochim. Biophys. Acta, **200**, 174–177 (1970)

10   PAULY, H., and H. P. SCHWAN: Mechanism of Absorption of Ultrasound in Liver Tissue   54
J. Acoust. Soc. Amer., **50**(2), Pt. 2, 692–699 (1971)

11   DUNN, F.: Temperature and Amplitude Dependence of Acoustic Absorption in Tissue   62
J. Acoust. Soc. Amer., **34**(10), 1545–1547 (1962)

12   DUNN, F., and J. K. BRADY: Absorption of Ultrasound in Biological Media   65
Biophysics, **18**, 1128–1132 (1974)

13   DUNN, F., and W. J. FRY: Ultrasonic Absorption and Reflection by Lung Tissue   70
Phys. Med. Biol., **5**(4), 401–410 (1961)

14   HAWLEY, S. A., and F. DUNN: Ultrasonic Absorption in Aqueous Solutions of Dextran   80
J. Chem. Phys., **50**(8), 3523–3526 (1969)

15   ZANA, R., and C. TONDRE: Ultrasonic Studies of Proton Transfers in Solutions of Poly(lysine) and Poly(ornithine). Implications for the Kinetics of the Helix-Coil Transition of Polypeptides and for the Ultrasonic Absorption of Proteins   84
J. Phys. Chem., **76**(12), 1737–1743 (1972)

16   KESSLER, L. W., and F. DUNN: Ultrasonic Investigation of the Conformal Changes of Bovine Serum Albumin in Aqueous Solution   91
J. Phys. Chem., **73**(12), 4256–4263 (1969)

17   O'BRIEN, W. D., JR., and F. DUNN: Ultrasonic Absorption Mechanisms in Aqueous Solutions of Bovine Hemoglobin   99
J. Phys. Chem., **76**(4), 528–533 (1972)

18   WHITE, R. D., and L. J. SLUTSKY: Ultrasonic Absorption and Relaxation Spectra in Aqueous Bovine Hemoglobin   105
Biopolymers, **11**(9), 1973–1984 (1972)

PART II. INTERACTION OF ULTRASOUND WITH BIOLOGICAL MEDIA IN SOLUTION AND IN SUSPENSION

Editors' Comments on Papers 19 Through 36   119

19   ESCHE, R.: Investigation of Vibratory Cavitation in Liquids   124
Translated from *Akust. Beih.*, No. 4, 217–218 (1952)

20	ELDER, S. A.: Cavitation Microstreaming *J. Acoust. Soc. Amer.*, **31**(1), 54–64 (1959)	126
21	DYER, H. J., and W. L. NYBORG: Ultrasonically-Induced Movements in Cells and Cell Models *IRE Trans. Med. Elec.*, **ME-7**, 163–165 (1960)	137
22	NYBORG, W. L., and H. J. DYER: Ultrasonically Induced Motions in Single Plant Cells *Medical Electronics, Proc. 2nd Intern. Conf. Med. Elec., June 24–27, 1959, Paris, 1960, pp. 391–396*	140
23	DYER, H. J., and W. L. NYBORG: Characteristics of Intracellular Motion Induced by Ultrasound *Proc. 3rd Intern. Conf. Med. Elec., London, 1960, pp. 445–449*	146
24	W. W. WILSON, F. J. WIERCINSKI, W. L. NYBORG, R. M. SCHNITZLER, and F. J. SICHEL: Deformation and Motion Produced in Isolated Living Cells by Localized Ultrasonic Vibration *J. Acoust. Soc. Amer.*, **40**(6), 1363–1370 (1966)	151
25	RAVITZ, M. J., and R. M. SCHNITZLER: Morphological Changes Induced in the Frog Semitendinosus Muscle Fiber by Localized Ultrasound *Exptl. Cell Res.*, **60**, 78–85 (1970)	159
26	NYBORG, W. L.: Mechanisms for Nonthermal Effects of Sound *J. Acoust. Soc. Amer.*, **44**(5), 1302–1309 (1968)	167
27	HUGHES, D. E., and W. L. NYBORG: Cell Disruption by Ultrasound *Science*, **138**(3537), 108–114 (1962)	175
28	PRITCHARD, N. J., D. E. HUGHES, and A. R. PEACOCKE: The Ultrasonic Degradation of Biological Macromolecules Under Conditions of Stable Cavitation: I. Theory, Methods, and Application to Deoxyribonucleic Acid *Biopolymers*, **4**(3), 259–273 (1966)	182
29	ROONEY, J. A.: Hemolysis near an Ultrasonically Pulsating Gas Bubble *Science*, **169**, 869–871 (Aug. 28, 1970)	197
30	WILLIAMS, A. R., D. E. HUGHES, and W. L. NYBORG: Hemolysis near a Transversely Oscillating Wire *Science*, **169**, 871–873 (Aug. 28, 1970)	199
31	ROONEY, J. A.: Shear as a Mechanism for Sonically Induced Biological Effects *J. Acoust. Soc. Amer.*, **52**(6), 1718–1724 (1972)	202
32	MACLEOD, R. M., and F. DUNN: Effects of Intense Noncavitating Ultrasound on Selected Enzymes *J. Acoust. Soc. Amer.*, **44**(4), 932–940 (1968)	209
33	COAKLEY, W. T., and F. DUNN: Degradation of DNA in High-Intensity Focused Ultrasonic Fields at 1 MHz *J. Acoust. Soc. Amer.*, **50**(6), Pt. 2, 1539–1545 (1971)	218

Contents

34 HRAZDIRA, I: Direct and Indirect Effect of Ultrasound on Bone Marrow Cell Suspensions — 225
Folia Biol. (Prague), **11**, 330–333 (1965)

35 CLARKE, P. R., and C. R. HILL: Physical and Chemical Aspects of Ultrasonic Disruption of Cells — 229
J. Acoust. Soc. Amer., **47**(2), Pt. 2, 649–653 (1970)

36 COAKLEY, W. T., D. HAMPTON, and F. DUNN: Quantitative Relationships Between Ultrasonic Cavitation and Effects upon Amoebae at 1 MHz — 234
J. Acoust. Soc. Amer., **50**(6), Pt. 2, 1546–1553 (1971)

PART III. INTERACTION OF ULTRASOUND WITH BIOLOGICAL TISSUES AND ORGANS

Editors' Comments on Papers 37 Through 53 — 244

37 FRY, W. J., V. J. WULFF, D. TUCKER, and F. J. FRY: Physical Factors Involved in Ultrasonically Induced Changes in Living Systems: I. Identification of Non-Temperature Effects — 249
J. Acoust. Soc. Amer., **22**(6), 867–876 (1950)

38 FRY, W. J., D. TUCKER, F. J. FRY, and V. J. WULFF: Physical Factors Involved in Ultrasonically Induced Changes in Living Systems: II. Amplitude Duration Relations and the Effect of Hydrostatic Pressure for Nerve Tissue — 259
J. Acoust. Soc. Amer., **23**(3), 364–368 (1951)

39 WULFF, V. J., W. J. FRY, D. TUCKER, F. J. FRY, and C. MELTON: Effects of Ultrasonic Vibrations on Nerve Tissues — 264
Proc. Soc. Exptl. Biol. Med., **76**(2), 361–366 (1951)

40 MAZOUÉ, H., P. CHAUCHARD, and R.-G. BUSNEL: Nerve Excitation Due to High-Frequency Ultrasound — 270
Translated from *J. Physiol. (Paris)*, **45**, 179–182 (1953)

41 LEHMANN, J. F., and R. BIEGLER: Changes of Potentials and Temperature Gradients in Membranes Caused by Ultrasound — 274
Arch. Phys. Med. Rehabil., **35**, 287–295 (1954)

42 FRY, W. J., J. W. BARNARD, F. J. FRY, R. F. KRUMINS, and J. F. BRENNAN: Ultrasonic Lesions in the Mammalian Central Nervous System — 283
Science, **122**, 517–518 (Sept. 1955)

43 WELKOWITZ, W., and W. J. FRY: Effects of High Intensity Sound on Electrical Conduction in Muscle — 285
J. Cell. Comp. Physiol., **48**(3), 435–457 (1956)

44 FRY, W. J., and F. DUNN: Ultrasonic Irradiation of the Central Nervous System at High Sound Levels — 308
J. Acoust. Soc. Amer., **28**(1), 129–131 (1956)

45	DUNN, F.: Physical Mechanisms of the Action of Intense Ultrasound on Tissue *Amer. J. Phys. Med.,* **37**(3), 148–151 (1958)	310
46	HUETER, T. F., H. T. BALLANTINE, JR., and W. C. COTTER: Production of Lesions in the Central Nervous System with Focused Ultrasound: A Study of Dosage Factors *J. Acoust. Soc. Amer.,* **28**(2), 192–201 (1956)	314
47	FRY, F. J., H. W. ADES, and W. J. FRY: Production of Reversible Changes in the Central Nervous System by Ultrasound *Science,* **127**, 83–84 (Jan. 1958)	324
48	ROBINSON, T. C., and P. P. LELE: An Analysis of Lesion Development in the Brain and in Plastics by High-Intensity Focused Ultrasound at Low-Megahertz Frequencies *J. Acoust. Soc. Amer.,* **51**(4), Pt. 2, 1333–1351 (1972)	325
49	HAWLEY, S. A., and F. DUNN: UHF Acoustic Interaction with Biological Media *Naturwiss.,* **51**(23), 555–556 (1964)	345
50	POND, J. B.: The Role of Heat in the Production of Ultrasonic Focal Lesions *J. Acoust. Soc. Amer.,* **47**(6), Pt. 2, 1607–1611 (1970)	347
51	FRY, F. J., G. KOSSOFF, R. C. EGGLETON, and F. DUNN: Threshold Ultrasonic Dosages for Structural Changes in the Mammalian Brain *J. Acoust. Soc. Amer.,* **48**(6), Pt. 2, 1413–1417 (1970)	352
52	DYSON, M., J. B. POND, B. WOODWARD, and J. BROADBENT: The Production of Blood Cell Stasis and Endothelial Damage in the Blood Vessels of Chick Embryos Treated with Ultrasound in a Stationary Wave Field *Ultrasound Med. Biol.,* **1**, 133–148 (1974)	357
53	TAYLOR, K. J. W., and J. B. POND: A Study of the Production of Haemorrhagic Injury and Paraplegia in Rat Spinal Cord by Pulsed Ultrasound of Low Megahertz Frequencies in the Context of the Safety for Clinical Usage *Brit. J. Radiol.,* **45**, 343–353 (May 1972)	373

Author Citation Index   385
Subject Index   393
About the Editor   411

CONTENTS BY AUTHOR

Ades, H. W., 324
Ballantine, H. T., Jr., 314
Barnard, J. W., 283
Bauld, T. J., III, 50
Biegler, R., 274
Brady, J. K., 65
Brennan, J. F., 283
Broadbent, J., 357
Busnel, R.-G., 270
Carstensen, E. L., 34, 38, 43
Chauchard, P., 270
Clarke, P. R., 229
Coakley, W. T., 218, 234
Cotter, W. C., 314
Dunn, F., 62, 65, 70, 80, 91, 99,
   209, 218, 234, 308, 310, 345, 352
Dyer, H. J., 137, 140, 146
Dyro, J. F., 50
Dyson, M., 357
Edmonds, P. D., 50
Eggleton, R. C., 352
Elder, S. A., 126
Esche, R., 124
Fry, F. J., 249, 259, 264, 283, 324, 352
Fry, W. J., 70, 249, 259, 264,
   283, 285, 308, 324
Goldman, D. E., 31
Hampton, D., 234
Hawley, S. A., 80, 345
Hill, C. R., 229
Hrazdira, I., 225
Hueter, T. F., 19, 28, 31, 314
Hughes, D. E., 175, 182, 199
Hussey, M., 50
Kessler, L. W., 91

Kossoff, G., 352
Krumins, R. F., 283
Lehmann, J. F., 274
Lele, P. P., 325
Li, K., 34
Ludwig, G. D., 23
Macleod, R. M., 209
Mazoué, H., 270
Melton, C., 264
Nyborg, W. L., 137, 140,
   146, 151, 167, 175, 199
O'Brien, W. D., Jr., 99
Pauly, H., 54
Peacocke, A. R., 182
Pohlman, R., 14
Pond, J. B., 347, 357, 373
Pritchard, N. J., 182
Ravitz, M. J., 159
Robinson, T. C., 325
Rooney, J. A., 197, 202
Schnitzler, R. M., 151, 159
Schwan, H. P., 34, 38, 43, 54
Sichel, F. J., 151
Slutsky, L. J., 105
Taylor, K. J. W., 373
Tondre, C., 84
Tucker, D., 249, 259, 264
Welkowitz, W., 285
White, R. D., 105
Wiercinski, F. J., 151
Williams, A. R., 199
Wilson, W. W., 151
Woodward, B., 357
Wulff, V. J., 249, 259, 264
Zana, R., 84

INTRODUCTION

The field of ultrasonic biophysics had its beginnings near the end of World War I when techniques for locating submarines were being developed. Among such pursuits were those of P. Langevin, who was investigating an acoustic method in which a piezoelectric transducer, in a circuit containing appropriate capacitors and inductors, was excited by a Poulsen arc converter to vibrate at the resonant frequency of the structure and emit ultrasound into the bay at Toulon. As the electric potentials applied to the quartz plate at times were as high as 40,000 V, the amplitude of the acoustic wave was appreciable and small fish and other marine animals were found dead in the vicinity of this radiation. Thus, some effects of ultrasound on living systems were apparent from the time ultrasound became available as a physical agent. Because of its inherent instability, the Poulsen arc was unsuitable for detailed investigation of these phenomena, and serious study awaited the development of the vacuum-tube oscillator for use as the piezoelectric transducer driver.

The first extensive investigation of the phenomena observed by Langevin was conducted by Wood and Loomis and their results were published in 1927. They described in some detail the electronic and acoustical aspects of their apparatus, various phenomena familiar from lower-frequency studies such as traveling- and standing-wave manifestations, heretofore unfamiliar phenomena such as some chemical and thermal effects, acoustic production of emulsions and fogs, and biological effects. Although the latter were exclusively of a thermal and mechanical

Introduction

destructive nature, they exhibited the demonstrable effects that this new form of energy could exert on living systems.

That portion of the Wood and Loomis paper which deals with the biological effects of ultrasound is reproduced in the following paragraphs.

> Though the effects of these waves upon living matter might more properly be discussed elsewhere, it may not be out of place to mention briefly a few of the observations which we have made as they have some bearing on the physical processes involved.
>
> In marked contrast to the flocculation, or driving together of small particles of suspended matter, which has been mentioned, we have fragmentation, or the tearing to pieces of small and fragile bodies. Filaments of living spirogyra were torn to pieces and the cells ruptured. Small unicellular organisms such as paramecium were rendered immobile by a short treatment to vibration of moderate intensity, subsequently recovering, but were killed by a longer exposure, many of them being torn open. The circumstance that all are not treated alike is doubtless due to the fact that those which manage to keep out of the nodes of the stationary wave system are less roughly handled by the vibrations. Bacteria apparently are able to survive owing to their small size, for the fragmentation of larger bodies is due to the fact that the forces applied to their surfaces vary in magnitude and direction at different points of the body, while in the case of a bacterium the whole body is subjected to the same treatment.
>
> Red blood corpuscles in physiological salt solution are rapidly destroyed, the turbid liquid becoming as clear as a solution of a red aniline dye.
>
> With vibrations of less intensity the destruction is less complete, a blood count made at the end of each 15 seconds of exposure showing that the percentage destroyed decreases, a point being reached at which no further destruction occurs unless the intensity of the radiation is augmented. This means of course that some of the corpuscles, the recently formed ones perhaps, are more hardy than those of greater age. Small fish and frogs are killed by an exposure of one or two minutes, an observation also made by Langevin at Toulon with his Poulsen arc oscillator (see Plate VIII.) Mice are less sensitive, a twenty-minute exposure not resulting in death, and though at the end of the treatment the animal was barely able to move, the recovery was fairly rapid. Blood counts made with a mouse during exposure showed a diminishing number of corpuscles, until a stationary state (about 60 per cent normal) was reached. The biologists inform us, however, that the blood count of a mouse is affected by fear, the corpuscles hiding in the liver until the danger is over! We made the count with drops taken from the tip of the tail.

Plate VIII Reproduced from *Phil. Mag.*, **6,** (1927); copyright © 1927 by Taylor & Francis Ltd.

Introduction

> We have not yet determined the cause of death in the case of the fishes and frogs. They were protected against rise of temperature as much as possible by ice fragments, dropped into the water from time to time, but this does not shield them from internal heating, which may be the cause of death, as in the case of small animals introduced into a high-frequency electric field. In the case of a mouse killed by an exposure of two minutes between the plates of an air condenser operated at about 1000 volts with a frequency of 100 million, we found that the temperature of the body cavity was over 113°F.
>
> With distilled water or a fairly strong solution of salt in a test tube between the plates of the condenser, little or no heating occurred; but for small concentrations the heating was very marked, the maximum being for 8 per cent, which is very nearly the amount found in mammalian blood. At lower frequencies the heating appears to be greater for distilled water, at least with high voltages. We found that one terminal of our 60,000-volt coil could be held in the fingers without the production of any sensation, but if dipped into the open end of a glass tube a metre long and filled with distilled water, caused the water to boil in less than 10 seconds. The introduction of a small amount of salt into the water prevented the heat entirely in this case, which explains why no thermal discomfort was felt when the wire was held in the hand. The wire must be seized, however, before the current is turned on, otherwise a very vicious arc jumps to the finger producing a burn which is very slow in healing.*

The reader is encouraged to examine the entire paper, as it is a particularly clear description of these first observed ultrasonic effects. [One source is the Benchmark volume *Physical Acoustics*, edited by R. B. Lindsay (Dowden, Hutchinson & Ross, Inc., Stroudsburg, Pa., 1974), pp. 240–266.]

Following the introduction of the piezoelectric element into acoustics and the subsequent rapid developments in electronics during the next three decades, it became feasible to construct instruments for precise measurement of the velocity of propagation and the absorption coefficient of ultrasonic waves in liquids and liquid-like media. An additional advance occurred after 1945 when the adaptation of radar techniques yielded pulsed ultrasonic instrumentation operating in the multimegahertz range. Numerous techniques have been developed since that permit utilization of ultrasound in the frequency range from about

---

* From *Phil. Mag.*, **6**, 434–436 (1927); copyright © 1927 by Taylor & Francis Ltd.

2×10^4 Hz (the arbitrary boundary with the "sonic" range) to 10^9 Hz; extension to beyond 10^{11} Hz is emerging with the continued development of Brillouin scattering techniques. Simultaneous development of piezoelectric materials, lens-focusing systems, and field-measuring schemes have allowed high-intensity ultrasonics to be employed for the precision production of reversible and irreversible effects in biological media. As a result, two distinct groups of investigators pursuing individual research interests have employed ultrasonic methods in biologically oriented studies.

First, ultrasonic techniques have been employed as a means of investigating fast biophysical and biochemical reactions, with the objective of obtaining detailed information about the time course of overall and intermediate reactions for the complete kinetic description of reaction mechanisms. Ultrasound currently offers access to the determination of time constants (ranging from 10^{-3} to 10^{-9} s) for the approach to equilibrium by physical–chemical reactions for which both reactants and products are present in comparable concentrations. In combination with stopped flow, pressure and temperature jump, and pulsed electric field methods, ultrasound and Brillouin scattering methods cover the considerable range in time constants that cannot be determined by the more conventional mixing procedures (greater than 10^{-3} s) and spectroscopic techniques (less than 10^{-10} s).

The second group of investigators requiring details of the propagation of ultrasound in, and the effects on, biological materials has been those concerned with the use of this form of energy as a tool for fundamental studies in cellular and organismic organization and function and as tools in medical practice. Ultrasound is employed clinically, for example, in medical diagnosis as a pulse-echo technique (time-averaged intensities less than 100 mW/cm^2 and 0.1 percent duty cycle) from obtaining information regarding the static and dynamic state of gross tissue structures. Recent advances exhibit promise that microscopic details will soon become available with instruments designed to operate in the several hundred megahertz region. As a therapeutic tool, continued wave ultrasound is employed as a deep-heating agent; low intensities (usually 1 W/cm^2 or less) are used. As a selective tissue-modifying agent, relatively high intensities (of the order of 10^3 W/cm^2) are used in the form of pulses of short duration. For the successful employment of all these techniques, details of the propagation properties and interaction mechanisms are essential for appropriate instrument design, for selection of

Introduction

the most efficacious procedures, and for accurate assessment of risk (see, for example, O'Brien et al., 1972).

The papers selected for inclusion in this volume exhibit, in somewhat historical fashion, the development of the two major portions of the field, that is, the two main groups of users of this knowledge, and simultaneously illustrate the experimental difficulties that have limited more rapid advance. The selections were based largely on their impact on the field and the continuing value of the contribution. This material can be readily catalogued into three groupings: Part I contains papers dealing exclusively with the ultrasonic propagation properties of biological materials, Part II covers interaction mechanisms at the macromolecular and cellular levels of biological structure, and Part III contains mechanism studies at the tissue and organ levels of structure.

REFERENCE

O'Brien, W. D., Jr., M. L. Shore, R. K. Fred, and W. M. Leach. 1972. "On the Assessment of Risks from Ultrasound." *Proc. 1972 Ultrasonics Symp.*, IEEE Catalog No. 72, CHO 708-8SU, New York.

Part I
ULTRASONIC PROPAGATION PROPERTIES OF BIOLOGICAL MEDIA

An initial motivation for serious study of the ultrasonic propagation properties of living materials was the early observation that ultrasound provided an opportunity for true deep heating in tissues and not simply the superficial heating that attended irradiation with infrared and the like. Thus, it became necessary to determine the accurate dosage for affecting various organs at different depths, and a knowledge of the absorption of sound by the intervening tissues was essential.

Editors' Comments on Papers 1 Through 18

1 **POHLMAN**
   On the Absorption of Ultrasound in Human Tissues and Their Dependence upon Frequency

2 **HUETER**
   Measurement of Ultrasonic Absorption in Animal Tissues and Its Dependence on Frequency

3 **LUDWIG**
   The Velocity of Sound Through Tissues and the Acoustic Impedance of Tissues

4 **HUETER**
   Ultrasonic Absorption Measurements in Human Skull Bone and Their Dependence upon Frequency

5 **GOLDMAN and HUETER**
   Tabular Data of the Velocity and Absorption of High-Frequency Sound in Mammalian Tissues

6 **CARSTENSEN et al.**
   Determination of the Acoustic Properties of Blood and Its Components

7 **CARSTENSEN and SCHWAN**
   Absorption of Sound Arising from the Presence of Intact Cells in Blood

8 **CARSTENSEN and SCHWAN**
   Acoustic Properties of Hemoglobin Solutions

9 **EDMONDS et al.**
   Ultrasonic Absorption of Aqueous Hemoglobin Solutions

10 **PAULY and SCHWAN**
   Mechanism of Absorption of Ultrasound in Liver Tissue

11 **DUNN**
   Temperature and Amplitude Dependence of Acoustic Absorption in Tissue

12 **DUNN and BRADY**
    Absorption of Ultrasound in Biological Media

13 **DUNN and FRY**
    Ultrasonic Absorption and Reflection by Lung Tissue

14 **HAWLEY and DUNN**
    Ultrasonic Absorption in Aqueous Solutions of Dextran

15 **ZANA and TONDRE**
    Ultrasonic Studies of Proton Transfers in Solutions of Poly(lysine) and Poly(ornithine). Implications for the Kinetics of the Helix-Coil Transition of Polypeptides and for the Ultrasonic Absorption of Proteins

16 **KESSLER and DUNN**
    Ultrasonic Investigation of the Conformal Changes of Bovine Serum Albumin in Aqueous Solution

17 **O'BRIEN and DUNN**
    Ultrasonic Absorption Mechanisms in Aqueous Solutions of Bovine Hemoglobin

18 **WHITE and SLUTSKY**
    Ultrasonic Absorption and Relaxation Spectra in Aqueous Bovine Hemoglobin

In Paper 1, by Pohlman, the astonishing, nearly linear dependence of the absorption coefficient on frequency is revealed, which encourages him to speculate on the mechanism for this nonclassical behavior. Pohlman observes that different tissues exhibit different rates of energy absorption.

Paper 2, by Hueter, verifies the earlier finding that the ultrasonic absorption coefficient is a nearly linear function of frequency. He further notes additions to the attenuation of this wave energy resulting from anisotropic structural features of striated muscle, observes that precautions should be taken in estimating heating when the nonuniformities of the diffraction field must be considered, and speculates on the absorption mechanism.

Editors' Comments on Papers 1 Through 18

With the successful development of ultrasonic nondestructive testing methods for solids, attention turned to the utilization of this schema in medical practice to diagnose abnormal conditions and disease states. It was soon recognized that, in addition to the absorption properties of tissues, the speed of sound in and the acoustic impedance of tissues were crucial parameters that needed to be made available with appreciable accuracy, if appropriate instrumentation and clinical procedures were to become universally available. Paper 3, by Ludwig, is the first significant undertaking in this area; he shows that the velocity and impedance values of high-water-content tissues do not differ greatly from those of water, and that anisotropic structural features do not contribute greatly to these parameters.

The possibility of extending the ultrasonic pulse-echo diagnostic tool to clinical observations of the central nervous system meant that mineralized tissues would also have to be traversed by acoustic energy. Paper 4, by Hueter, revealed the much higher absorption exhibited by bone and the different frequency dependence, as compared with those of soft tissues. He also identifies experimental problem areas, some of which have still not been disposed of.

Since Pohlman's 1939 publication dealing with absorption of ultrasound in human tissues, numerous measurements have been made of the ultrasonic propagation properties of tissues and organs. In 1956, Goldman and Hueter published a compilation of the then available data. This report, included here as Paper 5, still provides a useful qualitative picture of the frequency dependence and range of values of the important propagation parameters. Recent observations suggest that the linear frequency dependence of mammalian tissues prevails for about 100 MHz (Kessler, 1973).

One of the most important findings in the elucidation of the ultrasonic absorption process in biological media was the observation that protein constituents provide the main contribution and that protein solutions exhibit comparable absorption magnitudes and similar frequency dependencies as tissues. Papers 6 through 8 are from the laboratory that made this discovery. The first, Paper 6, by Carstensen, Li, and Schwan, deals with blood and various blood proteins in solution; they argue that a major fraction of the absorption in biological materials occurs at the molecular level. The next, Paper 7, by Carstensen and Schwan, shows that a small contribution to absorption in blood arises from the viscous interaction between the intact cells and their environment. The third, Paper 8, also by Carstensen and Schwan, shows that the absorption per wavelength and the velocity dispersion of hemoglobin solutions can be related through relaxation theory by assuming a broad distribution of relaxation times.

After approximately one decade, measuring techniques have progressed to the point where absorption can be determined in solutions of biopolymers to nearly 500 MHz. In Paper 9, Edmonds and associates summarize such data for aqueous hemoglobin solutions.

The question of whether the macromolecular origin of the ultrasonic absorption in blood also extended to solid tissues was first treated by Pauly and Schwan about 1957, although their results were published much later in the article reproduced as Paper 10. Their finding that approximately two thirds of the total absorption in liver arises at the macromolecular level implies that for tissues, to a large extent, the ultrasonic propagation properties are a reflection of their molecular composition.

Because of the strong influence associated with medical applications of ultrasound, the vast majority of investigations associated with the propagation properties of biological systems were conducted with homeothermic specimens, and the dependence of the absorption on temperature did not become evident. Papers 11 and 12 exhibit the importance of temperature on this property, as well as wave amplitude and frequency.

The difficulties of making ultrasonic measurements in tissues is nowhere better exemplified than in the lung. In addition to the usual variables that must be treated, the degree and state of inflation is of obvious importance. In Paper 13, by Dunn and Fry, the very high absorption in excised fresh tissue is noted and a loss mechanism is proposed.

Continuation of this work within the next few years (although published more than a decade later) verifies the unusually high absorption, provides information on the frequency dependence of the absorption and velocity, but shows the proposed absorption mechanisms to be erroneous (Dunn, 1974). A recent investigation on fixed lung tissue has confirmed these data and has also provided information on the dependence of absorption on inflation (Bauld and Schwan, 1974).

The important finding by Carstensen and co-workers that ultrasonic absorption in biological media occurs largely at the macromolecular level stimulated investigations to elucidate the mechanisms involved. Thus, the remaining papers of Part I deal with molecular occurrences that could require the extraction of energy from the acoustic wave process (absorption).

Random-coil polymers were studied to determine the importance of molecular weight on the ultrasonic absorption behavior of macromolecules in solution. In Paper 14, by Hawley and Dunn, which deals with dextran, it is shown that, beyond a molecular weight corresponding to approximately 100 monomer units, absorption becomes independent of molecular size. A similar result was obtained by Kessler et al. (1970) for aqueous polyethylene glycol solutions.

As the polysaccharide dextran is primarily a linear structure that assumes a random-coil configuration in aqueous solution and, for example, hemoglobin is a globular protein that displays a stabilized folded structure, the importance of higher-order structure could be examined by comparing the dextran and protein ultrasonic absorption data. This has been done with hemoglobin in Dunn et al. (1969), where it is argued that tertiary structure could be responsible for the excess absorption in hemoglobin. However, a direct attempt in which the higher-order structures of hemoglobin were denatured with guanidine hydrochloride did not produce the expected reduction in the absorption (O'Brien and Dunn, 1971).

The influence of the basic protein structure on ultrasonic absorption in aqueous solutions was treated by several investigators in two synthetic polyamino acids. The primary mechanisms proposed to explain the excess ultrasonic absorption in poly-L-glutamic acid solutions are solvent–solute interactions (Burke et al., 1965) and helix-coil transition (Schwarz, 1965; Wada et al., 1967); helix-coil transitions were important at 50 kHz and side chain dissociations at 3 MHz.

Parker et al. (1968) concluded that the observed ultrasonic absorption behavior in aqueous poly-L-lysine solutions can be attributed to the helix-coil transition.

In Paper 15, by Zana and Tondre, conditions are identified that favor helix-coil transitions in aqueous solutions of polypeptides.

Papers 16, 17, and 18, by Kessler and Dunn, O'Brien and Dunn, and White and Slutsky, deal with the processes contributing to excess ultrasonic absorption in aqueous protein solutions. While an understanding of the mechanisms involved at high and low pH values appears to have been achieved, such is not the case for proteins near physiological pH at megahertz frequencies.

An interesting suggestion that interactions between molecules may be of considerable importance has been proposed by Kremkau and Carstensen (1972). The absorption properties of nucleic acids in aqueous solutions have also received appreciable attention and the reader is urged to see Lang and Cerf (1969) and O'Brien et al. (1972) for details.

REFERENCES

Bauld, T. J., and H. P. Schwan. 1974. "Attenuation and Reflection of Ultrasound in Canine Lung Tissue." *J. Acoust. Soc. Amer.,* **56,** 1630–1637.

Burke, J. J., G. G. Hammes, and T. B. Lewis. 1965. "Ultrasonic Attenuation Measurements in Poly-L-glumatic Acid Solutions." *J. Chem. Phys. Chem.,* **42,** 3520–3525.

Dunn, F. 1974. "Attenuation and Speed of Sound in Lung." *J. Acoust. Soc. Amer.,* **56,** 1638–1639.

———, et al. 1969. "Absorption and Dispersion of Ultrasound in Biological Media." In H. P. Schwan (ed.), *Biological Engineering.* New York: McGraw-Hill, p. 205–332.

Kessler, L. W. 1973. "VHF Ultrasonic Attenuation in Mammalian Tissue." *J. Acoust. Soc. Amer.,* **53,** 1759–1760.

———, et al. 1970. "Ultrasonic Absorption in Aqueous Solutions of Polyethylene Glycol." *J. Phys. Chem.,* **74,** 4096–4102.

Kremkau, F. W., and E. L. Carstensen. 1972. "Macromolecular Interaction in Sound Absorption." In J. M. Reid and M. R. Sikov (eds.), *Interaction of Ultrasound and Biological Tissues,* DHEW/FDA 73-8008, Rockville, Md., pp. 37–42.

Lang, J., and R. Cerf. 1969. "Absorption ultrasonore dans des solutions d'acide desoxyribonucléique; étude de la dénaturation alcaline." *J. Chim. Phys.,* **66,** 81–87.

O'Brien, W. D., Jr., and F. Dunn. 1971. "Ultrasonic Examination of Hemoglobin Dissociation Process in Aqueous Solutions of Guanidine Hydrochloride." *J. Acoust. Soc. Amer.,* **50,** 1213–1215.

———, C. L. Christman, and F. Dunn. 1972. "Ultrasonic Investigation of Aqueous Solutions of Deoxyribose Nucleic Acid." *J. Acoust. Soc. Amer.,* **52,** 1251–1255.

Parker, R. C., J. J. Slutsky, and K. R. Applegate. 1968. "Ultrasonic Absorption and the Kinetics of Conformational Change in Poly-L-lysine." *J. Phys. Chem.,* **72,** 3177–3186.

Schwarz, G. 1965. "On the Kinetics of Helix-Coil Transition of Polypeptides in Solution." *J. Mol. Biol.,* **11,** 64–77.

Wada, Y., H. Sasabe, and M. Tomono. 1967. "Viscoelastic Relaxation in Solutions of Poly-(Glutamic Acid) and Gelatin at Ultrasonic Frequencies." *Biopolymers,* **5,** 887–897.

1
ON THE ABSORPTION OF ULTRASOUND IN HUMAN TISSUES AND THEIR DEPENDENCE UPON FREQUENCY

R. Pohlman

This article was translated expressly for this Benchmark volume by Floyd Dunn, University of Illinois at Urbana–Champaign, from "Über die Absorption des Ultraschalls im menschlichen Gewebe und ihre Abhängigkeit von der Frequenz," Physik Z., 40(5) 159–161 (1939)

The biological effects of ultrasound have been investigated extensively in a number of works.[1] The destructive effects were demonstrated almost exclusively and other results, under the imposed conditions, could scarcely be found.

In spite of the destructive effects on living organisms, it seemed to this writer that points of view exist which with correct dosage would allow therapeutic effects to be expected on nerves, tissue, and other parts. A detailed report has been made elsewhere[2] with regard to the trains of thought that led in this direction and the gratifying results which were achieved in the treatment of patients.

In connection with these investigations, the question of the absorption of ultrasound in human tissues, its frequency dependence, and the depth of penetration of the radiation was of major significance. The measurement of the absorption coefficient and its frequency dependence were also of interest from another point of view since we are concerned here with the propagation and absorption of ultrasound in an inhomogeneous medium for which the classical absorption formula for a homogeneous medium is no longer applicable. The possibility therefore exists that important differences from the classical relationships could appear.

The measurement of the absorption coefficients was undertaken in the following manner: a piece of tissue (gluteal musculature) in as fresh a condition as possible was placed into physiological sodium chloride solution in the absorption apparatus (Figure 1). In the schematic drawing, 1 indicates the piezoquartz for generating the ultrasound, which was glued under the thin metal foil (2). On a very thin paper screen (3) lies the tissue (4). Above the tissue is situated the diffuse-reflecting sound reflector (5),[3] which is attached to a very sensitive sound balance

---

[1] A Dognon and E. H. Biancani, *Ultrasons et Biologie,* Gauthier Villars, Paris, 1937.
[2] R. Pohlman, R. Richter, and E. Parow, *Deut. Med. Wochschr.,* 7 (1939).
[3] A detailed report concerning the influence of the vessel walls and reflectors during sound absorption measurements appears elsewhere.

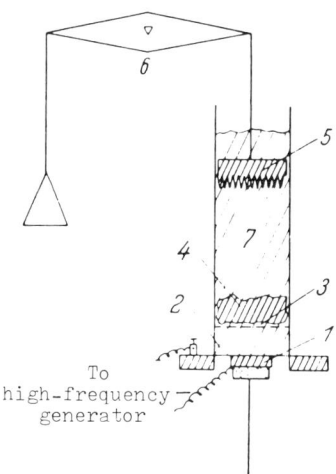

Figure 1 Reproduced from *Physik. Z.*, 40(5), 159 (1939); copyright © 1939 by S. Hirzel Verlag.

(6). The reflector and tissue are surrounded by a physiological sodium chloride solution (7). When the piezoquartz is excited, ultrasonic radiation penetrates the tissue and exerts on the reflector a sound pressure that is proportional to the intensity. The energy absorbed in the tissue can be calculated if measurements are made, at constant radiation, once with the tissue absent and a second time with the tissue present. If the layer of solium chloride solution traversed by the radiation is held constant for both comparative measurements, which could be affected by means of raising and lowering the balance by a vertical slide apparatus, then the simple relation

$$I_2 = I_1 e^{-\alpha(x_2 - x_1)} \tag{1}$$

is valid where $x_2 - x_1$ is the thickness of the tissue layer, α is the absorption coefficient, and I_1 and I_2 are, respectively, the entering and issuing radiation intensities. Reflection at the tissue boundaries has not been considered here. It can be eliminated if studies are carried out with tissue layers of different thicknesses. Such reflection is very small and can, in general, be neglected as it is of little importance since the uncertainty of the measurement, which depends upon the indefiniteness of the tissue-layer thickness, is much greater (about 10 percent).

In a series of experiments the absorption coefficient was determined by measuring the radiation pressure at different distances from the quartz for constant voltage on the piezoquartz. During this process different secondary phenomena (streaming phenomena, cavitation, etc.) can contribute to absorption. To avoid these errors, we proceeded in the following manner: the piezoquartz was excited with different voltages and the sound pressure associated with each voltage was ascertained at a determined distance from the quartz. Then the intensity

Absorption of Ultrasound in Human Tissues

(sound pressure) must be proportional to the square of the voltage; thus

$$I = \text{const. } V^2 \qquad (2)$$

(V = voltage). If the square of the applied voltage is plotted as the abscissa and the sound pressure as the ordinate for different distances from the quartz, straight lines must result for an acceptable series of measurements. Every disturbance, of whatever kind, must make itself noticeable by a deviation from the straight line. From the different slopes of the pair of straight lines, measured with and without the tissue present, the absorption coefficient α can be determined quite accurately from equations (1) and (2). Figure 2 shows an example of such curves, where a_1, a_2 is the corresponding curve pair at a frequency (ν) of 800 kHz. The difference in slope yields the absorption of an approximately 3.5 cm thick piece of tissue (gluteal region, containing skin, about 2 cm of subcutaneous tissue, and about 1.5 cm of musculature).

To determine the absorption coefficients as accurately as possible, different series of measurements were carried out. Measurements on the tissue of a child and of adults were undertaken.[4] No noteworthy variation resulted in the absorption within the limits of error. However, the absorption was different for fat tissue and muscle tissue; indeed, an equally thick layer of fat absorbed less than a corresponding layer of muscle. The investigation was undertaken in such a way that the absorption of the complete tissue was measured first; then the two separate layers were measured. The absorption coefficients are collected in the following table, where the tabulated values are averages of different series of measurements:

	Frequency (kHz)	Species	α/ν^2	α	Half-Value Thickness (cm)
1	800	Fat-musculative	22,000	0.141	4.9
2	800	Fat layer	16,000	0.102	6.8
3	800	Muscle layer	30,000	1.192	3.6
4	2,400	Fat-musculative	8,200	0.472	1.5
5	287	CS_2	19,200	0.016	—
6	870	CS_2	22,000	0.167	—

Source: *Physik. Z.*, **40**(5), 160 (1939); copyright © 1939 by S. Hirzel Verlag.

The data entered in row 1 correspond to the actual proportions in the human body. The absorption value α is remarkably small and corresponds approximately to that of a highly absorbing homogeneous fluid

[4] For kindly providing these materials, we express our thanks to the Kinderklinik des Kaiserin-Auguste-Viktoria-House and to the Martin-Luther-Krankenhaus, Berlin.

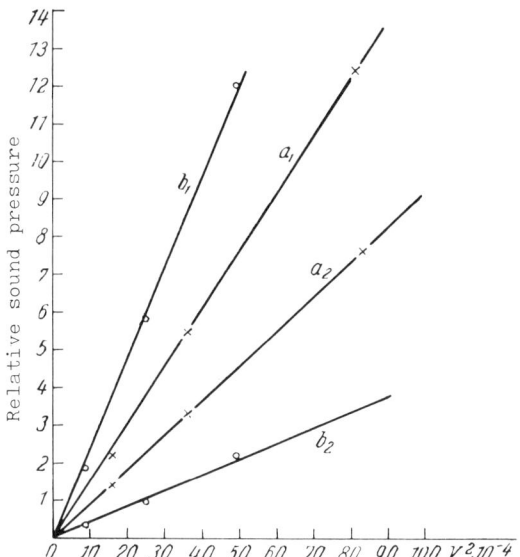

Figure 2 a_1, 800 kHz without tissue, b_1, 2400 kHz without tissue; a_2, 800 kHz with tissue, b_2, 2400 kHz with tissue. Reproduced from *Physik Z.*, **40**(5), 160 (1930); copyright © 1939 by S. Hirzel Verlag.

($CS_2 = 0.14$ for $\nu = 800$ kHz). The last column gives the layer thickness in which the radiation is decreased by one half. For a frequency of 800 kHz, the latter is, therefore, a penetration depth of 4.9 cm. Rows 2 and 3 show the absorption values separately for the same frequency. The relatively inhomogeneous muscle layer exhibits absorption nearly twice that of the fat layer. The corresponding half-value layers are in the last column.

Row 4 shows the dependence of the absorption and penetration depth upon frequency. The absorption α was calculated from curves b_1 and b_2 (Figure 2). It shows, as was suspected, that it is considerably greater at 2400 kHz than for 800 kHz. The radiation decrease results, therefore, in the substantially smaller half-value thickness, as the last column shows. It follows from this that a higher frequency must be chosen for the treatment of skin diseases and the like than for the treatment of deep-lying organs.

An interesting circumstance deserves to be emphasized. The values of α/ν^2, which are recorded in column 4, should, according to classical theory, be independent of frequency. This is generally true within the limits of error for homogeneous liquids. For example, recorded in rows 5 and 6 are the values for carbon disulphide from measurements by Claeys, Errera, and Sack.[5] Therefore, in our case the α/ν^2 value in row 4 ought to be the same as that in row 1, since we are dealing with the same tissue. As seen, this is by no means the case. If the value corresponding to the frequency of 2400 kHz was greater than that of 800 kHz,

[5] *Trans. Faraday Soc.*, **33**, 136 (1937).

this could easily be interpreted as an additional scattering in human tissue. However, since it is smaller, the interpretation runs into considerable difficulty. The answer can possibly be sought on the basis of intercellular cavitation. Investigation of the cavitation phenomena in homogeneous liquids has shown[6] that at high frequencies cavitation decreases noticeably with increasing frequency. However, since cavitation always contributes to absorption, so indeed a too small absorption would result with increasing frequency and the α/ν^2 value would turn out to be too small. However, it is certain that we are not dealing with the usual, well-known cavitation in homogeneous liquids, for this occurs in a discontinuous manner at high intensities, whereas a continuous process is apparently being dealt with here; otherwise the curves of Figure 2 would show deviation from the straight lines. Final clarification of this phenomena can result only from further studies.

SUMMARY

Ultrasonic absorption in human tissue is measured and its dependence on frequency exhibited; it does not follow the α/ν^2 law. The resulting absorption values correspond approximately to that of highly absorbent homogeneous liquids. Muscle layers show approximately twice the absorption as fat layers.

[6] A detailed report concerning this will be published shortly elsewhere.

2
MEASUREMENT OF ULTRASONIC ABSORPTION IN ANIMAL TISSUES AND ITS DEPENDENCE ON FREQUENCY

T. F. Hueter

This article was translated expressly for this Benchmark volume by Floyd Dunn, University of Illinois at Urbana–Champaign, from "Messung der Ultraschallabsorption in tierischen Geweben und ihre Abhängigkeit von der Frequenz." Naturwiss., 35(9), 285–287 (1948)

In recent years ultrasound has acquired increasing importance as a physical therapy agent. In this connection, knowledge of the absorption coefficient, or the half-value layer, in different tissues is of importance to enable one to estimate the effective dose at different body depths, for example, at the ischiatic nerve. For plane wave radiation, the intensity J at depth x is

$$J(x) = J(0) \exp(-\alpha x).$$

According to classical theory, the absorption coefficient α should increase with the square of the frequency f; consequently, $\alpha/f^2 = $ const. According to Pohlman's measurements (1) the validity of this relation was in doubt and a linear relationship was more likely. An exhaustive reexamination of the linear behavior of α in the frequency range from 1 to 4.5 MHz seemed appropriate. For that purpose Pohlman developed an optical method (2) which utilized photometric evaluation according to the Debye–Sears process of the diffraction of light by ultrasonic waves. Thus, the previous method of intensity measurement, which used a sound pressure balance was replaced by a more accurate method.

In the following discussion, results obtained by this method are reported for frequencies of 1.5, 2.4, and 4.5 MHz with tissue specimens of beef kidney, liver, heart, and tongue. After sacrificing the animals, the specimens were immediately transferred to a physiological sodium chloride solution, in a Dewar flask, at a temperature of 35°C and measured at once. A temperature influence on the magnitude of α could not be determined between 20° and 35°C. Measurements were carried out on tissue slices of different thicknesses (about 1, 2.3, and 3.6 cm) to eliminate possible reflections at the tissue boundaries. Since, from the known thickness and speed of sound in the tissue, the resulting reflection factor, relative to water, does not exceed a maximum of 10 percent for different layer thicknesses, approximately equal values of α were

Ultrasonic Absorption in Animal Tissues

expected. Nevertheless, if the tissue specimens were positioned immediately in front of the quartz in the measuring chamber in the interference field, which has a characteristic structure of alternating maxima and minima of intensity, lesser layer thicknesses consistently exhibited larger values of α than did large thicknesses (see Table 1). This astonishing result, which confirms Horvath's observations (3), can be interpreted in terms of an added shearing effect appearing in the interference field, which gives rise to additional internal friction. Since the structure of the interference field becomes homogeneous with increasing distance from the source, the thus limited additional portion of α disappears with greater tissue depth. This increase in α, owing to the structure of the sound field with increased energy transformation in the outer tissue layers, can be welcomed for dermatological uses, although it must be avoided for simple deep therapy (e.g., prostatitis treatment), for example, by means of an inserted water column, as Horvath has proposed.

Table 1

Frequency (MHz)	Layers Thickness (cm)	Kidney α	Kidney $\frac{\alpha}{f}\cdot 10^7$	Liver α	Liver $\frac{\alpha}{f}\cdot 10^7$	Heart α	Heart $\frac{\alpha}{f}\cdot 10^7$	Tongue (trans) α	Tongue (trans) $\frac{\alpha}{f}\cdot 10^7$	Tongue (long) α	Tongue (long) $\frac{\alpha}{f}\cdot 10^7$
1.5	1.0	—	—	0.48	3.2	0.76	5.0	1.31	8.8	0.67	4.8
	2.3	0.38	2.5	0.33	2.2	0.60	4.0	1.10	7.3	0.43	2.9
	3.6	—	—	—	—	—	—	0.9	6.0	—	—
2.4	1.0	—	—	0.62	2.6	0.92	3.8	3.1	12.4	—	—
	2.4	0.54	2.3	0.36	1.5	0.89	3.7	1.3	5.5	—	—
	3.6	—	—	—	—	—	—	—	—	0.64	2.7
4.5	1.0	—	—	1.4	3.2	—	—	4.2	9.2	—	—
	2.4	1.0	2.3	1.2	2.7	1.6	3.5	—	—	1.3	2.9
	3.6	—	—	0.8	1.8	—	—	—	—	—	—

Source: Naturwiss., **35**, 285 (1948); copyright © 1948 by Springer-Verlag.

The measured absorption coefficients are shown in Table 1. The quoted values are averages of numerous repeated measurements on tissue samples from different animals. For the present, this average representation could be extended only over a limited number of specimens for reasons of limited time; thus the values are scattered owing to the indefiniteness of the biological media. These preliminary values for absorption coefficients should be supplemented through further measurements on a greater number of tissue specimens.

The α values of the thickest layers (approximately 3.6 cm) give the best approximation to the absorption coefficients valid for homogeneous ultrasonic radiation (far field). In Figure 1, these values lead to the evident straight lines that largely demonstrate the constancy of α/f, the average values of which are given in Table 2.

The half-value layers, obtained by extrapolation from Figure 1, are given for the customary therapeutic frequency of 800 kHz in the second

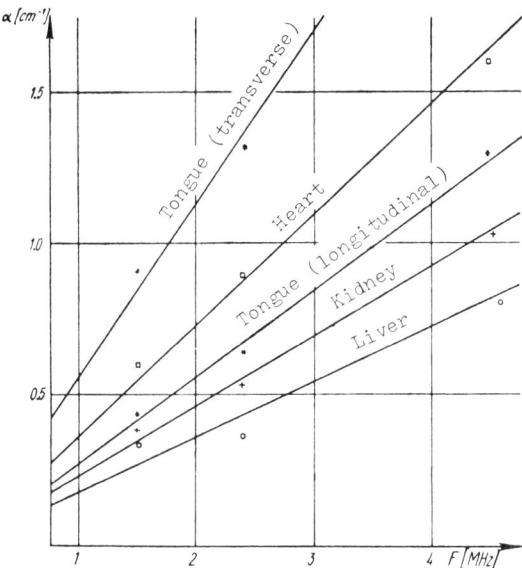

Figure 1 Reproduced from *Naturwiss.*, **35**(9), 286 (1948); copyright © 1948 by Springer-Verlag.

column of Table 2. The half-value layers resulting from 800 kHz agree with the values found by other measuring methods [radiation pressure (1), thermoelectric probes (4)]. It is worth noting here the behavior of the muscle tissue of tongue, which differs from the rest of the very homogeneous tissue materials owing to a large-scale superficial fibrous structure. In the case of the tongue, the nature of the anisotropy of the absorption dependence was determined by measuring not only across, but also along, the fiber direction.

The linear increase of the absorption with frequency is identical with the observed behavior of solids, for which Maxwell proposed a phenomenological theory. The absorption mechanisms were in this case determined first by the grain size or degree of inhomogeneity; here substantial shear and sliding effects were invoked as the mechanisms of internal friction. In the case of tissues the observed effect of the sound

Table 2

Tissue	Average Value $\alpha/f \cdot 10^7$ (cm^{-1} sec)	Half-Value Layer (cm)	
		800 kHz	2.4 MHz
Kidney	2.35	3.7	1.3
Liver	1.8	5.0	1.7
Heart	3.75	2.6	0.9
Tongue (trans)	5.75	1.7	0.6
Tongue (long)	2.8	3.5	1.2

Source: *Naturwiss.*, **35**, 286 (1948); copyright © 1948 by Springer-Verlag.

field structure and tissue inhomogeneity (tongue) underlines the analogy between tissues and solids, in which gum-like substances are most similar to tissue with regard to their elastic behavior and their absorption. Tissue layers of strongly elastic inhomogeneity will give rise to increased energy conversion, as observed at the boundaries by Horvath (5) and Pätzold (6), among others. This is due to the increased shearing effects attributable to distortion of the sound field in the layers.

In this context there will be a detailed report later on experiments on phantoms (which have commenced), on their theoretical significance, and on the practical consequences for dosimetry in ultrasonic therapy.

REFERENCES

1. R. Pohlman, *Physik. Z.*, **40**, 159–161 (1939).
2. Th. Hueter and R. Pohlman, *Z. Angew. Physik*, to appear shortly.
3. J. Horvath, *Klin. Prax.*, **108** (1946).
4. R. Pohlman, *Klin. Wochschr.*, **26**, 277–278 (1948).
5. J. Horvath, *Arzneimittel-Forsch.*, **1**, 357–364 (1947).
6. J. Pätzold and H. Born, *Strahlentherapie*, **76**, 486–492 (1947).

The Velocity of Sound through Tissues and the Acoustic Impedance of Tissues

George D. Ludwig*
Naval Medical Research Institute, Bethesda, Maryland
(Received August 11, 1950)

The velocity of sound through various animal organ tissues and through living human tissues is measured, using an ultrasonic pulse method, at 1.25 and 2.5 Mc. The effect of anisotropy (fiber direction) on velocity is determined with beef muscle. Values obtained with the beam traversing the tissue perpendicularly to the long axis of the muscle bundles do not differ significantly from those found with the energy directed parallel with the muscle fibers.

Velocity through living human tissues, consisting mostly of muscle, is measured by transmitting the ultrasound through various thicknesses of the arm, leg, and thigh.

Specific gravities of the tissues are measured. The characteristic acoustic impedances (ρc values), calculated from the density and velocity data, vary between 1.5×10^5 and 1.7×10^5 g/cm^2/sec. The imaginary component of tissue impedance is calculated and found to be negligible at the frequencies at which these measurements are made.

THE successful application of ultrasonic pulse techniques and the echo-ranging principle to underwater detection and ranging and to the localization of flaws in metals[1] prompted an investigation of the use of an analogous technique for diagnostic purposes in medicine and surgery.[2] The development of ultrasonic

---

* Now at Massachusetts General Hospital, Boston, Massachusetts, and Acoustics Laboratory, Massachusetts Institute of Technology, Cambridge, Massachusetts.

[1] F. A. Firestone, J. Acous. Soc. Am. **17**, 287 (1946).
[2] G. D. Ludwig and F. W. Struthers, "Considerations underlying the use of ultrasound to detect gallstones and foreign bodies in

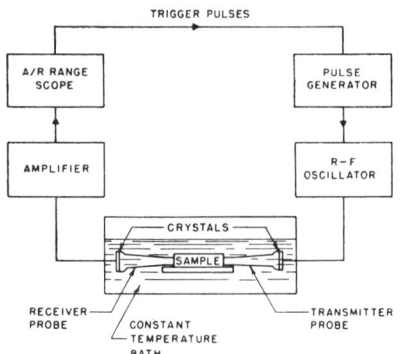

FIG. 1. Schematic diagram of apparatus for measuring velocity of sound through tissues.

instruments and techniques for medical applications requires a knowledge of some of the acoustic propagation characteristics of tissue. Values of the characteristic acoustic impedance of various tissues are needed to calculate reflection coefficients at interfaces such as those between dissimilar tissues, and between foreign bodies and tissue.

Sound velocity and attenuation measurements have been made through certain human and animal tissues. The specific gravity of each tissue was measured and the characteristic acoustic impedance (ρc values) calculated from the velocity and density data. This paper deals with the velocity and acoustic impedance data, some of which have been reported in an earlier paper.[2] The data on the attenuation of sound in tissue will be published in a subsequent paper.

When this work was begun, the literature contained no actual data on the velocity of sound through tissues or acoustic impedances. Pohlman reported measurements of the attenuation of sound in tissue[3] but gave only estimates of velocity and impedance.[4]

EXPERIMENTAL METHOD

For the velocity measurements an ultrasonic pulse technique similar to that described by Pellam and Galt[5] and used by Nolle and Mowry[6] is employed. Instead of a single transducer with a reflector, however, two transducers are used, one to transmit and the other to receive. Each contains an x-cut quartz crystal with the same resonant frequency. The time required for the pulse to travel through various thicknesses of each type of tissue is measured and the velocity is then calculated.

A schematic illustration of the apparatus is given in Fig. 1. The A/R range scope produces trigger pulses at the rate of several hundred per second. Each pulse initiates a new sweep of the oscilloscope and simultaneously triggers the external pulse generator which in turn delivers an r-f pulse to the crystal transducer. The ultrasonic pulse thus generated passes through the sample and is received by the second transducer whose output voltage is amplified and displayed on the oscilloscope screen. The scope is equipped with a delayed sweep that allows the received signal to be placed at the left-hand edge of the oscilloscope trace and the transmission time in "radar yards" to be read from a direct-reading dial. The dial is calibrated with respect to time by an internal crystal that places signal markers on the scope screen at precise intervals. Dial readings are converted to time (in microseconds) to obtain a plot of distance (thickness in centimeters) vs. time.

Various types of crystal holders were fashioned in an effort to achieve maximal transfer of the ultrasonic energy into tissue. Best results have been obtained with a hollow plastic probe. The crystal is mounted at one end. The probe is filled with water and the tip, which measures 1.5 cm in diameter, is closed by a Nylon diaphragm 3 mils thick.

The transducer probes are mounted in the lens holders of a standard optical bench. Movable supports are thus provided which maintain the probes in parallel alignment when they are moved in a horizontal plane. For the measurements on animal organ tissues the optical bench is inverted over a constant-temperature water tank. The transducers and the sample under investigation are immersed in the bath and the sample is supported on a movable tray. Good contact between the probes and the tissue sample is secured. Temperatures of the tissue samples are taken before and after each determination. For these particular experiments the temperature of the bath was maintained at 24°C and the tissue temperatures did not vary from this value by more than 1°C.

The choice of frequency for medical purposes involves many considerations. Since the attenuation in tissues is so great at higher frequencies, the frequency must be maintained low enough or the intensity increased to achieve deep penetration. The danger of tissue damage imposes an upper limit of intensity. Therefore, the frequency must be decreased to allow the desired amount of tissue penetration. However, lowering the frequency increases the wave-length with a resultant decrease in resolving power and beam directivity. In diagnostic applications, where resolution is of great importance, the choice of frequency must be a compromise, low enough to offset the increasing attenuation with increase in frequency and high enough to provide sufficient resolution. Previous experiments[2] had shown that the most desirable frequency range for a diagnostic instrument capable of detecting foreign bodies of the order of 0.5 cm diameter or larger at tissue depths up to 15 cm, is 1.0 to 2.5 Mc. Therefore, the velocity measurements reported here were made at frequencies in this range, namely 1.25 and 2.5 Mc.

---

tissue," Naval Medical Research Institute Project NM. 004 001 Report No. 4 (June, 1949).
[3] R. Pohlman, Physik. Zeits. 40, 159 (1939).
[4] R. Pohlman, Deut. med. Wochschr. 73, 373 (1948).
[5] J. R. Pellam and J. K. Galt, J. Chem. Phys. 14, 608 (1946).
[6] A. W. Nolle and S. C. Mowry, J. Acous. Soc. Am. 20, 432 (1948).

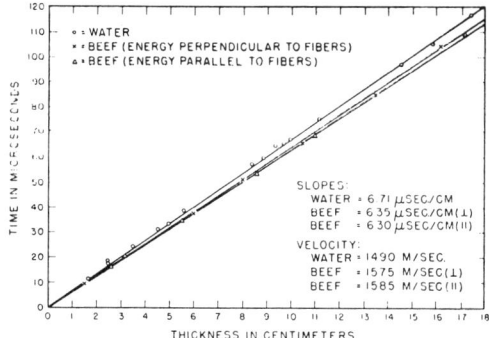

FIG. 2. Transmission time vs. thickness for distilled water and for beef. The velocity values calculated from the slopes of the lines are given in the lower right corner.

EXPERIMENTAL PROCEDURE

Zero time was determined by placing the transmitting and receiving probes in direct contact. A dial reading was taken with the leading edge of the received signal adjusted to coincide with the start of the oscilloscope trace. Samples of tissue of various thicknesses were successively interposed between the probes. The receiving probe position was adjustable to accomodate each sample while the transmitting probe remained fixed. For each sample thickness the received signal was adjusted to the start of the scope trace and a reading taken. The difference in readings between zero time and each sample reading was converted to time. A direct reading of the thickness was obtained from the amount of separation of the lens holders.

The transmission times through the various thicknesses of each tissue were measured and plotted against thickness. The velocity was calculated from the reciprocal of the slope of the line best fitting these data, determined by the method of least squares.

The specific gravity of each tissue was determined by means of the copper sulfate method.[7] Using an appropriate temperature factor, these values were converted to density values. The specific gravity of the test solutions varied between 1.000 and 1.100 in steps of 0.001. The specific gravity of representative sample bottles was checked from time to time on a specific gravity balance.

The characteristic acoustic impedance (ρc value) was calculated from the velocity and density data. The product of the velocity and density expresses the real (resistive) component of the impedance. In order to find the order of magnitude of the imaginary component of the impedance, measurements of the attenuation of sound in various tissues were made at 1.25 and 2.5 Mc. The methods used and the complete data will be re-ported at a later date. Suffice it to say that the attenuation values obtained ranged between 1.0 and 3.0 db/cm. The absorption coefficients calculated from these data agree closely with values reported by Hueter[8] and Hueter and Pohlman[9] who used an optical method. From these absorption values the imaginary component of the impedance has been calculated for tissues and has been found to be negligible. Therefore, in this paper all characteristic acoustic impedances of tissues are expressed as the ρc values.

The absolute accuracy of the velocity measurement is limited by the measurement of sample thickness. With the larger samples of animal tissues the ultrasonic path length in the tissue could be determined to about one part in 200 on the optical bench. The time measurements were accurate to approximately one part in 400 on the radar range scope. The probable error in the final results averaged from measurements on several samples is estimated to be one part in 200.

A check on the method was made by measuring the velocity through distilled water (Fig. 2). A value of 1490 m/sec. (± 0.7 percent) at 25°C was obtained. This agrees within ± 0.5 percent with the values given by Bergmann[10] and the *International Critical Tables*,[11] and with a value obtained at the Naval Research Laboratory with an interferometric method.[12]

EXPERIMENTAL RESULTS

The velocity through sections of boneless beef was measured at 1.25 and 2.5 Mc. The blocks consisted almost entirely of muscle and were cut from quarters of refrigerated beef so that the muscle fibers were oriented in one direction. To determine the effect of the fiber direction (anisotropy), measurements were made with the ultrasonic beam directed parallel to the axis of the muscle bundles and then perpendicular to their long axis. The transmission time through the largest dimension was measured first. Each succeeding measurement was made by cutting a few centimeters from the large block; the ultrasonic beam was directed through the same portion of the block as its thickness was decreased.

The data for the beef are given in Fig. 2 in which the values are compared with the data for distilled water. Average values of 1575 and 1585 m/sec. (temperature 24 to 25°C) were obtained for the transverse (perpendicular to fiber direction) and longitudinal (parallel to fiber direction) irradiations, respectively. These values are not significantly different and are within the experimental error of the method. These data are for 2.5 Mc;

---

[7] R. A. Phillips and D. D. Van Slyke, *Copper Sulfate Method for Measuring Specific Gravities of Whole Blood and Plasma* (from U. S. N. Research Unit, Rockefeller Institute for Medical Research, published by Josiah Macy, Jr. Foundation, New York, February, 1945).

[8] T. F. Hueter, Naturwiss. **9**, 285 (1948).
[9] T. F. Hueter and R. Pohlman, Zeits. f. angew. Physik I, 405 (1949).
[10] L. Bergmann, *Ultrasonics and their Scientific and Technical Applications* (John Wiley and Sons, Inc., New York, 1938), Hatfield translation.
[11] *International Critical Tables* (McGraw-Hill Book Company, Inc., New York, 1926), Vol. VI, p. 464, National Research Council.
[12] R. J. Urick, unpublished data, Naval Research Laboratory (1948).

data obtained at 1.25 Mc give approximately the same values and are not included.

In a similar fashion measurements were made of the velocity of sound through various organ tissues of dog and hog. Immediately the animal was killed the organs were removed and placed in normal saline solution. Measurements were made as soon as the temperature of the tissue came to equilibrium with room temperature (24 to 25°C). The temperature of the sample was held constant at 24 to 25°C during measurements by means of the large constant-temperature bath.

The data for brain, liver, spleen, and kidney are given in Fig. 3 and the velocity values are given in column 1 of Table II. The velocity value for dog and hog brain are approximately equal.

Average values of velocity through certain living human tissues were also obtained. The ultrasonic beam was directed through various thicknesses of calf muscles, thigh and biceps muscles. All these measurements were made with the beam perpendicular to the long axis of the muscles with care being taken to avoid the long bones. The data are given in Table I.

The data are more variable with this type of tissue than with water or beef. For the most part, this is probably attributable to the fact that variable amounts of fat, muscle, connective tissue, blood vessels, and nerves are traversed by the ultrasonic beam as it passes through the calf, thigh, or arm at different points. However, some of the variability can be attributed to the decreased accuracy of measurement of thickness of the living tissues.

TABLE I. Sound velocity at 2.5 Mc through living human tissues consisting mostly of muscle. The standard deviation for each value is approximately two percent.

Tissue	Transmission time μsec./cm	Velocity m/sec.
Leg (calf)		
G.L.	6.20	1610
R.U.	6.35	1575
T.C.	6.66	1500
J.B.	6.36	1565
Arm (biceps)		
G.L.	6.49	1540
R.U.	6.33	1580
T.C.	6.49	1540
J.B.	6.56	1515
A.L.	6.30	1587
Thigh (quadriceps)		
G.L.	6.40	1563
R.U.	6.51	1536
T.C.	6.64	1506
J.B.	6.65	1504
Mean value for human tissue (mostly muscle)	6.49	1540

A mean value for human tissue consisting principally of muscle was found by plotting all the data from measurements on the arms, legs, and thighs (Fig. 4); the best straight-line fit was determined by the method of least squares. The mean velocity, calculated from the

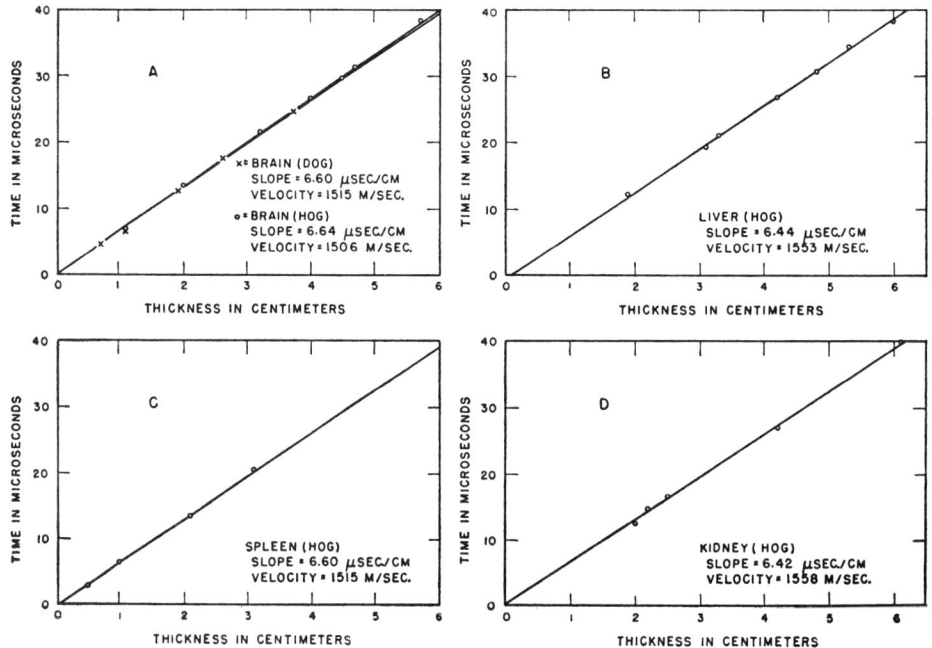

FIG. 3. Transmission time vs. thickness for various animal organ tissues.

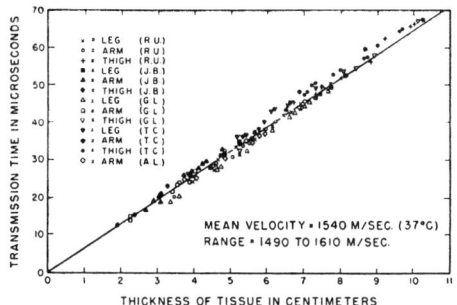

FIG. 4. Transmission time vs. thickness for living human tissue consisting principally of muscle. The line is the best straight-line fit of all data taken on living human tissues.

TABLE II. Velocity (c), density (ρ), and acoustic impedance (ρc) of tissue. The standard deviation in values for the animal tissues is ±1 percent.

Tissue	Velocity m/sec.	Density g/cm³	Acoustic impedance g/cm²/sec. ×10⁵
Brain (dog)	1515	1.028	1.56
Brain (hog)	1506	1.026	1.55
Spleen (hog)	1515	1.059	1.60
Liver (hog)	1553	1.064	1.65
Kidney (hog)	1558	1.040	1.62
Beef	1575–1585	1.068	1.68–1.69
Human tissue (mean value)	1490–1610 1540	1.06 1.06	1.58–1.70 1.63
Water	1490	1.00	1.49

slope of the line, was 1540 m/sec. Temperature for all the human tissue measurements was that of body temperature (37°C).

Ten small pieces, each approximately 1 mm³, were cut from each block of beef and organ tissue. The specific gravity of each was determined and a mean value obtained which was converted to a density value by appropriate temperature corrections. The density values agree closely with those obtained by other investigators.[13,14] Behnke[15] gives 1.06 for the over-all average specific gravity of human tissue. This was used in computing the characteristic acoustic impedance of living human tissue. The values for beef, organs and living tissue, are given in column 2 of Table II. The characteristic acoustic impedance of these tissues are given in column 3 of Table II.

SUMMARY

Sound velocity through tissues has been measured at frequencies of 1.25 and 2.5 Mc, using a pulse method. Values obtained at these frequencies are identical indicating that dispersion does not occur, at least in this range.

The effect of the anisotropy (fiber direction) of the tissue on the sound velocity was investigated with beef muscle. Values obtained with the energy traversing the tissue perpendicularly to the long axis of the muscle bundles do not differ significantly from those found with the irradiation directed parallel with the muscle bundles.

Values for brain, liver, kidney, and spleen of the dog and hog and for beef muscle vary between 1506 and 1585 m/sec. (24 to 25°C).

The velocity through living human tissue has been measured by transmitting the ultrasonic beam through the muscles of the leg, arm, and thigh of different individuals. A range of values between 1490 and 1610 m/sec. with a mean value of 1540 m/sec. is obtained.

The specific gravities of the animal tissues were measured; the values ranging from 1.026 to 1.068.

The characteristic acoustic impedances of these tissues were calculated. Values for impedance vary between 1.5×10^5 and 1.7×10^5 g/cm²/sec. These values, which were calculated from the velocity and density data express only the real component of the impedance. The imaginary component was calculated by utilizing data on the absorption of sound in tissue. In each case, the reactive component of the impedance has been found to be negligible, at the frequencies at which the measurements were made.

ACKNOWLEDGMENT

The author wishes to acknowledge the advice and criticism of Drs. J. P. Flynn, D. E. Goldman, and K. S. Cole of the Naval Medical Research Institute. He is also indebted to the General Precision Laboratory, Inc., Pleasantville, New York, for the generous use of its instruments and especially for the aid rendered by I. A. Greenwood of its staff. The kind cooperation of Dr. R. H. Bolt, of the Acoustics Laboratory, Massachusetts Institute of Technology, during the preparation of this report is greatly appreciated.

---

[13] H. Vierordt, *Anatomische, physiologische und physikalische Daten und Tabellen* (Gustav Fisher, Jena, 1906), third revised edition.

[14] Gersh, Hawkinson, Rahbun, and Behnke, "Changes in specific gravity of tissues, organs, and the animal as a whole resulting from rapid decompression of Guinea pigs from high pressure atmospheres," Naval Medical Research Institute, Project X-284 Report No. 2 (1944).

[15] A. R. Behnke, "Physiologic studies pertaining to deep sea diving and aviation, especially in relation to the fat content and composition of the body," The Harvey Lecture Series 37, 198–226 (1942).

4
ULTRASONIC ABSORPTION MEASUREMENTS IN HUMAN SKULL BONE AND THEIR DEPENDENCE UPON FREQUENCY

T. F. Hueter

This article was translated expressly for this Benchmark volume by Floyd Dunn, University of Illinois at Urbana–Champaign, from "Messung der Ultraschallabsorption im menschlichen Schädelknochen und ihre Abhängigkeit von der Frequenz," Naturwiss., 39(1), 21–22 (1952)

In an earlier publication (1), attenuation constants, determined by an optical method (2), were reported for different animal tissues. Recent investigations of the feasibility of using ultrasonic waves for diagnostic purposes, particularly for exhibiting the cerebral ventricles, have led to corresponding measurements on human skull bones with an impulse method (3).

The apparatus employed operated with impulses from 5- to 25-μs duration in the frequency range from 300 kHz to 3.5 MHz. Four measuring paths were arranged in a water-filled trough, each equipped with a sender–receiver pair made of barium titanate ceramic. The four measuring paths were excited by choice in their respective fundamental waves (0.3, 0.56, 1.15, and 2.36 MHz) or in various harmonics. Bone tissue specimens of different thicknesses were placed in the radiation path and the additional attenuation measured.

The bone specimens were obtained with a new kind of cylindrical boring tool (4). Thirty-five different specimens were measured at eight frequencies. These specimens were taken post mortem from seven autopsies and from different places in the skulls. During mechanical working (e.g., grinding thin specimens) of such fresh specimens, their consistency, and thereby their attenuation behavior, was greatly changed. Thus, there remained at one's disposal only the natural variation in thickness between the individual bone specimens for determination of the attenuation per unit thickness and, therewith, the reflection loss.

As a typical example of the scatter occurring with such measurements, Figure 1 shows the measured values for the frequency of 1.8 MHz. This scattering is primarily to be traced back to natural variations in the structure of the different bone specimens and, secondarily, to deficiencies in the parallelism of the specimens.

After determining the reflection losses from the ordinate point of intersection from Figure 1 for all investigated frequencies, the corresponding average straight line was determined (by the method of least squares); its slope yields the attenuation in decibels per centimeter.

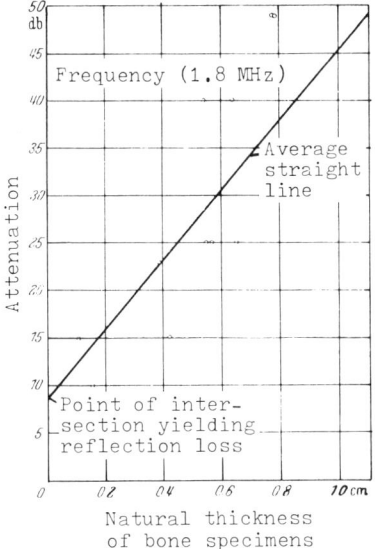

Figure 1 Example of the evaluation of the measured points. Reproduced from *Naturwiss.*, **39**(1), 21 (1952); copyright © 1952 by Springer-Verlag.

Figure 2 shows, in this way, the frequency behavior of the attenuation coefficients of skull bone, which exhibits strong deviation from a linear law. The only known value in the literature, by Theismann (5) at 800 kHz, lies essentially too high at 27 dB/cm. The general course of the curve shows strong similarity to the frequency dependence found by Mason (6) for attenuation in polycrystalline metals, which, for the investigated frequency range, permits the conclusion that the energy loss occurs

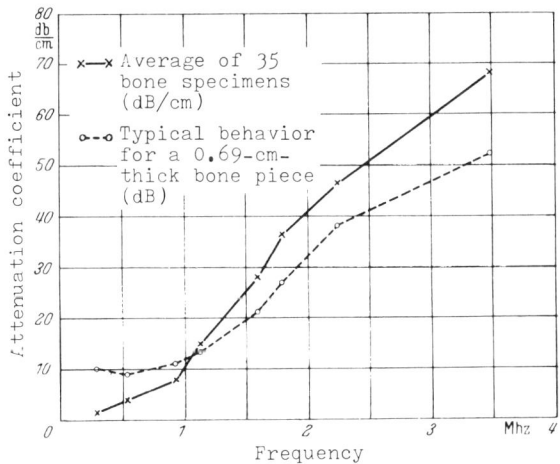

Figure 2 Frequency variation of the ultrasonic attenuation coefficient of human skull bone. Reproduced from *Naturwiss.*, **39**(1), 21 (1952); copyright © 1952 by Springer-Verlag.

through scattering. In distinction to bone, the acoustic attenuation in the soft body tissues is essentially determined by its viscous behavior (7).

Although severe variations appear among the attenuations of the different bone specimens, it has been shown that each specimen obeys a similar frequency rule. This is illustrated by the course of the dashed curve of attenuation in Figure 2 for a typical bone piece 0.69 cm thick. The observed nonlinearity of the attenuation frequency curve permits one to overcome the difficulties resulting from the ultrasound ventriculography experiment in regard to the separation of the brain portion and the bone portion of the received sound signal level (8). Results of such work, currently under way, that use frequency exploration will be extensively reported later.

This work was carried out in the Acoustics Laboratory, Massachusetts Institute of Technology, Cambridge, Massachusetts, USA (supported by a grant from the National Institutes of Health).

REFERENCES

1. Th. Hueter, *Naturwiss.* **35**, 285 (1948).
2. Th. Hueter and R. Pohlman, *Z. Angew. Phys.*, **2**, 75 (1949).
3. K. Th. Dussik and L. Wyt, *Wien. Med. Wochschr.*, **1947**, 425.
   ———, T. Ballentine, R. Bolt, T. Hueter, and G. Ludwig, *Science* (Lancaster (Pa.) **112**, 525 (1950).
4. Developed by A. Metcalf and T. Ballentine, Massachusetts General Hospital, Boston, Mass.
5. H. Theismann and F. Pfander, *Strahlentherapie*, **80**, 607 (1949).
6. W. P. Mason, *Piezoelectric Crystals and Their Application to Ultrasonics*, p. 427. New York: Van Nostrand Reinhold Company, 1950.
7. H. L. Ostreicher, *J. Acoust. Soc. Amer.* **23** (1951).
8. Th. F. Hueter and R. H. Bolt, *J. Acoust. Soc. Amer.* **23**, 160 (1951).

Tabular Data of the Velocity and Absorption of High-Frequency Sound in Mammalian Tissues

D. E. GOLDMAN,* *Naval Medical Research Institute, Bethesda, Maryland*

AND

T. F. HUETER, *Massachusetts Institute of Technology, Cambridge, Massachusetts*

(Received August 29, 1955)

This report is a condensed presentation of currently available data on the velocity and absorption of high-frequency sound in mammalian tissues.

IN the last several years, interest in the biological and medical applications of ultrasound has led to studies on the acoustic properties of cells and tissues. As interest continues to develop, it becomes useful for workers in the field to have readily available such data as have already been accumulated.

TABLE I. Velocity of high-frequency sound in mammalian tissues.

Species	Tissue	Condition	Temp °C	Frequency mcps	Velocity m/sec	Precision percent	Source
Man	muscle	refrig.	24	1.8	1568	0.5	e
	liver	refrig.	24	1.8	1570	0.5	e
	fat	...	24	1.8	1476	0.5	e
	limb	in vivo	body	2.5	1540	(1490–1610)	f
	meningioma	...	body	2.26	1540	...	g
	skull bone	fresh	body	0.8	3360	...	h
	breast carcinoma	refrig.	24	1.8	1573	0.5	e
Dog	muscle	fresh[a]	26	4, 12	1592	0.2[d]	i
	muscle	fresh[b]	26	4, 12	1576	0.2[d]	i
	brain	fresh	25	2.5	1520	1.	f
	liver	fresh	26	4, 12	1580	0.4	i
Pig	muscle	fresh	24	1.8	1580	0.5	e
	brain	fresh	24	1.8	1565	0.5	e
	brain	fresh	25	2.5	1510	1.	f
	liver	fresh	24	1.8	1585	0.5	e
	liver	fresh	25	2.5	1550	1.	f
	spleen	fresh	24	1.8	1576	0.5	e
	spleen	fresh	25	2.5	1520	1.	f
	kidney	fresh	24	1.8	1560	0.5	e
	kidney	fresh	25	2.5	1560	1.	f
	fat	fresh	24	1.8	1443	0.5	e
	fat	refrig.	37	1.6	1410	...	j
Cow	muscle	fresh	24	1.8	1580	0.5	e
	muscle	refrig.	25	2.5	1580	1.	f
	brain	fresh	24	1.8	1560	0.5	e
	liver	fresh	24	1.8	1590	0.5	f
	spleen	fresh	24	1.8	1577	0.5	e
	kidney	fresh	24	1.8	1568	0.5	e
	fat	fresh	24	1.8	1465	0.5	e
Horse	plasma	refrig.	37	1.0	1571	0.1	k
	blood	refrig.	37	1.0	1571	0.1	k
	muscle	fresh	24	1.8	1595	0.5	e
	brain	fresh	24	1.8	1560	0.5	e
	liver	fresh	24	1.8	1580	0.5	e
	spleen	fresh	24	1.8	1591	0.5	e
	kidney	fresh	24	1.8	1558	0.5	e
	fat	fresh	24	1.8	1443	0.5	e
Rabbit	muscle	fresh[a]	26	4, 12	1603	0.2[d]	i
	muscle	fresh[b]	26	4, 12	1587	0.2[d]	i
	liver	fresh	24	1.8	1599	0.5	e
	liver	fresh[c]	24	1.8	1607	0.5	e
	liver	fresh	26	4, 12	1575	0.2	i
Guinea Pig	liver	fresh	24	1.8	1575	0.5	e
	liver	fresh[c]	24	1.8	1589	0.5	e

[a] Sound traveling perpendicular to direction of fibers.
[b] Sound traveling in direction of fibers.
[c] Animal killed and then bled before removal of liver in order to reduce blood content.
[d] Difference in velocity is statistically significant at 1% confidence level.
[e] A. H. Frucht, Z. ges. exptl. Med. **120**, 526 (1953).
[f] G. D. Ludwig, J. Acoust. Soc. Am. **22**, 862 (1950).
[g] H. T. Ballantine *et al.*, M.I.T. Acoustics Lab. Quarterly Progress Report, January–March, 1951.
[h] H. Theismann and F. Pfander, Strahlentherapie **80**, 607 (1949).
[i] D. E. Goldman and J. R. Richards, J. Acoust. Soc. Am. **26**, 981 (1954).
[j] Schwan, Carstensen, and Li, Trans. Am. Inst. Elect. Engrs. Part I, **72**, 483 (1953).
[k] R. J. Urick, J. Appl. Phys. **18**, 983 (1947).

* The opinions expressed herein are those of the author and do not necessarily reflect the views of the U. S. Navy or the naval service at large.

TABLE II. Absorption of high-frequency sound in tissues.

Species	Tissue	Condition	Frequency mcps	Absorption cm^{-1} e d	Half-value layer cm^g	Source
Man	plasma	refrigerated	0.87	0.02	17.	h
	plasma	refrigerated	1.7	0.04	8.7	h
	plasma	...	1.0	0.007	100.0	i
	blood	...	1.0	0.02	35.	i
	blood	...	0.8	$0.002V^e$	...	j
	blood	...	1.2	$0.0045V^e$	...	j
	blood	...	2.4	$0.011V^e$	...	j
	muscle	...	0.80	0.1	3.6	k
	brain	fixed	0.30	0.09	4.1	h
	brain	fixed	0.87	0.14	2.5	h
	brain	fixed	1.7	0.18	1.9	h
	brain	fixed	3.4	0.37	0.9	h
	medulla oblongata	...a	1.7	0.14	2.5	h
	medulla oblongata	...a	3.4	0.34	1.0	h
	medulla oblongata	...b	1.7	0.22	1.6	h
	medulla oblongata	...b	3.4	0.46	0.75	h
	liver	20 hr post mortem	1.	0.15	2.4	l
	liver	20 hr post mortem	3.	0.23	1.5	l
	liver	20 hr post mortem	5.	0.35	1.0	l
	fat	...	0.80	0.05	6.9	k
	fat	melted	0.87	0.045	7.7	h
	fat	melted	1.7	0.09	4.1	h
	fat	melted	3.4	0.16	2.2	h
	skull bone	fresh or fixed	0.8	1.5	0.23	m
	skull bone	...	0.6	4.5	0.077	n
	skull bone	...	0.8	9.	0.038	n
	skull bone	...	1.2	17.	0.020	n
	skull bone	...	1.6	32.	0.011	n
	skull bone	...	1.8	42.	0.0083	n
	skull bone	...	2.25	53.	0.0065	n
	skull bone	...	3.5	80.	0.0043	n
	sciatic nerve	...a	3.4	0.35	1.0	h
	sciatic nerve	...b	3.4	0.55	0.63	h
Pig	brain	...	0.8	0.025	14.	o
	brain	...	2.4	0.075	4.6	o
	brain	...	0.35–2.4	$0.055f^f$	...	p
	brain (white matter)	fresh	2.5	0.19	1.8	q
	fat	...	1.6	0.035	10.	r
	fat	...	2.5	0.1	3.5	r
	fat	...	4.0	0.2	1.7	r
	fat	...	6.0	0.3	1.2	r
	fat	...	7.0	0.4	0.87	r
Cow	muscle (gluteal)	...a	0.3	0.1	3.5	h
	muscle (gluteal)	...a	0.87	0.2	1.7	h
	muscle (gluteal)	...a	1.7	0.25	1.4	h
	muscle (gluteal)	...a	3.4	0.6	0.58	h
	muscle (gluteal)	...b	0.3	0.075	4.6	h
	muscle (gluteal)	...b	0.87	0.05	6.9	h
	muscle (gluteal)	...b	3.4	0.25	1.4	h
	brain	fresh	0.87	0.1	3.5	h
	brain	fresh	1.7	0.15	2.3	h
	brain	fresh	3.4	0.35	1.0	h
	heart	...	0.8	0.2	1.7	s
	heart	...	1.5	0.35	1.0	s
	heart	...	2.4	0.4	0.87	s
	heart	...	4.5	0.8	0.43	s
	heart	...	0.35–4.5	$0.19f^f$	...	p
	liver	...	0.8	0.07	4.9	s
	liver	...	1.5	0.2	1.7	s
	liver	...	2.4	0.2	1.7	s
	liver	...	4.5	0.55	0.63	s
	liver	...	0.3	0.1	3.5	h

TABLE II.—(Continued)

Species	Tissue	Condition	Frequency mcps	Absorption cm^{-1} c d	Half-value layer cm^g	Source
	liver	...	0.87	0.1	3.5	h
	liver	...	1.7	0.15	2.3	h
	liver	...	3.4	0.25	1.4	h
	liver	...	0.35–4.5	0.10f^f	...	p
	liver	fresh	10.	1.37	0.25	t
Cow	liver	fresh	23.	3.0	0.115	t
	kidney	...	0.8	0.1	3.5	s
	kidney	...	1.5	0.2	1.7	s
	kidney	...	2.4	0.25	1.4	s
	kidney	...	4.5	0.5	0.69	s
	kidney	...	0.35–4.5	0.11f^f	...	p
	tongue	...^{b}	0.8	0.2	1.7	s
	tongue	...^{b}	1.5	0.55	0.63	s
	tongue	...^{b}	2.4	1.0	0.35	s
	tongue	...^{b}	4.5	2.1	1.6	s
	tongue	...^{a}	0.8	0.1	3.5	s
	tongue	...^{a}	1.5	0.25	1.4	s
	tongue	...^{a}	2.4	0.3	1.2	s
	tongue	...^{a}	4.5	0.65	0.53	s
Rat	spinal cord	fresh	0.98	0.09–0.12	3.6–3	u

^{a} Sound traveling in direction of fiber axis.
^{b} Sound traveling perpendicular to fiber axis.
^{c} Absorption coefficient α defined in $A = A_0 e^{-\alpha x}$, where A is amplitude of sound, and x is distance traveled through tissue.
^{d} Precision usually 10–25%.
^{e} V is volume concentration of cells in percent.
^{f} f is frequency in megacycles.
^{g} Half-value layer ($=(\ln)2/2\alpha$) is distance traveled to reduce *intensity* to half original value.
^{h} S. Colombati and S. Petralia, Riceica Sci. **20**, 71 (1950).
^{i} H. P. Schwan and E. L. Carstensen, J. Am. Med. Assoc. **149**, 121 (1952).
^{j} Carstensen, Li, and Schwan, J. Acoust. Soc. Am. **25**, 286 (1953).
^{k} R. Pohlman, Physik. Z. **40**, 159 (1939).
^{l} K. T. Dussik, Prog. Rept. National Institute of Health Project A-545, April 15, 1955.
^{m} H. Theisman and F. Pfander, Strahlentherapie **80**, 607 (1949).
^{n} T. F. Hueter, Naturwissenschaften **39**, 21 (1952).
^{o} Guttner, Fiedler, and Patzold, Acustica **2**, 148 (1952).
^{p} R. Esche, Akust. Beih. **2**, 71 (1952).
^{q} T. F. Hueter and R. H. Bolt, J. Acoust. Soc. Am. **23**, 160 (1951).
^{r} H. P. Schwan and E. L. Carstensen (unpublished data).
^{s} T. F. Hueter, Naturwissenschaften **35**, 285 (1948).
^{t} T. F. Hueter and M. S. Cohen, unpublished data presented at the Acoust. Soc. meeting in Cleveland, Ohio, October, 1953.
^{u} W. J. Fry and R. B. Fry, J. Acoust. Soc. Am. **25**, 6 (1953).

Tables I and II contain values of the velocity and absorption of high-frequency sound in a number of mammalian tissues of several species. The material is believed to be up to date and the tables give as much accessory information as has seemed to lend itself to condensed presentation. No critical discussion is indicated at this time, but it is anticipated that the accumulation of further data on the basis of more precise and extensive measurements should permit important generalizations on the acoustic characteristics of living matter.

We note from Table I that most measured velocities fall in the range from 1500 to 1600 m/sec, and that no significant frequency dispersion of velocity has been observed to date. Table II reveals that the absorption coefficients of most soft tissues lie in the range from 0.5 to 2 db/cm/Mc. Looking at the available data for each kind of tissue as a whole one finds that the absorption values lie within broad regions, as shown in Fig. 1, which indicate a constant loss per cycle (α/f=const) over the hitherto explored frequency range from 0.3 to 10 mcps. The significance of this finding in terms of basic acoustic loss mechanisms (relaxation, hysteresis, etc.) can only be determined by more accurate measurements over a wider frequency range and at different temperatures.

Comparing the various soft tissues amongst each other one notices that fat exhibits the lowest value of both sound velocity and absorption and that muscle shows the highest values in both respects.

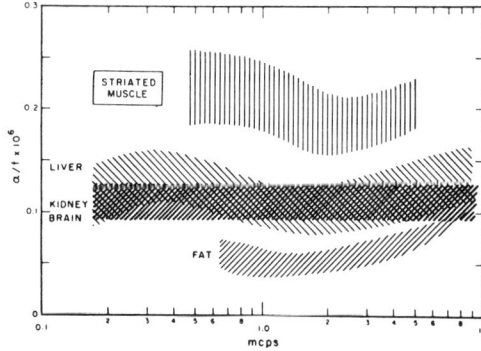

FIG. 1. Estimated relations between sound absorption and frequency for several mammalian tissues.

Determination of the Acoustic Properties of Blood and its Components*

EDWIN L. CARSTENSEN, KAM LI, AND HERMAN P. SCHWAN

Moore School of Electrical Engineering and Department of Physical Medicine, University of Pennsylvania, Philadelphia 4, Pennsylvania

Measurements of absorption and velocity of sound in blood, plasma, and solutions of albumin and hemoglobin have been carried out in the frequency range 800–3000 kc and temperature range 5–45°C. The absorption departs only slightly from a linear dependence upon frequency. Absorption for the various solutions is in direct proportion to protein content. It is concluded that the acoustic properties of blood are largely determined by the proteins which it contains.

CLINICAL interest in high frequency sound as a therapeutic agent is continually increasing. A wide range of applications have been suggested for its use[1]. From physical considerations, it appears that ultrasound fortuitously combines adequate depth of penetration in tissues with wavelengths short enough to permit sharp beaming, thus making it possible to provide a localized heating in the deep body tissues[2].

An investigation of the acoustic properties of the biological medium may be expected, first, to provide a quantitative basis for the phenomenological description of the heating processes and, second, to lead to an understanding of the mechanism of absorption.

Measurements of the absorption and velocity of sound in some of the solid tissues have been reported in the literature[3–6]. Blood was chosen to begin this investigation because it contains cells and in this sense is similar to body tissues. Yet, it is sufficiently homogeneous that it can be measured with a relatively high degree of accuracy.

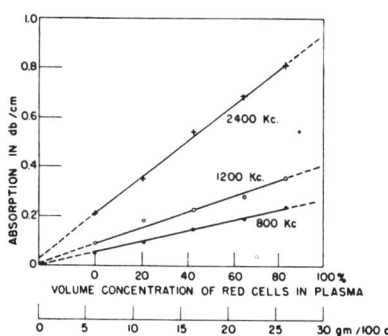

FIG. 1. Absorption of sound in human blood *versus* volume concentration of red cells.

---

* Aided by a grant from the National Foundation for Infantile Paralysis, Inc.
[1] *Der Ultraschall in der Medizin* (Hirzel, Zurich, 1949).
[2] H. P. Schwan, and E. L. Carstensen, J. Am. Med. Assoc. **149**, 121 (1952).
[3] R. Pohlman, Physik. Z. **40**, 159 (1939).
[4] T. Hüter, Naturwissenschaften **35**, 285 (1948).
[5] G. D. Ludwig, J. Acoust. Soc. Am. **22**, 862 (1951).
[6] R. Esche, Akust. Beih. **1**, 71 (1952).

EXPERIMENTAL PROCEDURE

Absorption measurements were carried out by a two-transducer pulse technique which has been described in detail previously.[7] The unusual feature of the measurement is that transducer separation is maintained constant, and continuously varying amounts of water are substituted for test liquid in the path between the transducers. This is accomplished by using a two-chamber test vessel with water on one end separated from the test liquid on the other end by a thin plastic window. The transducers are located on an assembly such that the source is in the water and receiver is in the test liquid. Substitution of water for test liquid is achieved by moving the entire transducer assembly along the axis of the test vessel. By maintaining constant transducer separation, it is possible to avoid difficulties which arise from complex variations in the field near a transducer.[8] This technique is particularly suited to measurements of water solutions or liquids with characteristic impedance approaching that of water. Errors arising from reflections at the interface between the two liquids, the transducer faces, and the sides of the test vessel are eliminated largely by use of pulsing techniques and directional transducers. Refraction of the sound beam at the interface between the liquids presents the possibility of error. However, experimental checks have shown that these errors can be made negligible, even for poorly matched liquids, by careful alignment of the transducers to provide normal incidence of the sound wave at the interface. The over-all error of the absorption determinations is estimated to be approximately ±10 percent or 0.02 db/cm, whichever is the larger.

To determine phase velocity, the wavelength in test liquid is measured by comparing the phase of a direct signal from the oscillator with that of the rf output of the receiver as its position is varied relative to the source. The accuracy of velocity measurements for biological substances was of the order of ±0.5 percent.

Both water and the test liquid were placed under vacuum before measurement to remove a part of the normally dissolved gases. The temperature was maintained constant to within a few tenths of a degree

---

[7] H. P. Schwan, and E. L. Carstensen, Electronics, July, 1952, p. 216–220.
[8] H. Born, Z. Physik **120**, 383 (1943).

centigrade. It was necessary to circulate blood during measurements, to prevent sedimentation of red cells.

EXPERIMENTAL RESULTS

Exploratory absorption measurements were made using the red cell residues of centrifuged human blood. The result of this work is illustrated in Fig. 1. The residues, containing about 85 percent red cells, were diluted with plasma and curves of absorption as a function of concentration of red cells in plasma were obtained. The curves are straight lines, indicating negligible interaction among the absorbers up to maximum concentration. These data are shown as a function of frequency[9] in Fig. 2. It is apparent that the absorption in plasma is not negligible but actually accounts for a significant portion of the absorption in whole blood. The high absorption in plasma is an indication that the cells, as such, are at least not solely responsible for the absorption. Other measurements, which were carried out, have shown that the cell membrane alone can account for a maximum of one-tenth of the absorption observed for red cells.

On the other hand, it is interesting to examine the data of Fig. 1 in terms of protein concentration of the solutions. Actual protein determinations were not made on the samples used in those measurements. However, the lower abscissa of Fig. 2 indicates roughly the protein concentration of the solution as estimated on the basis of generally accepted values of protein content of plasma and red cells. Extrapolation of the absorption curves gives to a first approximation zero absorption for zero protein concentration. Actually, if proteins were principally responsible for the absorption, it would be anticipated that the extrapolated curves would intercept the zero protein concentration axis at the value of the absorption for water. This is the case for the 800 kc and 1200 kc curves. The high intercept for the 2400 kc curve may be explained by the independent behavior of plasma at high frequencies as indicated in subsequent discussion. In reality, several different proteins are involved in this comparison. The red cell protein is almost entirely hemoglobin, while roughly 60 percent of the total plasma protein is albumin. These two proteins are similar. Both have molecular weights of the order of 70 000 and axis ratios between 1:5 and 1:9. All of the remaining plasma proteins, principally the globulins, are orders of magnitude heavier than albumin. Yet, the data of Fig. 1 provide a strong indication that the proteins of blood are responsible for its absorption.

A more extensive investigation was subsequently conducted to confirm the importance of the protein in the absorption. Plasma and red cell concentrate

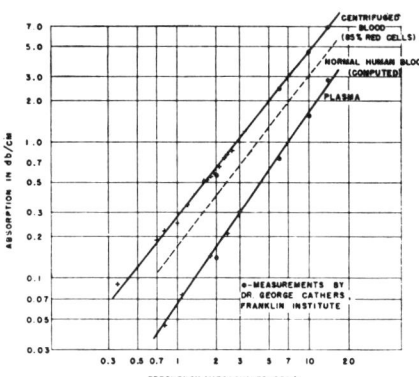

FIG. 2. Absorption of sound in human blood.

were obtained from horse blood by sedimentation. Sodium citrate solution was added to prevent coagulation. The protein concentration of each solution was determined by the Kjeldahl method. Protein concentration for plasma samples ranged from 5.4 to 6.0 g/100 cc, and for red cell concentrates from 32 to 33 g/100 cc. In addition, a 7 percent solution of human hemoglobin[10] in dextrose and water and a 12.5 percent solution of human serum albumin in distilled water were obtained. Absorption for these solutions was measured over the frequency range from 800 to 3000 kc/sec and for temperatures from 5° to 45°C. The data are summarized briefly in Figs. 3–5. The measured absorption, in db/cm, has been converted to absorption per unit quantity of protein present in the solution by dividing by the measured protein content. The contribution of water alone to the absorption is considered negligible.

The data for all the solutions are presented as a function of frequency at 10°C, 20°C, 40°C in Figs. 3, 4, and 5, respectively. At 40°C (Fig. 5) it is readily seen that there is no significant departure of any of the solutions from their average as indicated by the solid line. The same situation holds in general at 10°C and 20°C (Figs. 3 and 4). The agreement among the data for the various solutions shows that absorption is a direct function of protein concentration. The proportionality is the same for both albumin and hemoglobin within the limits of error. This relation appears to hold whether the protein is in solution or contained in the cells.

An interesting departure from the relation occurs in the case of plasma at low temperatures and high frequencies as shown in Figs. 3, 4, and 7. The excess absorption in plasma relative to the other solutions may be caused by the presence of the larger plasma proteins.

---

[9] The points above 3 mc, as shown in Fig. 3 were obtained at the Franklin Institute through the cooperation of Dr. George I. Cathers. Each circle represents a single measurement on one sample of red cells or plasma. This will provide an indication of behavior at frequencies higher than otherwise obtained in this investigation.

[10] R. B. Pennell, and W. C. Smith, J. Hematology 4, 380 (1949). This method uses a 6 percent dextrose solution as a medium for the hemoglobin. Six percent dextrose alone has negligible absorption.

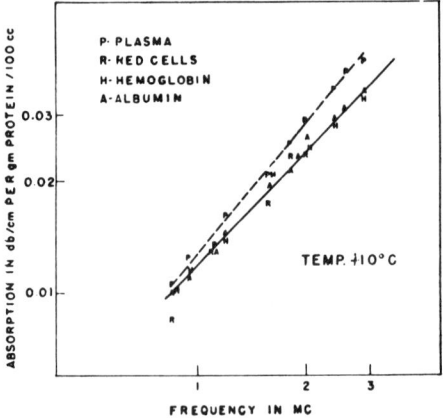

FIG. 3. Absorption of sound in protein solutions +10°C.

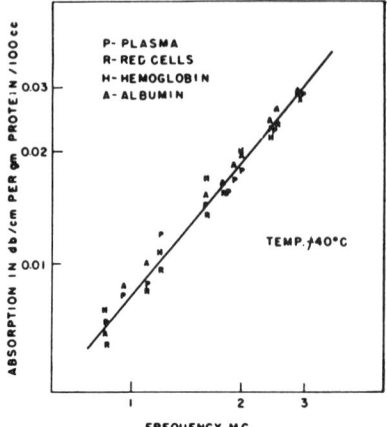

FIG. 5. Absorption of sound in protein solutions +40°C.

The absorption departs only slightly from a linear dependence upon frequency in the range measured. For clarity, the data of Figs. 3, 4, and 5 have been presented in Fig. 6 without the experimental points. The slope of the absorption *versus* frequency curves in log log presentation is approximately 1.2.

The temperature coefficient of absorption is small but definitely negative. Averages of absorption for all the solutions are plotted as a function of temperature for 1, 2, and 3 mc in Fig. 7. The independent behavior of plasma at low temperatures and high frequencies is also indicated.

The velocity of sound was determined for the various solutions described previously. Typical data are shown in Fig. 8, which gives the velocity of sound in 6.2 percent and 12.5 percent solutions of albumin in distilled water, as well as a curve for water alone. In general, addition of protein to a solvent has the effect of increasing velocity of sound. For all solutions measured, this increase was found to go in direct proportion to the protein concentration, averaging roughly 4 m/sec per gm/100 cc protein. Velocity measurements for all samples were performed at both 800 and 2400 kc. No dispersion was observed in this range.

SUMMARY

On the basis of this investigation, it can be concluded that the acoustic properties of blood are determined largely by the proteins which it contains. Absorption of sound has been shown to be directly proportional to the protein concentration whether in solution or contained within cells. Although the similar molecules

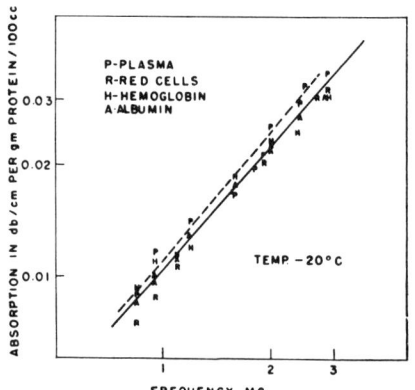

FIG. 4. Absorption of sound in protein solutions +20°C.

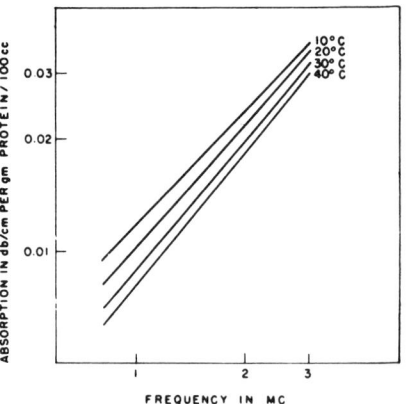

FIG. 6. Average absorption *versus* frequency for protein solutions (red cells, hemoglobin, albumin).

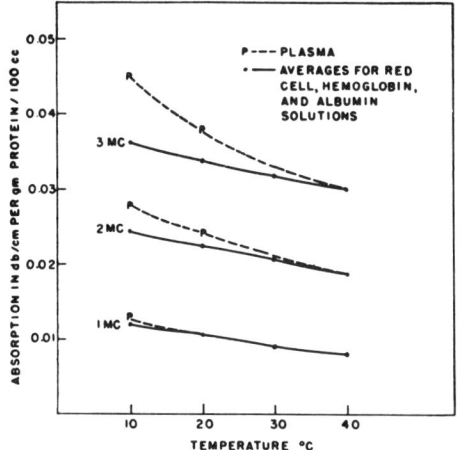

FIG. 7. Absorption *versus* temperature for protein solutions. Solid curve is average for red cells, hemoglobin and albumin. The independent behavior of plasma is indicated by dotted line.

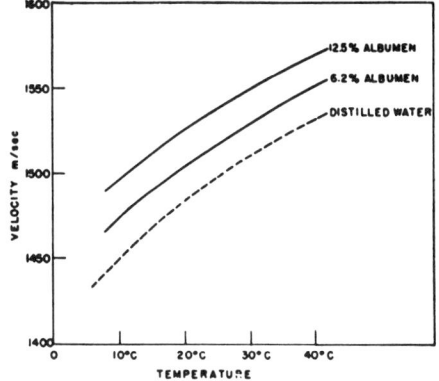

FIG. 8. Velocity of sound in albumin solutions.

albumin and hemoglobin have similar absorption, there is some indication that the larger plasma proteins have somewhat different characteristics.

The authors wish to acknowledge the generous help and guidance of Dr. Robert B. Pennell, of Sharpe and Dohme, Inc., Glenolden, Pennsylvania, who supplied the materials for this investigation and assisted through many helpful discussions; Dr. George I. Cathers, formerly of the Franklin Institute, who performed the preliminary measurements mentioned in the text and supplied helpful advice at the initiation of this work; Dr. John G. Reinhold, Hospital of the University of Pennsylvania, who performed the protein determinations on blood samples; and Dr. Eugene Ackerman and Mr. John Parnell for helpful discussions.

THE JOURNAL OF THE ACOUSTICAL SOCIETY OF AMERICA   VOLUME 25, NUMBER 2   MARCH, 1953

Effects of Sonic Vibration on the Proteolytic Activity of Pepsin*

GEORGE M. NAIMARK[†] AND WILLIAM A. MOSHER
Biochemical Research Foundation, Newark, Delaware
(Received December 3, 1952)

Pepsin preparations were sonically treated for various periods of time in a 9-kc magneto-striction oscillator with the temperature maintained at 13°C–16°C during treatment. After sonic treatment the residual proteolytic activity of the enzyme was determined by measuring the turbidity decrease during the digestion of an albumin substrate. It was found that dilute solutions of Merck U.S.P. pepsin were rapidly inactivated by sonic treatment whereas highly concentrated solutions were refractory to ultrasonic destruction. Sonic irradiation of a pure Armour crystalline pepsin solution yielded slight enzyme inactivation only. In no instance was enzyme activation observed in this study.

INTRODUCTION

A LIMITED number of papers have reported the effects of sonic and ultrasonic vibration on the biological activity of enzyme systems.[1] It has been shown that oxidases are usually inactivated by sonic treatment,[2–6] although Haas[7] successfully prepared an active cytochrome oxidase using sonic techniques Reductase and amylase, on the other hand, were found[4] to be highly resistant to inactivation by vibrational waves while catalases were unaffected by such treatment unless sufficiently dilute.[4] Chambers found[8] that

---
* Presented at the Symposium on Acoustics and Chemistry at Western Reserve University, May 21–23, 1952.
† Now at Strong Cobb and Company, Inc., 2654 Lisbon Road, Cleveland 4, Ohio.
[1] Naimark, Klair, and Mosher, J. Franklin Inst. **250**, 279–299, 402–408 (1951).
[2] R. J. Christensen and R. Samisch, Plant Physiol. **9**, 385–386 (1934).
[3] M. Matsudaira and A. Sato, Tohoku J. Exptl. Med. **22**, 412–416 (1934).
[4] M. Kasahara and T. Yoshinare, Z. Kinderheilk. **59**, 462–464 (1938).
[5] R. Wurmser and S. Filitti-Wurmser, Compt. rend. soc. biol. **128**, 475–476 (1938).
[6] Grabar, Voinovitch, and Prudhomme, Biochim. et Biophys. Acta **3**, 412–416 (1949).
[7] E. Haas, J. Biol. Chem. **148**, 481–493 (1943).
[8] L. A. Chambers, J. Biol. Chem. **117**, 639–649 (1937).

Absorption of Sound Arising from the Presence of Intact Cells in Blood*

Edwin L. Carstensen† and Herman P. Schwan

Electromedical Division, Moore School of Electrical Engineering, and Department of Physical Medicine, School of Medicine, University of Pennsylvania, Philadelphia 4, Pennsylvania

(Received July 10, 1958)

The absorption of sound in blood occurs primarily on a molecular level and is related to the presence of protein in the cells and plasma. However, a small contribution to the total absorption arises simply from the presence of intact cells in the blood.

Analysis based on Epstein's theory of scattering shows that the cellular absorption results from a viscous interaction between the fluid and cells when the latter, because of their greater density, fail to follow the oscillatory motion set up by the sound wave.

IT has been demonstrated earlier[1] that the absorption of sound in blood occurs primarily on a molecular level and can be attributed largely to the proteins present. A more recent paper[2] reports a study of the acoustic properties of hemoglobin, the most abundant of the blood proteins. These reports, however, do not give the complete picture, for in normal blood, where the red cells are dispersed in plasma, there appears to be a higher absorption than can be accounted for on the basis of protein concentration alone.

EXPERIMENTAL OBSERVATIONS

In general the techniques and precautions were the same as described earlier.[1] Citrated beef blood was used throughout the investigation. The red cells were separated from plasma and the lighter cellular fractions by centrifugation. For the measurements this red cell fraction was diluted either with plasma or with saline as noted. Measurements were made on the same day that the blood was collected to minimize the possibility of hemolysis.

The absorption of sound by blood and its components is summarized in Fig. 1. Absorption per wavelength in nepers is given for (1) the concentrated red cells of beef blood,‡ (2) beef plasma, and (3) a mixture of two parts red cells to 3 parts plasma, which is approximately the composition of the normal blood. Comparison of the absorption of packed red cells with hemoglobin solutions of similar concentration leads to the conclusion that when the cells are tightly packed the absorption must occur almost entirely on a molecular level. The absorption in plasma must also be of a molecular nature. If the same

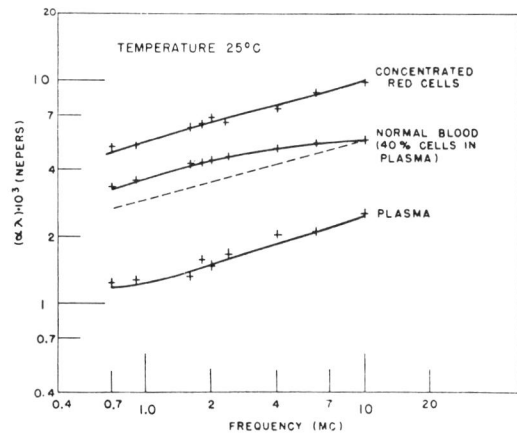

FIG. 1. Absorption of sound in beef blood.

---

* This work was supported by a contract from the Wright Patterson Air Force Base (AF33(616)2494).
† Now at Fort Detrick, Frederick, Maryland.
[1] Carstensen, Li, and Schwan, J. Acoust. Soc. Am. 25, 286 (1953).
[2] E. L. Carstensen and H. P. Schwan, J. Acoust. Soc. Am. (to be published).

‡ The cells were concentrated by centrifuging the blood at approximately 2000 g for thirty minutes. The residue contained roughly 10% intercellular fluid as indicated by high-speed hematocrit measurements.

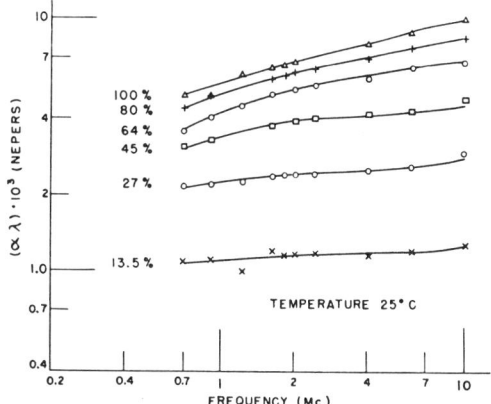

FIG. 2. Absorption of sound in suspensions of beef red cells in 0.9 g/100 cc NaCl solution.

were true for normal blood, it should be possible to predict the absorption of the suspension from a linear combination of the plasma and red cell absorptions. The dashed curve in Fig. 1 indicates this prediction. The observed values of absorption in normal blood are significantly higher than would be anticipated on the basis of linear combination of protein absorptions. We must conclude either (1) that the protein absorption per molecule has somehow been increased by the process of combining red cells and plasma or (2) that a mechanism, not operative in either packed cells or pure plasma, comes into play in the suspension and produces losses in addition to the protein absorption. The former appears to be very unlikely because the conditions at the molecular level are the same in all cases. There is no true mixing of the proteins, because in the absence of hemolysis the hemoglobin is confined in the cell. In addition, so long as the osmotic forces remain constant the cell volume should remain the same and hence the local concentration of hemoglobin in the cell is the same, whether the cells are in suspension or after packing by centrifugation.

To obtain a clearer picture of the problem a series of absorption measurements were performed on red cells diluted by saline (NaCl) solutions of varying tonicity. Using saline instead of plasma simplified the problem because in general the absorption of the saline could be neglected. Each series of measurements was started with concentrated red cells prepared by first centrifuging beef blood to remove the plasma, then washing the red cell residue with an equal volume of 0.9 g/100 cc NaCl solution and centrifuging again to remove the saline. The resulting preparation contained 90–95% cells by volume according to high-speed hematocrit measurements. The hemoglobin concentration of the cell preparations averaged about 32 g/100 cc, as indicated by colormetric measurements. These cells were diluted in progressive steps with solutions containing 0.0 to 2.0 g/100 cc of NaCl. The absorption was measured over a frequency range from 0.7 to 10.0 Mc for all concentrations. Figure 2 gives the data as measured for dilutions

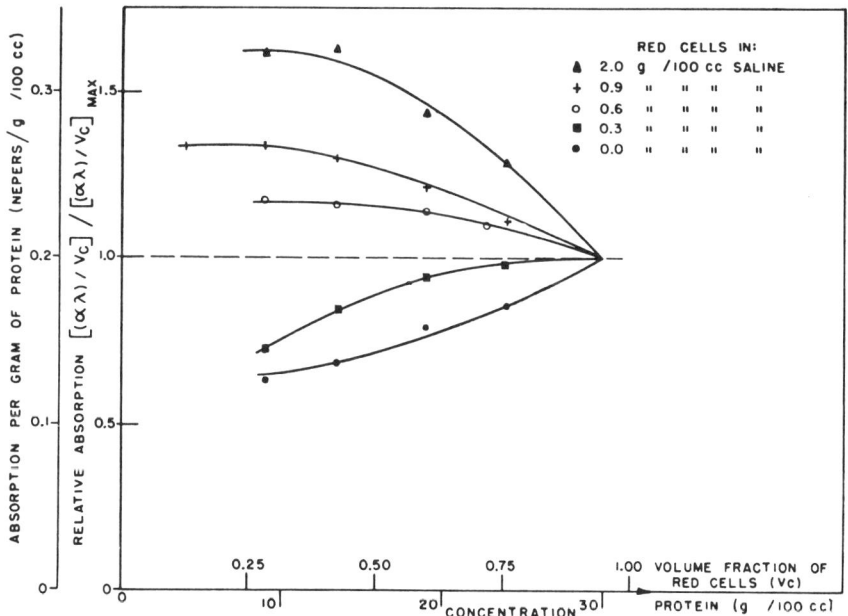

FIG. 3. Absorption of sound by suspensions of beef red cells in sodium chloride solutions (2 Mc). Absorption per wavelength per cc of cells relative to that observed at maximum concentrations is plotted against concentration.

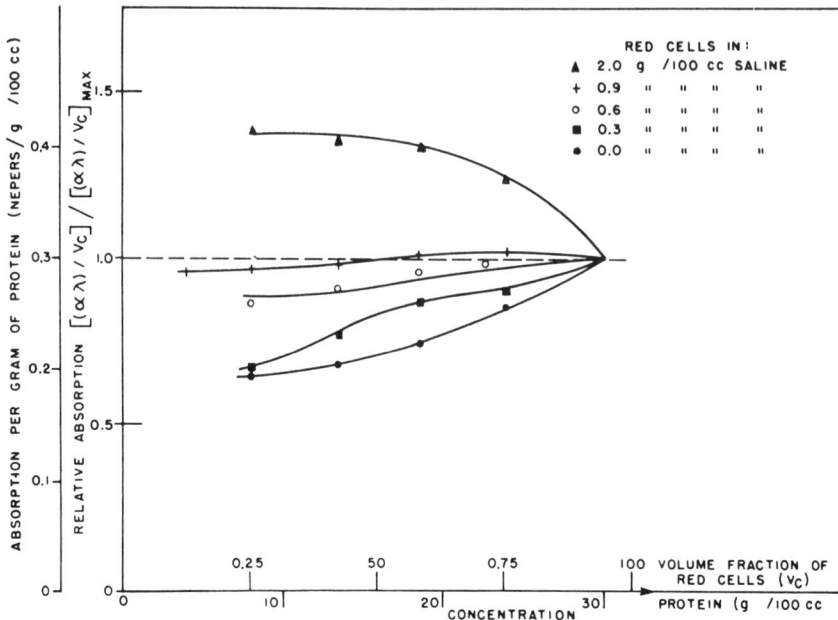

FIG. 4. Same as Fig. 3 except (10 Mc).

with isotonic (0.9 g 100 cc) saline. Note the change in slope of the curve as the dilution progresses. Below 25% volume concentration the shape of the curve becomes constant and the absorption goes in direct proportion to the concentration of erythrocytes. The results for all solutions are summarized in Figs. 3 and 4 for the frequencies of 2 Mc and 10 Mc, respectively. Here absorption per cell relative to that at maximum concentration is plotted as a function of concentration. The absorption of sound in the saline component has been subtracted out to show the behavior of the other components contributing to the absorption.

It is anticipated that in hypertonic solutions the cells will shrink causing the absorption per gram of protein to increase, while in hypotonic solutions, where the cell expands, there should be a small decrease in the absorption per gram of protein. (See Fig. 5, which has been taken from reference 2 in order that it may be compared with the present data.) When diluting with isotonic saline (0.9 g/100 cc), the cell volume and local protein concentration and hence absorption per gram of hemoglobin should remain constant. Thus, if the absorption is to be explained on the basis of protein content the data for dilution with 0.9 g/100 cc NaCl solution should be a horizontal straight line passing through the value measured at maximum concentrations. At 10 Mc this is approximately true, but at 2 Mc the observed absorption exceeds the protein absorption by roughly 30% at low cell concentrations. For the concentration of 0.6 g/100 cc saline, which is strongly hypotonic, the cells expand and there is actually a decrease in the absorption per gram of the hemoglobin (see Fig. 5). In spite of this, the net absorption for low cell concentrations is higher than would be predicted if the protein absorption per gram remained constant. A number of independent checks including hematocrit and conductivity measurements have been made which confirm the constancy of cell volume upon dilution with 0.9 g/100 cc NaCl solution and the increase in cell volume upon dilution with 0.6 g/100 cc saline. In

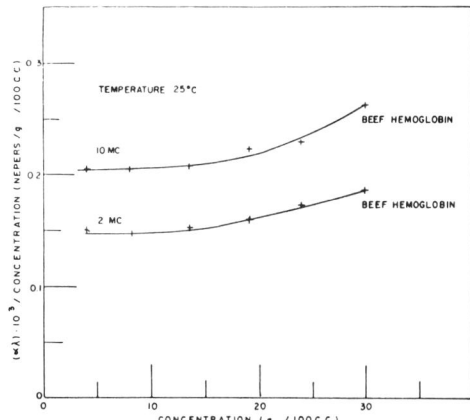

FIG. 5. Absorption of sound in solutions of beef hemoglobin. The water absorption has been subtracted out to show the behavior of the protein.

previous work[2] it was reported that even large concentrations of NaCl had a negligible effect on the absorption of hemoglobin. In other words, these dilution experiments could not have been influenced by chemical action of the diluent on the protein.

It must be concluded that, although the absorption of sound in the fluid and cellular components of blood taken separately is just that contributed by their constituent proteins, in dilute suspensions there is in addition absorption which arises simply from the presence of the intact cells.

THEORETICAL CONSIDERATIONS

We have then the problem of the absorption of sound by a medium containing a suspension of particles which are small in comparison with the wavelength of sound. This problem has been approached theoretically from various points of view by a number of investigators.[3-9] The most general approach is that of Epstein,[5] who considers the case of a longitudinal, progressive wave passing through a medium containing spherical particles which are small compared to the wavelength of the sound.§ The reflected waves, transverse and longitudinal, can be determined from the boundary conditions of the sphere. The absorption of sound energy which results under the assumption that the transverse wave is damped out within a very short distance from the particle is indicated to a first approximation by the absorption coefficient‖,¶

$$\alpha = \tfrac{2}{3}\pi a^3 n k (\delta-1) \operatorname{Re}\left\{\frac{i+b-ib^2/3}{\delta-i\delta b-(2+\delta)b^2/9}\right\}, \quad (1)$$

where a is the radius of the sphere, n the number of particles per cm^3, $k=\omega/c$ is the propagation constant for the suspending fluid, δ is the ratio of density of the suspending fluid to that of the particle,

$$b^2 = i\omega\frac{\rho}{\eta}a^2,$$

---

[3] C. J. T. Sewell, Phil. Trans. Roy. Soc. London, **B210**, 239–270 (1910).
[4] H. Lamb, *Hydrodynamics* (Dover Publications, New York, 1945), sixth edition.
[5] P. S. Epstein, *Theodore von Karman Anniversary Volume* (California Institute of Technology, Pasadena, California, 1941).
[6] Brandt, Freund, and Hiedemann, Z. Physik **104**, 511–533 (1937).
[7] Angerer, Barth, and Guttner, Strahlentherapie **84**, 601 (1951).
[8] W. J. Fry, J. Acoust. Soc. Am. **24**, 412–415 (1952).
[9] R. J. Urick, J. Acoust. Soc. Am. **20**, 283–289 (1948).
§ The compressibility of the erythrocytes is slightly less than that of water. Thus the assumption applies to all components of the medium.
‖ The absorption coefficient given here is defined in the usual way by the equation $I=I_0 e^{-2\alpha x}$ so that α is expressed in nepers per unit of distance. From Epstein's derivation, it is apparent that his absorption coefficient is actually 2α. Hence, the difference of a factor of 2 between Epstein's formula and Eq. (1).
¶ The equation for absorption is derived on the assumption that the ratio of viscosity of the suspending fluid to the shear modulus of the particle is very small. Very little is known about the mechanical properties of the erythrocytes. But, from the fact that the cell maintains a nonspherical (discoidal) shape, it may be inferred that it possesses a certain shear stiffness.

ρ being the density of the fluid, η its viscosity, and ω the angular frequency. In developing Eq. (1), it is assumed that the radius of the sphere is small in comparison to the wavelength of the longitudinal wave and that the velocity of sound for the longitudinal wave in the fluid is much greater than that for the transverse wave. Epstein gives correction terms which extend this expression in certain cases to somewhat higher frequencies or larger particles. It turns out that in the present case, these corrections become comparable in magnitude to Eq. (1) only at 100 Mc, where the nonprotein absorption is too small to be measured.

To make Eq. (1) more useful in the analysis of the present data, the frequency ω can be brought out explicitly by defining the quantity $\gamma^2 = \rho a^2/2\eta$, i.e., $b^2 = 2i\gamma^2\omega$. Then

$$\alpha = \frac{V_p}{c}\left(\frac{\delta-1}{\delta}\right)^2 \frac{2}{9}$$

$$\times \frac{\gamma^2\omega^2[1+\gamma\omega^{\frac{1}{2}}]}{[1+\gamma\omega^{\frac{1}{2}}]^2 + \gamma^2\omega\left[1+\frac{2}{9}\left(\frac{2}{\delta}+1\right)\gamma\omega^{\frac{1}{2}}\right]^2}, \quad (2)$$

when $V_p = \tfrac{4}{3}\pi a^3 n$ is the volume concentration of particles.

This equation may be written in the form usually used for a relaxation process. If we define the quantities**

$$R_e = 6\pi a\eta[1+\gamma\omega^{\frac{1}{2}}], \quad (3)$$

$$M_e = M + m\left[\frac{1}{2} + \frac{9}{4}\frac{1}{\gamma\omega^{\frac{1}{2}}}\right], \quad (4)$$

where $M = \tfrac{4}{3}\pi a^3 \rho_p$ is the mass of the particle and m is the mass of an equivalent sphere of the suspending fluid and

$$\omega_0 = R_e/M_e, \quad (5)$$

the expression for the absorption coefficient becomes

$$\alpha = \frac{V_p}{2c}\left(\frac{\delta-1}{\delta}\right)^2 \frac{m}{M_e}\omega_0 \frac{(\omega/\omega_0)^2}{1+(\omega/\omega_0)^2} \quad (6)$$

and the absorption per wavelength is

$$\alpha\lambda = V_p\pi\left(\frac{\delta-1}{\delta}\right)^2 \frac{m}{M_e}\frac{\omega/\omega_0}{1+(\omega/\omega_0)^2}. \quad (7)$$

This now has the form of the equation for a relaxation process, but in this case the coefficient as well as the relaxation frequency itself are slowly varying functions of frequency. As a result the curve of $(\alpha\lambda)$ vs frequency is similar to that for a simple relaxation process, but

---

** In discussing the case of a pendulum in a viscous fluid, Lamb[6] shows that these quantities have the meaning of resistance (modified Stokes resistance for sphere moving in a viscous fluid) and effective mass for the sphere.

somewhat broadened. For illustration Eq. (7) has been evaluated for conditions which apply to the present problem and is presented in Fig. 6. The dashed curve in Fig. 6 gives the shape of the ($\alpha\lambda$) curve for a simple relaxation, i.e., frequency independent coefficient and relaxation frequency.

A more restricted approach to the problem has been made by considering the relative motion between suspended particle and suspending fluid which arises because of inertial effects associated with the passage of the sound wave.[6-8] This leads to the same value for the velocity of the particle relative to the surrounding fluid as that predicted by the Epstein theory and gives an expression for absorption which is completely equivalent to Eq. (7). It may be concluded, as Urick[9] did, that the transverse scattered wave in Lamb's and Epstein's theories is a quantitative description of the thin layer of fluid which is undergoing shear in the transition from

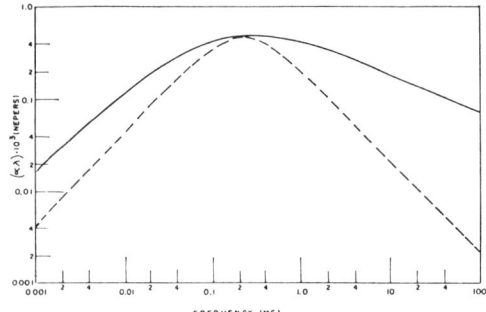

FIG. 6. Relative motion absorption. The solid curve has been computed from Eq. (7) using parameters similar to those which apply to the present experimental conditions. The dashed curve shows the shape of the absorption vs frequency curve which is obtained when the frequency dependence of the mass and resistance is neglected.

the surface of the more or less stationary particle to the moving fluid.

DISCUSSION

To compare the absorption predicted by Eq. (6) with the nonprotein absorption which is apparent in the data of Fig. 2 the following assumptions are made: (1) that the protein absorption and relative motion absorption are superimposed without interaction; (2) that as shown by physiological studies cells diluted in isotonic saline maintain a constant cell volume and hence the absorption per gram of protein remained constant throughout the series of measurements presented in Fig. 2; (3) that no relative motion is permitted when the cells are packed by centrifugation. The last is in reality an approximation, since the

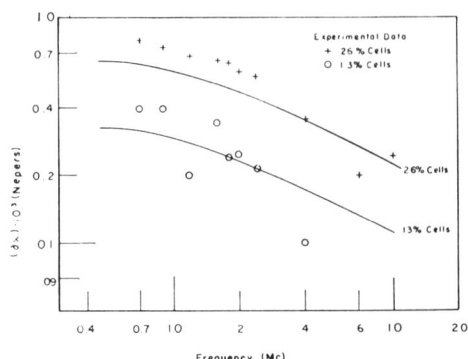

FIG. 7. Comparison of observed nonprotein absorption in suspensions of red cells in saline (points) with that predicted by Eq. (7) (solid curves).

medium is not homogeneous. However, with packing to 90% volume concentration the cells must be distorted so that they lie in close contact over a large part of their surfaces. The little intercellular fluid which remains will tend to be trapped in pockets among the cells and relative motion will be very largely inhibited.

The absorption observed in the "100%" curve of Fig. 2 is assumed to occur therefore on a purely molecular level. The nonprotein absorption as evidenced in the more dilute curves of Fig. 2 then is the observed absorption less the molecular absorption, which is determined by dividing the "100%" curve by the relative concentration of cells. This nonprotein absorption for 13 and 26% cell suspensions is shown by the experimental points in Fig. 7. The solid curves give the absorption predicted by Eq. (7) with the following values taken for its parameters: $\rho_p = 1.084$, $\eta = 0.01$ poise, $a = 2.3$ microns. The value taken for the radius of the particle is the radius of a sphere of 50 cubic microns volume, the value given by Ponder[10] for the volume of the beef cell. Actually, the effective radius of the nonspherical cells may be slightly larger than the value chosen, even so the agreement between theory and observation is within experimental error.

It is concluded that the absorption of sound in blood arises in part simply from the presence of intact cells in suspension and that this is added to the molecular absorption caused by the protein which they contain. The nonprotein absorption can be exlained quantitatively by Epstein's theory. Thus, it may be interpreted to result from a viscous interaction between the fluid and suspended particles when the latter, because of their greater density, fail to follow the oscillatory motion set up by the sound wave.

[10] E. Ponder, *Hemolysis and Related Phenomena* (Grune and Stratton, New York, 1948).

Acoustic Properties of Hemoglobin Solutions*

EDWIN L. CARSTENSEN† AND HERMAN P. SCHWAN

Electromedical Division, Moore School of Electrical Engineering, University of Pennsylvania, Philadelphia 4, Pennsylvania

(Received July 10, 1958)

> Absorption and velocity of sound in the range 0.5 to 10 Mc have been measured for solutions of hemoglobin under various conditions of temperature, concentration, and chemical environment. The hemoglobins of several species of mammals and different chemical forms of hemoglobin have been studied.
>
> In the absorption per wavelength only minor variations with frequency have been observed. Dispersion in the velocity of sound of these solutions has been measured. It has been possible to relate quantitatively the magnitude of the absorption and dispersion through relaxation theory by assuming a broad distribution of relaxation times.

IN an earlier paper[1] it was shown that the absorption of sound by blood could be attributed primarily to the proteins which it contains, both in the red cells and in solution in the plasma. To continue this investigation, hemoglobin, the most abundant of the blood proteins, has been chosen for intensive study. Measurements of the absorption and velocity of sound have been performed under a variety of conditions. The effects of temperature, concentration, pH, and presence of neutral salts have been studied. The measurements have been extended to the hemoglobins of several species of mammals and to different chemical forms of hemoglobin.

1. PREPARATION OF HEMOGLOBIN SOLUTIONS

Hemoglobin solutions, which were sufficiently pure for the purposes of these investigations, were obtained by treating washed, packed red cells with toluene to liberate the hemoglobin. In this process,[2] the cell membranes (stromata) become associated with the toluene so that when the mixture is centrifuged the lighter toluene-stromata fraction is separated from the heavier hemoglobin solution.

Hemoglobin solutions prepared in this manner have been shown to be essentially free of stromata. The concentration of stromata in the hemoglobin solutions was determined through a measurement of the electric capacitance of a sample of the solution. The presence of the very thin insulating cell membranes, whether a part of a normal red cell or simply an intact shell (ghost), causes an increase in the apparent dielectric constant of the solution depending on concentration up to values of the order of 3000 in the frequency range below 0.1 Mc, whereas the dielectric constant of the protein solution alone would be of the order of 120. Thus, a measurement of the capacitance of a sample of blood is one of the most sensitive methods for detecting the presence of stromata and estimating their concentration.

For certain applications the presence of stromata in a solution of hemoglobin does not affect the acoustic measurements significantly. In these cases a completely satisfactory hemoglobin solution can be made by diluting packed red cells with distilled water. The osmotic forces on the cell cause hemolysis, and by this process the hemoglobin is distributed throughout the solution. Preparations of this kind were used for the measurements shown in Figs. 1 and 5.

Spectrophotometric analysis showed that the hemoglobin solutions used in the present work were predominantly in the oxygenated form. Occasionally small amounts of methemoglobin were present (10–20%). However, a separate investigation showed that the absorption of sound in oxyhemoglobin and methemoglobin are substantially the same.

2. TECHNIQUES OF MEASUREMENT

Both the absorption[1] and velocity[3] of sound were measured by difference techniques. The sensitivity of these methods of measurement is high by comparison with conventional methods which are applicable to these experimental conditions. This is particularly true of the velocity difference measurements which are sufficiently sensitive that it has been possible to measure

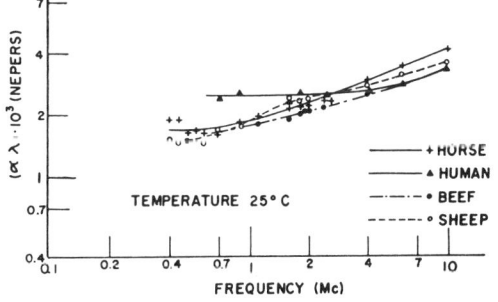

FIG. 1. Absorption of sound in solutions of four mammalian hemoglobins. Each solution contains approximately 45% of red cells in distilled water—an equivalent hemoglobin concentration of about 15 g/100 cc.

---

* This work was supported in part by a contract from the Wright Patterson Air Force Base [AF 33 (616) 2494] and in part by the Office of Naval Research, Contract 119-289.
† Now at Fort Detrick, Frederick, Maryland.
[1] Carstensen, Li, and Schwan, J. Acoust. Soc. Am. 25, 286 (1953).
[2] F. Haurowitz, Z. physiol. Chem. 186, 141 (1930).
[3] E. L. Carstensen, J. Acoust. Soc. Am. 26, 858 (1954).

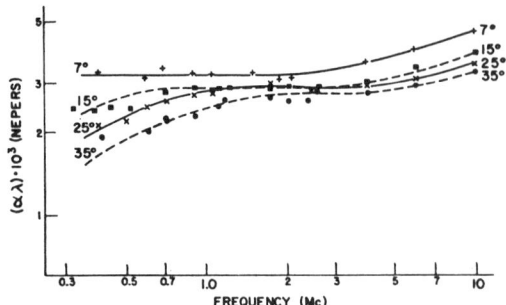

FIG. 2. Influence of temperature on the absorption of sound in solutions of human hemoglobin. Concentration of solutions was 16.5 g Hb/100 cc.

the dispersion in the velocity of sound in protein solutions.

For the purpose of detecting dispersion the errors in velocity measurement are estimated to be less than 5 parts in 10^5 of the velocity. Errors in absorption measurements are less than 5% in the frequency range above 1 Mc, but at the lowest frequencies of the investigation may go as high as 10%.

Absorption data are presented in terms of absorption per wavelength $(\alpha\lambda)$ in nepers throughout.

3. MAMMALIAN HEMOGLOBIN

A good general picture of the absorption data for hemoglobin solutions is given in Fig. 1. Here the absorption per wavelength is plotted as a function of frequency for the hemoglobin of man, cow, sheep, and horse. The experimental conditions were the same for each curve, i.e., washed, packed red cells were diluted with equal parts of distilled water. This produced optically clear solutions in which the distribution of hemoglobin was nearly uniform on a microscopic basis and at a concentration of approximately 15 g/100 cc. The measurements were carried out at a temperature of 25°C. Although there are small repeatable differences in absorption for different species, it is apparent from Fig. 1 that not only are the rough magnitudes of the absorption for all hemoglobins the same, but also that the $(\alpha\lambda)$ value depends upon frequency only to a small extent. This is neither the characteristic frequency dependence shown by classical absorption mechanisms nor that typical of a simple relaxation process. It might, however, be explained on the basis of a more complex process characterized by a broad distribution of relaxation frequencies. The specific frequency behavior characteristic of each type of hemoglobin is presumably just a reflection of the distribution of the energy of the relaxing elements in the complex system; e.g., it appears that in human hemoglobin the relative abundance of low-frequency, high-energy components is greater than in horse hemoglobin.

If the relaxation hypothesis is correct, there should be dispersion in the velocity of sound. Evidence on this point is given in Secs. 5 and 7.

4. TEMPERATURE EFFECTS

The data on human hemoglobin demonstrate particularly well the influence of temperature on the absorption of sound. At low frequencies absorption per wavelength increases gradually until in the middle of the range of observation it reaches a constant plateau. Because of its shape, it is easy to see shifts in the curve with temperature. The absorption of sound in a human hemoglobin solution with a concentration of approximately 16.5 g Hb/100 cc at temperatures 7°, 15°, 25°, and 35°C is plotted in Fig. 2. Note that as the temperature increases the curves appear to be shifted to the

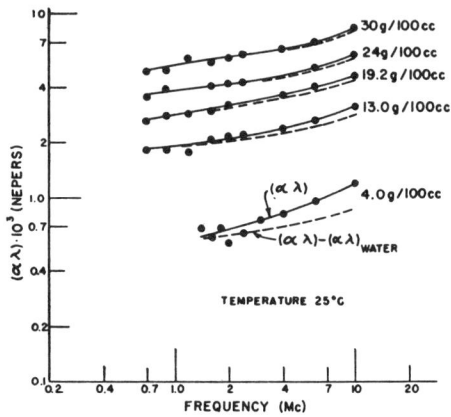

FIG. 3. Absorption of sound in solutions of beef hemoglobin. As the concentration of hemoglobin decreases, the influence of water on the total absorption becomes apparent. Dashed curves, representing the difference between total absorption and that contributed by water, indicate that the frequency dependence of the protein absorption is the same at all concentrations of Hb.

right and lowered slightly. Similar behavior is exhibited by the other types of hemoglobin.

5. CONCENTRATION DEPENDENCE

Solutions of beef hemoglobin were prepared with concentrations in excess of 30 g Hb/100 cc, a concentration which approaches that found in the normal red cell. Absorption of sound as a function of frequency in beef hemoglobin solutions is given in Fig. 3 for concentrations from 4 to 30 g Hb/100 cc. The absorption per wavelength per gram of hemoglobin at 2 Mc and 10 Mc is plotted in Fig. 4. The absorption coefficient is a linear function of concentration up to roughly 15 g Hb/100 cc. Above this concentration the absorption per gram increases until at 30 g Hb/100 cc the absorption per molecule is 20% greater than that found for dilute solutions.

ACOUSTIC PROPERTIES OF HEMOGLOBIN SOLUTIONS

The frequency dependence of the absorption is the same at all levels of concentration. This was observed for all species of hemoglobin studied. Figure 5 gives these data for human red cells diluted with distilled water.

Measurements of the velocity of sound have been performed on the solutions of beef hemoglobin described above (Fig. 3) and also at various concentrations for human hemoglobin. These data are given in Figs. 6 and 7, respectively. There is a strong dispersion over the entire frequency range, as would be anticipated for a relaxation process. If the limit of the distribution of relaxation frequencies were being approached in the range of measurement, the velocity curve should tend toward a constant value. Instead the slopes of the velocity data for beef hemoglobin show a tendency to increase at the highest frequencies, corresponding to the increasing values of $(\alpha\lambda)$ seen in Fig. 3. The indi-

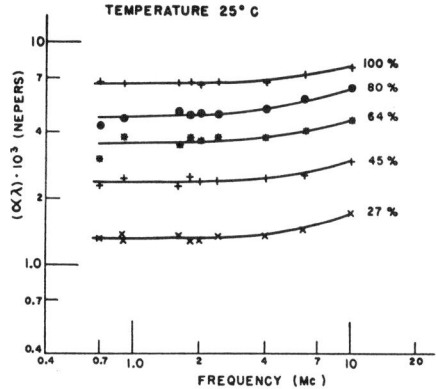

FIG. 5. Absorption of sound in solutions of human hemoglobin. The 100% solution consisted of packed red cells. The concentration of the other solutions is indicated in terms of percent of the original solution diluted in the distilled water. These data illustrate that the frequency dependence of the absorption is independent of concentration. The absorption of water has been subtracted from the data. 25°C.

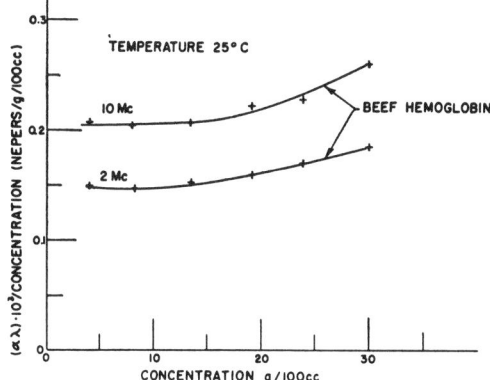

FIG. 4. Absorption of sound in solutions of beef hemoglobin. Absorption of sound per wavelength per gram of hemoglobin is plotted as a function of concentration. The absorption of the water in the solution has been subtracted from the observed absorption to show the behavior of protein.

cations are, therefore, that the relaxation frequencies outside the 0.3–10 Mc range of this investigation are involved.‡

6. ENVIRONMENT

Factors such as the solubility[4] of hemoglobin in water are strongly influenced by the presence of certain neutral salts and by changes in pH. There is some evidence[5,6] that the degree to which the protein is

‡ Subsequent to the investigation reported here, Dr. Hilde Gramberg (Doctoral dissertation, Johann-Wolfgang-Goethe-Universität, Frankfurt, Germany, 1956) has measured the absorption of sound in hemoglobin solutions at frequencies of 35 and 82 kc/cc. Her data indicate that relaxation frequencies as low as 30 kc are involved in the absorption.
[4] E. J. Cohn and J. T. Edsall, editors, *Proteins, Amino Acids and Peptides* (Reinhold Publishing Corporation, New York, 1943), p. 608.
[5] M. F. Perutz, Trans. Faraday Soc. 42, 137 (1946).
[6] McMeekin, Graves, and Hipp, J. Polymer Sci. 12, 309 (1954).

hydrated will depend upon salt concentration and pH as well. The absorption of sound was measured for hemoglobin solutions with pH values of 6, 7.4, and 9.4. No detectable effect of pH on the absorption could be observed.

The solubility of hemoglobin in concentrated NaCl

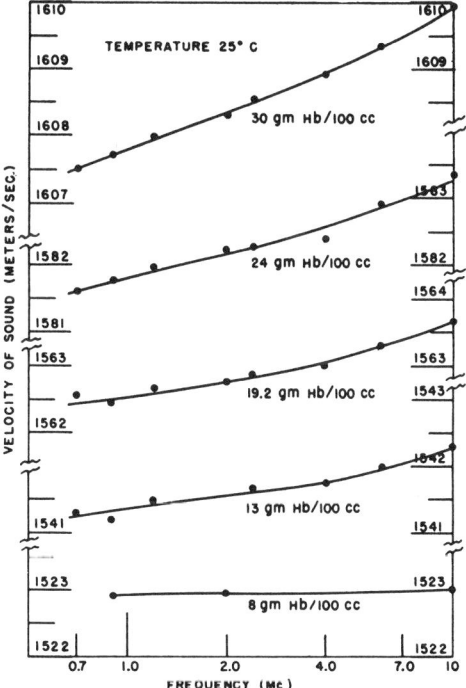

FIG. 6. Dispersion of the velocity of sound in solutions of beef hemoglobin. 25°C.

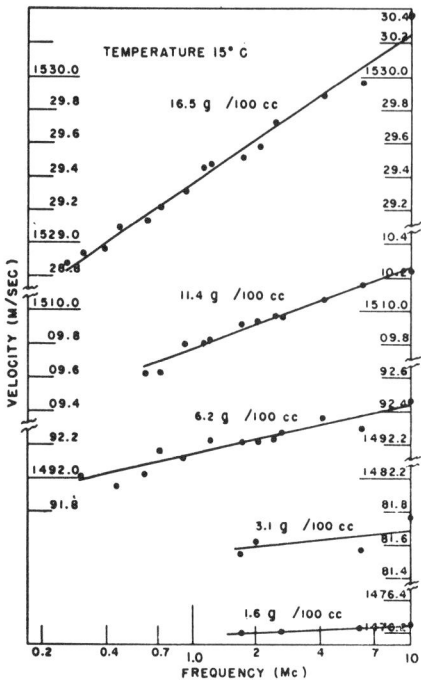

FIG. 7. Dispersion of the velocity of sound in solutions of human hemoglobin. 15°C.

solutions is somewhat greater than in distilled water.[4] In $(NH_4)_2SO_4$ solutions the solubility of hemoglobin is markedly decreased.[4] The absorption of sound by hemoglobin in molar solutions of NaCl and $(NH_4)_2SO_4$ was measured. Again no detectable effect of these salts on the absorption could be observed.

7. RELAXATION ABSORPTION

A number of qualitative observations indicate that some kind of relaxation process is responsible for the absorption of sound in hemoglobin solutions. First, the existence of dispersion in the velocity of sound is a strong indication of relaxation. Second, the departure of the absorption coefficient α from the classical proportionality with frequency squared is most easily explained by a distribution of relaxation times. Hysteresis as defined by Mason[7] gives rise to frequency independent values of $(\alpha\lambda)$, but this mechanism does not account for dispersion in the velocity of sound. Third, the dependence of the absorption on temperature is consistent with the behavior exhibited by other known relaxation mechanisms.

If a simple relaxation process is involved, the absorption and velocity dispersion are related quantitatively

[7] W. P. Mason, *Piezoelectric Crystals and Their Application to Ultrasonics* (D. Van Nostrand Company, Inc., Princeton, New Jersey, 1950), p. 482.

through relaxation theory. This relationship has been demonstrated experimentally for certain solutions.[8] Theoretically relating the two independent quantities, absorption and dispersion, is more difficult in the present case because of the lack of independent information regarding the distribution of relaxation frequencies. However, after introducing certain simplifying assumptions, it has been possible to apply relaxation theory to the observations for human and beef hemoglobin and show that the magnitudes of the absorption and dispersion agree within the limits permitted by the nature of the assumptions and experimental error.

The absorption for a medium in which a single-frequency relaxation process is controlling can be written as

$$\alpha\lambda = B\omega + 2(\alpha\lambda)'\frac{\omega/\Omega}{1+(\omega/\Omega)^2}, \quad (1)$$

where Ω is the relaxation frequency and B is related to the absorption resulting from the nonrelaxing components of the medium. If there are N relaxing elements with characteristic frequencies $\Omega_1, \Omega_2 \cdots, \Omega_N$ then, assuming superposition,

$$\alpha\lambda = B^*\omega + \sum_{n=1}^{N} 2(\alpha\lambda)_n*\frac{\omega/\Omega_n}{1+(\omega/\Omega_n)^2}. \quad (2)$$

Unfortunately, the data in the present case provide us with no detailed evidence regarding the values of Ω_n and the corresponding $(\alpha\lambda)_n*$. Because the influence of a single relaxation process extends over several octaves, it would be hard to say whether the observations resulted from a few discrete relaxation frequencies, two or more Gaussian type distributions spaced at wide intervals, or simply a continuous distribution.

To proceed with the analysis we make the most simple assumption consistent with the experimental data, namely, that the relaxing elements are distributed continuously in a simple power function of the frequency. The summation of Eq. (2) can be replaced by an integral,

$$\alpha\lambda = B^*\omega + \int 2(\alpha\lambda)*f(\Omega)\frac{\omega/\Omega}{1+(\omega/\Omega)^2}d\Omega, \quad (3)$$

where $f(\Omega)$ is the distribution function and $(\alpha\lambda)^*$ is a constant. As pointed out in Sec. 3, the absorption data give $(\alpha\lambda)$ values which are more or less independent of frequency over a wide range. Using the function $f(\Omega) = \Omega_0/\Omega$, which,§ as shown by Fig. 8, most closely

[8] E. L. Carstensen, J. Acoust. Soc. Am. **26**, 262 (1954).
§ This function corresponds to a uniform distribution of relaxing elements on an energy basis, i.e., the number of relaxing elements per unit interval of activation energy is a constant independent of the activation energy between the chosen limits.

approximates this condition, and integrating, Eq. (3) becomes

$$\alpha\lambda = B^*\omega + \frac{2}{\pi}(\alpha\lambda)^*\Omega_0 \tan^{-1}\frac{\omega[(1/\Omega_1)-(1/\Omega_2)]}{1+\omega^2/\Omega_1\Omega_2}, \quad (4)$$

where Ω_1 and Ω_2 are the limits of the distribution. By extending the limits Ω_1 and Ω_2 we can achieve a constant $(\alpha\lambda)$ over any desired frequency range.

A similar discussion applies to the dispersion in the velocity of sound which accompanies relaxation. For a single relaxation frequency the velocity c may be written

$$c^2 = c_0^2 + \frac{2}{\pi}(\alpha\lambda)'c_0 c_\infty \frac{(\omega/\Omega)^2}{1+(\omega/\Omega)^2}, \quad (5)$$

c_0 and c_∞ being the low- and high-frequency limits for the velocity. Because the dispersion is a small part of the total velocity, the approximation $c_0 c_\infty \sim c^2$ is very good in most practical applications. For a continuous distribution of relaxing elements,

$$c^2 = c_0^{*2} + \frac{2}{\pi}(\alpha\lambda)^* c^2 \int f(\Omega)\frac{(\omega/\Omega)^2}{1+(\omega/\Omega)^2}\alpha\Omega. \quad (6)$$

The same distribution must be taken for velocity as was chosen to represent the absorption data. With this

$$c^2 = c_1^{*2} + \frac{2}{\pi}(\alpha\lambda)^*\Omega_0 c^2\left[\frac{1}{2}\ln\frac{1+(\omega/\Omega_1)^2}{1+(\omega/\Omega_2)^2}\right]. \quad (7)$$

The result in this case is an increase in the velocity proportional to the logarithm of the frequency for $\Omega_1 \ll \omega \ll \Omega_2$. This agrees with observation in all cases investigated (see Figs. 6 and 7).

The quantities B^* and c_0^{*2} in Eqs. (4) and (7) cannot be determined from the present experimental data.∥ Before applying these equations to the data we eliminate the quantity B^* by taking the difference between values of $\alpha\lambda/\omega$ at two frequencies ω_1 and ω_2 where the ratio between the two frequencies is constant, i.e., $\omega_2/\omega_1 = n$. Similarly c_0^{*2} is eliminated by taking the difference between velocities at these same frequencies. We then can write

$$\frac{\left(\frac{\alpha\lambda}{\omega}\right)_1 - \left(\frac{\alpha\lambda}{\omega}\right)_2}{c_1^2 - c_2^2} = \frac{\pi}{\omega c_0 c_\infty}$$

$$\times \left\{\frac{\tan^{-1}\frac{\omega\left[\frac{1}{\Omega_1}-\frac{1}{\Omega_2}\right]}{1+\omega^2/\Omega_1\Omega_2} - \frac{1}{n}\tan^{-1}\frac{n\omega\left[\frac{1}{\Omega_1}-\frac{1}{\Omega_2}\right]}{1+n^2\omega^2/\Omega_1\Omega_2}}{\frac{1}{2}\left[\ln\frac{1+(\omega/\Omega_1)^2}{1+(\omega/\Omega_2)^2} - n\ln\frac{1+(n\omega/\Omega_1)^2}{1+(n\omega/\Omega_2)^2}\right]}\right\}, \quad (8)$$

where the subscripts 1 and 2 on the left side of the equation imply measurements at frequencies ω_1 and ω_2, respectively. Equation (8) can be written in a more convenient form for application to the experimental data.

$$\frac{\left(\frac{\alpha}{f^2}\right)_1 - \left(\frac{\alpha}{f^2}\right)_2}{2\Delta c} = \frac{\pi}{fc^2}F_{-1}(\omega,\Omega_1,\Omega_2,n), \quad (9)$$

where $\Delta c = (c_2 - c_1) \ll c_1$, $f = \omega/2\pi$, $c_0 c_\infty \sim c^2$; and $F_{-1}(\omega,\Omega_1,\Omega_2,n)$, the quantity in braces in Eq. (8), approaches unity in the dispersion region and is relatively insensitive to the choice of n, particularly for values of n near unity.

The solid curves in Fig. 9 represent the right-hand side of Eq. (8) for two conditions: (1) where the limits of distribution of relaxing elements $\Omega_1/2\pi$ and $\Omega_2/2\pi$ are taken as 0.2 Mc and 20 Mc and (2) where the

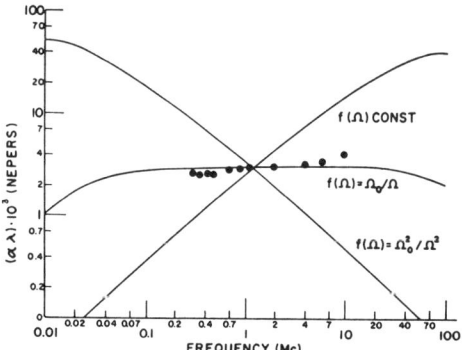

FIG. 8. Comparison of observed absorption of hemoglobin (experimental points taken from Fig. 2, 15°C curve) with that predicted by Eq. (3), making the indicated assumptions for the distribution function $f(\Omega)$. Best fit for the data is given by the function $f(\Omega) = \Omega_0/\Omega$. Limits of the integration in all cases were $\Omega/2\pi = 0.02$ Mc and 200 Mc.

∥ A similar problem might arise when dealing with a substance characterized by a single relaxation time and where, because of limited range of observation, B and c_0^2 could not be directly measured. In this case, differences in the absorption and dispersion data can be shown, by direct operation on Eqs. (1) and (5), to be related as follows:

$$\frac{(\alpha\lambda/\omega)_1 - (\alpha\lambda/\omega)_2}{c_1^2 - c_2^2} = \frac{\pi}{\Omega c_0 c_\infty}. \quad (8a)$$

This ratio should be a constant over the dispersion region and have a magnitude determined by the relaxation frequency of the process.

When two widely separated relaxation frequencies Ω_1 and Ω_2 are involved, the ratio on the left of Eq. (8a) will have two more or less constant values in the regions $\omega \sim \Omega_1$ and $\omega \sim \Omega_2$ with a transition between them, so that to a very rough approximation the ratio will go as $\pi/\omega c_0 c_\infty$ in the range between $\omega = \Omega_1$ and $\omega = \Omega_2$. Equation (8) is simply a refinement of this procedure based on the assumption of a particular continuous distribution of relaxing elements which conforms to a first approximation to the observed absorption data.

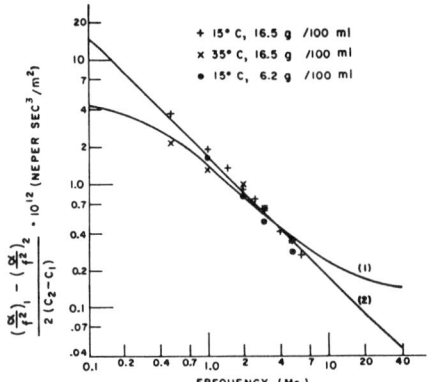

FIG. 9. Application of relaxation theory to data on human hemoglobin. Points represent ½ ratios of slopes of absorption (α/f^2) and velocity C curves for various temperatures and concentration. The solid curves are values for this ratio as computed from Eq. (9) for, a distribution of relaxation frequencies (1) from 0.2 to 20 Mc and (2) from 0.02 to 200 Mc.

limits are 0.02 and 200 Mc. The points in the same figure represent determinations of the ratio on the left of Eq. (8) from the experimental data for human hemoglobin at various concentrations and temperatures.

In the preceding discussion an attempt has been made to relate absorption and dispersion quantitatively through relaxation theory for a complex system about which little or no independent information of pertinent nature is available. To do this, it has been necessary to make assumptions about the nature of the distribution of relaxing elements. It is apparent that the function $F_{-1}(\omega,\Omega_1,\Omega_2,n)$ in Eq. (8) depends for its numerical value not only upon the range of the distribution but also upon the choice of the distribution function $f(\Omega)$ in Eqs. (3) and (6). However, some justification can be given for considering the excellent agreement between theory and observation in Fig. 9 to be more than fortuitous. First, the distribution function $f(\Omega) = \Omega_0/\Omega$ approximates the frequency dependence of both absorption (Fig. 8) and dispersion as observed. Second, the test applied through Eq. (9) is not strongly sensitive to the distribution function chosen. For example, the function $F_0(\omega,\Omega_1,\Omega_2,n)$ corresponding to the distribution $f(\Omega) = $ const behaves much the same as F_{-1} but has a value of 0.5 in the center of the distribution as opposed to a value of 1.2 for F_{-1}.

In light of the foregoing considerations, it may be concluded that the agreement between theory and observation shown in Fig. 9 is strong quantitative evidence that relaxation processes are responsible for the acoustic properties of hemoglobin solutions.

It was pointed out in Sec. 5 that the combined evidence of absorption and velocity data indicates that the distribution of characteristic frequencies of the relaxing elements extends from below 0.3 Mc to above 10 Mc. From the analysis presented in Fig. 8 it appears that the relaxation frequencies extend over at least three decades corresponding to a range of around 4 kcal per mole. The data on human hemoglobin (Fig. 2) give a rough indication of the activation energies associated with the relaxation process. From the shift in the knee of the absorption curve with temperature, it follows that activation energies[9] of between 8 and 9 kcal per mole control the relaxation time in the region of 1 Mc.

8. DISCUSSION

Very little of a positive nature may be said concerning the specific mechanism which is responsible for the relaxation absorption in hemoglobin solutions.

Hemoglobin solutions show a dielectric dispersion in the frequency range near 1 Mc. This has been explained by Oncley[10] as arising from a rotational relative motion between the hemoglobin molecule and water. The electric field provides a torque which tends to orient the polar molecules. In the acoustic case, nonspherical hemoglobin molecules might be oriented by flow if there were translational relative motion between the macromolecules and water. However, the ratio of the viscous to the inertial forces acting on these molecules suspended in water is so great at the frequencies of observation that no relative motion is possible.[11]

It is not possible to rule out completely all viscous processes. Measurements on hemoglobin solutions have shown that shear viscosity alone is much too small to account for the magnitude of the absorption observed. Furthermore, the temperature dependence of the shear viscosity is much greater than the temperature dependence of the absorption of sound observed here. Of course, it would be possible to explain the data mathematically by assuming a bulk viscosity which is orders of magnitude greater than the shear viscosity. In the absence of a precedent for this assumption, such a mathematical operation would add little to an understanding of the actual absorption process.

It is most likely that the sound wave disturbs the chemical or structural equilibrium either within the molecule itself or in the protein water complex.

9. SUMMARY

Absorption and velocity of sound measurements for hemoglobin solutions are presented. By quantitatively relating absorption and dispersion it has been possible to demonstrate that the absorption is controlled by a relaxation process. A broad spectrum of relaxation frequencies must be assumed corresponding to activation energies extending well above and below 8 kcal per mole.

---

[9] Glasstone, Laidler, and Eyring, *Theory of Rate Processes* (McGraw-Hill Book Company, Inc., New York, 1941).
[10] J. L. Oncley, Chap. 22 in reference 4.
[11] E. L. Carstensen and H. P. Schwan, J. Acoust. Soc. Am. 31, 185 (1959).

ACKNOWLEDGMENTS

The authors wish to express their appreciation to Dr. T. H. Wood for many helpful discussions; to Mr. G. Hausen and Mr. R. Gelfand, who assisted in the analysis of the data; to Miss A. M. Dych, who prepared some of the hemoglobin solutions which were used in the measurements; and to the Sharpe and Dohme Laboratories, Glenolden, Pennsylvania, who supplied the human red cells which were used in the research program.

For discussions leading to a better understanding of the biochemical aspects of the work the authors are also indebted to Dr. Hugo Fricke of the Biological Laboratory, Cold Spring Harbor, Long Island; Dr. Thomas L. McMeekin of the Eastern Regional Research Laboratory, Philadelphia; and within the University of Pennsylvania to Dr. David L. Drabkin, Graduate School of Medicine, Dr. Warner Love, Johnson Foundation for Medical Physics, Dr. M. H. Jacobs, School of Medicine, Dr. Brian E. Conway, Harrison Laboratory of Chemistry, and Dr. John C. Reinhold, Hospital of the University of Pennsylvania.

9

Copyright © 1970 by A.S.P. Biological and Medical Press, BV

Reprinted from *Biochim. Biophys. Acta*, **200**, 174–177 (1970)

Ultrasonic Absorption of Aqueous Hemoglobin Solutions

Peter D. Edmonds, Thomas J. Bauld III, Joseph F. Dyro, and Matthew Hussey

The frequency dependence of the absorption of ultrasound in aqueous hemoglobin solutions is pertinent to the kinetics of fast reversible reactions responding to perturbations of pressure[1]. This communication clarifies several aspects of data previously subject to uncertainty.

(1) The ultrasonic absorption coefficient of aqueous solutions of hemoglobin has been measured previously at 35 and 82 kHz[2], and in the ranges 0.4–10 MHz[3] and 32–232 MHz[4]. The measurements at highest frequencies were performed on solutions for which the concentration of hemoglobin was known only approximately and subsequent analysis has disclosed a probable error of 20% in the nominal concentrations

SHORT COMMUNICATIONS 175

reported earlier[4]. We have repeated the work in the high frequency range 10–130 MHz using solutions of known concentration, determined by weighing the crystalline solute, with subsequent verification by evaporation of the sample solution to dryness and reweighing.

(2) The previous measurements of GRAMBERG[2] exhibited strong dependence of the absorption coefficient upon the method of preparation of hemoglobin solutions (by hemolysis of erythrocytes with water, detergent, toluene or after crystallization of hemoglobin). We have compared solutions prepared by hemolysing bovine erythrocytes with toluene and by dissolving crystallized hemoglobin in water; the material was crystalline bovine hemoglobin (Sigma Chemical Co., Type I, twice recrystallized, dialyzed and lyophilized), dissolved in glass-distilled water. In the high frequency range we find no significant dependence upon preparative procedure.

(3) Previous measurements extending to 232 MHz did not adequately determine the asymptotic value for the parameter α/f^2 as $f \to \infty$, which is anticipated for a system exhibiting relaxational behavior (α = absorption coefficient, f = frequency). The measurements have been repeated and extended; strong evidence is found to support the estimate of the asymptotic value, $(\alpha/f^2) \simeq 38 \cdot 10^{-17}$ neper·cm^{-1}·sec^2 given previously[4]. The new data at high frequencies were obtained with use of pulse apparatus constructed in two laboratories[5,6,*].

(4) We have previously advanced alternative suggestions for the form of the spectrum of relaxation times required to describe the observed frequency dependence of the ultrasonic absorption coefficient[4,7]. It is now shown that the earlier, simpler hypothesis[4] adequately describes the available data, and that no evidence for specific structure within the relaxation spectrum can be adduced.

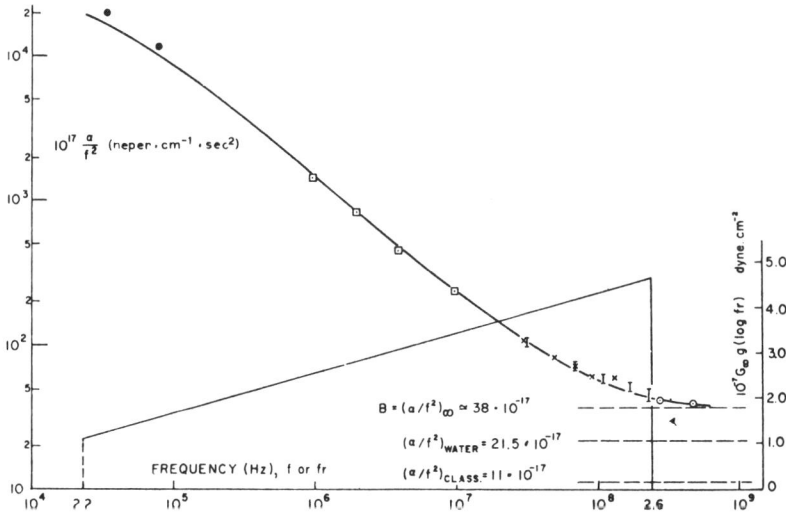

Fig. 1. Absorption parameter α/f^2 as a function of frequency for an aqueous solution of 15 g hemoglobin per 100 ml at 25° and neutral pH. ●, ref. 2, 20°; ⊡, ref. 3; I, ref. 4, revised; ○, J. L. HUNTER AND P. DURDEL (personal communication, 1966), crystalline hemoglobin; ×, this work, crystalline hemoglobin.

---

* We are indebted to Dr. J. L. HUNTER and Mr. PAUL DURDEL of John Carroll University for carrying out the two measurements at 270 and 470 MHz at our request.

Fig. 1 includes the revisions and additions to data for the measured absorption parameter a/f^2 as a function of frequency extending over a range of four decades, at 25° and a concentration of 15% (w/w) hemoglobin. This is the upper limit of proportionality between a and concentration; it is approximately one-half the concentration of hemoglobin in whole blood.

Relaxational behavior is exhibited throughout the entire frequency range of measurement. The high frequency asymptotic value of a/f^2 is significantly greater than either the value for the solvent water or the classical value calculated from the measured static viscosity of the hemoglobin solution.

Conclusions regarding the spectrum of relaxation times to be deduced from the measurements depend critically on the choice of high frequency asymptote. The empirical value of $38 \cdot 10^{-17}$ neper·cm^{-1}·sec^2 is shown to be a significant plateau in the data and it can be used as a base line for defining relaxational contributions at lower frequencies.

The spectrum of relaxation frequencies represented by the ramp function $G_\infty \cdot g(\log f_r)$ on Fig. 1 then adequately describes the measured data, where no assumption is made about the behavior of a/f^2 at frequencies above 470 MHz. The function $g(\log f_r)$ describes the density of unit step functions used to approximate relaxation functions of unit strength[8]. Iterative calculations[8] have been performed in an attempt to generate significant structure within the spectrum (in the sense of "ultrasonic spectroscopy"). No significantly closer approximation of recalculated values of a/f^2 to the experimental data is achieved by use of such a spectrum. Thus, the precision of the data does not warrant such refinement of the spectrum beyond the ramp function.

We have investigated the effects of a variety of conceivable changes to the ramp function shown in Fig. 1. An example is shown in Fig. 2, where the slope of the ramp

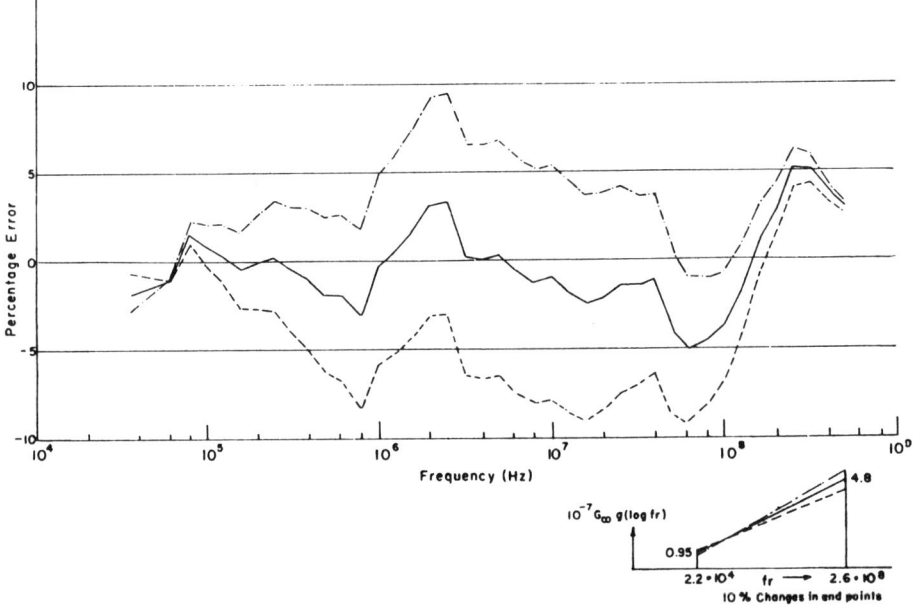

Fig. 2. Examples of the effects of varying the test ramp function $G_\infty g(\log f_r)$ from the optimal configuration.

has been changed by altering the heights of the end points by ± 10% (see insert). The ordinates are the percentage differences (at each frequency) between the measured value of a/f^2 and the values of a/f^2 recalculated from the three ramp functions, one optimal and two modified as stated. The optimal ramp generates an error curve lying entirely within the bounds at ± 5%, corresponding to the anticipated maximum error of individual measurements of a/f^2. The modified ramps generate the dashed error curves lying in part outside the bounds at ± 5%, thus entailing their rejection.

The upper bounding relaxation frequency $(f_r)_{max}$ for the optimal ramp is 260 MHz; the lower bounding relaxation frequency $(f_r)_{min}$ = 22 KHz is related only to the termination of data at 35 KHz and does not represent a lower bound in the same sense as the phenomenological upper bound. The range of associated relaxation times $\tau_r = 1/2\pi f_r$ is then $5 \cdot 10^{-10}$ to $> 5.5 \cdot 10^{-6}$ sec. The ordinates of the ends of optimal ramp are $G_\infty g (\log f_r) = 0.95 \cdot 10^7$ and $4.8 \cdot 10^7$ dyne·cm^{-2} at the lower and upper bounds, respectively. The possible errors in these results are less than ± 10%.

Accordingly we conclude that the available measured data on the absorption parameter a/f^2 are best fitted by a distribution function of relaxation frequencies, $g (\log f_r)$ shown in Fig. 1 and applicable to the expression:

$$\frac{a}{f^2} = \frac{\pi}{\varrho_0 v_1{}^3} \int_{\log f_{r_1}}^{\log f_{r_2}} G_\infty g(\log f_r) \cdot \frac{f_r}{f_r{}^2 + f^2} \, d(\log f_r)$$

where a is the amplitude absorption coefficient, f is the frequency of measurement, ϱ_0 is the density of the solution, v_1 is the velocity of longitudinal waves and f_r is a relaxation frequency, bounded by the limiting values f_{r_1} and f_{r_2}. The foregoing conclusion is strictly applicable only to hemoglobin solutions in distilled water at a concentration of 15 g hemoglobin per 100 ml and a temperature of 25°, but can be considered representative of likely behavior over a limited range of these parameters. Within the stated limitations this work serves by identification of the shortest relaxation time to define the fastest reaction rate associated with physicochemical reactions in these solutions, which respond to pressure changes; the relaxation spectrum demonstrates that relevant reaction rates probably extend over at least 4 orders of magnitude.

This work was supported by the U.S. Public Health Service under grants GM-12299 and HE-01253 from the National Institute of Health.

Biomedical Engineering Department, PETER D. EDMONDS
Moore School of Electrical Engineering, THOMAS J. BAULD III
University of Pennsylvania, JOSEPH F. DYRO
Philadelphia, Pa. 19 104 (U.S.A.) MATTHEW HUSSEY

1 M. EIGEN AND G. G. HAMMES, *Advan. Enzymol.*, 35 (1963) 1.
2 H. GRAMBERG, Doctoral Dissertation, Goethe Universität, Frankfurt, 1956.
3 E. L. CARSTENSEN AND H. P. SCHWAN, *J. Acoust. Soc. Am.*, 31 (1959) 305.
4 P. D. EDMONDS, *Biochim. Biophys. Acta*, 63 (1962) 215.
5 J. L. HUNTER AND H. D. DARDY, *J. Acoust. Soc. Am.*, 36 (1964) 1914.
6 P. D. EDMONDS, *Rev. Sci. Inst.*, 37 (1966) 367.
7 P. D. EDMONDS AND J. F. DYRO, *Proc. 19th Ann. Conf. Eng. Med. Biol.*, 8 (1966) 217.
8 A. J. BARLOW AND J. LAMB, *Proc. Roy. Soc. London, Ser. A*, 253 (1959) 52.

10

Copyright © 1971 by the Acoustical Society of America

Reprinted from *J. Acoust. Soc. Amer.*, 50(2), Pt. 2, 692–699 (1971)

Mechanism of Absorption of Ultrasound in Liver Tissue

H. Pauly* and H. P. Schwan

*Electromedical Division, Moore School of Electrical Engineering, University of Pennsylvania,
Philadelphia, Pennsylvania 19104*

The dominant part of the acoustic absorption of liver tissue and its components results from macromolecular relaxation processes. The absorption has been investigated over the frequency range 1–10 MHz and the following results have been obtained: (1) About two-thirds of the total absorption arises at the macromolecular level, with the remainder caused by macroscopic structure. (2) The specific absorption of tissue macromolecules, as expressed in absorption per weight percent, varies considerably from one biopolymer to another. (3) The absorption is related to the structure of the biological macromolecule or its hydration and changes with heat denaturation and pH. (4) A similar frequency dependence results for all materials investigated. This dependence is to be expected if one assumes that the molecular processes of absorption are characterized by a broad spectrum of relaxational time constants and activation energies extending over a range of at least 1:7.

INTRODUCTION

It appears that significant advances of ultrasonic applications in medicine and biology will depend on a thorough understanding of the mode of propagation of ultrasound through tissue. Since the earlier work on tissue absorption in the 1950s, little has been done which sheds light on the mode of propagation of sound through tissues.[1,2] In particular, no explanation for the observed high ultrasonic absorption coefficient in tissues has been given. The investigations of the absorption of ultrasound in hemoglobin and blood by Carstensen and Schwan[3,4] had shown that in this case absorption was largely due to macromolecular process and that hemoglobin and albumin have similar specific absorption values (expressed in decibels/cm per weight percentage protein). However, these specific absorption values were too small to suggest an explanation of tissue absorption. Clearly, tissue absorption must arise from other than macromolecular processes, or the specific absorption of tissue proteins is considerably larger than that of the blood proteins hemoglobin and albumin.

We have attempted to contribute to this problem by investigating on what level of cellular, subcellular, and molecular structure most of the sonic absorption takes place. To this end, liver tissue samples have been fractionated by a modification of the method of Schneider and Hogeboom,[5] resulting in fractions and preparations from solid tissue to homogenates where even subcellular organelles were largely destroyed.

Ultrasonic absorption coefficients have been determined over a wide frequency range in all cases and found to be largely independent of the degree of cellular and subcellular destruction, suggesting that more than two-thirds of liver tissue absorption originates on a macromolecular level including submicroscopic particles such as ribosomes and aggregations of macromolecules.

The specific absorption of the most abundant tissue proteins and its different and largely unknown forms of submicroscopic associations is considerably higher than values previously reported for hemoglobin and albumin. Additional material obtained with gelatin is presented, which further supports that the specific absorption of various proteins varies by at least a factor of 10. Furthermore, it is shown that factors such as pH and denaturation strongly affect the specific ultrasonic absorption of proteins. Thus sonic absorption data are shown to respond strongly to variation of protein structure.

The frequency dependence of proteins is explained in terms of a broad spectrum of activation energies. The similarity in the character of the frequency dependence of all protein and tissue samples investigated is explained by the assumption of relaxation processes with activation energies extending over a broad range.

I. MATERIALS AND TECHNIQUE

Beef and lamb liver were obtained from the slaughterhouse and kept at 4°C until used. "Ground liver" was obtained utilizing a meat grinder with coarse blades and consisted of pieces of solid liver tissue 1–5 mm in size. Homogenate was obtained by blending one part of ground liver with two parts of physiological saline solution in a Waring blender. The preparation was performed at 0°C. The tissue of homogenization depended on the degree of destruction of the cells desired. The pH value of the supernatant and the gelatin[6] solution were adjusted to the desired value by addition of hydrochloric acid or sodium hydroxide. pH values were determined with a Cambridge pH meter model L and glass electrodes.

Specific gravity was obtained from weight determinations. Solid content was determined from dry weight. The composition of beef liver was obtained from literature.[7] The data for the amount of protein, total lipid, and RNA vary somewhat. Larger variations were found in the content of metabolites, but these components are present in small amounts compared to protein or lipids. We can assume the following composition (percent by weight): 70% water, 20% protein, 5% total lipid, 2.5% carbohydrates and metabolites, 1.5% ash, and 1% nuclei acids. Therefore, the main component is protein accounting for about 67% of the dry weight, while the total lipids account for only 17%. Most of the lipids are found in the cytoplasmic membranes and the endoplasmatic reticulum. Therefore, the lipid content of mitochondria and the microsomal fraction is up to 30%–50% of the dry weight of these fractions.[7] The

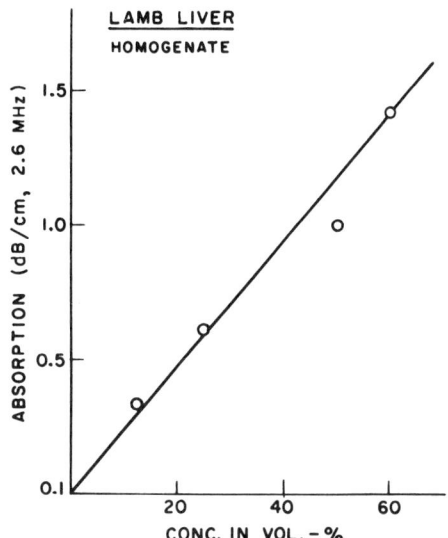

FIG. 2. Absorption of lamb liver homogenate as function of concentration of liver material. Increasing viscosity did not permit measurements at concentration values above 60%.

total lipid content of the nuclei is about 10%–17% of the dry weight and therefore comparable to the lipid content of liver tissue. Because of the different distribution of the membraneous material in the preparations used in this investigation, the lipid content should be accordingly different. The lipid content of the supernatant is smaller than 5%, and the lipid content of the sediment should be somewhat higher than 5%.

Acoustic absorption coefficients were determined using a substitution technique previously described by Carstensen and Schwan.[8] In this technique the power transmitted from one transducer to another is monitored while the test substance between the transducers is replaced by an electrolyte. The electrolyte concentration was determined by the salts added to the liver fraction tested. The precise amount of salt is unimportant since the absorption of the substitute electrolytes is small compared to that of the substances of interest in this study.

II. RESULTS

Figure 1 shows the frequency dependence of the absorption of ground beef liver. The material was diluted with an equal volume of physiological saline solution in order to transform the originally rather viscous semisolid material into the solution needed to apply the substitution technique for absorption measurements listed above. The reduced viscosity made it also possible to separate gas bubbles, introduced into the sample during the process of homogenization, either under vacuum or with a centrifuge. Determina-

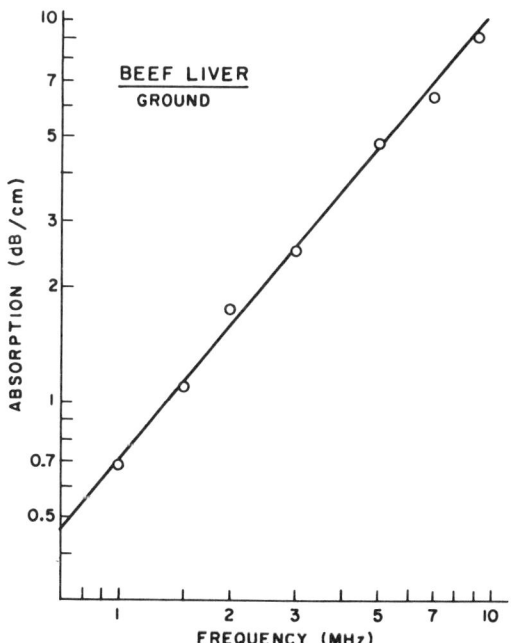

FIG. 1. Absorption of ground beef liver as function of frequency. Temperature 25°C.

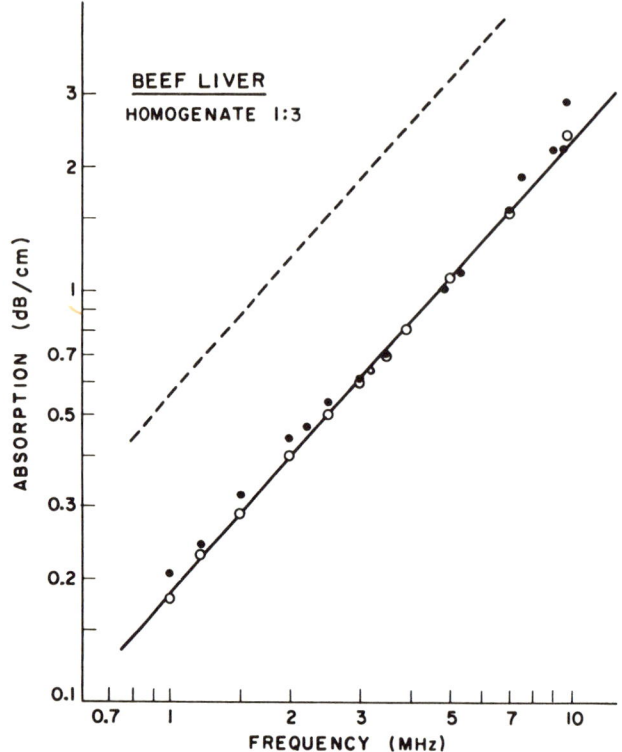

Fig. 3. Frequency dependence of the absorption of beef liver homogenate. Ground liver was diluted with 2 volumes of physiological saline and homogenized for 1 min (○) or 25 min (●) at °C. Measured after degassing at 25°C. Dashed curve corrects for dilution factor (see text).

tion of the solid content of the homogenates was carried out from dry weight and specific weight. The data in Fig. 1 are corrected for the dilution factor, i.e., pertain to solid liver. This correction is allowed since the absorption was observed to depend linearly on the concentration of liver material as demonstrated in Fig. 2. We consider these data characteristic of "idealized" liver tissue of random orientation of cells and not affected by major structural disturbances of large size (pieces of connective tissue, arteries, etc). The curve is the result of several measurements, averaging points taken repeatedly at each frequency. Individual variation up to 20% from the curve were in part due to sedimentation. The data are almost identical with our results on whole lamb liver[9] and published absorption data[1,2]; i.e., ground tissue has the same absorption as solid tissue.

Figure 3 shows the frequency dependence of the absorption of beef liver homogenate. Microscopic examination of the material obtained after 1-min homogenization time revealed about 90% intact cell nuclei, and other small particles such as mitochondria, but almost complete destruction of the cells (>95%). The sample obtained after 25 min of preparation in the blender had all its cells and nuclei and a large part of its mitochondria destroyed. The absorption of both samples is identical, indicating that the absorption processes do not arise on a cellular level and a nuclear level. That is, neither cells nor subcellular organelles contribute as structural units noticeably to the total absorption. The dashed curve corrects for the dilution factor, i.e., pertains to undiluted homogenate and is obtained from the experimental data by linear transformation. Its frequency dependence is nearly the same as given in previous figures for solid tissue. Its absolute level corresponds to absorption coefficients which are about 30% smaller than those pertaining to solid and ground tissue. We conclude that the major part of the total absorption arises on a level of organization smaller than that defined by cells, cell nuclei, and mitochondria.

Different molecular components of tissue have different specific absorption values. This is indicated in Fig. 4 by the results obtained on different liver fractions with different chemical components, different types of proteins, and different types and amounts of lipids. The components were obtained by centrifuging the homogenate for 1 h at 2500 g. The supernatant consisted of soluble proteins, microsomes, and a small fraction of mitochondria. The volume of the sediment was 20%–25% of the total homogenate and consisted predominantly of unbroken cells, nuclei, mitochondria, and connective tissue. From specific weight of supernatant (1.026 g/cm³ at 20°C) and dry weight, it was estimated that the specific absorption of the supernatant (absorption per gram solid in 100 cm³) compares with that of hemoglobin. On the other hand, the specific absorption of the sediment is considerably higher, indicating that different tissue components have different specific absorption values.

If the specific absorption value is different for the different molecular components, we may expect changes in absorption accompanying changes in molecular structure. Such a change is accomplished, for example, by heat denaturation of the protein molecule. Indeed we were able to find a 30% increased absorption after heat denaturation of the proteins in a liver homogenate as shown in Fig. 5. The data support the belief that the specific absorption per molecule depends on molecular properties such as structure or hydration.

Another property of the molecule, the dissociation of the acid and basic groups, will change the pH value. Closely related with the electric charge of the macromolecule is the amount of bound water. Indeed we found the specific absorption to depend on the pH value as shown in Fig. 6. The pH was changed by addition of concentrated NaOH. The addition of NaOH itself does not affect the absorption, since all absorption data are obtained with a substitution technique where the homogenate and its components are compared against the dilutent, thus canceling

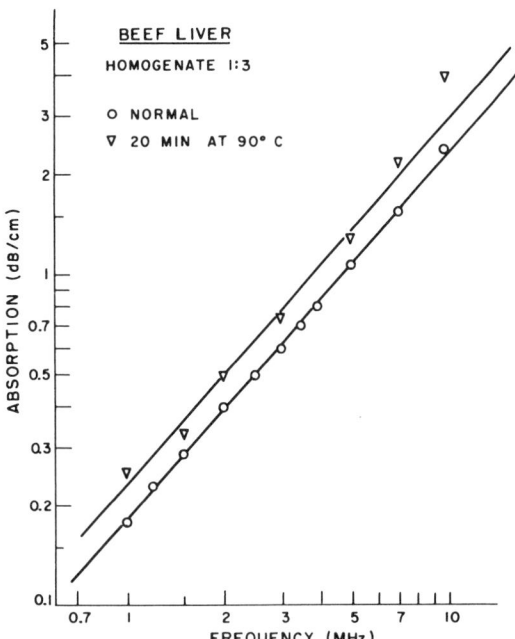

Fig. 5. Effect of heat denaturation on the absorption coefficient of beef liver homogenate. The beef liver homogenate was heated up to 90°C for 20 min. Measurement at 25°C.

the small effects due to salt content. In Fig. 6, the absorption is shown at two different pH values. However, the slope of the frequency-dependent curve remains practically unchanged. Figures 7 and 8 show in greater detail how the absorption depends on pH. The absorption changes strongly with pH, passing through a pH maximum near 12. On the other hand, the frequency dependence does not change significantly over the total range of pH values investigated.

The pH dependence of the absorption appears to be a very general feature of biopolymers and proteins. Figure 9 displays the absorption of a 25% gelatin solution versus pH. There is a broad minimum of the absorption in the region of the isoelectric points of the various protein components of the gelatin solution. To the acid and to the alkaline side, the absorption increases and passes through pronounced maxima. Proton transfer processes are probably largely responsible for this behavior.[10,11] Clearly, the acoustic behavior of the gelatin solution displays the same general pH dependence as the liver preparations. Since the liver preparations do contain some lipids while the gelatin solution is free of it, it appears that the specific pH dependence of the acoustic properties is due to the protein component.

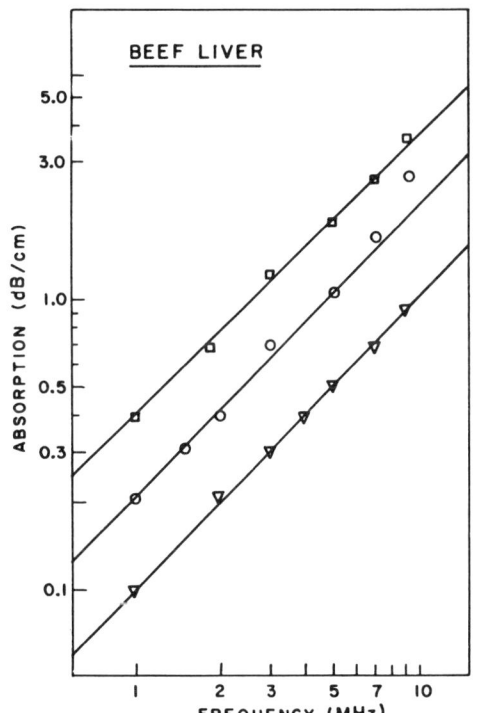

Fig. 4. Frequency dependence of the absorption of beef liver homogenate 1:3 (○), sediment (□), and supernatant (▽). Temperature 25°C.

III. MATHEMATICAL MODEL

All results reported above as well as previously published ultrasonic absorption data obtained with

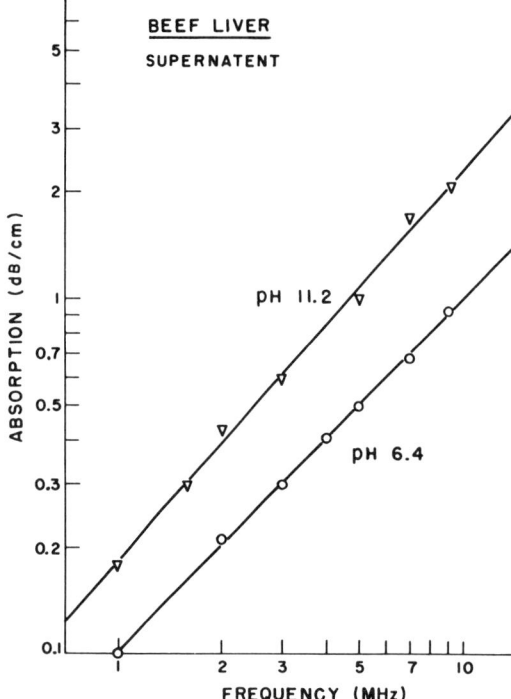

FIG. 6. Frequency dependence of beef liver supernatant absorption at two different pH values. Supernatant from a 33% homogenate. Temperature 25°C.

$\alpha\lambda$ is fairly independent of frequency. This behavior can occur if a large number of relaxation processes contribute to the absorption, as is shown next.

In the presence of a single relaxational mechanism, characterized by one time constant T, the absorption per wavelength $\alpha\lambda$ varies with frequency: $f=1/2\pi\omega$.

$$\alpha\lambda/\omega = B + 2(\alpha\lambda)^* T/[1+(\omega T)^2], \quad (1)$$

where B is a constant characteristic of classical contributions to absorption and relaxational mechanisms which occur at frequencies much higher than $1/T$; $(\alpha\lambda)^*$ is the relaxation contribution to absorption found for $\omega = 1/T$ and equal to the highest value of the relaxation absorption per wavelength; T is a time constant $= 1/2\pi f_0$, where f_0 is the relaxation frequency.

If many time constants are involved, representative of many relaxational processes and extending from a smallest value T_1 to a large value T_2, a continuous distribution of time constants is conveniently considered and Eq. 1 is to be replaced by

$$\frac{\alpha\lambda}{\omega} = B + 2\int_{T_1}^{T_2} \frac{(\alpha\lambda)^* T}{1+(\omega T)^2} dT, \quad (2)$$

where $(\alpha\lambda)^*$ is a function of T characterizing the dis-

tissues, cell suspensions, and large biological macromolecules indicate that the absorption α changes approximately proportionally with frequency. The precise dependence is characterized by a power function f^m, where m varies between 1 and 1.2 depending on the data chosen. Hence, the absorption per wavelength

TABLE I. Absorption per wavelength $\alpha\lambda$ is calculated for various distribution functions of time constants $F(T) = aT^n$. a is chosen for equal unit area $\int_0^\infty F(T)dT$.

n	$\int \dfrac{z^{n+1}}{1+z^2}dz$	$\alpha\lambda$	a
-2	$\ln\dfrac{z}{(1+z^2)^{\frac{1}{2}}}$	$\dfrac{a}{\omega}\ln\dfrac{1+(1/\omega T_1)^2}{1+(1/\omega T_2)^2}$	$\dfrac{T_1 T_2}{T_2-T_1}$
-1	$\tan^{-1}z$	$2a\tan^{-1}\dfrac{\omega(T_2-T_1)}{1+\omega^2 T_1 T_2}$	$\dfrac{1}{\ln T_2/T_1}$
0	$\ln(1+z^2)^{\frac{1}{2}}$	$\dfrac{a}{\omega}\ln\dfrac{1+(\omega T_2)^2}{1+(\omega T_1)^2}$	$\dfrac{1}{T_2-T_1}$
$+1$	$z-\tan^{-1}z$	$\dfrac{2a}{\omega}(T_2-T_1)-\dfrac{2a}{\omega^2}\tan^{-1}\dfrac{\omega(T_2-T_1)}{1+\omega^2 T_1 T_2}$	$\dfrac{2}{T_2^2-T_1^2}$

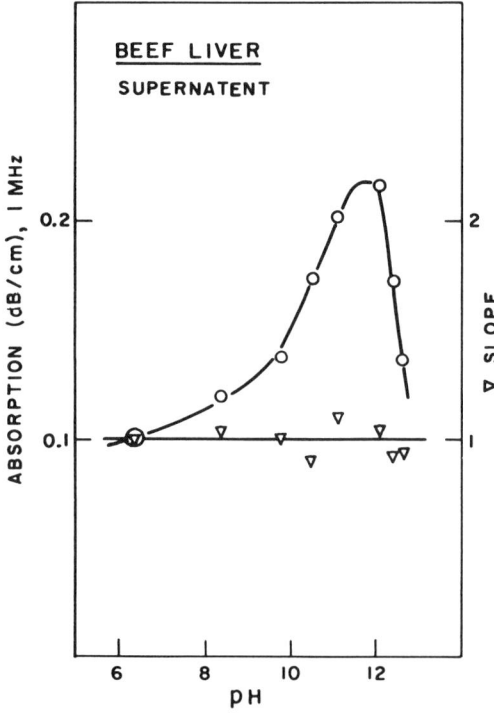

FIG. 7. Absorption at 1 MHz (left ordinate) and slope of the absorption of beef liver supernatant as a function of pH. Temperature 25°C.

tribution of time constants. Suppose the time-constant distribution function is a power function of T. Then, with $z = \omega T$,

$$\frac{\alpha \lambda}{\omega} = B + \frac{2a}{\omega^{2+n}} \int_{z_1}^{z_2} \frac{z^{n+1}}{1+z^2} dz, \quad (3)$$

where a is a frequency-independent constant. Table I gives some of the exponent values which permit ready integration of Eq. 3 and corresponding expressions for $\alpha\lambda$. In order to illustrate the acoustic behavior predicted by these functions, Fig. 10 shows evaluations for time-constant spectra extending over T ranges $T_2/T_1 = 1, 3, 10, 100,$ and 1000. The constants B are neglected, as they characterize that part of the total absorption which is not of interest here.

It is apparent that only for the case of $n = -1$ can a substantial broadening of the peak of the $\alpha\lambda$-versus-frequency curve be achieved. The absorption per wavelength $\alpha\lambda$ varies less than 33% with frequency over nearly 3 frequency decades for a T spread of $1:1000$. It appears justified, therefore, to conclude that for any frequency dependence where $\alpha\lambda$ does not vary appreciably with frequency, the distribution of time constants does not differ greatly from the Eq. 3 with $n = -1$, i.e., the distribution function of time constants

$$f(T) = (K/T) dt, \quad (4)$$

where K is a constant. This function can be rewritten

$$f(T)/d(\ln T) = K, \quad (5)$$

i.e., precisely that to be anticipated if the relaxational behavior is characterized by a spectrum of activation energies $A \sim \ln T$, which can be approximated by a box

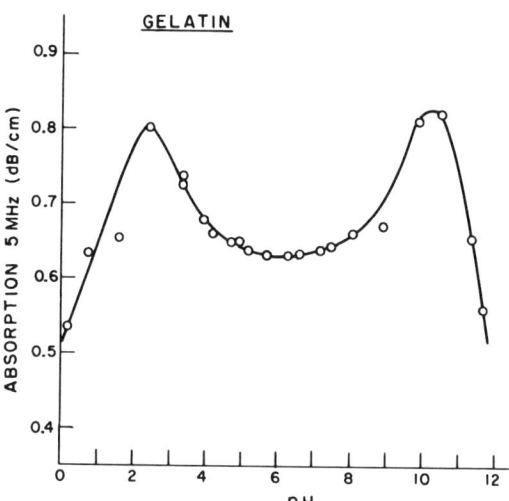

FIG. 9. Absorption at 5 MHz of a 25% gelatin solution as a function of pH. Temperature 35°C.

function.[12] For a time-constant range $1:1000$, the activation energies range $1:7$.

IV. DISCUSSION

The results presented above show that the absorption of liver tissue and its subcellular and macromolecular components changes approximately proportionally with frequency. In Sec. III, this behavior was shown to result if a broad distribution of relaxational processes is assumed to exist. One can anticipate that all biological materials which contain complex macromolecules giving rise to such a relaxational behavior exhibit a similar frequency dependence extending over several frequency decades. Indeed, a survey of all results presented so far in the literature is in accord with this behavior.[13-16] We therefore conclude that the similarity of all the slopes of $\alpha\lambda$ versus frequency, as reported

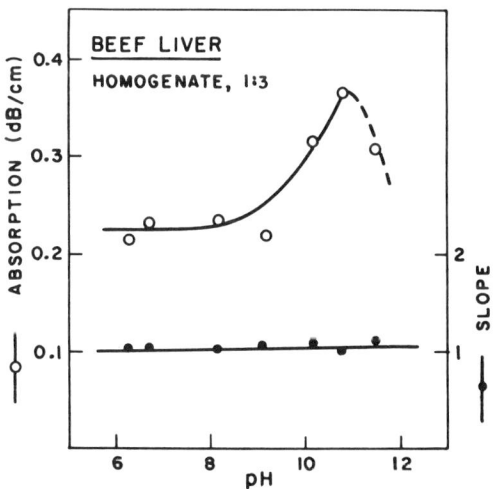

FIG. 8. Corresponding representation as in Fig. 7 for 33% beef liver homogenate.

TABLE II. Summary of specific absorption (in decibels/cm per gram solid component in 100 cm³) at 1 Mc. Values at other frequencies are obtained approximately by simple multiplication with the frequency in megacycles.

Specific absorption for various substances	
Liver (Solid tissue)[a]	0.027
Liver—Ground	0.023
—Homogenate	0.020
—Sediment	0.027
—Supernatant	0.011
—Nuclei (Ref. 17)	0.040
Hemoglobin (Ref. 5)	0.010
Gelatin	0.005
Albumin (Ref. 5)	0.010
Blood plasma (Ref. 5)	0.010

[a] Obtained by insertion of slices of tissue, 1–2 cm thick, between transducers. The value agrees within 10% with published data.

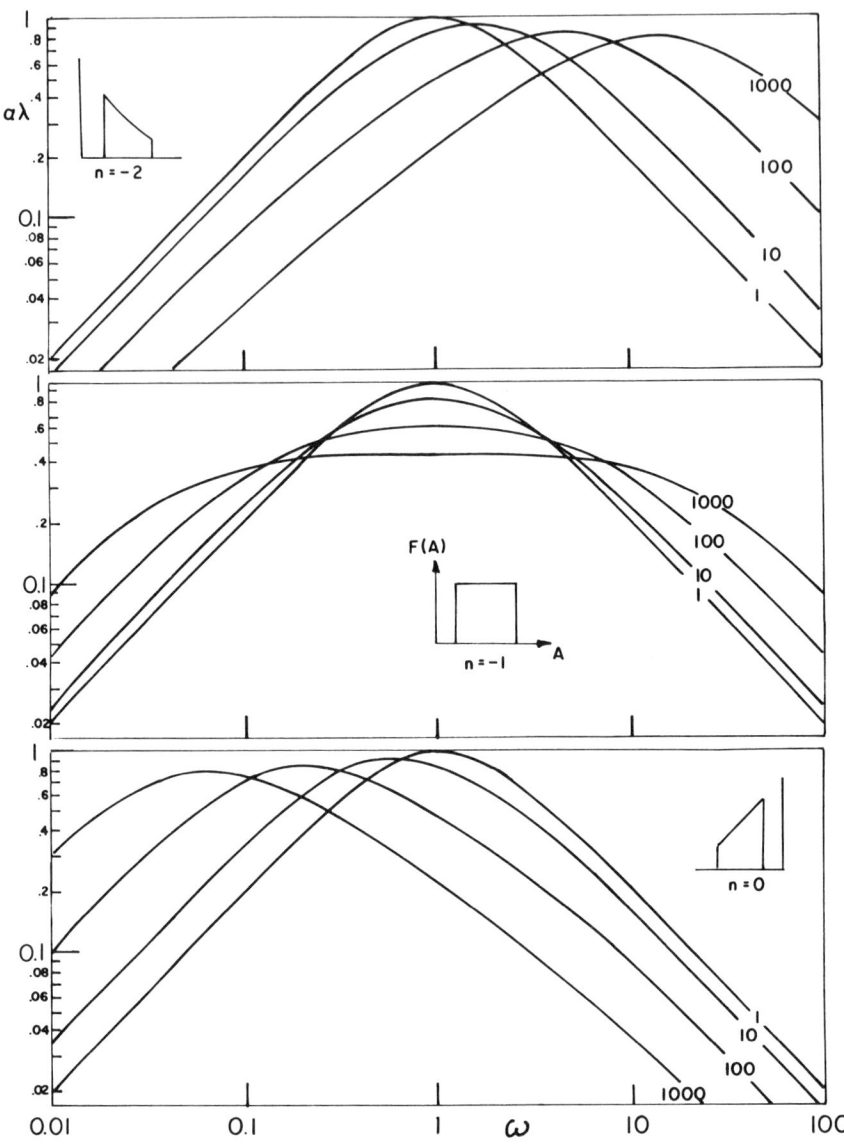

FIG. 10. Absorption per wavelength $\alpha\lambda$ as function of frequency $f=\omega/2\pi$ for various distribution functions of time constants $F(T)=aT^n$. All curves are calculated for a unit value of the sum of all time constants $\int_{T_1}^{T_2} F(T)dt$. The parameter indicates the ratio of largest to smallest time constant T_2/T_1. T_1T_2 is set equal to 1. Inserts show the distribution functions with $A=\ln T$. The curves $n=-2$ are symmetrical to the curves $n=0$, with the symmetry axis at $\omega=1$. All curves demonstrate that only in the case $n=-1$ may a fair degree of frequency independence of $\alpha\lambda$ be achieved.

above, is not surprising and is to be anticipated. However, this does not imply that the amplitude of the individual contributing relaxational processes is the same. Indeed, the results presented here clearly show that the specific absorption varies almost by one order of magnitude from one type of substance to another (Table II and Fig. 10).[17]

The results presented show, furthermore, that the destruction of gross tissue structure results only in a moderate reduction of the absorption. Comparison of the data in Figs. 1 and 2 demonstrates that about two-thirds of the total absorption is apparently generated at a submicroscopic level, probably at the macromolecular level. The different preparations investigated do contain mainly proteins and some lipids. It is difficult to separate the effect of the protein component from the effect of the lipid component. The pH dependence of the acoustic properties of the liver preparations and the lipid-free gelatin solution suggest that the specific absorption is mainly due to the protein component. The comparable specific absorption of a hemoglobin solution and of the liver supernatant which does contain some lipids provides additional support that the lipid component does not noticeably affect the acoustic properties. The fact that liver tissue is so strongly sound absorbing in comparison with blood is not due to a different mechanism of interaction originating at a higher structural level. It merely represents higher specific absorption values for the abundant liver tissue macromolecules and, perhaps, their different kinds of submicroscopic associations.

The macromolecular absorption is shown to change with macromolecular denaturation such as caused by temperature increase and with pH. The mechanism responsible for this behavior is not clarified in this paper. However, a similar dependence of absorption on pH in various biological macromolecules has been recently investigated in greater detail by Dunn and his associates[18,19] and by Hussey and Edmonds.[10,11] A recent summary of various possible mechanisms and absorption data of biological and other molecules has been given by Dunn, Edmonds, and Fry.[2]

ACKNOWLEDGMENTS

This study was supported by Air Force Contract AF 33(616)-2494 and NIH Grant RO1 HE-01253.

---

* Present address: Institute f. Phys. u. Med. Strahlenkunde, Univ. of Erlangen-Nürnberg, Erlangen, West Germany.

[1] D. E. Goldman and T. F. Hueter, "Tabular Data of the Velocity and Absorption of High-Frequency Sound in Mammalian Tissues," J. Acoust. Soc. Amer. **28**, 35 (1956); **29**, 655 (L) (1957).

[2] Floyd Dunn, P. D. Edmonds, and W. J. Fry, "Absorption and Dispersion of Ultrasound in Biological Media," in *Biological Engineering*, H. P. Schwan, Ed. (McGraw–Hill, New York, 1969).

[3] E. L. Carstensen and H. P. Schwan, "Acoustic Properties of Hemoglobin Solutions," J. Acoust Soc. Amer. **31**, 305–311 (1959).

[4] E. L. Carstensen and H. P. Schwan, "Absorption of Sound Arising from Presence of Intact Cells in Blood," J. Acoust. Soc. Amer. **31**, 185–189 (1959).

[5] W. C. Schneider and G. H. Hogeboom, "Cytochemical Studies of Mammalian Tissues: The Isolation of Cell Components by Differential Centrifugation: A Review," Cancer Res. **11**, 1–22 (1951).

[6] Difco Bacto gelatin was used.

[7] W. S. Spector, Ed., *Handbook of Biological Data* (Saunders, Philadelphia, 1961), Tables 52 and 54.

[8] E. L. Carstensen, K. Li, and H. P. Schwan, "Determination of the Acoustic Properties of Blood and its Components," J. Acoust. Soc. Amer. **25**, 286–289 (1953).

[9] H. P. Schwan, H. Pauly, and A. Smith, "Research Studies on the Effect of Sound on Biological Material," Wright Air Develop. Center, Wright–Patterson Air Force Base, Ohio, 3rd Tech. Rep. (1957).

[10] M. Hussey and P. D. Edmonds, "Ultrasonic Examination of Proton Reactions at the α-Amino and Side Chain Groups of Arginine and Lysine in Aqueous Solution," J. Acoust. Soc. Amer. (to be published).

[11] M. Hussey and P. D. Edmonds, "Ultrasonic Examination of Proton Transfer Reactions in Aqueous Solutions of Glycine," J. Acoust. Soc. Amer. (to be published).

[12] For n values 0 and -2, the elevated ramp functions indicated in Fig. 10 result. The box function appears a more reasonable approximation of any flat distribution function of activation energies.

[13] Only in the case of blood have measurements been extended to sufficiently low and high frequencies (see Refs. 14–16) to recognize the frequency limits of this behavior.

[14] H. Gramberg, PhD thesis, Johann Wolfgang Goethe Univ., Frankfurt (1956).

[15] P. D. Edmonds, "Ultrasonic Absorption of Hemoglobin Solutions," Biochim. Biophys. Acta **63**, 216 (1962).

[16] P. D. Edmonds, T. J. Bauld, J. F. Dyro, and M. Hussey, "Ultrasonic Absorption of Aqueous Hemoglobin Solutions," Biochim. Biophys. Acta **200**, 174–177 (1970).

[17] A. Smith and H. P. Schwan, "The Acoustic Properties of Cell Nuclei," J. Acoust. Soc. Amer. (to be published).

[18] W. D. O'Brien, Jr., "The Absorption of Ultrasound in Aqueous Solutions of Biological Polymers," PhD thesis, Univ. Illinois, Urbana, Illinois (1970).

[19] L. W. Kessler and F. Dunn, "Ultrasonic Investigation of the Conformal Changes of Bovine Serum Albumin in Aqueous Solutions," J. Phys. Chem. **73**, 4256 (1969).

11

Copyright © 1962 by the Acoustical Society of America

Reprinted from J. Acoust. Soc. Amer., 34(10), 1545–1547 (1962)

Temperature and Amplitude Dependence of Acoustic Absorption in Tissue

FLOYD DUNN

Biophysical Research Laboratory, University of Illinois, Urbana, Illinois

(Received June 21, 1962)

The acoustic intensity absorption coefficient of tissue of the central nervous system has been determined at the sound frequency of 1 Mc/sec in the temperature range from 2° to 28°C at incident sound intensities ranging from 5 to 200 W/cm². The absorption coefficient exhibits an increase with increasing temperature and no variation with the acoustic intensity.

IT is well established that properly controlled, high-intensity ultrasound produces unique effects on the central nervous system[1] and provides a versatile method for modifying brain structures.[2] In order that the full potentialities of this methodology be realized for fundamental biological research and for medicine, it is essential that the physical mechanism(s) of the interaction of intense ultrasound and biological structures be understood. An important adjunct to the elucidation of these physical mechanisms is a basic understanding of the absorption processes occurring when biological materials are irradiated with ultrasound. The frequency dependence of the absorption coefficient of most investigated tissues can be described by a power function whose exponent varies between 1 and 1.3.[3,4] Knowledge of the temperature dependence of the ultrasonic absorption coefficient of tissue is lacking, owing to the fact that investigations have been restricted largely to adult mammals, which are thermally homeostatic, and thereby essentially excluding the possibility of making absorption measurements over a wide temperature range. The present study was undertaken to fill, in part, this existing gap in knowledge and to provide necessary data for the accurate determination of the dosage parameters to affect specific biological systems.

The method employed for determining the value of absorption coefficients in tissue has been described previously.[3,5,6] Briefly, a small calibrated thermocouple probe is imbedded in the tissue and the specimen is exposed to rectangular acoustic pulses of known intensities. The transient thermoelectric output produced in response to a pulse is recorded on a magnetic oscillograph employing a galvanometer with a time constant of approximately 0.02 sec. The rise time of the pulse is of the order of 10^{-3} sec. The transient temperature change detected by the thermocouple imbedded in the tissue possesses two distinct phases. The first phase, which reaches an equilibrium value rapidly (in about 0.1 sec), results from the conversion of acoustic energy into heat by the viscous forces acting between the wire and the immediately surrounding medium.[7] This phase is, of course, not present in the tissue when the thermocouple is absent. The second phase exhibits an almost linear characteristic (for a pulse duration of approximately 1 sec) and results from acoustic energy converted into heat by absorption in the tissue surrounding the thermocouple junction. If the thermocouple wires are sufficiently small in diameter, the initial time rate of change of temperature from the second phase is related to the acoustic intensity absorption coefficient per unit path length by the relation

$$\mu = (\rho C_p K/I)(dT/dt)_0, \quad (1)$$

where ρC_p is the heat capacity per unit volume of the tissue (cal/cm³°C), I is the acoustic intensity (W/cm²), and K is the mechanical equivalent of heat. The acoustic intensity of the plane traveling wave field is determined by a thermocouple probe which has been calibrated against a radiation-pressure detector.[3,7,8]

The young mouse, 24 h after birth, is a convenient preparation for the study described herein for a number of reasons,[5,9,10] one of the more important being that it is an essentially poikilothermic animal which readily allows temperature cycles to as low as 0°C to be carried out without producing permanent changes in the animal. The following procedure is followed in preparing the animals for ultrasonic irradiation and the associated measurements of the concomitant transient temperature rise.[5] The animal is cooled to render it dormant so that it can be properly positioned in the mouse holder and to ensure that it will remain in that position until it is placed in the sound tank and irradiated. The traveling-wave sound field is produced by a 1-Mc, unfocused, X-cut quartz plate. The acoustic transmission medium is degassed mammalian Ringer's solution. When the mouse is sufficiently cool, it is placed in the mouse holder which secures the head, hind limbs, and tail firmly. The previously fabricated and calibrated copper–constantan thermocouple is then inserted in the spinal

---

[1] W. J. Fry, in *Advances in Biological and Medical Physics*, edited by J. H. Lawrence and C. A. Tobias (Academic Press Inc., New York, 1958), Vol. VI, p. 281.
[2] W. J. Fry, R. Meyers, F. J. Fry, D. F. Schultz, L. L. Dreyer, and R. F. Noyes, Trans. Am. Neurol. Assoc. 16 (1958).
[3] W. J. Fry and F. Dunn, in *Physical Techniques in Biological Research*, edited by W. L. Nastuk (Academic Press Inc., New York, 1962), Vol. IV, Chap. 6, p. 261.
[4] D. E. Goldman and T. F. Hueter, J. Acoust. Soc. Am. 28, 35 (1956).
[5] F. Dunn, Ph.D. thesis, University of Illinois, Urbana, Illinois (1956).
[6] W. J. Fry and R. B. Fry, J. Acoust. Soc. Am. 25, 6 (1953).
[7] W. J. Fry and R. B. Fry, J. Acoust. Soc. Am. 26, 294, 311 (1954).
[8] F. Dunn and W. J. Fry, IRE Trans. on Ultrasonic Engineering PGUE-5, 59 (1957).
[9] F. Dunn, Am. J. Phys. Med. 37, 148 (1958).
[10] W. J. Fry and F. Dunn, J. Acoust. Soc. Am. 28, 129 (1956).

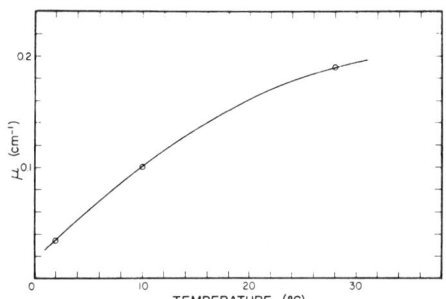

FIG. 1. Acoustic intensity absorption coefficient vs temperature at 1 Mc for spinal cord of young mice.

cord at the level of the third lumbar vertebra. (There is very nearly a 1:1 spatial correspondence between vertebral segments and cord segments of these animals.) An optical arrangement is employed to position the animal accurately, and hence the junction of the thermocouple, in the predetermined site in the sound field. The method of inserting the thermocouple into the tissue (threading) does not yield information about the precise position of the junction with respect to the anatomical structure of the animal. This is determined after the animal is sacrificed. The soft tissues are cleared in a 1% aqueous solution of KOH after which the oseous material is stained with Alizarin Red-S.[11] The specimen can then be viewed under a microscope and the position of the thermocouple junction is located accurately with respect to the vertebral structures.

The thermocouples are fabricated from 0.003-in.-diameter commercially available wires, which are etched in acid to reduce the diameter to approximately 0.0005 in. in the vicinity of the junction. Assembly of the thermocouple is accomplished by soldering. Both lapp and butt joints have been used without any observed difference in the ensuing results.

Table I shows the results at the three base temperatures of the animals considered in this study. The base temperatures of the animals were known to ±0.1°C. The quantity $\mu/\rho C_p$ is computed using Eq. (1) and a knowledge of the incident sound intensity and the experimentally determined initial temperature rise measured by the inserted thermocouple (the temperature rise associated with the viscous forces being subtracted away). The acoustic intensity absorption coefficient per unit path length, μ, is computed using the value[12] $\rho C_p = 0.84$ cal/cm³ °C for the heat capacity per unit volume ($\rho = 1.03$ g/cm³, $C_p = 0.81$ cal/g°C). In the absence of more specific information, the heat capacity per unit volume is considered to be constant within the temperature range of these experiments. The over-all uncertainty in the numerical values of $\mu/\rho C_p$ is thought to be of the order of 10–15% so that a small dependence upon temperature of ρC_p does not add appreciably to the percentage uncertainty in the results. It should be noted that in computing the values of $\mu/\rho C_p$ and μ, account was taken of the reduction in the incident intensity caused by absorption of energy in the tissue. This small correction was accomplished by using a value of μ obtained by inserting the value of the incident intensity in Eq. (1) and this is considered to be sufficiently accurate for the purposes of this study. Figure 1 is a graphical representation of the intensity absorption coefficient versus temperature.

The procedure for acquiring the data included the exposure of each specimen to different levels of incident sound. Thus data is available relating the initial time rate of change of temperature, as observed by the inserted thermocouple probe, as a function of the intensity of the incident acoustic wave and is illustrated graphically in Fig. 2. On the assumption that the heat capacity per unit volume, ρC_p, is not dependent upon the intensity, it can be concluded [see Eq. (1)] that, in the range of incident sound intensities from zero to approximately 200 W/cm², the absorption coefficient is independent of the intensity of the incident wave for the experimental preparations of this study. Concerning the observed lack of dependence of the absorption coefficient of nerve tissue upon the sound intensity in these experiments, the following statements can be made: The propagation distance from sound source via degassed mammalian saline to tissue was, in these experiments, approximately 5 cm. The discontinuity distance in water for the highest sound amplitudes employed in these experiments is approximately 8 to 10 cm. The

TABLE I. Absorption data.

T(°C)	$\mu/\rho C_p$	μ(cm⁻¹)
2	0.040	0.034
10	0.12	0.10
28	0.23	0.19

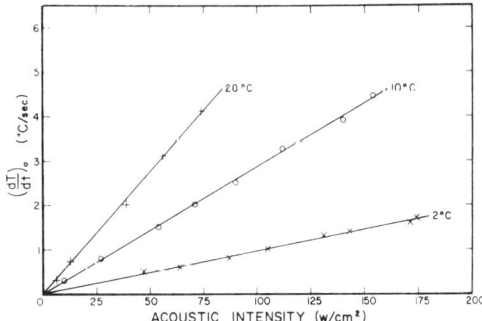

FIG. 2. Time rate of change of temperature in spinal cord of young mice produced by absorption of 1 Mc sound vs incident acoustic intensity.

[11] H. J. Conn and M. A. Darrow, *Staining Procedures used by the Biological Stain Commission* (Biotech Publications, Geneva, New York, 1946), Part I, Page ID₃–13.

[12] W. Guttner, Acustica 4, 547 (1954).

product αL (the infinitesimal amplitude absorption coefficient × the discontinuity distance) is of the order of 10^{-3}.[3] Using the results of recent computations,[13] it is seen that the amplitude of the fundamental frequency component of the propagated sound wave at the tissue–saline interface is at most a few percent less than that initially at the surface of the vibrating element. Thus the wave incident at the tissue is relatively undistorted. On the assumption that the ratio B/A for tissue is approximately the same as that for water, and since the propagation distance in the tissue is relatively short (approximately 0.5 mm), the transfer of energy from the incident fundamental wave to its harmonics is negligible.

The form of the temperature dependence of the absorption coefficient (Fig. 1) eliminates shear viscosity as the absorption mechanism in tissue. On the basis of the work reported herein the following empirical relation describes the temperature dependence of the amplitude absorption coefficient in the range from 0° to 35°C:

$$\alpha = \tfrac{1}{10}[2 - e^{0.016(35-T)}], \qquad (2)$$

where T is the temperature of the tissue in °C. The temperature dependence of the absorption coefficient appears to resemble that of high viscosity liquids above the main relaxation frequencies of the liquid.[14] Indeed, it has been estimated that the bulk viscosity of muscle relaxes near 40 kc and the shear viscosity near 400 kc.[15]

ACKNOWLEDGMENT

This research was supported by Contract US Ph B-1017, Institute of Neurological Diseases and Blindness, National Institutes of Health, U. S. Public Health Service.

---

[13] W. W. Lester, J. Acoust. Soc. Am. **33**, 1196 (1961).

[14] T. A. Litovitz and T. Lyon, J. Acoust. Soc. Am. **26**, 577 (1954).

[15] T. F. Hueter, WADC Tech. Rept. 57-706 (August 1958).

ABSORPTION OF ULTRASOUND IN BIOLOGICAL MEDIA

F. Dunn and J. K. Brady

Bioacoustics Research Laboratory, University of Illinois, Illinois 61801, U.S.A.

(*Received* 20 *November* 1972)

Measurements have been made of the linear absorption coefficient of ultrasound in tissues of the c.n.s. of mammals over the temperature range from 0 to 50°C at frequencies of 0·5, 0·7 and 1·0 Mc/s. The spinal cord of newborn mice (24 hr after birth) is a convenient object for the investigation. The character of the change in absorption is similar to the changes found close to the main relaxation frequencies in highly viscous fluids.

ULTRASOUND is becoming ever more accessible to biochemists and biophysicists for studying fast reactions [1] and physicians for diagnostic and therapeutic purposes [2, 3]. However, the full possibilities of these methods are realized neither in fundamental biological research nor in medicine since the physical mechanisms of the interaction of ultrasound with biological media are insufficiently understood. In addition, the presence of information on the interaction of ultrasound with biological media is important for practical purposes in order to avoid clinically dangerous conditions. An important element for determining these physical mechanisms is an understanding of the basic processes of absorption of ultrasound occurring on insonation of biological material.

The frequency dependence of the absorption coefficient of a number of tissues and organs has been studied by many authors and in most cases may be evaluated by a power function with exponent from 1·0 to 1·3 [1]. Exceptions are the lungs [4] and bones [5]. Less attention has been paid to the temperature dependence of the absorption coefficient since investigators have chiefly worked with the tissues of adult mammals and were seriously restricted in the possibility of carrying out measurements over a very wide temperature range. The investigation carried out on newborn mice [6] showed that at a frequency of 1 Mc/s the absorption coefficient rises with increase in the temperature inversely to the expected by analogy with liquid media in which shear viscosity predominates. Such a phenomenon was also observed in the brain of adult cats [7] on measurement immediately after death when it was found that absorption of ultrasound undergoes an insignificant change up to a temperature of about 50°C above which it substantially increases. The present paper continues investigations on the newborn mice in order to bring in an additional parameter – the frequency of the ultrasound.

The methods used to determine the value of the absorption coefficient in tissues has been described previously [8, 9]. The method consists in the following: a small calibrated thermocouple pick-up is fixed in the tissue and the sample insonated with ultrasonic pulses of known intensity. The temperature changes measured with the thermocouple fixed in the tissue have two distinctly marked phases. The first phase which rapidly (about 0·1 sec) reaches equilibrium value is due to the transformation of the acoustic energy into heat by viscous forces acting between the wire and the medium directly surrounding it [10]. This phase of course does not appear in tissue in the absence of a thermocouple. The second phase has an almost linear character (for a pulse duration of about 1 sec) and is due to conversion of acoustic energy into heat as a result of absorption in the tissue surrounding the thermocouple contact. If the diameter of the thermocouple wires is sufficiently small then the initial rate of change in temperature in the second phase is connected with the linear absorption coefficient of ultrasound by the following relation:

$$\alpha = \frac{\rho C_p K}{2I} \left(\frac{dT}{dt}\right)_0, \tag{1}$$

where ρC_p—thermal capacity of unit volume of tissue (cal/cm³·degree); I—intensity of ultrasound (W/cm²); K—mechanical equivalent of heat.

The acoustic intensity in the field of the flat running wave is determined by a thermoelectric probe which was initially calibrated against a radiation pressure meter [9–11].

Newborn mice 24 hr after birth are a convenient object for investigations similar to the present one for a number of reasons [12–14] of which the most important is the substantial poikilothermal power of such a mouse readily permitting variation in temperature down to 0°C without any residual changes in the animal. The animal was anaesthetized in order to ensure its necessary position in the holder and the invariability of this position before the completion of the procedure of insonation. When the measurements were made at temperatures below 25°C cooling of the animal itself caused it to fall asleep. At experimental temperature above 25°C the animal was immobilized by adding a small quantity of CO_2 to the respiratory mixture.

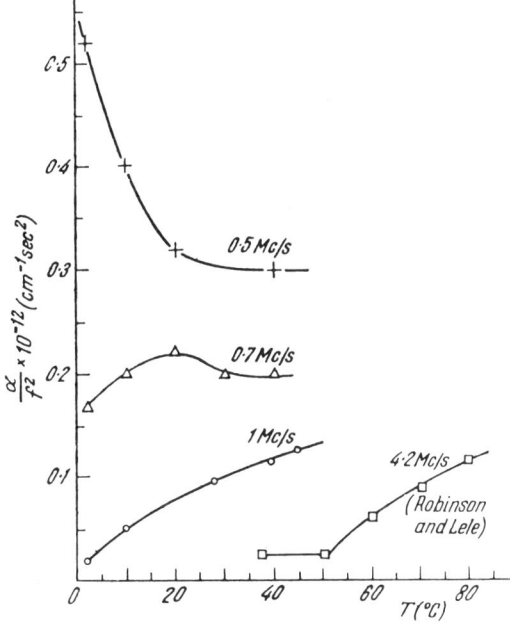

FIG. 1. Temperature dependence of α/f^2 in spinal cord of newborn mouse.

The field of the sonic running wave was produced by a quartz plate of X-section of such a construction that the inhomogeneity of distribution of the intensity of ultrasound along the spinal cord was less than 5 per cent (the tissue samples were checked). As transmitting acoustic medium we used degassed mammalian Ringer solution. After the corresponding immobilization, the mouse was placed in a holder in which the head, hind limbs and tail were firmly fixed. Then, into the spinal cord at the level of the third lumbar vertebra was inserted a copper–constantan thermocouple prepared and calibrated previously (between the vertebral and spinal segments of this animal an almost exact 1 : 1 spatial correspondence existed). For the precise location of the animal and correspondingly, the thermocouple junction at a set point of the sonic field we used an optical device. The method of inserting the thermocouple into the

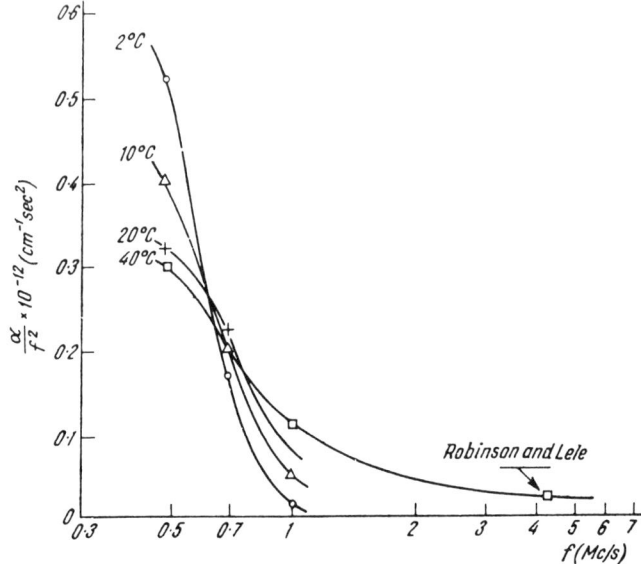

FIG. 2. Frequency dependence of α/f^2 in spinal cord of newborn mouse.

tissue did not allow us to obtain information on the precise location of the thermocouple junction in relation to the anatomical structure of the animal. Precise localization of the thermocouple junction was determined on sectioning the tissue under a microscope after decapitating the animal and freezing with liquid nitrogen.

The thermocouples were prepared from wire 7·62 mm diameter which was pickled in acid to reduce the diameter close to the junction approximately to 1·27 mm. The junction of the thermocouple was achieved by soldering.

The absorption coefficient of the ultrasound was estimated from equation (1) using the value $\rho C_p = 0.84$ cal/cm^3·degree [15] for the thermal capacity per unit volume ($\rho = 1.03$ g/cm^3, $C_p = 0.81$ cal/g·degree). In view of the absence of more detailed information the thermal capacity per unit volume is considered to be constant within the limits of the entire temperature range of the investigation. The total error in the numerical value of the absorption coefficient was of the order of 10 per cent so that the weak dependence of ρC_p on temperature insignificantly increased the relative error of the results. In calculating the value of the absorption coefficient we took into account the reduction in the intensity of the incident ultrasound due to absorption in the tissue. This minor correction was achieved by using the value α obtained on introducing the value of the incident intensity of the ultrasound into equation (1) which is presumed to be sufficiently accurate for the purposes of our investigation.

Figure 1 shows the dependence of the frequency-independent absorption coefficient α/f^2 on temperature for the spinal cord of the newborn mouse at frequencies of 0·5, 0·7 and 1 Mc/s. It also presents the findings obtained by Robinson and Lele [7] for the brain of the adult cat.

Figure 2 shows the dependence of α/f^2 on frequency. It is clear that the results obtained are similar to the findings for highly viscous fluids close to the regions of the main relaxation frequencies.

It is now necessary to obtain information at lower frequencies to determine the width and form of the relaxation region in the hope of identifying the mechanism of absorption. It is also important to determine whether the character of the change in absorption shown in Figs. 1 and 2 is typical of all soft tissues, in general, with the possible exception of pulmonary tissues [5].

The authors are grateful to the National Institute of Public Health for financing this work with grant GM 12281.

REFERENCES

1. DUNN, F., EDMONDS, P. D. and FRY, W. J., Biological Engineering, p. 205, (Ed. H. P. Schwan). McGraw-Hill, N. Y. 1969
2. BOCK, J. and OSSOINIG, K. (Eds.) Ultrasono Graphia Medica, Proc. lst. World Congress Ultrasonic Diagnostics in Medicine (Verlag der Weiner Medizinischen). Akademie, Veinna, Vol. I, II and III, 1971
3. LEHMANN, J. F., De LATEUR, B. J. and SILVERMAN, D. R., Arch. Phys. Med. Rehabilit 47: 331, 1966
4. DUNN, F. and FRY, W. J., Phys. Med. Biol. 5: 401, 1961
5. HUETER, T. F., Naturwissenschaften 39: 21, 1952
6. DUNN, F., J. Acoust. Soc. Amer. 34: 1545, 1962
7. ROBINSON, T. C. and LELE, P. P., J. Acoust. Soc. Amer. 51: 1333, 1972
8. FRY, W. J., and FRY, R. B., J. Acoust. Soc. Amer. 25: 6, 1953
9. FRY, W. J. and DUNN, F., Physical Techniques in Biological Research p. 251-394, (Ed. W. L. Nastuk). Acad. Press, N. Y., Vol, IV, Chap. 6, 1962
10. FRY, W. J. and FRY, R. B., J. Acoust. Soc. Amer. 26: 294, 311, 1954
11. DUNN, F. and FRY, W. J., IRE Trans. Ultrasonic Engr. PGUE-5, 59, 1957
12. DUNN, F., Ph. D., Thesis, University of Illinois, Urbana, 1956
13. FRY, W. J. and DUNN, F., J. Acoust. Soc. Amer. 28: 129, 1966
14. DUNN, F., Amer. J. Phys. Med., 37: 148, 1958
15. GUTTNER, W., Acustica 4: 547, 1954

13

Copyright © 1961 by the Institute of Physics, London

Reprinted from *Phys. Med. Biol.*, 5(4), 401–410 (1961)

Ultrasonic Absorption and Reflection by Lung Tissue

By F. Dunn, Ph.D. and W. J. Fry, M.Sc.

Biophysical Research Laboratory, University of Illinois,
Urbana, Illinois, U.S.A.

§ 1. Introduction

It is well known that ionizing radiation affects biological systems in a cumulative fashion (Blum 1959). Each exposure to the radiation increases the probability that pathologic growth will become apparent in the irradiated tissue structure in some specified time interval. X-ray procedures have been long employed as diagnostic tools but it is only relatively recently that an awareness of the effects of even very small doses has become generally known. The radiation dose can be reduced in lengthy diagnostic procedures such as cardiac catheterization or examination of the GI tract, for example, by the employment of fluoroscopic gastro-intestinal image intensifiers (Mallams and Miller 1956). However, it is of considerable importance to devise procedures to accomplish the purpose of X-ray methods whenever possible, further reducing hazards of ionizing radiation. In addition, a method which reveals soft tissue structure would have considerable application in medicine. One such procedure which suggests itself is the ultrasonic method of visualization of soft tissue structures developed by Howry (1957), Kikuchi, Uchida, Tanaka and Wagai (1957), and Wild and Reid (1957). To examine this possibility for lung tissue, which is especially difficult because of the presence of undissolved gas and which represents a tissue structure that annually receives considerable diagnostic radiation attention, experiments were carried out to determine the acoustical properties of excised dog lung.

§ 2. Experimental Method and Results

The tissue was used immediately after excision and was kept continuously in normal physiological saline. For the measurements, the tissue was encapsulated in a stainless steel frame having thin polyethylene windows (0·003 in. thick), the same diameter as that of the lung sample and sufficiently large to prevent interference with the acoustic beam. Care was taken not to compress the tissue from its original thickness and it is estimated that two-thirds of the residual air remained in the lung sample. The slight surface curvature of the major faces of the sample was eliminated by the confining capsule. The capsule was filled with degassed physiological saline to minimize possible interface difficulties which would result if gas bubbles were occluded on the tissue surface. The volume of the degassed saline was a small fraction (approximately 5%) of the volume of the lung tissue in the capsule so that only a small fraction (a maximum of 0·2%) of the gas present within the lung could be dissolved by the degassed saline. This estimate assumes that the lung contains residual air.

A thermocouple probe (Fry and Fry 1954, Dunn and Fry 1957) was used as the detector of the pulsed acoustic energy. All measurements were carried out at a temperature of $35 \cdot 0 \pm 0 \cdot 5 °C$. The acoustic source was a circular X-cut quartz plate $1\frac{1}{2}$ in. in diameter and having a fundamental thickness resonant frequency of 0·98 Mc/sec. The experimental configuration was arranged so that essentially plane waves of sound were propagated over the path lengths of interest in this study, as shown by investigation of the acoustic field distribution normal to the direction of propagation. Fig. 1 schematically illustrates the experimental arrangement. Determinations of the fraction of the incident acoustic energy

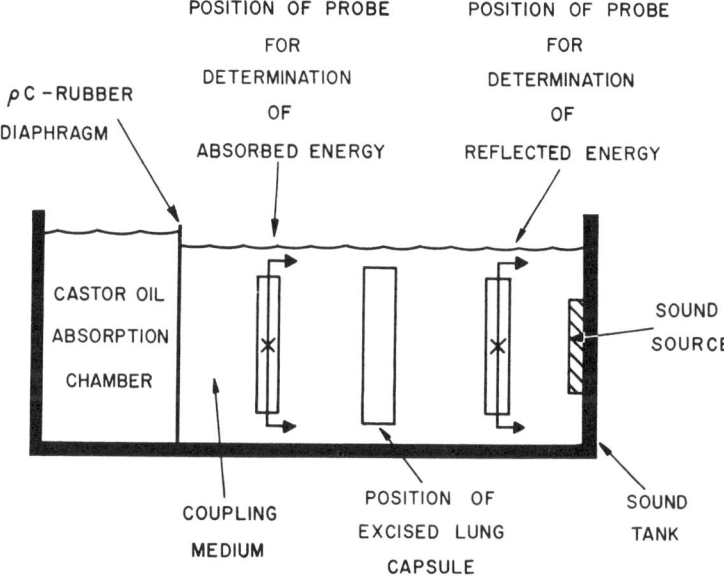

Fig. 1. Schematic illustration of the experimental arrangement.

reflected from the encapsulated lung and that propagated through the same sample were made. The coupling medium through which the ultrasound propagated from source through probe assembly and sample to the absorber was degassed physiological saline. The absorption chamber absorbed sound incident upon it to the extent that no reflected component interfered with the measurements.

For the determination of the fraction of the energy reflected at the lung-saline interface, the thermoelectric probe was placed between the encapsulated lung and the sound source. The lung sample was held in a fixed position while the probe was moved along the axis of propagation of the acoustic field. The transient response of the probe to 0·1 sec pulses of sound was observed as a function of the probe position (0·1 mm intervals) and yields a well defined standing wave pattern. The 'standing wave' pattern measured in this fashion can be used to deduce values of reflection coefficients by the use of an appropriate analysis.† In this way, the standing wave pattern existing in the lung–saline–probe system was determined quantitatively. Similar measurements of the standing wave pattern were made with the capsule filled with degassed saline and air to check the method. The experimentally determined reflection coefficient obtained in this fashion (with two samples of dog lung) is 0·71 implying that 50% of the incident energy is reflected at the lung-saline interface.

The density of normal lung tissue, determined from measurements of the weight and volume, is 0·40 g/cm³. The sound propagation velocity in lung tissue can be obtained from the following relation relating the acoustic impedances (ρv), in the two media and the reflection coefficient, r (the ratio of the amplitudes of the reflected to incident acoustic pressure),

$$(\rho v)_{\text{lung}} = (\rho v)_{\text{saline}} \left(\frac{1-r}{1+r}\right),$$

where ρ is the density and v is the sound velocity. Using the values given above for the reflection coefficient and the density of lung tissue and $1·53 \times 10^5$ g/cm² sec for the acoustic impedance of saline, the sound propagation velocity in the lung tissue is found to be $6·5 \times 10^4$ cm/sec. The wavelength of the sound in the lung at 0·98 Mc/sec (the frequency at which the experiments were carrried out) is thus 0·66 mm.

The acoustic absorption coefficient per unit path length within lung tissue was determined by placing the encapsulated lung between the sound source and the thermoelectric probe (see fig. 1), i.e. the probe was used to investigate the intensity of the sound wave after it traversed the lung sample. The air-filled capsule was used in place of the encapsulated lung to demonstrate that within the limits of detectability (approximately one part in 10^5) no energy was received at the probe. This configuration was used to determine whether leakage of acoustic energy around the capsule would be of sufficient magnitude to interfere with the measurements. Values of the acoustic amplitude absorption coefficient per unit path length were determined from a knowledge of the energy reflected at the two lung-saline interfaces, the thickness of the sample, and the acoustic intensity detected by the probe in accordance with the relation

$$I_1 = I_0 \exp(-2al),$$

where I_0 and I_1 are, respectively, the acoustic intensitites at the lung-saline interfaces nearest to and farthest from the source, l is the thickness

---

† The 'standing wave' ratio determined by the thermocouple is a function of both the acoustic pressure and particle velocity amplitudes and their spatial distributions.

of the lung sample, and α is the acoustic amplitude absorption coefficient per unit path length. The attenuation of the acoustic energy is sufficiently great such that energy reflected need not be considered. The data are tabulated in table 1. Fig. 2 shows the total absorption as a

Table 1. Tabulation of Experimental Data and Results

Sample	ρ (g/cm^3)	l (cm)	$(I_1/I_0) \times 10^3$	αl	α (cm^{-1})	$v \times 10^{-4}$ (cm/sec)
normal	0·40	0·76	0·68	3·6	4·8	6·5
normal	0·40	0·60	4·1	2·8	4·6	6·5
pneumonitis	0·76	0·99	1·08	3·4	3·5	3·4

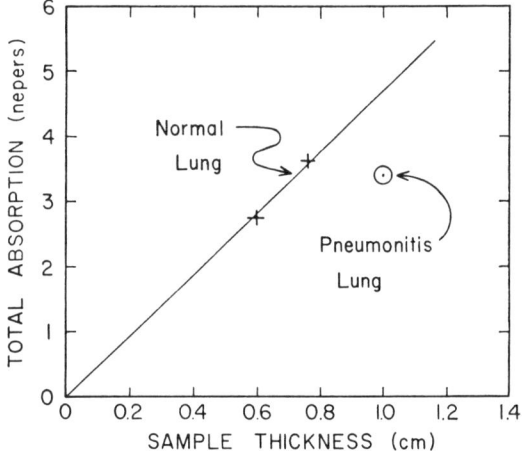

Fig. 2. Total acoustic amplitude absorption *vs* lung sample thickness.

function of sample thickness. The average value for the amplitude absorption coefficient for normal dog lung (excised) is found to be 4·7 cm^{-1}. One of the animals had a pneumonitis at the time the lung was excised. This resulted in an increased density by approximately a factor of two, a decrease in the velocity of propagation in the lung by the same factor, but a decrease in the absorption coefficient by only 25%.

§ 3. Acoustic Model of Lung Tissue

The unusually high absorption displayed by lung tissue requires comment. For example, the value obtained is more than an order of magnitude greater than that found for dry oxygen or nitrogen (Beranak 1949). In order to account for the absorption greatly in excess of that displayed by soft tissue, an acoustic model based upon the gross structure of lung tissue is postulated.

Ultrasonic Absorption and Reflection by Lung Tissue

The structure of lung tissue pertinent to the present discussion can be described as follows (Best and Taylor 1945). The bronchioles branch and rebranch repeatedly as they proceed toward the periphery of the lung. The first branchings are approximately 1·5 mm in length and 0·4 mm in diameter. The terminal and respiratory bronchioles are approximately 0·5 mm in length, but the diameters are approximately the same as that of earlier branchings. That is, even though the bronchioles become shorter, virtually no decrease in diameter takes place as they pass toward the periphery. Each respiratory bronchiole gives rise to five or six alveolar ducts each of which in turn rebranch a number of times and each gives rise to approximately five alveolar sacs. The hemispherical protuberances in the walls of the sacs constitute the

Table 2. Tabulation of Quantities Appearing in Eqn. (2) and Associated Numerical Values

Quantity	Definition	Numerical Value
ω_0	$\frac{1}{R_0}\left\{\left(\frac{3\gamma P_0}{\rho}\right)\left(\frac{g}{\epsilon}\right)\right\}^{\frac{1}{2}}$, angular resonant frequency of bubble	
g	$1 + \frac{2\sigma}{P_0 R_0}\left(1 - \frac{1}{3h}\right)$	
ϵ	$1 + \frac{3(\gamma-1)}{2\Phi R_0}\left\{1 + \frac{3(\gamma-1)}{2\Phi R_0}\right\}$	
Φ	$\left(\frac{\omega_0 \rho_g C_p}{2K}\right)^{\frac{1}{2}}$	
v	acoustic velocity in liquid	1.5×10^5 cm/sec
γ	ratio of specific heats of gas	1·4
p_0	static pressure	10^6 dyne/cm^2
ρ	density of liquid	1·0 g/cm^3
ρ_g	density of gas	1.29×10^{-3} g/cm^3
σ	surface tension	75 dyne/cm
h	γ/ϵ	$1 < h < \gamma$
C_p	heat capacity at constant pressure of gas	0·24 cal/g
K	thermal conductivity coefficient of gas	5.6×10^{-6} cal/cm sec °C
η	viscosity of liquid-like medium	1.5×10^2 poise*
N	number of bubbles per unit volume	$0.143\, R_0^{-3}$
b	total dissipation parameter, $b_t + b_r + b_v$	
b_t	thermal dissipation parameter (does not contribute appreciably)	
b_r	radiation dissipation parameter, $\frac{\rho\omega^2}{4\pi v}$	
b_v	viscous dissipation parameter, $\frac{\eta}{\pi R_0^3}$	
R_0	radius of gas bubble	
ω	angular frequency of sound field	6.28×10^6 radians/sec

* von GIERKE, OESTREICHER, FRANKE, PARRACK and VON WITTERN (1952)

pulmonary alveoli which are approximately 0·1 mm in diameter. The alveolar sacs have dimensions considerably greater than that of both the respiratory bronchiole and the alveolar duct.

The very large number of spheroid-like and cylindrical gaseous elements with dimensions comparable to the wavelength of sound in the lung at the frequency of the measurements (0·98 Mc/sec), suggests a model composed of a uniform distribution of spherical gas bubbles imbedded in a liquid-like medium. It is considered that sound energy excites the bubbles to pulsate and that the bubbles dissipate their energy by radiating spherical sound waves, by polytropic compressions and expansions of the enclosed gas, and by viscous dissipation attributed to viscous forces acting at the gas-liquid interface. Assume that the liquid-like material has properties similar to water (except as noted in table 2) and that the gas has properties similar to air. In order to be consistent with the experimentally determined density of lung tissue (0·40 g/cm³), the number of bubbles per unit volume, N, is related to the mean bubble radius as

$$N = \frac{0 \cdot 6}{4/3 \pi R_0^3} = 0 \cdot 143\, R_0^{-3}, \quad \ldots \ldots \quad (1)$$

i.e., this requires a packing of the bubbles approximately midway between hexagonal closest packing (each sphere touched by twelve neighboring spheres) and cubic packing (each sphere touched by six neighboring spheres). The acoustic amplitude absorption coefficient per unit path length for such a system is (Devin 1959; Fry and Dunn 1961)

$$\alpha = \frac{bNv}{4} \left[\frac{\frac{3\gamma P_0}{R_0^2} + \omega^2 \rho}{\left\{ \frac{1}{4\pi R_0} \left(\omega^2 \rho - \frac{3g\gamma P_0}{\epsilon R_0^2} \right) \right\}^2 + b^2 \omega^2} \right] \quad \ldots \ldots \quad (2)$$

The quantities appearing in eqn. (2) are defined in table 2 and values for these quantities useful in making the computation are also tabulated.

Fig. 3 shows the absorption coefficient described by eqn. (2) as a function of frequency for a bubble radius of 0·3 mm, a value consistent with the dimensions of the gaseous elements of lung, as discussed above. Fig. 4 shows the absorption coefficient as a function of bubble radius at a frequency of 1 Mc/sec. In the range of values chosen for the frequency and the bubble radius in the computations, radiation of spherical sound waves contributes the greatest source of energy dissipation; dissipation attributed to viscous forces acting at the gas-liquid interface is approximately an order of magnitude less and thermal conduction is negligible. On the assumption that the bubble radius is 0·3 mm, giving $5 \cdot 3 \times 10^3$ bubbles/cm³ from eqn. (1), fig. 3 shows that the absorption coefficient at 1 Mc/sec is 5·7 cm⁻¹ as compared with the experimentally determined value of 4·7 cm⁻¹. Further, fig. 4 shows that, at 1 Mc/sec, the experimentally determined value of the absorption coefficient is obtained for a bubble radius of 0·32 mm corresponding to a bubble population of $4 \cdot 4 \times 10^3$ bubbles/cm³.

The agreement obtained between the experimentally determined value of the absorption coefficient and that of the model is considered to be sufficiently good to lend support to the view that the mechanism of ultrasonic absorption in lung tissue at the sonic frequency of 1 Mc/sec is primarily the result of radiation of sound waves by pulsating gaseous structures.

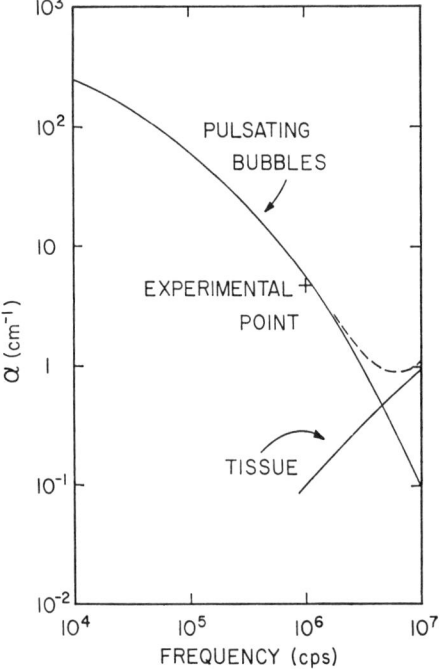

Fig. 3. Acoustic amplitude absorption coefficient per unit path length frequency for bubble radius of 0·3 mm.

§ 4. Discussion

Fig. 3 shows that as the sound frequency increases, the acoustic amplitude absorption coefficient per unit path length resulting from pulsating bubbles decreases and has the value of approximately $0·1 \text{ cm}^{-1}$ at 10 Mc/sec (for bubbles of 0·3 mm radius). It is reasonable to assume that a contribution to the total absorption coefficient of lung also results from absorption of acoustic energy in the tissue. An examination of the literature (Goldman and Hueter 1956) indicates that with the exception of bone and lung, virtually all other mammalian tissues thus far investigated exhibit a linear dependence of the absorption coefficient per unit path length upon frequency and, at a frequency of 1 Mc/sec, the values lie between the limits of $0·025 \text{ cm}^{-1}$ and $0·25 \text{ cm}^{-1}$. Let it be assumed that the total absorption coefficient, α_T, is the algebraic sum of that due to pulsating bubbles, α_b, and that due to tissue, α_t, i.e.,

$$\alpha_T = \alpha_b + \alpha_t \qquad \ldots \ldots \ldots \quad (3)$$

Let it be assumed further that $\alpha_t = 0.1$ cm^{-1} at 1 Mc/sec and that it increases linearly with frequency as shown in fig. 3. Then α_T, determined in accordance with eqn. (3) and illustrated in fig. 3 exhibits a minimum which (for the numerical values chosen) occurs at approximately 6 Mc/sec and has the value of 0.9 cm^{-1}.

The lung samples used in this study contained approximately one-third the average resting respiratory air of normal lung *in vivo*. If it is assumed that inflation of the lung to normal respiratory level has the effect of increasing the bubble radii (without altering the bubble population), then the acoustic absorption decreases, as evident from fig. 4. For such a case, the curve for pulsating bubbles would appear in fig. 3 below the one shown, and consequently would intersect the tissue absorption curve at a lower frequency yielding smaller values for the total absorption coefficient over the frequency range where the absorption due to pulsating bubbles is appreciable compared with that of tissue.

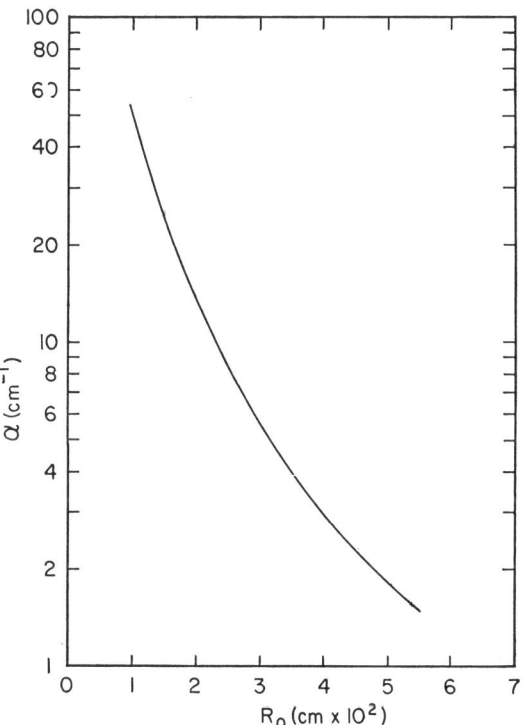

Fig. 4. Acoustic amplitude absorption coefficient per unit path length *vs* bubble radius at a frequency of 1 Mc/sec.

It thus appears that if ultrasound is to be used as a diagnostic tool to examine lung tissue *in vivo*, it is desirable to operate at frequencies near that of minimum absorption per unit path length and, in all probability, also to increase considerably the sensitivity of the receiver for a given depth of penetration and incident sound intensity over that commonly employed for the examination of soft tissue structures of comparable thickness. Using the numerics obtained above, if internal lung structure visualization is to be accomplished comparable with the visualization of structures at the same depth in other soft tissue, it would be necessary to increase the sensitivity of the receiving system by approximately a factor of two for each centimeter depth as compared with that employed for the examination of other soft tissues. Further, there is a loss of 3 db on passing through the lung interface. With systems presently available, ultrasound could be used to examine the surface of the lung for uniformity and the immediate sub-surface tissue (the first few millimeters) for pathologic changes.

Acknowledgments

This work was aided by a grant from the American Thoracic Society. The authors are indebted to Professor F. J. Fry and Mr. R. C. Eggleton for contributions to this research.

Summary

The acoustic reflection and absorption coefficients of both normal and diseased (pneumonitis) excised lung tissue (dog) were experimentally determined at a frequency of 0·98 Mc/sec. It is found that the physiological saline-lung interface reflects 50% of the sound energy falling on it at normal incidence. The acoustic amplitude absorption coefficient per unit path length of lung tissue is 4·7 cm^{-1}. The very high absorption exhibited can be explained as caused by radiation of acoustic energy by the pulsating gaseous structures in the lung tissue. The theory indicates that the absorption coefficient of lung tissue should approach a minimum as the frequency is increased above 1 Mc/sec and should then increase at still higher frequencies. The diseased lung exhibited an acoustic absorption coefficient approximately 25% less than that of normal lung specimens.

Résumé

On a determiné expérimentalement pour la fréquence de 0·98 Mc/sec les coéfficients de réflexion et absorption acoustique pour les tissus pulmonaires normaux et morbides (pnéumonie) d'une chien. On a trouvé que la surface-frontière entre la solution physiologique et le poumon réflète 50% de l'énergie acoustique à l'incidence normale. Le coéfficient d'absorption d'amplitude acoustique par longueur unitaire de trajectoire dans le tissu pulmonaire s'élève à 4,7 cm^{-1}. La très grande absorption observée peut être expliquée par l'émission de l'énergie acoustique par les structures gaseuses vibrantes dans le tissus pulmonaires. La théroie indique que le coéfficient d'absorption du tissu pulmonaire doit s'approcher d'un minimum quand la fréquence surpasse 1 Mc/sec et qu'il doit augmenter aux fréquences plus hautes. Le poumon morbide avait un coéfficient d'absorption acoustique s'élevant à environ 75% de celui des échantillons du poumon normal.

Zusammenfassung

Die Schallreflexions- und Absorptionskoeffizienten sind bei einer Frequenz von 0,98 MHz für gesunde und kranke (Pneumonia) herausgeschnittene Hundeslungengewebe experimentell bestimmt worden. Es wurde gefunden, dass die Grenzfläche zwischen der physiologischen Salzlösung und der Lunge 50% der normal einfallenden Schallenergie reflektiert. Der Schallamplituden-Absorptionsbeiwert für Einheitsweglänge im Lungengewebe beträgt

4,7 cm^{-1}. Die äusserst hohe Absorption kann dadurch erklärt werden, dass die pulsierenden gasförmigen Gebilde im Lungengewebe Schallenergie ausstrahlen. Die Theorie deutet an, dass der Absorptionsbeiwert des Lungengewebes sich einem Minimum nähern dürfte, wenn die Frequenz über 1 MHz gesteigert wird, worauf er bei noch höheren Frequenzen ansteigt. Die kranke Lunge zeigte einen um etwa 25% kleineren Schallabsorptionsbeiwert als die normalen Lungenausschnitte.

REFERENCES

BERANAK, L. L., 1949, *Acoustic Measurements* (New York: John Wiley & Sons, Inc.), p. 68.

BEST, C. H., and TAYLOR, N. B., 1945, *The Physiological Basis of Medical Practice* (Baltimore: Williams & Wilkins Co.), p. 295.

BLUM, H. F., 1959, *Science*, **130**, 1545.

DEVIN, C., 1959, *J. acoust. Soc. Amer.*, **31**, 1654.

DUNN, F., and FRY, W. J., 1957, *I.R.E. Trans. on Ultrasonic Engr.*, *PGUE*-**5**, 59.

FRY, W. J., and FRY, R. B., 1954, *J. acoust. Soc. Amer.*, **26**, 294, 311.

FRY, W. J., and DUNN, F., 1961, *Physical Techniques in Biological Research* (New York: Academic Press), Vol. IV, ed. by W. L. Nastuck (to be published).

GOLDMAN, D. E., and HUETER, T. F., 1956, *J. acoust. Soc. Amer.*, **28**, 35.

HOWRY, D. H., 1957, *Ultrasound in Biology and Medicine* (Washington, D.C.: Amer. Inst. Biol. Sci.), ed. by E. Kelly, p. 49.

KIKUCHI, Y., UCHIDA, R., TANAKA, K., and WAGAI, T., 1957, *J. acoust. Soc. Amer.*, **39**, 824.

MALLAMS, J. T., and Miller, J. E., 1956, *Radiology*, **67**, 877.

VON GIERKE, H. E., OESTREICHER, H. L., FRANKE, E. K., PARRACK, H. O., and VON WITTERN, W. W., 1952, *J. appl. Physiol.*, **4**, 886.

WILD, J. J., and REID, J. M., 1957, *Ultrasound in Biology and Medicine* (Washington, D.C.: Amer. Inst. Biol. Sci.), ed. by E. Kelly, p. 30.

Ultrasonic Absorption in Aqueous Solutions of Dextran*

S. A. HAWLEY† AND F. DUNN

Biophysical Research Laboratory, University of Illinois, Urbana, Illinois 61801

(Received 8 July 1968)

The ultrasonic absorption measured in aqueous solutions of dextran (1–6 polyanhydroglucose) as a function of molecular weight (3.4×10^3–3.7×10^6) and frequency (3–195 MHz) reveals a relaxational behavior related to the calculated viscoelastic properties of free-draining random coils.

INTRODUCTON

In recent years there have been a number of investigations involving aqueous solutions of biopolymers which employ ultrasonic spectoscopy as a method of analysis.[1] The particular interest in the acoustic properties of these substances is related both to biomedical applications of ultrasound as well as to the relevance of the propagation parameters to characterization of equilibrium kinetics of reactions involving biologically interesting marcomolecules.

It has been observed that in solutions of hemoglobin,[2] bovine serum albumin,[3] and poly-L-glutamic acid[4] the frequency dependence of the absorption coefficient α is compatible with a distribution of relaxation processes, viz.,

$$\frac{\alpha}{f^2} = \sum_{p=1}^{N} \frac{A_p}{1+(f/f_p)^2} + B, \quad (1)$$

in which $f_p = 1/2\pi\tau_p$ where τ_p is the characteristic relaxation time of the pth process and B represents nonrelaxing absorption contributions. Little information has been provided, however, which serves to reveal the molecular origins of acoustic-macromolecular interactions. Part of the difficulty resides in the complexity of the macromolecular species, in particular the uncertainties concerning dynamic viscoelastic interactions that may be present for a polyelectrolytic molecular species.

In this study, ultrasonic absorption measurements are presented for aqueous solutions of dextran, a quasi-linear uncharged polymer, as a function of molecular weight as well as frequency. Since there is available both experimental and theoretical information that delineates the shear viscoelastic behavior of linear polymer solutions, the choice of such a material for ultrasonic absorption measurements provides an opportunity to examine any interrelations that may exist.

EXPERIMENTAL

The dextran (1–6 polyanhydroglucose, 5% non 1–6 linkages) employed in this investigation is type NRRL B512 obtained in fractionated form.[5] Molecular-weight and intrinsic viscosity information provided by the manufacturer is listed in Table I. Solutions were prepared with distilled water and clarified via membrane filtering prior to use. Concentration was determined via optical rotation based on a measured $[\alpha]_{\pm}^{20}$ of $+199°$.

---

* This work represents in part, the Ph.D. thesis in biophysics of the first named author, University of Illinois, Urbana, Ill., 1967.
† Present address, Lyman Laboratory, Physics Department, Harvard University, Cambridge, Massachusetts 02138.
[1] A. detailed bibliography will be found in F. Dunn, P. D. Edmonds, and W. J. Fry, in *Bioelectronics*, H. P. Schwan, Ed. (McGraw-Hill Book Co., New York, 1968).
[2] P. Edmonds, Biochem. Biophys. Acta **63**, 216 (1962).
[3] S. A. Hawley and F. Dunn (unpublished).
[4] J. Burke, G. Hammes, and T. Lewis, J. Chem. Phys. **42**, 3250 (1965).

[5] Obtained from Pharmacia Fine Chemical Co., Piscataway, N.J.

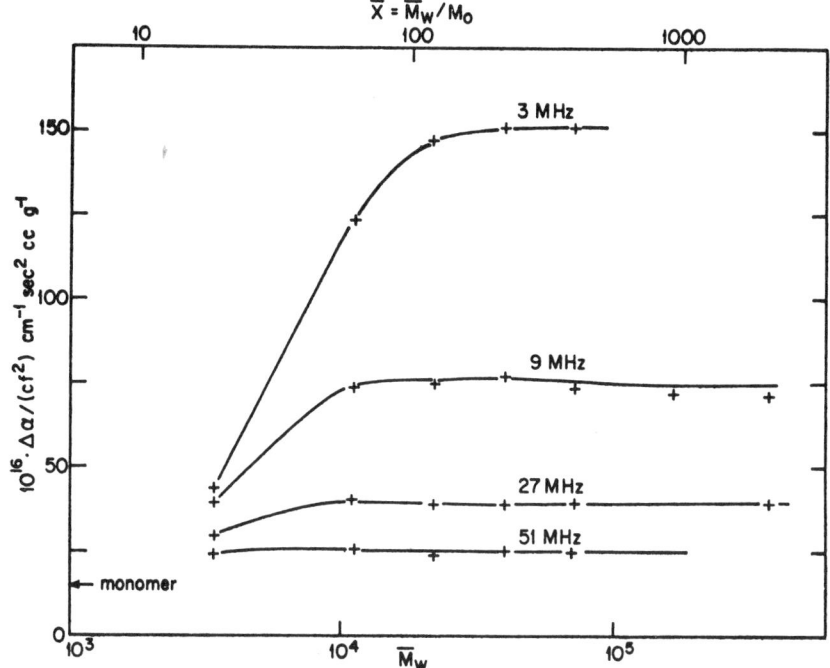

FIG. 1. $\Delta\alpha/(cf^2)$ as a function of weight average molecular weight at four frequencies, 20.0°C. The average degree of polymerization $\bar{X}$ is denoted on the superior abscissa.

A semiautomatic version of the conventional pulse technique was employed over the frequency range 3–195 MHz. The instrumental precision has been determined in the couse of several hundred determinations of α in distilled water conducted throughout the course of the investigation. Diffraction corrections are made at 3 and 9 MHz assuming that these losses are additive and noninteracting with the intrinsic absorption of the sample. The fractional standard deviation has been found to vary between 3.72% at 9 MHz to 18.0% at 195 MHz, with a minimum deviation of 2.46% observed at 21 MHz. Throughout the frequency range the mean values of α obtained in water agreed in all instances within 3% of accepted values.[1] Sample temperature was determined and maintained at 20°± 0.01°C.

The measurement of the speed of sound in the dextran solutions revealed that the presence of the solute decreases the over-all compressibility, with solutions having a slightly higher sound speed than water. For most solutions considered in this study, the net change is of the order of measurement error (0.3%) so that a systematic examination of this data will not be attempted here.

RESULTS

Ultrasonic absorption values have been determined for at least three concentrations at each molecular weight and frequency. Within the experimental error, the absorption was found to be a linear function of concentration to about 10% throughout the molecular-weight and frequency range, extrapolating to the absorption coefficient for water at zero concentration.

The results are depicted in Figs. 1 and 2 in terms of the concentration-free relaxation parameter $\Delta\alpha/(cf^2)$

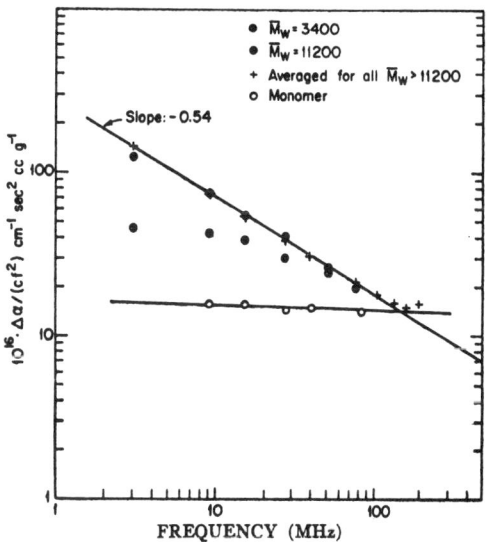

FIG. 2. $\Delta\alpha/(cf^2)$ vs frequency for the monomer glucose and dextran of molecular weights listed in Table I.

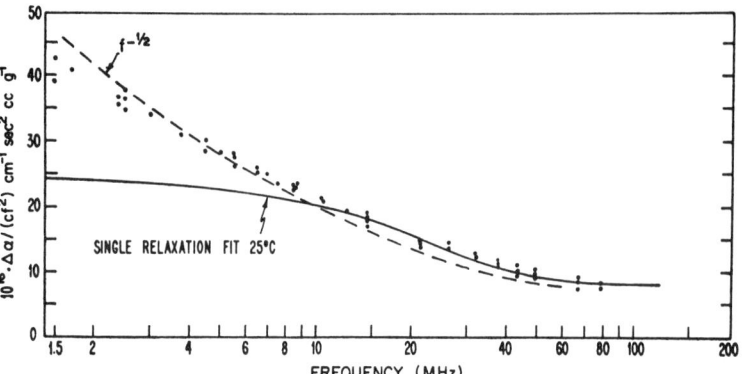

FIG. 3. $\Delta\alpha/(cf^2)$ vs frequency for polyethylene glycol, molecular weight=20 000. Solid line represents authors' (Ref. 10) single relaxation fit to their data in frequency range 10–185 MHz, 25°C; points correspond to measurements in frequency range 1.5–80 MHz, 20°C (Ref. 11).

in which $\Delta\alpha$ represents the absorption difference between that of solution and of solvent and c is the concentration. The important features to note are: (1) except at the highest frequencies of observation there is an absorption contribution associated with polymerization, (2) there is a molecular-weight dependence of $\Delta\alpha/(cf^2)$ for molecular weights less than about 25×10^3 which becomes less pronounced as the frequency is increased; (3) at molecular weights above 25×10^3 the absorption is independent of molecular weight within the frequency range considered; and (4) there is an apparent limiting dependence corresponding to $\Delta\alpha/(cf^2) \propto f^{-1/2}$.

These observations are suggestive of results obtained for shear viscoelastic properties of polymer solutions,[6] and one can proceed along this line of analysis by considering the anticipated viscous losses, viz.,

$$\alpha/f^2 = (2\pi^2/\rho V^3)(\eta_v' + \tfrac{4}{3}\eta_s'), \qquad (2)$$

where ρ is the density, V is the sound velocity, and η_v' and η_s' are the dynamic volume and shear viscosities, respectively. Static hydrodynamic and viscosity measurements in Pharmacia NRRL B512 dextran have been observed to be compatible with random-coil behavior with a small excluded-volume effect attributable to branching.[7] If this behavior is maintained under dynamic conditions, one may reasonably formulate an expression for η_s'. For this purpose the limiting slope suggests that the Rouse formulation[8] for the free-draining coil, which predicts a limiting viscosity proportional to $f^{-1/2}$, may be appropriate. One can then easily show that if the concentration dependence of the speed of sound and of the density are neglected, an assumption which introduces an uncertainty of the order of the experimental error, the excess concentration-free absorption attributable to shear viscosity, $\Delta\alpha_s/(cf^2)$, in terms of the free-draining coil becomes

$$\frac{\Delta\alpha_s}{cf^2} = \frac{4\pi RT}{3M\rho V^3} \sum_{p=1}^{N} \frac{1/f_p}{1+(f/f_p)^2}. \qquad (3)$$

One would expect, on the basis of the random coil model, that[9] for $f<f_1\approx\pi RT/(12M\eta_{s,0}[\eta]_0)$, the shear viscosity, and consequently $\Delta\alpha/cf^2$, will be independent of frequency, passing to an $f^{-1/2}$ dependence in the region above f_1. Employing the above relation for f_1 to the two lowest molecular weights ($\bar{M}_w=3.4\times10^3$ and 11.2×10^3) considered in this investigation, one expects the transitions to occur at about 35 and 6 MHz, respectively. Transitions of the correct form are in fact observed in both cases which suggest that the absorption may be intimately related to the normal modes involved in shear viscosity (see Fig. 2).

This approach is not without frailty, however, inasmuch as it does not explain the magnitude of the absorption. Consider, for example, the predicted behavior of the highest-molecular-weight solute employed ($M_w=3.7\times10^5$) which will yield the greatest theoretical shear viscosity. For a 5% solution at 9 MHz ($f/f_1\approx270$), the observed excess absorption will be

TABLE I. Molecular weights and intrinsic viscosities of dextran samples employed in this investigation, 20°C.

$\bar{M}_w\times10^{-3}$	$\bar{M}_n\times10^{-3}$	$[\eta]_0$ (cc/g)	$\bar{M}_w/\bar{M}_n$
3.4	2.0	5.4	1.7
11.2	5.7	9.8	1.97
21.8	14.5	14.7	1.50
39.8	25.6	19.8	1.55
72.0	40.5	25.6	1.77
167	105	38	1.59
370	185	50	2.00

[6] W. Philippoff, in *Physical Acoustics*, W. P. Mason, Ed. (Academic Press Inc., New York, 1965), Vol. II, Part B, pp. 1–90.

[7] K. Granath, J. Colloid. Sci. **13**, 308 (1958).
[8] P. Rouse, J. Chem. Phys. **21**, 1272 (1953).
[9] $\bar{M}_w$ is employed in all relevant computation. $\eta_{s,0}$ represents the viscosity of the solvent.

approximately 25×10^{-17} cm^{-1} sec^2 or nearly that of water. The excess attenuation anticipated from the model is less than 2×10^{-17} cm^{-1} sec^2 so that it becomes evident from Eq. (2) that a disproportionately high volume viscosity ($\eta_v'/\eta_s' > 10$) would be required for adherence to the simple approach presented here.

DISCUSSION

At this juncture it is worthwhile to compare the results obtained in dextran to measurements obtained by others in aqueous solutions of another linear polymer, viz., polyethylene glycol. Measurements[10] over the frequency range 10–185 MHz, obtained for molecular weights of 7500 and 20 000 have been interpreted in terms of a single relaxation, i.e., $N=1$ in Eq. (1). While this formulation does provide a somewhat reasonable fit to the data, the total relaxation occurring over this frequency range is somewhat less than that usually necessary to distinguish a single process. That this procedure can be somewhat perilous may be seen in Fig. 3 in which the solid line depicts the single relaxation fit to these data[10] for $M_w = 20\,000$ at 25°C, and the solid points correspond to more recent measurements[11] at 20°C from 1.5 to 80 MHz in the same material, suggesting the presence of multiple relaxations. The relaxation frequency corresponding to the primary mode at this molecular weight, calculated as above for dextran, is 0.95 MHz and it is evident from Fig. 3 that for polyethylene glycol, as well as for dextran (Fig. 2), the absorption behavior above this frequency is in reasonable agreement with the $f^{-1/2}$ dependence.

The absorption behavior in solutions of polyethylene glycol has been measured as a function of molecular weight to a lower limit of 600 by Hammes and Schimmel,[12] and a single relaxation analysis was again employed to characterize the relaxation behavior in the frequency range 14–175 MHz. Although the complete absorption spectra are not presented, they observe that the apparent relaxation frequency decreases with increasing molecular weight to a molecular weight of about 7000 and is thereafter independent of molecular size. This may be interpreted in the following way: At high molecular weight (above 7000), the primary relaxation frequency is below the frequency range of observation and single relaxation analysis will be insensitive to variations of the primary relaxation frequency. As the molecular weight is lowered, however, the relaxation frequency increases and becomes important in determining the apparent single relaxation parameters as it approaches and enters the experimental range and thus provides a useful qualitative indication of the relaxational properties. Their observations are in this sense compatible with those indicated above, together suggesting the possibility that, in solutions of linear polymers, the normal modes involved in shear viscoelastic behavior may become intimately involved in the ultrasonic relaxation.

In general, the origin of a volume viscosity in associated liquids such as water is believed to be closely related to that of shear viscosity,[13] both involving structural rearrangements of the liquid lattice. In addition, for simple hydrogen-bonded liquids the ratio η_v'/η_s' is usually of the order of 3 or less. For polyethylene glycol, and as indicated above for dextran, the magnitude of the observed absorption would require a substantially higher volume contribution than would be warranted by straightforward predictions from the random-coil model. Thus it appears likely that normal modes of the molecule merely determine the mode of coupling to some other relaxation process involving the solvent. One may speculate, for example, that acoustic perturbation of normal modes accomodates rearrangement of water molecules engaged in weak bond interactions along the polymer backbone. More substantial delineation, however, is not presently available and remains the subject of work in progress.

ACKNOWLEDGMENTS

This investigation was supported in part by the Institute of General Medical Sciences, National Institute of Health, Public Health Service, U.S. Department of Health, Education, and Welfare, and in part by the U.S. Office of Naval Research, Acoustics Programs.

---

[10] G. Hammes and T. Lewis, J. Phys. Chem. **70**, 1610 (1966).
[11] W. O'Brien, M. S. thesis in electrical engineering, University of Illinois, 1968.
[12] G. Hammes and P. Schimmel, J. Am. Chem. Soc. **89**, 442 (1967).

[13] T. Litovitz and C. Davis, in *Physical Acoustics*, W. P. Mason, Ed. (Academic Press Inc., New York, 1965), Vol. II, Part A, pp. 281–349.

Ultrasonic Studies of Proton Transfers in Solutions of Poly(lysine) and Poly(ornithine). Implications for the Kinetics of the Helix–Coil Transition of Polypeptides and for the Ultrasonic Absorption of Proteins

by R. Zana* and C. Tondre

C.N.R.S.—Centre de Recherches sur les Macromolécules, 67 Strasbourg, France (Received January 21, 1972)

Publication costs borne completely by The Journal of Physical Chemistry

Ultrasonic absorption measurements have been performed at pH ranging from 5 to 13 and in the frequency range 1–115 MHz on solutions of poly(L-lysine) and poly(DL-lysine) (PLL and PDLL) and of poly(L-ornithine) (PLO) in H_2O–0.3 M NaBr and of PLO in 85 parts H_2O–15 parts methanol–0.3 M NaBr. The curves absorption vs. pH go through a maximum at pH 11.4–11.5. From the study of the effect of pH on the excess ultrasonic absorption of these solutions it is concluded that the absorption maxima are due to proton-transfer reactions on the side-chain amino group of the polypeptides and not to the helix–coil transition because (1) PLL and PDLL give essentially identical absorption results and (2) methanol has only very little influence on the absorption of PLO solutions. The effect of frequency on the excess absorption of PLL and PLO solutions has enabled us to determine the rate constants and volume changes for the proton-transfer reactions. Finally, the implications of the results obtained in this study on the possibility of observing the helix–coil equilibrium using ultrasonic absorption in the megahertz range and on the interpretation of the ultrasonic absorption data obtained with protein solutions are examined.

I. Introduction

In a recent study, Chou and Scheraga[1] showed that if poly(L-lysine) (PLL) undergoes a helix–coil transition at pH 10.3, such a conformational change does not occur with poly(DL-lysine) (PDLL) over the whole pH range. This result prompted us to undertake new ultrasonic absorption measurements as a function of pH on both PLL and PDLL. It was hoped that the comparison of the curves absorption vs. pH relative to these two polymers would provide us with direct evidence as to whether the increase of absorption found at pH >9 by Parker, et al.,[2] for PLL is due to the helix–coil transition, as postulated by these authors, or to a proton-transfer reaction on the ϵ-NH_3^+ group of PLL, as was advanced in a recent paper.[3] If the first process prevails, one can readily predict that the increase of absorption at pH >9 will be strongly affected in going from PLL to PDLL. On the contrary, if this increase is due to a proton transfer, it should not be changed much in going from PLL to PDLL since our previous studies[4,5] have shown this process not to be greatly affected by the overall conformation of the molecule under study.

As part of this work, absorption measurements have also been performed on poly(L-ornithine) (PLO). This polypeptide is very similar to PLL, its side chain being one CH_2 shorter than that of PLL, and the pK_a values of the side-chain NH_2 groups of PLO and PLL are quite close (within 0.1–0.2 pH unit[1,6]). However, the shorter side chain of PLO results in lesser stability of its helical conformation: in water the maximum helical content of PLO found at high pH is 25%, as compared with a value close to 100% for PLL;[6] moreover, the midpoint of the helix–coil transition is found at pH 11 for PLO[6] and 10.3 for PLL.[1] A comparative study of the effect of pH on the absorption of PLL and PLO may therefore also provide some evidence as to whether proton transfer or helix–coil transition is responsible for the excess absorption observed with aqueous solutions of these two polymers.[2,7] PLO also gives a second manner to approach this problem: Chaudhuri and Yang[6] have shown that the helical content of PLO is greatly increased by relatively small additions of methanol. For instance, the helical content is increased from 25% in water to 60% in H_2O–methanol (85:15, v/v). This mixture is the solvent used by Hammes and Roberts[7] in their ultrasonic study of the helix–coil transition of PLO induced by pH changes. These authors interpreted their results in terms of a helix–coil equilibrium. As will be seen now, the com-

---

(1) P. Chou and H. A. Scheraga, *Biopolymers*, **10**, 657 (1971).
(2) R. Parker, L. Slutsky, and K. Applegate, *J. Phys. Chem.*, **72**, 3177 (1968).
(3) R. Zana and C. Tondre, *Biopolymers*, **10**, 2635 (1971).
(4) J. Sturm, J. Lang, and R. Zana, *Biopolymers*, **10**, 2639 (1971).
(5) J. Lang, C. Tondre, and R. Zana, *J. Phys. Chem.*, **75**, 374 (1971); R. Zana and J. Lang, *ibid.*, **74**, 2734 (1970).
(6) S. Chaudhuri and J. Yang, *Biochemistry*, **7**, 1379 (1968).
(7) G. Hammes and P. Roberts, *J. Amer. Chem. Soc.*, **91**, 1812 (1969).

parison of the results obtained in this work for PLL, PDLL, and PLO, the latter being studied both in H$_2$O and 85:15 H$_2$O–methanol leaves no doubt that the excess absorption found in PLL, PDLL, and PLO is entirely due to proton transfer on the side-chain amino group.

II. Experimental Section

Poly(L-lysine) and poly(L-ornithine) hydrobromides with polymerization degrees 480 and 540 were purchased from Pilot Chem. Co., while poly(DL-lysine) hydrobromide (polymerization degree 200) was obtained from Miles-Yeda, Israel. The three polymers were used without further purification.

The ultrasonic absorption measurements were carried out on solutions in H$_2$O–0.3 M NaBr (PLL, PDLL, and PLO) or in 85 parts H$_2$O–15 parts methanol–0.3 M NaBr (PLO), using the standard pulse technique[8] or the ultrasonic interferometer.[9]

While PDLL and PLO remained dissolved in the whole pH range, precipitation of PLL occured slowly at pH >10.3. To further slow down this process, the concentration of the PLL solution was brought down to 0.06 mol of monomer per liter (M in the following) as compared to 0.116 and 0.156 M in the study of Parker, et al.[2] It was then observed that the pH of the 0.06 M solution could be increased above 10.3 by addition of NaOH, with the solution remaining relatively clear during the time required for the measurement of its absorption coefficient α (15 min). Moreover, no effect of time on α could be measured during the hour which followed the increase of pH, although a considerable increase of turbidity occurred meanwhile. For further measurements the turbid solution was clarified by bringing its pH down to 9 by addition of concentrated HBr with stirring. The pH was then raised to the desired value and the measurement performed. Such a procedure was repeated for each of the experimental results at pH >10.3. These successive pH cycles increased the NaBr concentration, C_{NaBr}, of the solution. We found, however, that an increase of C_{NaBr} from 0.3 M (initial concentration) to 0.5 M (final concentration after six pH cycles) resulted in a negligible change of absorption. Similarly Parker, et al.,[2] found practically no effect of C_{NaCl} on the ultrasonic absorption of PLL solutions.

Optical rotation measurements were performed on 0.06 M PLL and PDLL solutions in H$_2$O–0.3 M NaBr using a Zeiss polarimeter at 589 nm. No change of optical rotation with pH was detected with PDLL, while the rapid change of optical rotation due to the helix–coil transition of PLL was observed at pH 10.5. The values 10.3 and 10.4 have been reported by other workers for PLL in H$_2$O–0.1 M KCl[1] and H$_2$O–0.2 M KCl,[6] respectively. For PLO, the measurements were performed both in H$_2$O–0.3 M NaBr and in 85 parts H$_2$O–15 parts methanol–0.3 M NaBr at 365 nm. In both solvents the helix–coil transition was found to occur at pH close to 11.0, in good agreement with the findings of others.[7] Moreover, the change of optical rotation was two times smaller in water than in the H$_2$O–methanol mixture, thereby indicating a higher helical content in this mixed solvent, as previously reported.[6]

III. Effect of pH on the Excess Ultrasonic Absorption of PLL, PDLL, and PLO Solutions

Figure 1 depicts the curves α/N^2 vs. pH, N being the frequency of the ultrasonic radiation, for PLL and PDLL. The α/N^2 value for H$_2$O–0.3 M NaBr was found to be very close to that of water (horizontal dotted line on Figure 1) over the whole pH and frequency range. In their work on PLL, Parker, et al.,[2] did not observe the absorption maximum which appears on curve at pH 11.5 because their solutions were rather concentrated and PLL readily precipitated at pH >10.2. However, at pH 7 their results are in good agreement with ours when concentrations are taken into account. The results of Figure 1 demand the following remarks: (1) the absorption maximum for PLL occurs at pH 11.5 and not at pH 10.5, i.e., at the midpoint of the transition as would be expected if it is assumed that this maximum is due to the transition;[10] (2) an absorption maximum is also observed at pH 11.5 with PDLL, which does not present any helix–coil transition when the pH is changed;[1] and (3) curves 1 and 2 on Figure 1 can be made coincident by a translation of 110 × 10^{-17} cm^{-1} sec^2 (the dotted curve, very

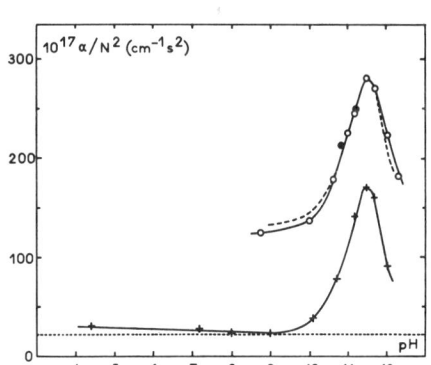

Figure 1. Variation of α/N^2 with pH at 6.49 MHz and 25° for 0.06 M solutions of PLL (+) and PDLL (○, direct titration; ●, reverse titration) in H$_2$O–0.3 M NaBr. The curve shown by the broken line practically coincident with the curve for PDLL was obtained by translating the curve relative to PLL. (· · · · · · ·), α/N^2 for water.

(8) J. Andreae, R. Bass, E. Heasell, and J. Lamb, *Acustica*, **8**, 131 (1958).
(9) R. Musa, *J. Acoust. Soc. Amer.*, **30**, 215 (1958).
(10) G. Schwarz, *J. Mol. Biol.*, **11**, 64 (1965).

close to that for PDLL, was obtained by such a translation of the curve relative to PLL; the difference of $5\text{--}10 \times 10^{-17}$ sec^2 which then appears in the pH range where α/N^2 is almost constant is within the experimental accuracy). These results strongly suggest that the absorption maxima on Figure 1 are not due to the helix–coil transition but to proton transfer on the amino groups of PLL and PDLL. As will be seen now, the results obtained with PLO lead to the same conclusion. However, before considering these results some comments must be made on the shift of absorption which appears in Figure 1 between equimolecular solutions of PLL and PDLL. This shift is not modified when PLL undergoes the helix–coil transition at pH 10.5 and is practically constant over the whole pH range. Therefore, it cannot be due to the difference of conformation between helical PLL and randomly coiled PDLL.   We have tentatively assigned this shift to a small difference of composition between the PLL and PDLL samples used in this work. The two samples were of different origins. It is thought that an unknown but small percentage of residues of the PDLL sample may not have been decarbobenzoxylated in the process of preparation of this polymer [poly(lysine) is always obtained by decarbobenzoxylation of poly(ϵ-carbobenzoxylysine)]. Evidence for this may be found in the fact that the PDLL sample was not totally soluble in water at neutral pH; the insoluble part was removed by filtration. However, this fact affects in no way the conclusions which have been drawn from the results of Figure 1 because in this work PDLL was used as a nonhelical model compound for PLL.

Figure 2 shows the results obtained with PLO solutions. The α/N^2 value for the solvent 85 parts H$_2$O–15 parts methanol–0.3 M NaBr was found to be $2\text{--}5 \times 10^{-17}$ cm^{-1} sec^2 over that for water in the whole pH and frequency range. As for PLL and PDLL, absorption maxima can be seen in Figure 1 at pH close to 11.5. The amplitude of the absorption maximum is slightly larger in water than in the mixed solvent in which the helical content of PLO is more than twice that in water. In the assumption of the helix–coil equilibrium, one would have expected an opposite result.[10] The effect of methanol on the excess absorption of PLO is more thoroughly studied in the next paragraph. In addition to the effect of methanol, it is found that when plotted on the same graph the two curves α/N^2 vs. pH relative to PLL and PLO in H$_2$O–0.3 M NaBr are practically coincident at $N = 6.49$ MHz, despite the quite different stabilities of the helical forms of these two polymers. This also suggests that as for PLL and PDLL the absorption peaks in Figure 2 are not due to the helix–coil transition but to the proton-transfer reaction

$$\text{NH}_3^+ + \text{OH}^- \underset{k_\text{B}}{\overset{k_\text{F}}{\rightleftarrows}} \text{NH}_2 + \text{H}_2\text{O} \qquad (1)$$

This assignment will now be substantiated in a quantitative manner.

IV. Study of the Kinetics of the Proton-Transfer Reaction in PLL and PLO Solutions

The expressions for the pH corresponding to the absorption maximum (pH$_\text{M}$), for the amplitude A of this maximum, and for the relaxation frequency characterizing the excess absorption associated with reaction 1 have been given in several papers[11–15] and will be used in what follows.

(1) Position of the Absorption Maximum. For a solution of simple base, pH$_\text{M}$ is given by eq 2, where pK_a has its usual meaning and C is the concentration in

$$\text{pH}_\text{M} = (14 + pK_\text{a} + \log C)/2 \qquad (2)$$

moles per liter. For a solution of polybase a complication arises because pK_a depends on the state of ionization of the polybase.[16] However, both PLL and PLO are weak polybases with a degree of ionization of about 5% at pH 11.5. Therefore, in eq 2 pK_a can be replaced by the intrinsic pK_a, pK_0.

Chou and Scheraga[1] have reported the value $pK_0 = 10.26$ for both PLL and PDLL in H$_2$O–0.1 M KCl. Our work was performed at a higher ionic strength (0.3 M NaBr), but the change of pK_0 with ionic strength is expected to introduce in the calculation of pH$_\text{M}$ an error well within the experimental accuracy on pH$_\text{M}$ (± 0.15 pH unit). The above value of pK_0 together with $C = 0.06$ M yields pH$_\text{M} = 11.5 \pm 0.1$, in excellent agreement with the experimental result.

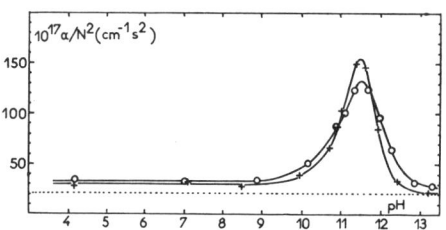

Figure 2. Variation of α/N^2 with pH at 6.49 MHz and 25° for 0.056 M solutions of PLO in H$_2$O–0.3 M NaBr (+) and in 85 parts H$_2$O–15 parts methanol–0.3 M NaBr (O). (. . . . . . .), α/N^2 for water.

(11) K. Applegate, L. Slutsky, and C. Parker, *J. Amer. Chem. Soc.*, **90**, 6909 (1968).

(12) R. White, L. Slutsky, and S. Pattison, *J. Phys. Chem.*, **75**, 161 (1971).

(13) M. Hussey and P. Edmonds, *J. Acoust. Soc. Amer.*, **49**, 1309, 1907 (1971).

(14) H. Inoue, *J. Sci. Hiroshima Univ.*, Ser. A2, **34**, 17 (1970).

(15) M. Emara, G. Atkinson, and E. Baumgartner, *J. Phys. Chem.*, **76**, 334 (1972).

(16) A. Katchalsky, N. Shavit, and H. Eisenberg, *J. Polym. Sci.*, **13**, 69 (1954).

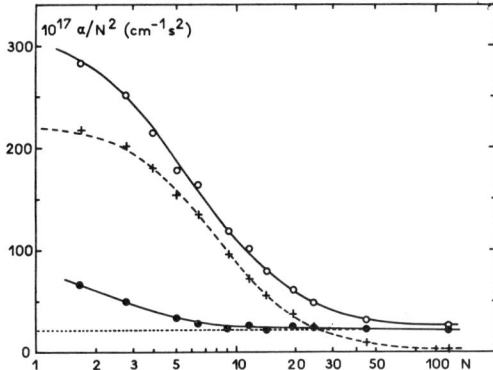

Figure 3. Ultrasonic absorption spectra of PLL in H₂O–0.3 M NaBr at 25° and at pH 7.0 (●) and 11.5 (○). The crosses show the values of $\delta\alpha/N^2$ (see text) and the broken line curve is the single relaxation time curve giving the best fit with the values of $\delta\alpha/N^2$. Concentration, 0.06 M. (·······), α/N^2 for the solvent.

For PLO in H₂O–0.2 M KCl the "minimum value" $pK_0 = 10.1$ has been reported by Chaudhuri and Yang,[6] thus yielding pH_M 11.45 ± 0.05, in excellent agreement with the experimental result on Figure 2. On the other hand, the results compiled by Robinson and Stokes[17] indicate that the pK_a's of weak bases are increased by less than 0.2 pH unit in the presence of 15% methanol. If it is assumed that methanol has a similar effect on PLO, the maximum increase of pH_M would then be less than 0.1 pH unit, i.e., practically within the experimental accuracy.

(2) *Determination of the Rate Constants k_F and k_B and of the Volume Change Associated with Reaction 1.* For this purpose the relaxation spectra of PLL and PLO solutions at pH 7.0 and 11.5 have been determined. The spectra relative to PLL are shown in Figure 3, in which are also given the values of $\delta\alpha/N^2 = (\alpha/N^2)_{pH\ 11.5} - (\alpha/N^2)_{pH\ 7.0}$. These values of $\delta\alpha/N^2$ have been obtained by drawing smooth curves through the experimental points at pH 11.5 and 7.0 and taking the difference between the values of α/N^2 at pH 11.5 and 7.0, read on these curves, at each of the frequencies where the absorption had been measured. This procedure averages the errors. The difference $\delta\alpha/N^2$ can be safely assumed to represent the contribution of reaction 1 since, as can be seen in Figure 2, the values of α/N^2 at pH 7.0 and 13.0 are practically identical. In Figure 3 the broken line curve represents the single relaxation time curve fitting the best with our results. This curve obeys the equation

$$\frac{\delta\alpha}{N^2} = \frac{A}{1 + (N^2/N_R^2)} + B \quad (3)$$

The values of the relaxation parameters are given in Table I.

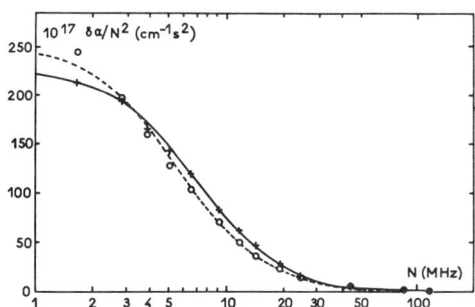

Figure 4. Variation of $\delta\alpha/N^2$ vs. N for 0.056 M solutions of PLO in H₂O–0.3 M NaBr (+) and in 85 parts H₂O–15 parts methanol–0.3 M NaBr (○) at 25°. The curves shown by the solid line and the broken line are single relaxation time curves giving the best fit with the results.

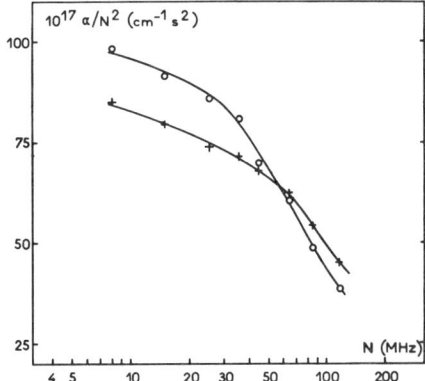

Figure 5. Ultrasonic absorption spectra of 0.1 M solutions of diethylamine in H₂O (+) and in 85% H₂O–15% methanol (○).

The ultrasonic absorption spectra of PLO at pH 7.0 and 11.5, in H₂O–0.3 M NaBr and in 85 parts H₂O–15 parts methanol–0.3 M NaBr have been determined. The results are represented in Figure 4 in terms of $\delta\alpha/N^2$. The lines are single relaxation time curves fitting the results. The relaxation parameters characterizing these curves are given in Table I. The results in Figure 4 show that the introduction of methanol brings about a decrease of relaxation frequency from 7 to 5.5 MHz. A similar decrease of N_R upon addition of methanol can also be seen in Figure 5 for a 0.1 M solution of diethylamine. Thus, this slowing down of proton-transfer reactions upon introduction of methanol appears to be quite general. Our measurements also indicate that at neutral pH the α/N^2 value is smaller for PLO than for PLL, particularly at $N < 5$ MHz. At 1.6 MHz the α/N^2 of PLL and PLO were,

(17) R. Robinson and R. Stokes, "Electrolyte Solutions," 2nd ed, Butterworths, London, 1970, p 541.

Table I

Polypeptide (solvent)	N_R, MHz	$10^{17}A$, cm^{-1} sec^2	$10^{17}B$, cm^{-1} sec^2	$10^{-9}k_F$, M^{-1} sec^{-1}	$10^{-6}k_B$, sec^{-1}	ΔV_0, cm^3/mol
PLL (H$_2$O–0.3 M NaBr)	8	220	2	7.35	1.3	24
PLO (H$_2$O–0.3 M NaBr)	7	225	0	8.1	1.0	25.7
PLO (85 parts H$_2$O–15 parts methanol–0.3 M NaBr)	5.5	250	0	6.35	0.8	24

respectively, 60 and $40 \cdot 10^{-17}$ cm^{-1} sec^2. This difference is to be compared to the increased excess absorption with increasing alkyl chain length observed in aqueous solutions of tetraalkylammonium salts.[18]

In the calculations of k_F, k_B, and ΔV_0 from the relaxation parameters A and N_R, the following equations were used[2]

$$pK_a = 14 + pK = 14 + \log(k_B/k_F) \quad (4)$$

$$\tau^{-1} = 2\pi N_R = k_F \left(\frac{c}{1 + Kc_{OH^-}} + c_{OH^-} + \frac{1}{K}\right) \quad (5)$$

$$A = \frac{2\pi^2 \rho v}{RT} \frac{(Kc)c_{OH^-}}{Kc + (Kc_{OH^-} + 1)^2} \Delta V_0^2 \tau \quad (6)$$

$$pH = 14 + \log c_{OH^-} \quad (7)$$

where τ is the relaxation time and ρ and v are, respectively, the density and velocity of ultrasound for the solvent. The activity coefficients were assumed to be unity. The results of these calculations are given in Table I. The values of the rate constants in Table I are the first ones reported for proton-transfer reactions on polymers. For PLL the value $\Delta V_0 = 24$ cm^3/mol obtained in this work compares well with that determined by Noguchi,[19] 22.9 cm^3/mol at ionic strength 0.2, using a dilatometric technique. On the other hand, some insight can be gained by comparing the value of k_F for PLL to that of the ϵ-amino group of its monomer, L-lysine. For this last compound two independent determinations of k_F have been reported.[13,14] The average value, $k_F = 1.9 \times 10^{10}$ M^{-1} sec^{-1} is larger by a factor of 2.5 than for PLL. This is a relatively small difference; one would have expected electrostatic effects to be of major importance in the case of PLL, as this polymer is also a polyelectrolyte. However, as stated above, only 1 out of every 20 residues is ionized at pH 11.5. This, together with the fact that the ϵ-amino side chain is more than 6 Å long, yields an average distance between charged residues over 30 Å. Therefore, in the vicinity of the polymer chain, the interaction of one OH$^-$ ion with the chain can be approximated by the interaction of this ion with the closest –NH$_3^+$ group on the chain; the smaller the OH$^-$–NH$_3^+$ distance, the better this approximation. Another factor which must be taken into account in the comparison of the k_F values for lysine and PLL is the diffusion coefficient. The Debye–Eigen[20] equation giving k_F contains the sum of the diffusion coefficients of the reacting species. In going from lysine to PLL, the diffusion coefficient per –NH$_3^+$ group will decrease, thereby yielding a lower rate constant. Also, steric effects are likely to be more important with PLL than with its monomer because of the hindrance due to the polymer chain, thus resulting in a smaller rate constant. The three effects invoked above are difficult to take into account quantitatively. However, the results on lysine and PLL are to be compared with the findings of Atkinson, et al.,[15] for the effect of molecular weight on k_F for the four first oligomers of polyethylenimine. These authors observed a decrease of k_F with increasing molecular weight from 2.8×10^{10} M^{-1} sec^{-1} for ethylenediamine to 2.3×10^{10} M^{-1} sec^{-1} for tetraethylenepentamine.

At the end of this paragraph, it may be stated that proton-transfer reactions account quantitatively for the excess absorption observed in PLL, PDLL, and PLO solutions in the frequency range 1–155 MHz. This conclusion has important implications as regards the kinetics of the helix–coil transition in solutions of polypeptides and to the excess ultrasonic absorption of protein solutions.

V. Considerations on the Kinetics of the Helix–Coil Transition in Solutions of PLL and PLO

Unlike PLL and PLO, a small though measurable ultrasonic absorption associated with the helix–coil equilibrium has been found in aqueous solutions of poly(L-glutamic acid) (PLGA) in the megahertz range.[21,22] However, PLGA solutions in 2:1 (v/v) water–dioxane–0.2 M NaCl did not show any such excess absorption.[23,24] The first and, in our mind, most important reason for these differences is to be found in the value of nucleation parameter σ, characterizing the

(18) M. Blandamer, M. Foster, N. Hidden, M. Symons, *Trans. Faraday Soc.*, **64**, 3247 (1968).
(19) H. Noguchi, *Biopolymers*, **4**, 1105 (1966).
(20) M. Eigen and L. De Maeyer, *Tech. Org. Chem.*, **8**, 1032 (1963).
(21) T. Saksena, B. Michels, and R. Zana, *J. Chim. Phys. Physicochim. Biol.*, **65**, 597 (1968).
(22) R. Zana, *J. Amer. Chem. Soc.*, **94**, 3646 (1972).
(23) R. Zana, S. Candau, and R. Cerf, *J. Chim. Phys. Physicochim. Biol.*, **60**, 869 (1963).
(24) J. Burke, G. Hammes, and T. Lewis, *J. Chem. Phys.*, **42**, 3520 (1965).

helix–coil equilibrium,[25] which changes according to the nature of the polypeptide and of the solvent.[26,27] In fact, Schwarz[10] has shown that for the helix–coil transition

$$N_R = 2\sigma k_H/\pi \qquad (8)$$

where k_H is the rate constant for the elementary step of growth of a helical region, i.e., for the formation of a hydrogen bond. N_R has been found to be 0.15 MHz for aqueous solutions of PDGA[28] and PLGA.[29] If we assume that in water k_H is independent of the nature of the polypeptide, N_R will still depend on this parameter through σ. The values $\sigma = 3\text{–}5 \times 10^{-3}$, 10^{-3}, and 7×10^{-4} have been determined for aqueous solutions of PLGA,[30] PLO,[31] and PLL.[32] One can therefore expect smaller values of N_R for PLL and PLO than for PLGA, and in the megahertz range the condition $N^2/N_R^2 \gg 1$ to be fullfilled for these three polymers. Equation 3 then indicates that $\delta\alpha/N^2$ is proportional to AN_R^2, that is to $\Delta V_0^2 N_R$ since A contains the ratio $\Delta V_0/N_R$,[20] where ΔV_0 is the volume change associated with the helix–coil transition. Dilatometric measurements have shown[19,33] that in water ΔV_0 depends only little on the nature of the polypeptide. Therefore, to a first approximation $\delta\alpha/N^2$ is proportional to N_R, i.e., to σ. For PLGA, $\delta\alpha/N^2$ has been found[34] to be 17 and 10×10^{-17} cm^{-1} sec^2 at 2.8 and 5 MHz, respectively. From the above one should expect for PLL and PLO excess absorptions three to seven times smaller than for PLGA, i.e., $2\text{–}5 \times 10^{-17}$ cm^{-1} sec^2 at 2.8 MHz and $1\text{–}3 \times 10^{-17}$ cm^{-1} sec^2 at 5 MHz. Such excess absorptions would be barely detectable had the helix–coil transition been the only process giving rise to the absorption. That proton-transfer reactions contribute to the excess absorption in the same pH range as the helix–coil transition renders practically impossible the observation of this process. This is well illustrated by the following example: Figure 1 shows that at pH 10.5, midpoint of the helix–coil transition for PLL, the absorption due to reaction 1 is $40 \pm 5 \times 10^{-17}$ cm^{-1} sec^2; at the same pH, frequency, and concentration the helix–coil transition would contribute by about $1\text{–}2 \times 10^{-17}$ cm^{-1} sec^2.

Differences in σ values also permit us to explain the different behaviors of PLGA in H$_2$O[21,22] and in 2:1 H$_2$O–dioxane.[23,24] In these solvents σ has the values $3\text{–}5 \times 10^{-3}$ [30] and 1.2×10^{-3},[27] respectively. With the same assumptions as above, this would result in an excess absorption too small to be detected in the solvent mixture.

In addition to the larger value of σ, two other facts make the observation of the helix–coil equilibrium easier with aqueous PLGA than with PLL and PLO. First, the pH at the midpoint of the transition (5.1) is over two pH units higher than that at which the contribution of the proton transfer is maximum (2.8 for a 0.1 M PLGA solution), while the pH's characterizing both processes are 1 and 0.5 pH units apart for PLL and PLO, respectively. Second, the contribution of the proton-transfer reaction on a carboxylic group (reaction 9) such as that of PLGA is much smaller than that for reaction 1 in the megahertz range[5,13] and becomes quite small at the pH corresponding to the mid transition.

$$-\text{CO}_2\text{H} \rightleftharpoons -\text{CO}_2^- + \text{H}^+ \qquad (9)$$

In conclusion, it appears that for aqueous solutions of polypeptides the helix–coil equilibrium gives rise to a measurable excess absorption in the megahertz range only when the cooperativity of the transition is low (high σ values) and when the pH's characterizing the transition and the proton transfer reactions are well separated on the pH scale (at least two pH units).

VI. Considerations on the Ultrasonic Absorption of Protein Solutions in the Alkaline Range

Studies of the ultrasonic absorption of protein solutions in the alkaline range have shown that the excess absorption goes through a well-defined maximum at pH 11.1–11.3.[5,35,36] We have shown[5] that this maximum is due to proton-transfer reactions which may involve all of the residues bearing on their side chains functional groups which titrate in the alkaline range, i.e., lysine, arginine, histidine, and tyrosine. Thus, at first sight the situation appears quite complicated, since each type of residue is characterized by a different pK_0. However, upon examination of the pK_0 values,[37] histidine and tyrosine can be readily eliminated, as their pK_0's are such that they would result in calculated values of pH$_M$ too different from the experimental ones. Moreover, the tables of the amino acid content of the proteins[38] whose ultrasonic absorption has been investigated reveal that lysine residues constitute 65–75% of the residues of interest. This led us as a first approximation to try to interpret the results obtained for protein solutions in the alkaline range in terms of reaction 1 involving only lysine residues. From the tables of the amino acid content and molecular weight

(25) B. Zimm and J. Bragg, J. Chem. Phys., **31**, 256 (1959).
(26) O. Ptitsyn and A. Skvortsov, Biofizika, **10**, 909 (1965).
(27) G. Hagnauer and W. Miller, Biopolymers, **9**, 589 (1970).
(28) H. Inoue, J. Sci. Hiroshima Univ., Ser. A2, **34**, 37 (1970).
(29) A. Barksdale and J. Stuehr, J. Amer. Chem. Soc., **94**, 3334 (1972).
(30) J. Rifkind and J. Applequist, J. Amer. Chem. Soc., **86**, 4207 (1964); R. Snipp, W. Miller, and R. Nylund, ibid., **87**, 3547 (1965).
(31) M. Gourke and J. Gibbs, Biopolymers, **10**, 795 (1971).
(32) G. Hagnauer, Ph.D. Thesis, University of Iowa, 1970.
(33) H. Noguchi and J. T. Yang, Biopolymers, **1**, 359 (1963).
(34) B. Michels and R. Cerf, C. R. Acad. Sci. (Paris), Ser. D, **274**, 1096 (1972).
(35) F. Dunn and L. Kessler, J. Phys. Chem., **74**, 2736 (1970).
(36) I. Elpiner, F. Braginskaya, and O. Zorina, 7th International Congress on Acoustics, Communication 25 M7, Vol. II, Akademiai Kiado Ed., Budapest, 1971, p 153.
(37) C. Tanford, Advan. Protein Chem., **17**, 69 (1962).
(38) C. Tristram and R. Smith, ibid., **18**, 227 (1963).

of proteins,[38] one can calculate the concentration C of lysine residues which appears in eq 2 for a given weight–volume concentration of the protein solution. Then, using eq 2 together with the experimental values of C and pH_M, one can obtain the value of pK_0 which is to be used in the calculations of k_F and ΔV_0. The value 10.6 was thus obtained for bovine serum albumin (BSA). This value is close to the pK_0 for PLL, but it is higher than that obtained from potentiometric titration of BSA,[37] thus revealing the contribution of the arginine residues which have been completely neglected in this calculation. The ultrasonic absorption spectra of a 0.0116 g/cm³ solution of BSA have been determined at pH 7.0 and 11.3 (absorption maximum).[39] Here again, the curve $\delta\alpha/N^2 = (\alpha/N^2)_{11.3} - (\alpha/N^2)_{7.0}$ vs. N was found to be fitted by a single time relaxation curve with $N_R = 4.7$ MHz, $A = 225 \times 10^{-17}$ cm^{-1} sec^2, and $B = 0$. From these data and the above value of pK_0, we obtained $k_F = 0.7 \times 10^{10}$ M^{-1} sec^{-1} and $\Delta V_0 = 26$ cm³ mol^{-1}. These values are very close to those obtained for PLL (see Section III), thereby justifying *a posteriori* the above simplifications. Contrary to the opinion advanced by other workers,[11,14] the example of BSA shows that despite the complexity of proteins the ultrasonic absorption data in the alkaline range can be interpreted simply but also quantitatively by considering only lysine residues.

Acknowledgment. We gratefully acknowledge the technical assistance of M. Pister and M. Krauskopf in building the pulse equipment.

(39) R. Zana, J. Lang, C. Tondre, and J. Sturm, Proceedings of the Workshop on the Interactions of Ultrasound with Biological Tissues, Seattle, Nov 1971, in press.

16

Copyright © 1969 by the American Chemical Society

Reprinted from *J. Phys. Chem.*, 73(12), 4256–4263 (1969)

Ultrasonic Investigation of the Conformal Changes of Bovine Serum Albumin in Aqueous Solution[1a]

by L. W. Kessler[1b] and F. Dunn

Bioacoustics Research Laboratory, University of Illinois, Urbana, Illinois 61801 (Received May 13, 1969)

The excess ultrasonic absorption and the speed of sound were measured in aqueous solutions of bovine serum albumin (a globular protein which undergoes marked configurational change with pH) at 20° over the frequency range 0.3 to 163 MHz and over the pH range 2.3 to 11.8. A sharp increase in the excess absorption is found outside the range 4.3 < pH < 10.5. The effect is reversible throughout this range and is more pronounced at the lower frequencies. The increase in the absorption below pH 4.3 appears to be correlated with the intermediate N–F' transition discussed by Foster and the change above pH 10.5 is thought to correspond with expansion of the molecule. At neutral pH, the ultrasonic absorption spectrum and the velocity dispersion are indicative of a broad distribution of relaxation processes. The magnitude of the ultrasonic absorption over the range 4.3 < pH < 10.5 is attributed to the perturbation of the solute–solvent equilibrium by the sound wave. Based on data at 20°, between 2.4 and 50 MHz, the frequency spectrum of the absorption increase at pH 3.5 over that at pH 7.0 may be described by a single relaxation process whose characteristic frequency is 2.2 MHz. Based on measurements of the velocity of sound at pH 7.0, the bulk modulus of BSA has been found to be 3.86×10^9 n/m².

Introduction

The spatial configuration assumed by a macromolecule within the environment of solvent molecules plays an important role in determining the hydrodynamic properties of the solution,[2a] as well as the chemical activity of the solute molecule. It appears that in cellular processes, for example, the transport of molecules across the protoplasmic membrane is related, in some complex manner, to the spatial geometry of the transported molecule as well as to the spatial arrangement of the lipoprotein complex constituting the cell membrane.[2b] Thus, it is within the realm of profitable investigation to explore methods which have the potential of providing information about molecular configuration and changes in molecule structure. Ultrasonic techniques have already been employed successfully to observe the dynamic equilibrium between multiple isomeric forms of molecules and to arrive at more complete kinetic descriptions of chemical and structural reactions whose relaxation times are comparable to the period of the ultrasonic wave.[3,4] The adiabatic propagation of a longitudinal acoustic wave through a fluid medium results in time-varying, localized changes in pressure, density, and temperature. Thus, the wave motion may perturb molecular equilibria at rates which depend upon the sound frequency.[5] For a nonideal fluid, this leads to a time lag between an applied pressure and the ensuing change in density. Consequently, molecular energy level populations are perturbed at the expense of acoustic wave energy, and the process is referred to as absorption. Ultrasonic absorption spectroscopy is useful for studying fast reactions having rate constants over the range from 10^{-9} sec to 10^{-4} sec.[6] The reader unacquainted with ultrasonic technology may wish to consult ref 4 and 6 for details of theory and experimental methods for determining absorption and velocity, and for current reviews of measurements on liquids, including biological media.

Only a few biologically important macromolecules have been subjected to thorough ultrasonic examination. For both hemoglobin,[7] a globular protein, and dextran,[8,9] a random coil molecule, the interaction between solvent and solute is probably the principal mechanism of acoustic absorption. The pressure variation associated with the sound wave perturbs the equilibrium distribution of solvent molecules that are weakly bonded to the solute, and, since rearrangement

---

(1) (a) Portions of this work were extracted from the Thesis submitted by the first-named author in partial fulfillment of the requirements for the Ph.D. degree in Electrical Engineering, University of Illinois, 1968. (b) Research Department, Zenith Radio Corporation, 6001 W. Dickens, Chicago, Ill. 60639.

(2) (a) C. Tanford, "Physical Chemistry of Macromolecules," John Wiley & Sons, Inc., New York, N. Y. (1961); (b) A. C. Giese "Cell Physiology," W. B. Saunders Co., Philadelphia, Pa., 1962.

(3) J. Lamb "Physical Acoustics," Vol. II, Part A, W. P. Mason, Ed., Academic Press, New York, N. Y., 1965, Chapter 4.

(4) M. Eigen and L. deMayer, "Technique of Organic Chemistry," Vol. 8, Part 2, S. L. Friess, E. S. Lewis, and A. Weissberger, Ed., Interscience Publishers, New York, N. Y., 1963.

(5) K. F. Herzfeld and T. A. Litovitz, "Absorption and Dispersion of Ultrasonic Waves," Academic Press, New York, N. Y., 1959.

(6) See for example, F. Dunn, P. D. Edmonds, and W. J. Fry, *Absorption and Dispersion of Ultrasound in Biological Media*, in "Bioelectronics," H. P. Schwan, Ed., McGraw-Hill Book Co., Inc., New York, N. Y., 1969.

(7) E. L. Carstensen and H. P. Schwan, *J. Acoust. Soc. Amer.*, 31, 305 (1959).

(8) L. W. Kessler, M.S. Thesis, Univ. of Illinois, Urbana, Ill., 1966.

(9) S. A. Hawley and F. Dunn, *J. Chem. Phys.*, 50, 3523 (1969).

of the solvent molecules does not occur instantaneously, absorption results. Of those biomacromolecules studied, hemoglobin can be made to undergo configurational changes,[2a] although ultrasonic spectroscopy has not been employed to study them. Polyglutamic acid, a synthetic polypeptide that can undergo a configurational change from helix coil to random coil in aqueous solution, has been examined ultrasonically by several investigators.[10,11,12] Schwarz[10] has shown that the ultrasonic absorption shows a sharp maximum at the midpoint of the helix coil transition and that theoretically this effect should be most pronounced at the relaxation frequency. According to his estimated value of 10^{-7} sec as the mean relaxation time at the midpoint of the transition, the mean relaxation frequency is 1.6 MHz. Lewis,[11] on the other hand, was not able to observe the absorption maximum corresponding to the helix coil transition and attributed the observed absorption to solvation phenomena. Saksena, et al.[12] have observed the absorption maximum but have calculated that the relaxation time is smaller than that predicted by Schwarz by a factor of approximately 10. Zana, et al.,[13] have investigated the absorption in nonaqueous solutions of several synthetic polypeptides that were made to undergo helix coil transitions, and, although the results obtained do not indicate that the helix coil transition is the principal mechanism of absorption, definite changes in the absorption are observed with changes in the molecular configuration.

Bovine serum albumin, a globular protein which undergoes a complex conformal change with pH in aqueous solution, was chosen for study because many of its physical and chemical properties have been rather well characterized by numerous investigations.[14] Briefly, the bovine serum albumin molecule has a compact structure within the pH range 4.3 < pH < 10.5 and an expanded structure outside that region. Although the nature of the conformal change has not yet been determined exactly, it is reasonable that the unfolding of the BSA molecule which results in the expanded structure may somehow involve a helix–coil transition. This is suggested by optical rotation and dispersion experiments in which the apparent helix content of bovine serum albumin changes with pH.[14]

Experimental Section

Bovine serum albumin (BSA) Fraction V powder, Lot 82268, was obtained from General Biochemicals and maintained at $-7°$ until used. Fraction V grade material was used for this investigation since it is more readily available than the purest grade in the large quantities necessary for this investigation. The solution was prepared by placing the BSA on the surface of a quantity of singly distilled water sufficient to prepare a solution of concentration 0.04 to 0.1 g/cc. The flask was then placed in the refrigerator until mixing of the two components was complete, usually accomplished overnight. The solution was filtered twice through type A glass fiber filters (Gelman Instrument Co.) in order to remove foreign particles larger than 0.3μ diameter, and after filtration the solution was maintained at $8°$ until used. Generally, the acoustic experiments were started within a few hours after the solutions were prepared. The concentration of each protein solution was determined with a Beckman Model DU spectrophotometer using the extinction coefficient determined by Cohn, et al.,[15] viz., $[E]_{1\text{ cm}}^{1\%} = 6.6$ at 280 mμ.

The pH of each solution was changed in steps ranging from 0.2 to 0.5 pH unit by the addition of known quantities of standard volumetric solutions of HCl and KOH. During this procedure, the solutions were stirred gently with a magnetic stirring bar to minimize pH gradients. Measurements of pH were made to within ± 0.1 pH unit, in the temperature range 19 to 23°, with a Beckman Model H-2 glass electrode pH meter which was standardized with accurate buffer solutions at pH 4, 7, and 9. Ultrasonic absorption and velocity measurements were not begun on the BSA solutions until at least 15 min after a pH change was made, which allowed sufficient time for the temperature to reach the desired value and also allowed the BSA molecules to reach configurational equilibrium.[16] The titrations were carried out in two stages using separate solutions, i.e., the pH was varied from neutral to about 2.3 for one set of solutions and from neutral to about 11.8 for the second set of solutions. This was considered the maximum pH range allowable to avoid possible damage to the sample chamber. For a particular set of experiments, the neutral solution was examined first, and then, after the investigation at either pH 2.3 or pH 11.8 the solution was titrated back to neutral for comparison with the first measurement.

The amplitude of a plane, progressive sinusoidal wave decays exponentially as it propagates through a lossy, homogeneous, infinitely extended medium according to

$$P(x,t) = P_0 \exp(-ax) \exp i(\omega t - kx) \quad (1)$$

where P is the instantaneous value of the acoustic pressure amplitude as a function of distance x and time t, a is the absorption coefficient, ω is the angular frequency, and k is the wave number. The pulse technique employed in this study to measure the absorption coefficient in liquids simulates the free field condition

---

(10) G. Schwarz, J. Mol. Biol., 11, 64 (1965).
(11) T. B. Lewis, Ph.D. Thesis, M.I.T., Cambridge, Mass., 1965.
(12) T. K. Saksena, B. Michels, and R. Zana, J. Chim. Phys., 65, 597 (1968).
(13) R. Zana, R. Cerf, and S. Candau, ibid., 60, 869 (1963).
(14) J. F. Foster "Plasma Proteins," F. W. Putnam, Ed., Academic Press, New York, N. Y., 1960, Chapter 6.
(15) E. J. Cohn, W. L. Hughes, and J. H. Weare, J. Amer. Chem. Soc., 69, 1753 (1947).
(16) C. Tanford, J. G. Buzzel, D. G. Rands, and S. A. Swanson, ibid., 77, 6421 (1955).

expressed by eq 1 for finite sample sizes, provided that the pulse length in the medium is short compared with the acoustic path length. In addition, the error in the absorption coefficient due to the spectrum of frequencies associated with a pulse train is negligible if the pulse is at least $30\pi/\omega$ sec in length.[17]

Two techniques were employed to measure the absorption coefficient to within 5% over the frequency range from 0.3 to 163 MHz. The first technique,[18] an automated version of that described by Pellam and Galt,[17] can be employed for frequencies greater than 9 MHz, for the transducer diameter available (1 in.), where diffraction effects are small. Over the frequency range from 9 to 69 MHz, two matched 3 MHz fundamental frequency X-cut quartz transducers are set parallel and coaxial to each other in the liquid to be studied. Each transducer is edge mounted with its front face in direct contact with the liquid and its back face exposed to an air-filled cavity. One transducer emits pulses of ultrasonic energy while the other transducer detects acoustic pulses. The acoustic path length is varied by displacing one transducer relative to the other at constant velocity. The amplitude of the received acoustic pulse varies according to eq 1 (where x is the instantaneous acoustic path length) and the electrical pulse from the receiving transducer is recorded on a logarithmic chart recorder whose paper displacement is slaved to the moving transducer. Over the frequency range 75 to 165 MHz, a pair of 15 MHz fundamental, X-cut quartz transducers bonded to fused quartz delay rods are substituted for the 3 MHz transducers. Velocity measurements are performed by electronically measuring the length of time required to change the acoustic path length by 100 wavelengths of sound. Details of this system can be found in ref 18 and 19.

At frequencies below 9 MHz, the technique described above for measuring the absorption coefficient and velocity of sound requires unreasonably large diameter transducer elements to minimize diffraction effects. These effects arise because the requirement of plane waves, as described by eq 1, is only approximated by a finite size transducer and the approximation deteriorates as the wavelength of sound approaches the diameter of the transducer. A comparison technique, described by Carstensen,[20] in which the acoustic properties of the sample liquid are determined relative to those of a known reference liquid, minimizes diffraction effects. In the present study, water served as a convenient reference liquid since its absorption coefficient and velocity of sound have already been determined accurately.[21,22] Two compartments of a double chamber tank are separated by an acoustically transparent window and are filled, respectively, with the reference liquid and the sample liquid. Two 3 in. diameter 0.3 MHz fundamental frequency ceramic transducers are placed in the respective chambers and face each other through the window. The transducers are mounted coaxially parallel and are supported a fixed distance apart on a sliding carriage. The acoustic measurements are made by varying the relative amounts of sample and reference liquids in the acoustic path by moving the carriage along the axis of sound propagation. If the velocity of sound in the sample liquid is within a few per cent of that in the reference liquid, then varying the relative amounts of each within the acoustic path produces little change in the acoustic path length and consequently only a negligible diffraction effect. Details of this technique are to be found in ref 19 and 20.

Results

The ultrasonic absorption in aqueous solutions of BSA was measured as a function of pH from 2.3 to 11.8 at six frequencies ranging from 2.39 MHz to 50.25 MHz at 20°. The data are presented in terms of the excess frequency-free absorption per unit concentration, *i.e.*

$$A = \Delta a/cf^2 \qquad (2)$$

where c is the concentration of solute in grams per cubic centimeter of solution and Δa is the difference in the absorption coefficients between the solution and the solvent. The absorption coefficient of the solvent found by Pinkerton[21] has been used to evaluate Δa. In aqueous solutions of BSA, the excess absorption has been shown to increase linearly with concentration at least to about 0.1 g/cc in the pH region where expansion of the molecule is known not to occur,[18] Figure 1.

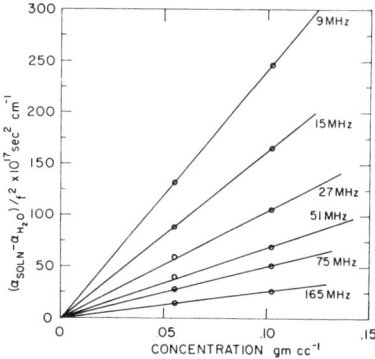

Figure 1. Concentration dependence of acoustic absorption of BSA at 20.0° in an aqueous solution of 0.15 M KCl.

(17) J. R. Pellam and J. K. Galt, *J. Chem. Phys.*, **14**, 608 (1946).
(18) S. A. Hawley, Ph.D. Thesis, Univ. of Illinois, Urbana, Ill., 1966.
(19) L. W. Kessler, Ph.D. Thesis, Univ. of Illinois, Urbana, Ill., 1968.
(20) E. L. Carstensen, *J. Acoust. Soc. Amer.*, **26**, 858 (1954).
(21) J. M. M. Pinkerton, *Proc. Phys. Soc.*, **B62**, 129 (1949).
(22) M. Greenspan and C. E. Tschiegg, *J. Acoust. Soc. Amer.*, **31**, 75 (1959).

Ultrasonic Investigation of the Conformal Changes of Bovine Serum Albumin

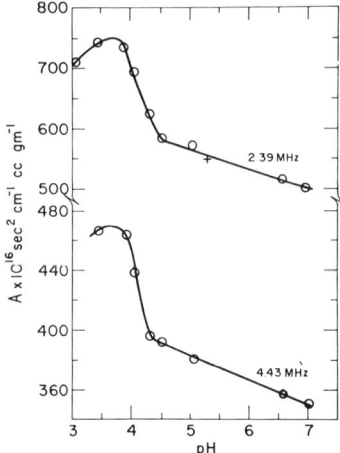

Figure 2. Ultrasonic absorption titration of BSA in an aqueous solution at 19.9° (+ indicates back titration).

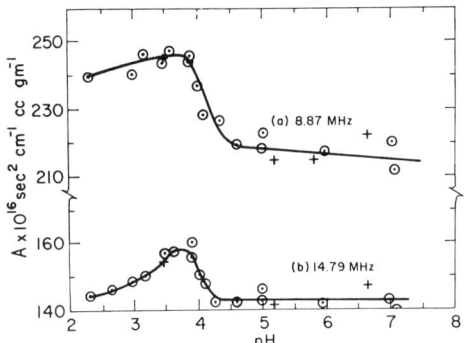

Figure 3. Ultrasonic absorption titration of BSA in an aqueous solution at 20.0° (+ indicates back titration).

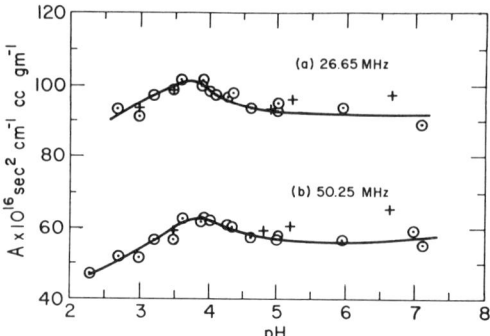

Figure 4. Ultrasonic absorption titration of BSA in an aqueous solution at 20.0° (+ indicates back titration).

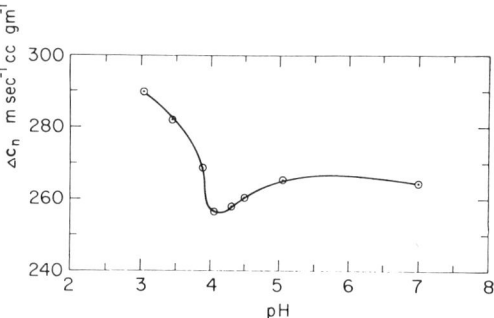

Figure 5. Ultrasonic velocity titration of BSA in an aqueous solution at 19.9° (BSA concentration 0.092 g/cc).

This implies that contributions to the absorption arising from possible intermolecular interactions are not important, and that in this region, as far as the ultrasonic wave is concerned the solution may be considered equivalent to one which is infinitely dilute.

The excess frequency-free absorption is plotted as a function of pH for the acid titration in Figures 2, 3, and 4 and it is clear that within experimental error the ultrasonic absorption changes reversibly with changes of pH. This is reasonable on the basis of the previously observed reversibility of other properties of BSA in aqueous solution over the same pH range.[14] The very small effect on the absorption of the formation of KCl as a result of back titration[19] has been corrected in all the figures. As the BSA solution is made acidic, the absorption increases by a small amount until the pH reaches about 4.3. Beyond this point, there is an abrupt increase in A which is markedly greater at lower frequencies than at higher frequencies. The velocity of sound in the solution was also measured as a function of pH at 2.39 MHz and the result for the acid titration is presented in Figure 5. In this figure, Δc_n is the difference between velocity of sound in the solution and that in the solvent, divided by the solute concentration. For the purpose of this calculation, changes in the velocity of sound in water were assumed to be negligible with changes of pH compared with changes in the velocity of sound in the solution.[19] The resulting curve exhibits a minimum at about pH 4.1 which is also the approximate midpoint of the abrupt absorption increase.

Ultrasonic titration on the alkaline side shows a somewhat greater increase in the absorption coefficient than at acid pH, an example of which is shown in Figure 6. As the alkalinity of the solution is increased from pH 7, the variation of A with pH is small up to about pH 10 where A increases rapidly. The alkaline effect is also reversible over the pH range investigated. Further, as with the acid titration, the effect is more pronounced at lower frequencies. The velocity of sound, on the other hand, was found to increase monotonically with increasing alkalinity.

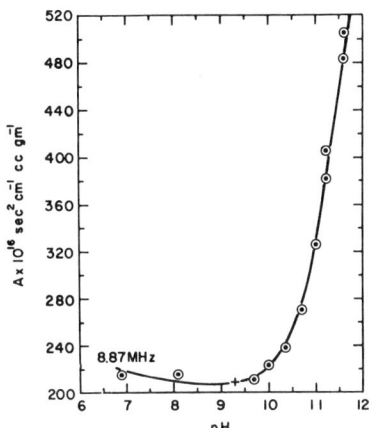

Figure 6. Ultrasonic absorption titration of BSA in an aqueous solution (+ indicates back titration).

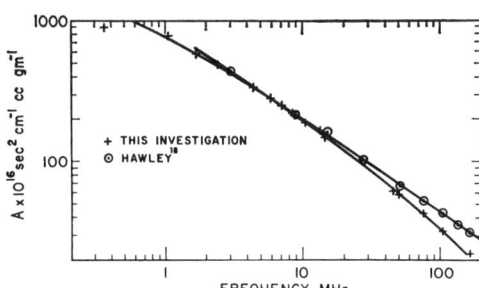

Figure 7. Ultrasonic absorption spectrogram at pH 7 and 20.0°.

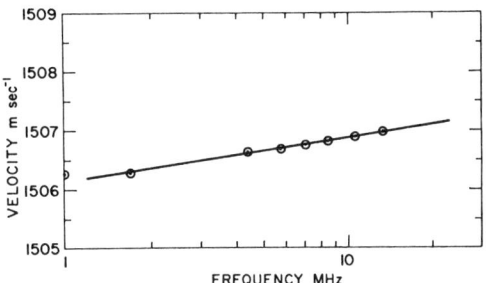

Figure 8. Velocity dispersion in an aqueous solution of BSA at pH 7 and 19.9°. BSA concentration, 0.092 g/cc. The velocity of sound in pure solvent at 19.9° is 1482.35 m/sec.

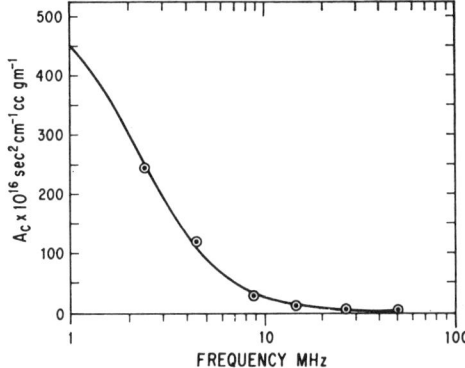

Figure 9. Spectrogram of absorption difference between pH values 3.5 and 7.0, compiled from data in Figures 2, 3, and 4.

The frequency dependence of A was measured over the frequency range from 0.3 to 163 MHz by the techniques indicated above and a composite of all the data, at neutral pH, is shown in Figure 7. Over the frequency range covered, a broad distribution of relaxation times may be necessary to characterize the absorption behavior. The presence of velocity dispersion further indicates the relaxational behavior of the absorption over this frequency range, Figure 8. In order to relate changes in the relaxational properties of BSA in aqueous solution with molecular conformal changes, the variation of the absorption coefficient as a function of frequency is examined. It is evident from Figures 2–6 that there are two sharp transition regions, one which is completely delineated in the acid pH region, and a second, in the alkaline pH region, which extends beyond the range of these measurements. If the following definition is made, $A_c = A(\text{pH } 3.5) - A(\text{pH } 7.0)$, then the frequency spectrum of A_c is described, approximately, by a single relaxation process as shown in Figure 9. The best fit single relaxation curve was determined by the method of "least squares" for an equation of the form

$$A_c = \frac{a_c}{1 + (f/f_0)^2} + b_c \quad (3)$$

where a_c, b_c, and f_0 are constants, f_0 being the characteristic relaxation frequency. For eq 3, $f_0 = 2.2$ MHz, $a_c = 543 \times 10^{-16}$ sec² cc/cm g and $b_c = 2.2 \times 10^{-16}$ sec² cc/cm g.

The absorption coefficient was determined as a function of temperature from 15 to 38° and at two pH values viz., 7.0 and 2.9. Higher temperatures were avoided since above 40° irreversible denaturation of the molecule can occur.[23] For absorption due to either shear or bulk viscosity, η, the temperature dependence of the absorption coefficient is given by Eyring[24] as

$$\alpha = \frac{2\omega^2}{3\rho_0 c_0^3}\left[\eta \exp\frac{\Delta F}{RT}\right] \quad (4)$$

(23) J. F. Foster and J. T. Yang, J. Amer. Chem. Soc., **77**, 3895 (1955).
(24) H. Eyring, J. Chem. Phys., **4**, 263 (1936).

where ρ_0 and c_0 are the density of the medium and velocity of sound respectively, ΔF is the activation energy, R is the gas constant and T is the absolute temperature. This relation is valid if the period of the acoustic wave does not approach a characteristic time constant associated with a relaxation process. Table I lists the apparent activation energies determined for a solution of 0.05 g/cc of BSA in water. In all cases, the apparent activation energy is less than that for pure water. At pH 7.0, ΔF is independent of frequency, within the experimental error, as it is for water. However, for low pH values, ΔF is a strong function of frequency which indicates the presence of additional relaxation phenomena.

Table I: Apparent Activation Energy of BSA in Aqueous Solution (Concn, 0.05 g/cc)

pH	f(MHz)	ΔF, kcal/mol
7.0	8.87	2.19
7.0	14.79	2.23
7.0	26.65	2.26
2.9	8.87	0.671
2.9	14.79	1.35
2.9	26.65	1.70
Water	All	4.23

Discussion

In order to account for the magnitude of the excess absorption coefficient at neutral pH, quantitative consideration is given to (1) dynamic shear viscosity of prolate ellipsoidal molecules in solution; (2) frictional losses associated with relative motion between the solute and solvent particles; (3) scattering of the acoustic waves by the solute particles; and (4) mode conversion of the longitudinal acoustic wave into rapidly decaying transverse acoustic waves at the solute–solvent interfaces. Such consideration[19] shows that, collectively, the above mechanisms of absorption account for only a small percentage of that observed. An investigation of the dynamic shear viscoelasticity of BSA in glycerol and water mixtures was reported by Allis and Ferry[25] in the frequency range from 0.04 to 400 Hz and the dynamic shear viscosity observed did not agree with theories of rotational relaxation. It is suggested by these authors that the origin of the viscoelasticity in BSA does not arise from orientation of the molecule by the shear stresses but primarily from an intramolecular flexibility not otherwise observable. However, a reexamination of the contribution of shear viscosity, taking into account this discrepancy, does not alter significantly the results presented here.

As a result of the above considerations it is concluded that the structural (bulk) viscosity of the solution is the principal factor responsible for the ultrasonic absorption. The distribution of relaxation times and the velocity dispersion indicate structural relaxation[26] and the same mechanism is thought to be present in aqueous solutions of other biomacromolecules, viz., hemoglobin,[7] and dextran.[27] A reasonable description of the observed phenomena has been given by Andreae, et al.,[28] for a nonelectrolyte dissolved in water. The short range structure of water rapidly breaks down as solute molecules are dissolved. Water molecules, which bind to the solute molecules in hydration layers, reach an equilibrium state with the unbound solvent molecules and it is this equilibrium which is perturbed by the sound wave and gives rise to absorption.

In order to understand the observed changes of the ultrasonic absorption coefficient that occur with changes in pH, a brief resumé of the physical-chemical properties of BSA in aqueous solution is presented. It has been generally recognized that serum albumin molecules undergo marked reversible structural changes as the pH of the environment is altered, although the exact nature of these changes remains unclear.[14] Within the range 4.3 < pH < 10.5, each BSA molecule behaves as an undeformable solid particle whose shape can be approximated by a prolate ellipsoid. Outside this range, it was thought that a simple swelling of the compact globular structure was responsible for the observed increase in optical rotation,[29] in viscosity,[30] and in other physical parameters until Tanford, et al.,[16] discovered a distinct stepwise change in the intrinsic viscosity, $[\eta]$, as the pH decreased below 4.3. Specifically, between pH 4.3 and 4.0, $[\eta]$ increases sharply by about 22% and between pH 3.5 and 2.8 an 84% increase occurs. This two-step process was observed when the BSA was suspended in 0.15 M KCl and was not observed at low ionic strengths. It should be noted that the ionic strength of the Fraction V material supplied by the manufacturer is high. In nearly the same pH region as the smaller increase in $[\eta]$, A increases abruptly between pH 4.4 and 3.8. There is no change in A corresponding to the 84% increase in $[\eta]$ below pH 3.5. This suggests that if the same mechanism is responsible for both the first increase in A and the increase in $[\eta]$ at pH 4.3, then a separate mechanism which does not affect the ultrasonic properties of the solution is responsible for the larger increase in $[\eta]$. Tanford, et al.,[16] propose that the complete expan-

(25) J. W. Allis and J. D. Ferry, J. Amer. Chem. Soc., **87**, 4681 (1965).
(26) T. A. Litovitz and C. M. Davis, "Physical Acoustics," Vol. II, Part A, W. P. Mason, Ed., Academic Press, New York, N. Y., 1965, Chapter 5.
(27) S. A. Hawley, L. W. Kessler, and F. Dunn, J. Acoust. Soc. Amer., **38**, 521 (1965).
(28) J. H. Andreae, P. D. Edmonds, and J. F. McKellar, Acustica, **15**, 74 (1965).
(29) B. Jirgensons, Arch. Biochem. Biophys., **39**, 261 (1952).
(30) J. T. Yang and J. F. Foster, J. Amer. Chem. Soc., **76**, 1588 (1954).

sion of BSA occurs in at least three distinct stages. As the pH is decreased a structural change occurs from a compact form to an "expandable form." At lower pH, the molecule expands physically, and for pH < 4, a small time-dependent increase in the viscosity occurs. This last stage is attributed to possible slow aggregation of the molecules.

Aggregation has been observed in aqueous BSA solutions at acid pH by many investigators. According to Williams and Foster,[31] the principal aggregate is the dimer and it has a maximum concentration at pH 3.3. Below pH 3.0 or above pH 3.5, the rate and extent of dimerization is diminished and these investigators conclude that the turbidity, or cloudiness of the solution that occurs below the isoelectric point which is usually attributed to aggregation, is due to the liberation of a lipid impurity that is carried along with the BSA molecule. Since the abrupt change of the ultrasonic absorption coefficient does not occur in the neighborhood of pH 3.3, it is unlikely that the acoustic disturbance is associated with dimerization. If the impurity itself was involved in the absorption mechanism, then some correlation should have been observed with liberation of this material, which occurred above pH 4.3.

The stepwise expansion process and more recent evidence of a two-step change in the rotational relaxation time occurring at pH 4.1 and pH 3.6[32] support the N–F transformation theory proposed by Aoki and Foster[33,34] and Foster.[14] The N state is the compact rigid form of the molecule which exists between pH 4.3 and 10.5. It is thought to be composed of two pairs of globular subunits held together tightly with each member of a pair held to its partner by hydrophobic bonds. The two pairs of subunits are then bonded to one another electrostatically to make up a four unit globule. As the pH of the BSA solution is reduced below its isoelectric point, electrostatic repulsion forces the molecule to separate into two units linked by flexible chains. This is the so-called F' state or "intermediate F" form. This state, which corresponds to Tanford's "expandable form," occurs in the same region as the observed ultrasonic effect. As the pH is reduced below 3.6, electrostatic forces become strong enough to overcome the hydrophobic bonds and each unit pair separates. This configuration of the molecule, which consists of four subunits interconnected by flexible linkages, is known as the F state.

Weber and Young[35] subjected acidified BSA to short enzymatic digestion and found that the BSA molecule was split into one large and two smaller globular fragments, a total of three instead of four subunits. With this evidence and with the hydrodynamic properties obtained by others, Bloomfield[36] determined a suitable three subunit model for BSA which consists of a central sphere, of radius 26.6 Å and two flanking spheres of radius 19 Å each. The hydrodynamic properties of this model agree well with those experimentally observed at pH 3.6. However, this model does not account for any stepwise expansion process. It is possible, however, that a combination of Bloomfield's model and Foster's could do so. For example, assume that the N state is composed of a single pair of hydrophobically bonded subunits which is electrostatically bonded to a third, larger subunit. The F' state and F state would correspond to the flanking spheres of Bloomfield's model.

At this point it is proposed that the observed absorption increase at acid pH over that at neutral pH is related to the N ⇌ F' transformation and that the best estimate of the relaxation time of this reaction that can be made on the basis of this study is $\tau_{NF'} = 0.72 \times 10^{-7}$ sec. (This was calculated from the simple relation $\tau_{NF'} = 1/2\pi f_0$.) The transformation F' ⇌ F is too slow to be observed with the ultrasonic frequencies employed in this investigation.

Figure 5 shows that the reduced velocity exhibits a minimum at pH 4.1. If the expansion of BSA occurs by separation of the globular units without certain changes to the globular units themselves, as implied by the fragmentation experiments of Weber and Young,[35] then it may be assumed, to a first approximation, that the elastic moduli and densities of these units remain constant. Urick,[37] under the simplifying assumption that a solution can be considered homogeneous to the sound wave, derived the following equation for the velocity of sound in a solution, C_x.

$$C_x = C_0 \left(\frac{1}{(1 + \phi \sigma_K)(1 + \phi \sigma_\rho)} \right)^{1/2} \quad (5)$$

where

$$\sigma_K = \frac{K_0 - K_1}{K_0}; \quad \sigma_\rho = \frac{\rho_1 - \rho_0}{\rho_0}$$

In eq 5, the subscripts 0 and 1 correspond to the solvent and solute, respectively, K is the bulk modulus, ρ is the density and ϕ is the volume fraction of the solution occupied by solute. At 2.39 MHz, pH 7.0 and 20°, $C_0 = 1482.4$ m/sec and $C_x = 1506.5$ m/sec when the solute concentration is 9.18%. If a reasonable value for ρ_1 is taken as 1.33 (see page 307 of ref 1) then $\sigma_\rho = 0.33$ and $\sigma_K = -0.77$. If K_0 is 2.18×10^9 n/m² then the bulk modulus of BSA is 3.86×10^9 n/m². It follows from eq 5, under the assumption stated above, that the minimum at pH 4.1 corresponds to a net decrease in the volume fraction of molecules in solution,

(31) E. J. Williams and J. F. Foster, *J. Amer. Chem. Soc.*, **82**, 3741 (1959).
(32) C. L. Riddiford and B. R. Jennings, *J. Chem. Soc.*, **88**, 4359 (1966).
(33) K. Aoki and J. F. Foster, *J. Amer. Chem. Soc.*, **79**, 3385 (1957).
(34) K. Aoki and J. F. Foster, *ibid.*, **79**, 3393 (1957).
(35) G. Weber and L. B. Young, *J. Biol. Chem.*, **239**, 1424 (1964).
(36) V. Bloomfield, *Biochem.*, **5**, 684 (1966).
(37) R. J. Urick, *J. Appl. Phys.*, **18**, 983 (1947).

which could result from a reduction of the protein hydration. This is reasoned physically as follows. In the N state, each gram of protein is associated with approximately 0.2 g of bound water molecules. If, when the molecule assumes the intermediate F' state, the hydration layer remains intact, then the internal rotational freedom possessed by the separated globular units would be suppressed and a decrease in the rotational relaxation time would not be observed. As pH decreases below 4.1, the velocity of sound increases monotonically due, in part, to the increased velocity in the solvent alone, and, in part, to the changes in the chemical nature of the solvent and solute at low pH.[38]

The change in the ultrasonic absorption coefficient for pH > 10.5 is more difficult to correlate with molecular events than it is for pH < 4.3 since very little experimental work on the nature of the alkaline expansion has been reported in the literature. Weber[39] and Tanford, et al.,[16] suggest that both the acid and alkaline expansions are similar; however, no other experimental evidence has been presented to substantiate this. Abrupt decreases of the rotational relaxation time of BSA were observed by Weber[39] at pH 3.6 and pH 11.2. It has been observed[14] that at pH 3.6, BSA begins to expand rapidly with decreasing pH, and thus it is reasonable to assume that expansion is also rapid with pH changes above 11.2. The ultrasonic titrations at acid pH and alkaline pH are similar in that the absorption coefficient begins to increase above pH 3.6 and below pH 11.2, i.e., before the molecule actually expands. For pH < 3.6, no correspondence is evident between expansion and A; however, for pH > 11.2, A is increasing sharply which implies that the mechanisms responsible for the absorption increase may be different in each region. Further evidence to support this is the sharp contrast between the velocity titrations at acid and alkaline pH. Tanford and Buzzell[40] have measured $[\eta]$ to pH 10.5 and found that no observable change occurs between pH 9.3 and pH 10.5 whereas here it has been shown that A increases significantly. Further correlation of the ultrasonic absorption increase for pH > 10.5 must await further details of the nature of the alkaline expansion.

Acknowledgment. The authors acknowledge gratefully the support of this research in part by a grant from the Institute of General Medical Sciences, National Institutes of Health, and in part by the Office of Naval Research, Acoustics Program. The authors also wish to extend their appreciation to G. Weber for his numerous and helpful discussions, to S. A. Hawley for his valuable suggestions regarding preparation of this manuscript, and to P. D. Edmonds for the use of his computer program for determining the best fit single relaxation curve.

(38) H. A. Saroff, *J. Phys. Chem.*, **61**, 1364 (1957).
(39) G. Weber, *Biochem. J.*, **51**, 155 (1952).
(40) C. Tanford and J. G. Buzzell, *J. Phys. Chem.*, **60**, 225 (1956).

Ultrasonic Absorption Mechanisms in Aqueous Solutions of Bovine Hemoglobin[1]

by W. D. O'Brien, Jr., and F. Dunn*

Bioacoustics Research Laboratory, University of Illinois, Urbana, Illinois 61801 (Received June 21, 1971)

Publication costs assisted by the Institute of General Medical Sciences, National Institutes of Health

In order to study further the principal loss mechanisms of ultrasonic energy in biological media, the absorption and velocity were determined in aqueous solutions of hemoglobin, largely at 10°, over the frequency range 1–50 MHz. A distribution of relaxation processes is necessary to characterize the absorption spectra. Interaction of the acoustic wave with the hydration layer of the macromolecule, and not direct interaction with macromolecular configuration, appears as a dominant mechanism. The ultrasonic absorption titration curves exhibit maxima around pH 2–4 and pH 11–13, in addition to possessing a broad peak in the pH range 5–9. The peaks in the absorption titration and the similar ones observed for bovine serum albumin (*J. Phys. Chem.*, **73**, 4256 (1969)), are attributed to the proton-transfer reaction occurring between particular amino acid side chain groups and the solvent. The broad peak is partially attributed to the proton transfer resulting from the imidazolium function of the histidine residue.

Introduction

Although ultrasonic spectroscopy has been available, in principle, for at least two decades, to date only a few biological macromolecules have been examined. Probably the most extensively studied biopolymer is the globular protein hemoglobin, the oxygen carrier in the red blood cells of vertebrates.[2-4] The earliest work of importance is that of Carstensen, *et al.*,[5] who investigated the ultrasonic absorption and velocity in the blood, plasma and solutions of albumin, and hemoglobin, and concluded that the acoustical properties of blood are largely determined by the protein concentration. In addition they showed that the absorption coefficient of hemoglobin is approximately the same as that of serum albumin within the frequency range 0.8–3 MHz in the temperature range 10–40°.

Within the neutral pH region, aqueous solutions of hemoglobin have now been examined over the extended frequency range 35–1000 MHz[2,3] and it has been shown that it is possible to approximate the entire spectrum with four appropriately selected discrete relaxation processes.

More recently other globular proteins have been examined. Kessler and Dunn[6] studied aqueous solutions of bovine serum albumin and attributed the ultrasonic absorption in the neutral pH region to solvent–solute interactions. Outside of the neutral pH range,

---

(1) (a) Portions of this work were extracted from the Thesis submitted by W. D. O'Brien, Jr., in partial fulfillment of the requirement for the Ph.D. degree in electrical engineering, University of Illinois. (b) A preliminary report of this work was presented at the 78th meeting of the Acoustical Society of America in San Diego, Calif., Nov 1969 [*J. Acoust. Soc. Amer.*, **47**, 98 (1970)].

(2) F. Dunn, P. D. Edmonds, and W. J. Fry, in "Biological Engineering," H. P. Schwan, Ed., McGraw Hill, New York, N. Y., 1969, Chapter 3, p 205.

(3) F. Schneider, F. Muller-Landay, and A. Mayer, *Biopolymers*, **8**, 537 (1969).

(4) P. D. Edmonds, T. J. Bauld, J. F. Dyro, and M. Hussey, *Biochem. Biophys. Acta*, **200**, 174 (1970).

(5) E. L. Carstensen, K. Li, and H. P. Schwan, *J. Acoust. Soc. Amer.*, **25**, 286 (1953).

(6) L. W. Kessler and F. Dunn, *J. Phys. Chem.*, **73**, 4256 (1969).

the absorption behavior was thought to correlate with conformation changes. Wada, et al.,[7] investigated the ultrasonic absorption of gelatin at 3 MHz as a function of pH which had a peak in the absorption spectrum around pH 4 and was interpreted to be a dissociation type reaction of the protein side chains. Also it has been reported that the absorption magnitude of gelatin solutions within the frequency range 0.7–10 MHz is approximately half that in hemoglobin and albumin solutions.[8] For each of these protein solutions the ultrasonic absorption spectra exhibit a distribution of relaxation times.

Two synthetic polyamino acids have been examined under varying environmental conditions with respect to their ultrasonic absorption. The primary mechanism proposed to explain the excess ultrasonic absorption in poly-L-glutamic acid solutions[9] is that of the solvent–solute interaction whereas Schwarz[10] attributes the excess absorption to the helix–coil transition. The examination by Wada, et al.,[7] revealed that at 50 kHz the absorption mechanism is that of helix–coil transition while at 3 MHz the absorption is attributed to side chain dissociation. It has been shown[11] that the observed ultrasonic absorption behavior in aqueous poly-L-lysine solutions can be associated with the helix–coil transition.

The carbohydrate, dextran, a linear α (1–6) anhydroglucose polysaccharide, assumes a random coil conformation in solution whereas most proteins exist as a compact, rigid molecule. The ultrasonic absorption spectra of dextran, in the frequency range 3 to 69 MHz, can be represented by a distribution of relaxation times.[12] The absorption magnitude is considerably less than that exhibited by proteins, and has been attributed to the proteins possessing a secondary and tertiary structure while dextran does not. In addition, the protein gelatin, which does not possess a tertiary structure, also exhibits a lesser absorption magnitude than those proteins with higher ordered structure. Thus the suggestion that the tertiary structure may be responsible for some of the excess ultrasonic absorption observed in protein solutions appears to have received attention.[13,14]

It is apparent from above that the mechanisms mainly responsible for the ultrasonic absorption are unsettled. The present study was undertaken to provide additional data from observations at extreme pH values.

Experimental Techniques

Two distinct systems, described in detail elsewhere,[15,16] were utilized. The high frequency system, an automated version of that of Pellam and Galt,[17] utilizes a transmitting and a receiving transducer and has a lower frequency limit of approximately 9 MHz because of the apparent attenuation due to diffraction effects. The upper frequency limit is well beyond the 50 MHz employed here. At the lower frequencies, a comparison method is used which minimizes the difficulties due to diffraction,[5,18] and is capable of measuring ultrasonic absorption and velocity over the frequency range 0.3–20 MHz. Both systems employ the standard pulse techniques. These measurement techniques assume the absorption behavior of the fluid under investigation to be exponential in nature, viz.

$$p = p_0 e^{-\alpha x} \quad (1)$$

where p_0 is the pressure amplitude at $x = 0$, x is the acoustic path length over which the acoustic pulse travels, and α is the amplitude absorption coefficient per unit length. Speed of sound measurements are accomplished by adding algebraically the received signal to a reference signal and recording the time required to change the acoustic path length by 100 wavelengths.[16] The temperature of the liquid under investigation was maintained to $\pm 0.05°$.

Two grades of methemoglobin (Hb), obtained from Nutritional Biochemicals Corp., Cleveland, Ohio, were investigated, viz., uncrystallized Bovine Hemoglobin (Hb-OX), control no. 3099, and Hemoglobin Bovine 2X Crystalline (Hb-2X), control no. 1480, 8647, and 8995. Singly deionized and distilled water, testing to at most 0.15 ppm impurities[19] was used throughout. The protein solutions were prepared by placing the proper amount on top of a measured volume of water and refrigerating until mixing was complete, usually 2–5 hr. The uncrystallized hemoglobin solution, which contained some red blood cell structures, was centrifuged at 20,000g for 2 hr to remove heavier particles. This supernatant, as well as other protein solutions, were filtered twice through type A glass fiber filters (Gellman Inst. Co., Ann Arbor, Mich.) to remove particles larger than 0.3 μ in diameter and stored at 7° until used, usually not more than a few hours. The weight concentrations of the acoustically measured solutions were determined, to an accuracy of better

(7) Y. Wada, H. Sasaba, and M. Tomono, *Biopolymers*, 5, 887 (1967).
(8) H. P. Schwan and H. Pauly, *J. Acoust. Soc. Amer.*, 50, 692 (1971).
(9) J. J. Burke, G. G. Hammes, and T. B. Lewis, *J. Chem. Phys.*, 42, 3520 (1965).
(10) G. Schwarz, *J. Mol. Biol.*, 11, 64 (1965).
(11) R. C. Parker, L. J. Slutsky, and K. R. Applegate, *J. Phys. Chem.*, 72, 1968 (1968).
(12) S. A. Hawley and F. Dunn, *J. Chem. Phys.*, 52, 5497 (1970).
(13) S. A. Hawley, L. W. Kessler, and F. Dunn, *J. Acoust. Soc. Amer.*, 38, 521 (1965).
(14) W. D. O'Brien, Jr., and F. Dunn, *J. Acoust. Soc. Amer.*, 50, 1213 (1971).
(15) W. D. O'Brien, Jr., Ph.D. Thesis, University of Illinois, Urbana, Ill., 1970.
(16) L. W. Kessler, S. A. Hawley, and F. Dunn, *Acustica*, 23, 105 (1971).
(17) J. R. Pellam and J. K. Galt, *J. Chem. Phys.*, 14, 608 (1946).
(18) E. L. Carstensen, *J. Acoust. Soc. Amer.*, 26, 858 (1954).
(19) Tested at the Illinois State Water Survey, Champaign, Ill.

than ±0.3% at room temperature, by evaporating 15 ml in a tared beaker over air until dry and placing in a vacuum desiccator for 24 hr.

The pH of the hemoglobin solution was altered by the addition of standard volumetric solutions of either 1.0 N HCl or KOH. The acid or base was introduced slowly, to minimize pH gradients, into the circulating fluid within the acoustic chamber in steps to produce changes ranging from 0.1 to 1.5 pH units. The pH readings were obtained to within a relative accuracy of ±0.01 pH unit with a Beckman Century SS pH Meter using a Beckman pH combination electrode (39013) which fitted directly into the acoustic chamber. The pH meter was standardized at pH 2.01, 4.01, 7.00, 9.18, and 12.45.

In order to obtain the entire ultrasonic titration spectrogram, two measurement procedures were performed, each starting at neutral pH. A single procedure took from 8 to 20 hr, depending upon the number of points to be determined. Typically, the test liquid remained at a fixed pH for 1 hr, of which 15–20 min was allowed for the test liquid to equilibrate and the balance of time devoted to determining the ultrasonic absorption and velocity at the desired frequencies. Two complete titration spectrograms were obtained for each grade of hemoglobin in order to verify the results, the data being available in ref 15. The solutions, at the terminal pH values, were stored for several days at 7°, following the measurement procedure, with no observable precipitation occurring.

Results

The ultrasonic absorption of the two grades of hemoglobin was examined as a function of solute concentration to determine the onset of finite concentration effects. Deviation from linearity of Hb-OX solutions occurs around 0.10 g/cm³ while that for the purer Hb-2X deviates around 0.16 g/cm³, the latter agreeing with earlier reports.[20] Such deviation from linearity is commonly attributed to interactions among the solute molecules. Thus equating the volume of an assumed spherical particle and the volume per molecule, $4/3\pi R^3 = cN_A/M$, where M is the molecular weight of the biopolymer and N_A is Avogadro's number, yields a molecular radius $R = 54$ Å at the concentration, c, of 0.16 g/cm³, i.e., where the ultrasonic absorption begins to deviate from linearity. This value is in substantial agreement with that determined from X-ray diffraction techniques wherein the Hb molecule was found to be roughly spherical with overall dimensions of 64 Å by 55 Å by 50 Å.[21] In all the work reported herein, the concentration was maintained well below 0.16 g/cm³.

The ultrasonic absorption data are presented in terms of the excess frequency-free absorption per unit concentration parameter

$$A = \Delta\alpha/cf^2 \qquad (2)$$

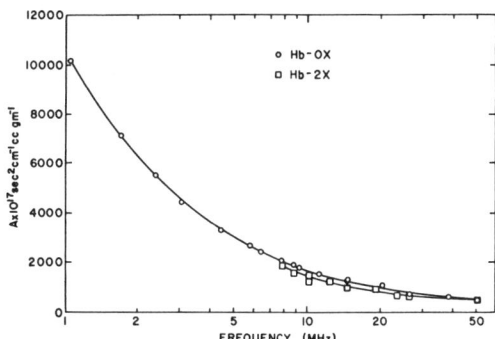

Figure 1. Ultrasonic absorption spectrogram in aqueous solutions of bovine hemoglobin ($T = 10°$).

where $\Delta\alpha$ is the difference between the absorption of the solution and that of the solvent, c is the biopolymer concentration in grams per cubic centimeter, and f is the ultrasonic frequency. The absorption parameter is shown in Figure 1 as a function of frequency for aqueous solutions of Hb-OX and Hb-2X at their isoelectric point, at 10.0°. The excess absorption exhibited by the uncrystallized Hb over the purer grade (about 10%) possibly reflects impurities contained in the former, and not present in the latter, which may also possess relaxational behavior. A similar situation has been reported for bovine serum albumin.[22]

The composite ultrasonic absorption titration curve is shown in Figure 2 for the frequency range 8.9–50.5 MHz and over the pH range 1.5–12.3. Similar shaped curves were also determined for Hb-OX down to 2.4 MHz.[15] The titration curves are similar in shape and magnitude to those for aqueous solutions for bovine serum albumin,[6] where the increase in A below 4.3 was associated with the intermediate N–F′ transition of Foster.[23] Hemoglobin, however, does not exhibit this transition. Similarly, in the alkaline pH region, both BSA and Hb show excess absorption peaking beyond pH 10.5.

The ultrasonic absorption coefficient and velocity were determined as functions of temperature over the range 10–37° in a neutral aqueous solution of Hb-2X at a concentration of 0.0349 g/cm³ in order to provide information on apparent activation energies, shown in Table I with that of water. In all cases ΔF is less than that of water though it increases with increasing frequency. A similar experiment reported for BSA[6] indicates no frequency dependence within the neutral

(20) E. L. Carstensen and H. P. Schwan, *J. Acoust. Soc. Amer.*, **31**, 185 (1959).

(21) M. F. Perutz, M. G. Rossmann, A. F. Cullis, H. Muirhead, G. Will, and A. C. T. North, *Nature (London)*, **185**, 416 (1960).

(22) L. W. Kessler, private communication, 1968.

(23) J. F. Foster, in "Plasma Proteins," F. W. Putnam, Ed., Academic Press, New York, N. Y., 1960, Chapter 6.

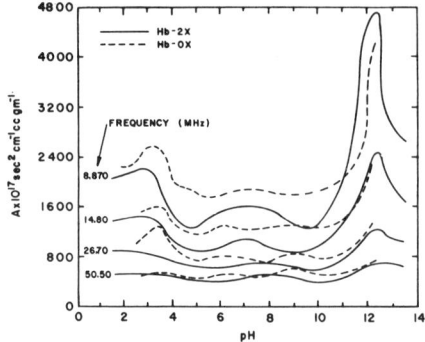

Figure 2. Composite ultrasonic absorption titration curve in aqueous solutions of bovine hemoglobin ($T = 10°$).

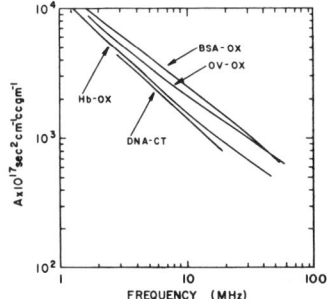

Figure 3. Composite ultrasonic absorption spectrogram, $10°$: Hb-OX, uncrystallized bovine hemoglobin; BSA-OX, uncrystallized bovine serum albumin; OV-OX, uncrystallized ovalbumin; DNA-CT, calf thymus deoxyribose nucleic acid.

pH region and strong frequency dependence at pH 2.9, which was attributed to additional relaxation processes occurring in the acidic region. Such processess, resulting in strong frequency dependence of the activation energy, could be proton-transfer reactions occurring at neutral pH values in Hb and at pH 2.9 in BSA aqueous solutions which are absent in neutral pH BSA solutions.

Discussion

Figure 3 shows the absorption parameter A for aqueous solutions of Hb-OX, BSA, ovalbumin-uncrystal-

Table I: Apparent Activation Energy of Aqueous Solution of Hb-2X (Concentration = 0.0349 g/cm^3)

pH	f, MHz	ΔF, kcal/mol
6.9	8.870	1.2
6.9	14.80	2.0
6.9	26.70	2.6
6.9	50.50	3.4
Water	All	4.48

lized,[24] and calf thymus DNA[25] at $10°$. All four curves possess approximately the same frequency dependence although differences in magnitude are apparent. The former suggests that the mechanism(s) of ultrasonic absorption, for these four solutions, at their isoelectric point, may be the same. One universal feature of aqueous solutions of globular proteins and nucleic acids is the existence of the ubiquitous structuring of water about the macromolecule. It thus seems reasonable to speculate that the magnitude differences result from differing degrees of hydration, depending upon detailed molecular differences. Interaction of the solvent and solute has already been invoked to describe the excess ultrasonic absorption in aqueous solutions of biological molecules.[6,9] The present consideration is that the mechanism is the perturbation of the hydration layer, which is essential for maintaining the integrity of the biopolymer, by the acoustic wave process.

Hemoglobin is a globular protein composed of four subunits, two α polypeptide chains and two β polypeptide chains, with each consisting of approximately 144 amino acid residues and conjugated to a heme moiety (an iron-containing porphyrin derivative). No disulfide cross-links exist which would prevent each of the polypeptide chains from assuming a completely extended or helical configuration and each chain possesses approximately 65% helix content.[26] Under normal conditions approximately 6% of the human hemoglobin molecules in aqueous solution (pH 7; no denaturing agents present) are dissociated in half without loss of tertiary structure of the individual polypeptide chains.[27] This indicates the unique nature of the hemoglobin molecule to dissociate into its half subunits under very mild conditions without hydrolysis or pigment denaturation occurring. The nature of the bonds, or noncovalent links, which connect the individual polypeptide chains is unknown although it is assumed that the contacts are between α and β chains resulting in $\alpha\beta$ half molecules upon dissociation rather than α and β half molecules.[28] All forms of hemoglobin undergo major conformational changes when exposed to low pH. As the pH is lowered from the isoelectric point to 4.5, the hemoglobin molecule dissociates in half without appreciable change in the conformation of the resulting $\alpha\beta$ polypeptide chains.[29] As the pH is lowered from 7 to 3.5 to 2.9, the intrinsic viscosity increases from 3.5 cm^3/g[30] to 13.55 cm^3/g[30] to 17 cm^3/

(24) W. D. O'Brien, Jr., and F. Dunn, unpublished results.
(25) W. D. O'Brien, Jr., and F. Dunn, unpublished results.
(26) C. Tanford, "Physical Chemistry of Macromolecules," Wiley, New York, N. Y., 1961.
(27) K. Kawahara, A. G. Kirschner, and C. Tanford, *Biochemistry*, **4**, 1203 (1965).
(28) A. F. Cullis, H. Muirhead, M. F. Perutz, M. G. Grossman, and A. C. T. North, *Proc. Roy. Soc. Ser. A*, **265**, 161 (1962).
(29) E. C. Field, and J. R. P. O'Brien, *Biochem. J.*, **60**, 656 (1955).
(30) D. Tanford, *J. Amer. Chem. Soc.*, **79**, 3931 (1957).

g,[31] respectively. Thus at pH values less than 4, the Hb molecule clearly shows a marked expansion. Within the pH range 4.0–2.2, Polet and Steinhardt[32] reported that the heme is expelled from the apoprotein and the globin is unfolded, but the specific pH at which these two events occur is still unknown. Finally, Reichmann and Colvin[33] determined that around pH 2.0–1.8, the Hb molecule splits into four fragments of nearly equal size.

The hemoglobin molecule also dissociates at alkaline pH, *viz.*, the sedimentation coefficient decreases above pH 10, where it is 4.2×10^{-13} sec, reaching a value of 2.55×10^{-13} sec at pH 11.[34] Within the pH range 7–11, the diffusion coefficient remains relatively constant, indicating that the hemoglobin molecule dissociates in half as the pH is increased from 10 to 11, while the individual polypeptide chains retain their tertiary structure. As the pH is increased from 11, the sedimentation coefficient, along with the diffusion coefficient, decreases showing that the two subunits are expanding but not necessarily dissociating further.

Repulsion of like charges has been discounted as a mechanism of dissociation since at pH 6 the molecule possesses a charge of $+5$ whereas at pH 10 it is -30.[34] Thus the denaturation of Hb must be attributed to different mechanisms in the acidic and alkaline pH regions.

The effect of pH on the ultrasonic absorption characteristics of a number of aqueous solutions of amino acids have been investigated, *viz.*, serine and threonine,[35] glycine,[36,37] glutamic acid, aspartic acid and alanine,[36] and arginine and lysine,[38] and absorption maxima were observed in the acidic and alkaline pH regions. This absorption peaking, as a function of pH, has been described quantitatively for the above amino acids, assuming the proton-transfer reaction dominates the absorption process. However, the task of dealing with the similar peaking for macromolecules such as proteins which contain a large number of side chains able to participate in such reactions is far more difficult since detail of reaction coupling among sub-chain groups is not currently available. Thus a more qualitative discussion of the proton-transfer reaction for proteins must suffice for the present. For this reaction, which is a chemical relaxation mechanism, it is considered that the propagating acoustic wave disturbance perturbs the proton from the solvent (water) to the solute, an amino acid side chain, and *vice versa*. The energy necessary to drive the reaction is extracted from the acoustic field. The pH values at which the peaks in Figure 2 are maximum can be correlated with the pK values of the individual amino acid side chains which participate, both in the acidic and alkaline pH regions, *viz.*

$$pH_{acid} = 1/2(pK + pC) \quad (3)$$

and

$$pH_{base} = 1/2(14 + pK - pC) \quad (4)$$

where C is the molar concentration of the particular side chain group. Table II lists the amino acid side chains for Hb, along with their quantity in moles per 100,000 g of protein which can participate in the proton transfer reaction. Within the acidic pH range, the μ-carboxyl group (pK range on protein of 3.0–4.7) of aspartic acid (ASP) and glutamic acid (GLU) is primarily responsible for absorption peak. At the alkaline pH values, the ϵ-amino group (pK range 9.4–10.6) of lysine (LYS), the guanidinium group (pK range 9.8–10.4) of tyrosine (TYR), and the sulfhydryl group (pK range 9.4–10.8) of cysteine (CYS) contribute primarily to the absorption maximum. The diffuse peak, in Figure 2, occurs within the same pH range (5–9) in which about one-third, or 22, of the histidine groups (HIS) per Hb molecule titrate[39] suggesting that the broad peak around neutral pH may be a result of the imidazolium group (pK range 5.6–7.0), although the majority of the excess absorption still results from the

Table II: Proton-Transfer Side-Chain Groups (in mol/100,000 g of Protein)

Amino acid side chain	Human Hb[a]	Bovine Hb[b]	BSA[c]
ASP/GLU (ω-carboxyl)	162	194	152
LYS (ϵ-amino)	133	133	88
HIS (Imidazolium)	115	103	24.6
ARG (Guanidinium)	36.4	42.4	33.8
TYR (Phenolic hydroxyl)	36.4	36.4	29.3
CYS (Sulfhydryl)	18.2	6.1	0
α-Carboxyl	12.1	12.1	1.54
α-Amino	12.1	12.1	1.54

[a] G. Braunitzer, K. Hilse, V. Rudloff, and N. Hilshmann, *Advan. Protein Chem.*, **19**, 1 (1964). [b] W. B. Gratzer and A. C. Allison, *Biol. Rev. Cambridge Phil. Soc.*, **35**, 459 (1960). [c] See ref 23.

(31) D. Tanford, *Advan. Protein Chem.*, **23**, 121 (1968).
(32) H. Polet and J. Steinhardt, *Biochemistry*, **8**, 857 (1969).
(33) M. E. Reichmann and J. E. Colvin, *Can. J. Chem.*, **34**, 411 (1956).
(34) U. Hasserodt and J. Vinograd, *Proc. Nat. Acad. Sci. U. S.*, **15**, 19 (1959).
(35) R. D. White, L. J. Slutsky, and S. Pattison, *J. Phys. Chem.*, **75**, 161 (1971).
(36) K. Applegate, L. J. Slutsky, and R. C. Parker, *J. Amer. Chem. Soc.*, **90**, 6909 (1968).
(37) M. Hussey and P. D. Edmonds, *J. Acoust. Soc. Amer.*, **49**, 1309 (1971).
(38) M. Hussey and P. D. Edmonds, *ibid.*, **49**, 1907 (1971).
(39) C. Tanford and Y. Nozaki, *J. Biol. Chem.*, **241**, 2832 (1966).

interaction of the hydration layer with the acoustic wave.

Bovine serum albumin consists of a single polypeptide chain with molecular weight of about 68,000. The conformation of the molecule remains unchanged at pH values between 4.3 and 10.5 and sedimentation coefficients decrease, with their ratios remaining constant,[40] implying that the molecule expands but does not dissociate. The intrinsic viscosity of the BSA molecule expands in steps, from 3.6 to 4.5 cm^3/g around pH 4 and from 4.5 to 8.4 cm^3/g around pH 3.[41] At $[\eta]$ = 8.4 cm^3/g, the BSA molecule is not highly expanded, and this can be attributed to the large number of disulfide bonds the molecule possesses. Tanford, et al.,[41] and Weber[42] have speculated that the alkaline expansion of the BSA molecule is similar to that in the acid pH region but this has neither been confirmed nor denied[23] by experiment.

The effect of the protonation reaction of the imidazolium group of the histidine amino acid to the results of serum albumin[6] appears to be minimal on the ultrasonic absorption within the pH range 5–9, possibly because of the small number of histidine amino acids in serum albumin (Table II). For both Hb and BSA, there is approximately the same ratio of amino acid groups which can partake in the proton transfer reaction between the alkaline (LYS, ARG, TYR, CYS) and acidic (ASP, GLU) pH region and as shown in Table II, these concentration ratios are 217/194 for Hb and 151/152 for BSA. The similarity of these ratios is considered to lend support to the view that the similar ultrasonic absorption magnitude of the peaks of these two globular proteins results from the same mechanism. Previously[6] the peaks in the BSA spectrogram were discussed in terms of conformational changes such as Foster's[23] N–F' transition. It now appears that a more acceptable explanation for such peaks in the ultrasonic absorption titration curves of Hb and BSA may be the proton-transfer reaction. However, the role of conformation cannot be completely discounted at this time as the evidence for each mechanism (1) is less convincing than one would like and (2) does not exclude the other from being present, simultaneously, to some degree of effectiveness.

Acknowledgment. The authors acknowledge gratefully the support of this research by a grant from the Institute of General Medical Sciences, National Institutes of Health.

(40) W. F. Harrington, P. Johnson, and R. H. Ottewill, *Biochem. J.*, **62**, 569 (1956).
(41) C. Tanford, J. G. Buzzel, D. G. Rands, and S. A. Swanson, *J. Amer. Chem. Soc.*, **77**, 6421 (1955).
(42) G. Weber, *Biochem. J.*, **51**, 155 (1952).

ERRATUM

Page 530, column 1, line 11 under "Results": the equation should read

$$\tfrac{4}{3} \pi R^3 = \frac{M}{cN_A}$$

18

Copyright © 1972 by John Wiley & Sons, Inc.

Reprinted with permission from *Biopolymers*, 11(9), 1973–1984 (1972)

Ultrasonic Absorption and Relaxation Spectra in Aqueous Bovine Hemoglobin

R. D. WHITE and L. J. SLUTSKY, *Department of Chemistry, University of Washington, Seattle, Washington 98195*

Synopsis

Measurements of the frequency and pH dependence of acoustic absorption at 0°C in aqueous solutions of freshly prepared bovine oxyhemoglobin are reported. The role of ionization and possible direct proton-transfer between proximal pairs in determining the characteristic times for the relaxation of the internal charge distribution is discussed. It is concluded that treatments which consider various classes of residues as ionizing independently will not give approximately correct relaxation spectra. A model which takes into account the coupling between the degrees of ionization of the various residues is found to give rough agreement with the observed acoustic absorption in the pH range in which the native conformation is stable.

INTRODUCTION

Since Carstensen and Schwan's work on aqueous solutions of hemoglobin there have been a number of studies of excess acoustic absorption in solutions of biologically important macromolecules[2-7] and their synthetic analogs.[8-10] Relaxation times or relaxation spectra adequate to characterize the observed frequency dependence of the ultrasonic velocity and attenuation have been deduced,[1-3,8,9] but the variety of chemical and structural equilibria possible in a solution of flexible polyelectrolytes is great enough to render the unambiguous identification of the microscopic process reponsible for the excess absorption a matter of some difficulty. In individual cases arguments have been advanced nominating solvation,[1,4,8] ionization,[4,6] and conformational change[5,9] as the principal contributor to the observed relaxation spectrum in aqueous protein solution.

There exists a fair body of experimental information on the kinetics[11-14] and thermodynamics[15] of the acidic and basic ionization reactions of amino acids and simple polypeptides. In these systems the Debye-Smoluchowsky[16,17] theory of diffusion-controlled reaction rates has given a good account of the kinetic parameters. The data on simple systems constitutes a basis for the estimation of the contribution of perturbation of acid–base and internal charge-transfer equilibria to the acoustic absorption in protein solutions. We wish to explore here the extent to which it is possible to account for the frequency and pH dependence of the excess acoustic absorption in aqueous bovine hemoglobin on this basis, and to briefly discuss some convenient mathematical procedures for computing relaxation spectra in complex systems.

IONIZATION REACTIONS OF HEMOGLOBIN

In a system of c components with r chemical or structural equilibria it will be possible to write r independent equations of the form

$$\sum_i \mu_{ij} B_i = 0 \tag{1}$$

where μ_{ij} is the coefficient of the ith chemical species in the jth reaction (μ_{ij} is here taken to be positive for products negative for reactants) and B_i designates the ith component. The change in the concentration of the ith component when the jth reaction is displaced from equilibrium serves to define the "degree of advancement," dx_j, of the jth reaction

$$\mu_{ij} dx_j = dc_i \tag{2}$$

and the total change in the concentration of the ith component when all reactions are displaced from equilibrium is

$$dc_i = \sum_j \mu_{ij} dx_j \tag{3}$$

The degrees of advancement of the autoionization of water and the reactions

$$-\text{AH} + \text{H}_2\text{O} \underset{k_b}{\overset{k_f}{\rightleftarrows}} \text{A}^- + \text{H}_3\text{O}^+ \tag{4a}$$

$$-\text{BH}^+ + \text{OH}^- \underset{k_b}{\overset{k_f}{\rightleftarrows}} \text{B} + \text{H}_2\text{O} \tag{4b}$$

where —AH represents the protonated form of an acid residue (aspartyl, glutamyl, terminal carboxyl) and BH+ the protonated form of a basic residue (histidyl, lysyl, arginyl, etc) constitute a basis set sufficient to describe both the internal and overall state of ionization of the hemoglobin molecule. The volume changes for the proton-transfer reactions of amino acids and simple polypeptides are known. Orttung[18] has made a detailed calculation based on the Kirkwood-Tanford[19] theory from which the degree of ionization of individual residues or groups of residues in hemoglobin can be estimated.

The volume changes for the basis reactions are given in Table I as are the degrees of protonation of each residue as a function of pH.

Measured values[11] of k_f in Eq. (4a) for the carboxylic acids range from 3.5×10^{10} M^{-1} sec^{-1} for benzoic acid to 5×10^{10} M^{-1} sec^{-1} for formic acid. We have measured k_f in Eq. (4b) for a number of amino acids and simple peptides.[13] Typical results in units of 10^{10} M^{-1} sec^{-1} are 1.9, 2.0, 2.1, 2.2, and 2.2 for glycine, threonine, triglycine, serine, and piperidine, respectively. If the effective radius for reaction, r_d, is taken to be equal

TABLE I
Volume and Enthalpy Changes for the Ionization of Acid Residues. Fraction (f) of Residues Protonated as a Function of pH

Residue	Base	ΔV in cc/mole	ΔH^d in kcal	Fraction of residues protonated[e]				Number of residue/molecule
				pH				
				6	7.5	9	12	
Arginyl$^+$	OH$^-$	25[a]	0	1.0	1.0	1.0	0.959	14
Lysyl$^+$	OH$^-$	25[a]	−3	1.0	1.0	0.996	0.863	44
Cystinyl	OH$^-$	6[b]	−6	1.0	0.997	0.979	0.720	4
Tyrosyl	OH$^-$	4[c]	−6	1.0	0.99	0.840	0.314	12
n-Terminal amino$^+$	OH$^-$	25[a]	−3	0.930	0.87	0.711	0.058	4
Histidyl$^+$	OH$^-$	25[a]	−6	0.784	0.67	0.529	0.022	38
CCOH	H$_2$O	−10[d]	1.5	0.112	0.013	5 × 10^{-4}	0	72

[a] 9.
[b] 20.
[c] 21.
[d] 15.
[e] 18.

to an N–H—O hydrogen bond distance (2.7 Å) and a geometrically reasonable steric factor is chosen (0.6), Debye's result

$$k_{12} = \frac{\sigma 4\pi N z_A z_B e_0^2 (D_A + D_B)}{\epsilon kT [\exp(z_A z_B e_0^2 / \epsilon r_d kT) - 1]} \quad (5)$$

where N is Avogadro's number, σ a steric factor, e_0 is the electronic charge, z_A and z_B are the algebraic charges of the ions, ϵ is the dielectric constant of the solvent, D_A and D_B are the diffusion coefficients of the reacting ions, and r_d is an effective radius for reaction predicts values between 2.0×13^{10} M^{-1} sec^{-1} (triglycine) and 2.2×10^{10} (glycine).

A plausible approach to the computation of the contribution of ionization and internal charge-transfer reactions to the relaxation spectrum of hemoglobin is then to estimate the forward rate constants of Eq. (4) from the conventional theory of diffusion-controlled reaction rates, to obtain the reverse rate constants and equilibrium concentrations at any pH from Orttung's computed degree of ionization, and to solve the system of coupled first-order equations obtained by linearizing the kinetic equations (one for each acidic or basic residue) implied by Eq. (4).

The eigenvalues of the kinetic matrix directly give the relaxation times. In order to calculate the acoustic absorption (α/f^2) it is necessary to compute the standard volume change, (ΔV), standard enthalpy change (ΔH), and chemical factor $\Gamma_r = dx_r/d \ln K_r$ (where K_r is the equilibrium constant for the rth normal reaction) for each of the normal reactions from the eigenvectors of the kinetic matrix and ΔV, ΔH, and K for the basis reactions. The absorption is then calculated by summing over the independent normal reactions

$$\frac{\alpha}{f^2} = A + \sum_j \frac{C_j \tau_j}{1 + (2\pi\tau_j f)^2}$$

$$C = 2\pi^2 \rho C_0 \bar{V}^2 RT \Gamma_M [(\beta \Delta H / C_p RT) - (\Delta V / \bar{V} RT)]^2 \quad (6)$$

where the ρ is the density, C_0 the velocity of sound, $\bar{V}$ the volume per mole of solution, β the coefficient of thermal expansion, C_p the molar heat capacity at constant pressure.

The results of such a computation for a $0.0023M$ solution of hemoglobin at pH 12 and 0°C is compared with experiment in Figure 1. The computed relaxation frequencies and the corresponding values of C are given in Table II. The important contributors to the excess absorption over most of the range displayed in Figure 1 are normal reactions 6 (qualitatively the reaction between protonated cystyl residues and lysyl or OH$^-$) and 7 (principally proton transfer between lysyl and tyrosyl residues). Normal reaction 1 (ionization of the carboxyl group) contributes nothing to the relaxation spectrum at high pH. Since it presumed here that ΔV for the protonation of all-NH$_2$ groups is the same, normal reaction 3 (principally proton exchange between lysyl and arginyl residues) makes only a small contribution; ΔV for this transfer being zero. At the higher fre-

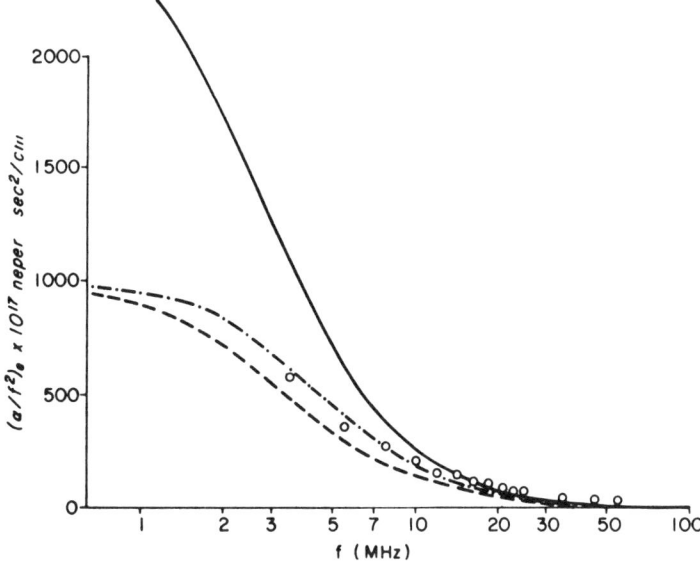

Fig. 1. The frequency dependence of the excess acoustic absorption of $0.0023M$ oxyhemoglobin at 0°C and pH 12. The open circles represent our data while the broken curve (---) shows the calculated absorption due to dissociation only. The curve (-.-.) represents the effect of including direct transfers between tyrosyl–lysyl and tyrosyl–cystinyl pairs assuming independent probabilities for simultaneous proton occupation of a pair of neighboring residues, and the solid curve shows the calculated result when direct transfers are included and the joint probabilities are computed in such a way as to maximize the chance that a pair is only single ionized.

quencies ($\geq$ 100 MHz) normal reaction 2 (hydrolysis of lysyl and arginyl residues) becomes the principal contributor to the excess absorption. The remaining eigenvectors do not have simple character approximate chemical equations being: $-$ (4. $CYS^- + 3HIS + NTA + 4LYSH^+ + ARGH^+ \rightarrow CYSH + 3HISH^+ + NTAH + 4LYS + ARG$); 5. $3CYS^- + LYSH^+ + NTAH^+ + 2HIS + 3H_2O \rightarrow 3CYSH + 3OH^- + LYS + NTA + 2HISH^+$; 8. $3COOH + OH^- + 2LYS \rightarrow 3COO^- + H_2O + 2LYSH^+$.

TABLE II
Relaxation Spectrum of Hemoglobin, pH 12 and 0°C

	f_R in MHz	C, sec cm^{-1}	$C\tau$, sec^2 cm^{-1}
1	1800	1.2×10^{-17}	1.1×10^{-27}
2	179	1.66×10^{-7}	1.5×10^{-17}
3	15	3.4×10^{-10}	0.38×10^{-17}
4	5.115	3.3×10^{-10}	1.0×10^{-17}
5	5.110	0.27×10^{-10}	0.08×10^{-17}
6	4.6	1.3×10^{-7}	464×10^{-17}
7	2.9	0.88×10^{-7}	483×10^{-17}
8	0.004	0.012×10^{-10}	4.6×10^{-17}

Hussey and Edmonds[3b] and O'Brien and Dunn[5c] have given simplified versions of this treatment in which the off-diagonal terms in the kinetic matrix are arbitrarily set equal to zero. The results of this calculation suggest that within the framework of the Debye-Smoluchowsky theory and Orttung's group-average pK_a's, the normal reactions are not simply individual group dissociations and that the off-diagonal terms cannot be justifiably eliminated from the calculation. At pH 12 this calculation gives quite a good account of the experimentally determined excess absorption, however, the procedure outlined above cannot account for the absorption near neutral pH.

O'Brien and Dunn[5c] have suggested that dissociative proton transfers of the histidyl residues contribute to a broad peak in a acoustic absorption at megacycle frequencies in the neighborhood of neutral pH. However, at neutral pH, 10^{-7} is a strict upper limit to Γ_r, $2\pi^2\rho c_0(\Delta V/RT)^2$ is approximately 4×10^{-5}, C in Equation (6) is then $\approx 4 \times 10^{-12}$, and even for a 1 megacycle relaxation frequency the excess value of α/f^2 will not exceed 10^{-18} neper sec^2/cm and thus it seems unlikely that simple ionization of histidyl residues can contribute significantly at neutrality.

Near neutrality the important eigenvectors of the kinetic matrix will correspond to reactions of the form,

$$-AH + B = BH^+ + A^- \quad (7)$$

that is, to intramolecular proton transfers with —AH a carboxyl and B a histidyl group. However, intramolecular proton transfers based on Eq. (4) proceed of necessity by the mechanism

$$-AH + H_2O = A^- + H_3O^+$$

$$H_3O^+ + B = BH^+ + H_2O$$

or the equivalent with hydroxyl ion acting as the base. At neutral pH, where the concentration of hydronium and hydroxide ions are in the neighborhood of 10^{-7} M even the most optimistic assumptions about the reaction radii and steric factors in the Debye theory of diffusion-controlled dissociation and recombination reactions predict relaxation frequencies of the order of 10 KHz and, while internal transfers by a dissociative mechanism may contribute to the very low-frequency spectrum observed by Carstensen,[1] they do not appear to be important at frequencies higher than 1 MHz.

There remains the possibility of a more direct protont-ransfer mechanism. The histidyl residues G19β, E13β, H5α, D6α, EF1α, all have carboxyl neighbors within a 5 Å radius, tyrosine H23α has a cystine neighbor at 5.5 Å. There is some reason to expect that proton exchange between near neighbors is significantly faster than dissociation and recombination in dilute solution; but there is not a large body of information on the rates of direct proton exchange. The isomeric aminobenzoic acids constitute a possible set of model compounds. Ultrasonic attenuation in 0.5M o-amino

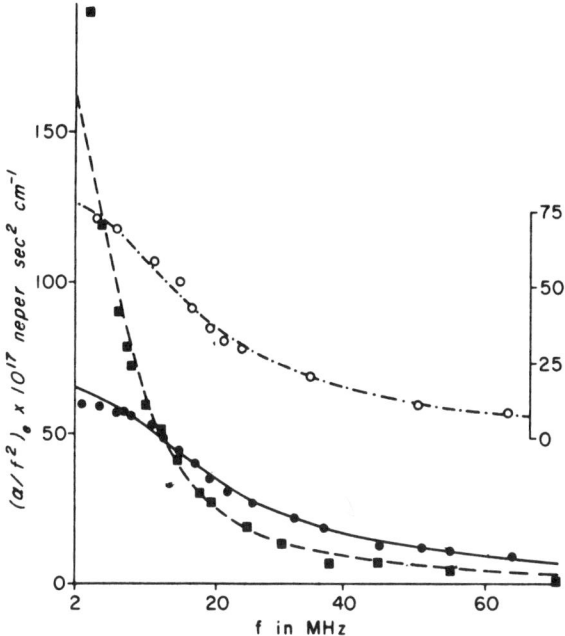

Fig. 2. Frequency dependence of the excess acoustic absorption in 0.5M solutions of o-aminobenzoic acid in methanol at 25°C (solid circles) and 0°C (squares). The open circles and righthand scale give the experimental results for o-aminobenzoic acid 1M in acetone at 25°C.

benzoic acid in methanol is given in Figure 2. The relaxation frequency is 20 MHz, presumably corresponding to direct transfer in a cyclic hydrogen-bonded system. The relaxation frequencies in methanolic solutions of the meta and para isomers are, as would be expected, too low to be conveniently determined by ultrasonic techniques.

Although there is justification for the assignment of megacycle relaxation frequencies to proton exchange between proximal amino and carboxyl groups the exact dependence of the rate on distance is still a matter of conjecture. Moreover, the set of basis reactions given in Eq. (4) is complete and introduction of the equations for direct proton-transfer reactions results in a system of linearly dependent kinetic equations. Before considering the possible role of direct transfer mechanisms in the relaxation spectrum of proteins we wish to briefly discuss the development of a convenient algorithm for the computation of α due to chemical relaxation in a system of kinetically distinct but thermodynamically redundant reactions.

ABSORPTION DUE TO CHEMICAL RELAXATION IN COMPLEX SYSTEMS

Starting with all species in Eq. (1) at their equilibrium concentration ^0C_i and introducing the definition $R_j = k_{fj} \Pi_i {}^0C_i {}^{\mu_{ij}}$, the mass-action expression for the rate of the jth reaction becomes, for small displacements from equilibrium,

$$\frac{dx_j}{dt}(t) = -R_j \sum_{ij} \frac{\mu_{ij}\mu_{ij}'}{C_i} x_j'(t) \tag{8}$$

or, in matrix notation,

$$\frac{d\mathbf{x}}{dt}(t) = -\mathbf{RAX}(t) \tag{9}$$

where $\mathbf{R}$ is a diagonal matrix with elements R_j and $A_{jj'} = \sum_i \mu_{ij}\mu_{ij'}/{}^0C_i$. With the usual substitution $\mathbf{X}(t) = \mathbf{X}e^{-\lambda t}$, Eq. (9) becomes

$$\lambda \mathbf{X} = \mathbf{RAX} \tag{10}$$

For computational purposes it is convenient to symmetrize via the substitution[23] $\mathbf{Y} = \mathbf{R}^{-1/2}\mathbf{X}$ and obtain the eigenvalues λ_i and diagonalizing matrix $\mathbf{S}$ of

$$\lambda \mathbf{Y} = (\mathbf{R}^{1/2}\mathbf{A}\mathbf{R}^{1/2})\mathbf{Y} \tag{11}$$

The ultrasonic absorption is given by[25]

$$\frac{\alpha}{f^2} = \frac{2\pi^2 \rho C_0}{RT(1000)} \bar{V}^2 \sum \left(\frac{\Delta V_k^*}{\bar{V}} - \frac{\beta}{C_p}\Delta H_k^*\right)^2 \frac{\lambda_k^{-2}}{1+\left(\frac{2\pi f}{\lambda_k}\right)^2} \tag{12}$$

If the volume and enthalpy changes for the basis reactions are represented as column vectors $\Delta \mathbf{V}$ and $\Delta \mathbf{H}$ with components ΔV_j and ΔH_j then[24]

$$\Delta \mathbf{V}^* = (\mathbf{S}^{-1}\mathbf{V}^{1/2})\Delta \mathbf{V}$$
$$\Delta \mathbf{H}^* = (\mathbf{S}^{-1}\mathbf{V}^{1/2})\Delta \mathbf{H} \tag{13}$$

Equations (12) and (13) are suitable for the computation of α when a linearly dependent basis-set is employed, the zero eigen values obtained in such a case make no contribution to the absorption.

To proceed further it is necessary to make some assumption about the variation in the rate of proton transfer between neighboring groups l and m as a function of distance. The rate in o-aminobenzoic acid is presumably an upper limit. For want of more detailed information we have assumed that the rate varies as the inverse cube of the distance as would be be expected if one were dealing with a diffusion-controlled rate enhanced by a locally high concentration of hydrogen ion due to the presence of a neighboring acid group.

Potentially the acoustic absorption associated with a reaction of the form of Eq. (7) is quite high. One transfer per α–β dimer with an equilibrium constant of one and a relaxation frequency of 3 MHz could account for about $1/4$ of the observed excess absorption in neutral solution. Most of the 38 histidyl residues have at least one carboxyl neighbor within 8 Å.

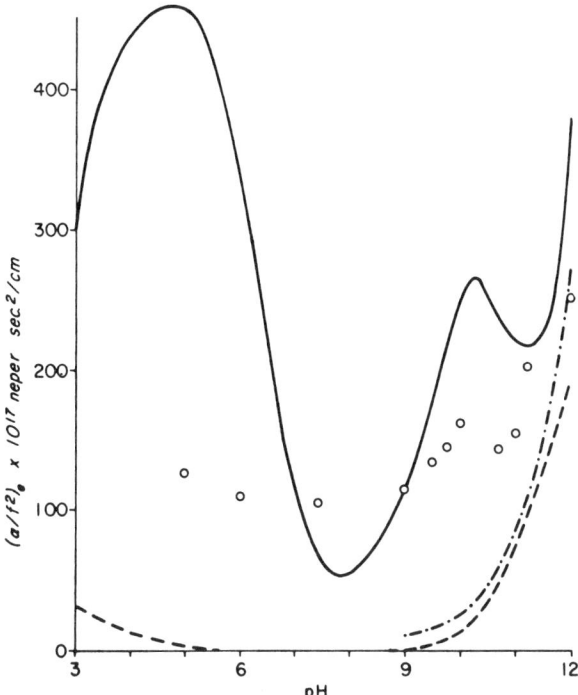

Fig. 3. Experimental and theoretical pH dependence of the acoustic absorption in 0.0023M bovine hemoglobin at 0°C and 7.9 MHz. The broken curve represents the effect of dissociative transfer only, -.-. includes direct transfers between tyrosyl–lysyl and histidyl–cystyl pairs calculated assuming independent probabilities of protonation of adjoining residues. The solid curve includes all neighboring pairs assuming that no protonated residue has a protonated neighbor.

However, in any analysis based on the ionization of a single group in the average electrostatic potential of its neighbors, the fraction f_{lm} of protonated residues l with unprotonated neighbors m will be given by $f_{lm} = f_l(1 - f_m)$. Groups of similar pK_a will be either simultaneously protonated or simultaneously unpopulated and the concentration of neighboring pairs with an approximately even distribution of protonated and unprotonated forms will be high only in the narrow range of pH in which $f_l \approx f_m \approx 1/2$. Groups of dissimilar p$K_a$ will make no important contribution at any pH.

The effect of including direct transfers subject to the approximations discussed in the foregoing on the calculated acoustic absorption is shown in Figures 3–5. At high pH the important transfers are between the tyrosine–lysine $H_{22\alpha}$–$H_{23\alpha}$ pair and the tyrosine–cystine $H_{23\beta}$–$F_{9\beta}$ pairs. At lower pH, histidine–tyrosine pairs (B1α–B5α, $F_{8\alpha}$–$H_{23\alpha}$) and $E_{20\beta}$–$H_{8\beta}$) become important contributors. Near neutrality a large number of carboxyl–histidine pairs must be considered.

These calculations are expected to underestimate the contribution of internal transfers. Protonating the near neighbor of a given residue obviously reduces the probability that the residue itself is protonated and

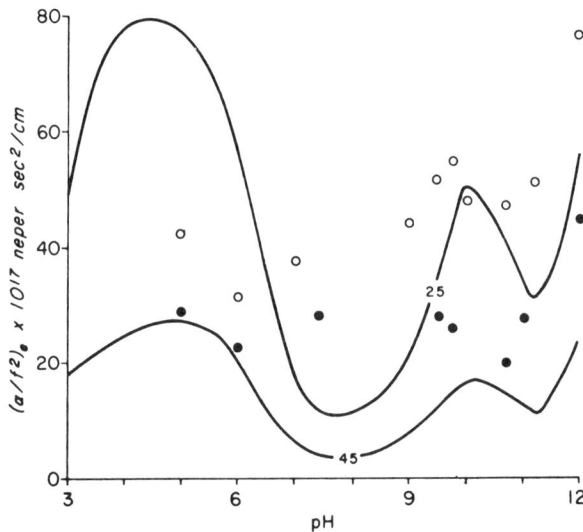

Fig. 4. Experimental and predicted pH dependence of the acoustic absorption in 0.0023M bovine hemoglobin at 0°C at 25 and 45 MHz. The solid curves are calculated assuming no protonated residue has a protonated nearest neighbor. Open circles represent the experimental results at 25 MHz, solid circles the experimental results at 45 MHz.

taking the joint probability to be the product of the individual probabilities does not properly count the number of pairs of unoccupied sites near protonated sites. Near neutral pH most carboxyl sites are not protonated. At separations of 5–10 Å the electrostatic potential due to a proton charge is sufficiently great so that it is more reasonable to assume that none of the few protonated carboxyl sites is adjacent to a positively charged histidine than to assume random mixing of occupied and unoccupied sites for proton binding. In Table III, the probabilities of the four possible states of ionization of a carboxylhistidyl pair calculated assuming zero probability of simultaneous protonation of adjacent residues (column II) are contrasted with the result of random mixing (column I). The value of the chemical factor for the intramolecular transfer calculated in this way, $f_C f_H/(f_C + f_H)$, will be approximately equal to the fraction of protonated carboxyl groups.

TABLE III
Probability of States of Protonation of Proximal Carboxyl–Histidyl Pairs

State of pair		Probability of state	
		I	II
—COOH	$^+$H—N	$f_C f_H$	0
—COO$^-$	$^+$H—N	$(1 - f_C)f_H$	f_H
—COOH	N	$f_C(1 - f_H)$	f_C
—COO$^-$	N	$(1 - f_C)(1 - f_H)$	$1 - (f_C + f_H)$

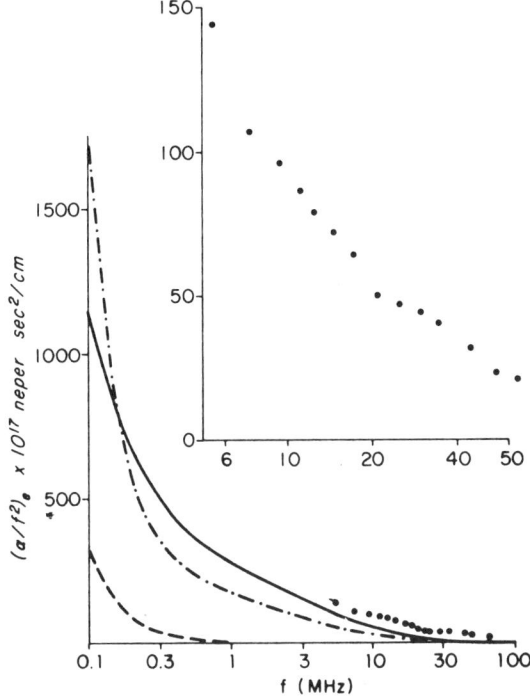

Fig. 5. Excess ultrasonic absorption in $0.0023M$ aqueous bovine hemoglobin at 0°C and neutral pH. The lowest solid curve represents the effect of dissociation, the broken curve includes the effect of direct intramolecular transfers calculated assuming independent probability of protonation of adjacent groups. The upper solid curve is calculated on the basis of the assumption that no carboxyl group immediately adjacent to a protonated histidyl is protonated at neutral pH.

Much the same argument is applicable to tyrosyl–histidyl pairs at higher pH.

The solid curves in Figures 1, 3, 4, and 5 are calculated on the basis of the populations given in column II. Such a calculation will necessarily overestimate the absorption at low pH where the probability of simultaneous protonation of a histidyl residue and a carboxyl near neighbor can not be legitimately neglected. Finally, the volume changes associated with ionization reactions in nonaqueous media are very much larger than those in water.[15,22] If one were to assign substantially larger ΔV's to transfers between pairs of interior residues which are in an essentially nonaqueous environment, the calculated absorption would be considerably larger.

CONCLUSIONS

Dissociative ionization reactions are not likely to make an important contribution to the relaxation spectrum in protein solutions at megacycle frequencies in the neighborhood of physiological pH, but they are likely to be important contributors to the relaxation spectrum at high and low pH. If one accepts the premise that charge transfer between proximal acidic and basic residues occurs at megacycle frequencies, then internal ionization reactions involving histidyl–carboxyl pairs are likely to be significant sources of absorption at megacycle frequencies even near physiological pH. There is not at present a sufficient body of experimenta results on simple compounds to allow one to arrive at a kinetic model with any degree of confidence; the isomeric aminobenzoic acids have so far been the only model compounds which have proven amenable to investigation by ultrasonic techniques. Our results indicate that internal ionization reactions are a significant source, but not the sole source, of acoustic absorption in aqueous hemoglobin solutions.

References

1. E. L. Carstensen and H. P. Schwan, *J. Acoust. Soc. Amer.*, **31**, 305 (1959).
2. F. Schneider, F. Müller-Landau, and A. Mayer, *Biopolymers*, **8**, 537 (1969).
3. a) P. D. Edmonds, *Biochim. Biophys. Acta*, **63**, 216 (1962).
   b) M. Hussey and P. D. Edmonds, *J. Phys. Chem.*, **75**, 4012 (1971).
4. R. Zana and J. Lang, *J. Phys. Chem.*, **74**, 2735 (1970).
5. a) F. Dunn and L. W. Kessler, *J. Phys. Chem.*, **74**, 2736 (1970).
   b) L. W. Kessler and F. Dunn, *J. Phys. Chem.*, **73**, 4256 (1969).
   c) W. D. O'Brien, Jr., and F. Dunn, *J. Phys. Chem.*, **76**, 528 (1972).
6. R. D. White and L. J. Slutsky, *J. Col. Inter. Science*, **37**, 727 (1971).
7. S. A. Hawley and F. Dunn, *J. Chem. Phys.*, **50**, 3523 (1969).
8. J. Burke, G. Hammes, and T. Lewis, *J. Chem. Phys.*, **42**, 3520 (1965). G. G. Hammes and P. B. Roberts, *J. Am. Chem. Soc.*, **91**, 1812 (1969).
9. R. Parker, L. J. Slutsky, and K. Applegate, *J. Phys. Chem.*, **72**, 3177 (1968); **70**, 3018 (1966).
10. R. Zana and C. Tondre, *Biopolyjers*, **10**, 2635 (1971).
11. M. Eigen and L. de Maeyer, *Technique of Organic Chemistry*, Vol. VIII, Part II, A. Weissberger, Jr., Ed., John Wiley & Sons, Inc., New York, 1961.
12. M. Eigen and E. Eyring, *J. Am. Chem. Soc.*, **84**, 3254 (1962).
13. K. Applegate, L. J. Slutsky, and R. C. Parker, *J. Am. Chem. Soc.*, **90**, 6909 (1968); R. D. White, L. J. Slutsky, and S. Pattison, *J. Phys. Chem.*, **75**, 161 (1971).
14. R. D. White and L. J. Slutsky, *J. Phys. Chem.* (in press).
15. E. J. Cohn and J. T. Edsall, *Proteins, Amino Acids, and Peptides as Ions and Dipolar Ions*, Reinhold Publishing Corp., New York, 1943.
16. P. Debye, *Trans. Electrochem. Soc.*, **82**, 265 (1942).
17. M. V. Smoluchowski, *Z. Physik. Chem.*, **92**, 129 (1917).
18. W. H. Orttung, *J. Am. Chem. Soc.*, **91**, 162 (1969). *Biochem.*, **9**, 2394 (1970).
19. C. Tanford and J. G. Kirkwood, *J. Am. Chem. Soc.*, **79**, 5333 (1957).
20. A. Elis, *J. Chem. Soc.*, **1961**, 4678.
21. S. Hamman and S. Lim, *Austral. J. Chem.*, **1**, 329 (1934).
23. G. Castellan, *Ber. Bunsenges. Physik. Chem.*, **67**, 898 (1963).
24. P. R. Schimmel, *J. Chem. Phys.*, **54**, 4136 (1971).
25. R. D. White (to be published).

Part II
INTERACTION OF ULTRASOUND WITH BIOLOGICAL MEDIA IN SOLUTION AND IN SUSPENSION

In an attempt to understand the mechanisms of the interaction of ultrasound with biological materials, experimental studies have been conducted at various levels of biological complexity. Part II is devoted to the biophysical interaction studies of solutions of macromolecules and suspensions of microorganisms and cells. In such studies, a principal question deals with the necessity for the presence of cavitation to affect the biological end point. By experimental design, thermal mechanisms are generally minimized in these systems.

Cavitation is the general term used to describe the growth and subsequent behavior of cavities in an acoustically perturbed medium. It is useful to think in terms of two types of cavitation. The violent type, called *transient* or *collapse cavitation*, produces intense hydrodynamic forces within the vicinity of the collapsing bubble, which are capable of severely disrupting biological structures. Highly reactive free radicals can also be by-products of transient cavitation.

For the less violent type, called *stable cavitation*, the bubble (or cavity) does not collapse but rather grows to a resonant size and oscillates or pulsates under the influence of the ultrasonic field. The hydrodynamic forces in the vicinity of the oscillating bubble are considered responsible for affecting biological materials.

Two excellent reviews (Flynn, 1964, and Nyborg, 1965) have been prepared on the physics of cavitation and the reader is urged to consult them for further details of this phenomena.

REFERENCES

Flynn, H. G. 1964. "Physics of Acoustic Cavitation in Liquids." In W. P. Mason (ed.), *Physical Acoustics*, Vol. I, Part B. New York: Academic Press, pp. 57–172.

Nyborg, W. L. 1965. "Acoustic Streaming." In W. P. Mason (ed.), *Physical Acoustics*, Vol. II, Part B. New York: Academic Press, pp. 265–331.

Editors' Comments on Papers 19 Through 36

19 **ESCHE**
Excerpt from *Investigation of Vibratory Cavitation in Liquids*

20 **ELDER**
Cavitation Microstreaming

21 **DYER and NYBORG**
Ultrasonically-Induced Movements in Cells and Cell Models

22 **NYBORG and DYER**
Ultrasonically Induced Motions in Single Plant Cells

23 **DYER and NYBORG**
Characteristics of Intracellular Motion Induced by Ultrasound

24 **WILSON et al.**
Deformation and Motion Produced in Isolated Living Cells by Localized Ultrasonic Vibration

25 **RAVITZ and SCHNITZLER**
Morphological Changes Induced in the Frog Semitendinosus Muscle Fiber by Localized Ultrasound

26 **NYBORG**
Mechanisms for Nonthermal Effects of Sound

27 **HUGHES and NYBORG**
Cell Disruption by Ultrasound

28 **PRITCHARD et al.**
The Ultrasonic Degradation of Biological Macromolecules Under Conditions of Stable Cavitation: I. Theory, Methods, and Application to Deoxyribonucleic Acid

29 **ROONEY**
Hemolysis near an Ultrasonically Pulsating Gas Bubble

30 **WILLIAMS et al.**
Hemolysis in a Transversely Oscillating Wire

31 **ROONEY**
   Shear as a Mechanism for Sonically Induced Biological Effects

32 **MACLEOD and DUNN**
   Effects of Intense Noncavitating Ultrasound on Selected Enzymes

33 **COAKLEY and DUNN**
   Degradation of DNA in High-Intensity Focused Ultrasonic Fields at 1 MHz

34 **HRAZDIRA**
   Direct and Indirect Effect of Ultrasound on Bone Marrow Cell Suspension

35 **CLARKE and HILL**
   Physical and Chemical Aspects of Ultrasonic Disruption of Cells

36 **COAKLEY et al.**
   Quantitative Relationships Between Ultrasonic Cavitation and Effects upon Amoebae at 1 MHz

Paper 19 is an extract from an article by Esche that describes the first attempt to examine cavitation thresholds in mammalian blood and tissue near the megahertz frequency range. Until this time, such studies had been performed a decade or more lower in frequency. The portions of the paper not included deal with determination of the cavitation threshold as a function of frequency from 0 to 3.3 MHz in gassy and degassed water, the results of which are collected in Figure 13, which is included here.

Paper 20 descibes the type of streaming fields and forces that develop around stable cavitation bubbles under the varying conditions of sound pressure amplitude and kinematic viscosity.

Harvey and Loomis (1928) and Schmitt (1929) observed intracellular eddying motions when a vibrating needle was brought in contact with a biological cell wall. Papers 21 through 24 deal with these phenomena. Papers 21 and 22 describe studies in which a localized divergent sound source is brought into contact with the cell wall; the authors suggest that the intracellular motions are related to acoustic microstreaming and can be explained in terms of acoustic streaming theory.

In Paper 23, Dyer and Nyborg expand upon this work and apply acoustic streaming theory to explain two cases of eddying motion. In addition, they note that cytoplasmic particles appear to be attracted toward the sound source and suggest that radiation pressure theory may explain this attraction phenomenon.

Wilson et al., in Paper 24, apply a special case of Embleton's theory of the mean force acting on a rigid sphere (discussed in this paper) to explain the attraction phenomenon of intracellular particles toward the vibrating needle source. An order of magnitude calculation is within experimental observations. Additionally, it is pointed out that this experimental technique is useful in gaining insight into the physical state of cytoplasm.

Paper 25, by Ravitz and Schnitzler, represents the first application of the localized divergent sound source to intact tissue. It is shown that irreversible changes can occur in tissue in the absence of heat and transient cavitation.

In Paper 26, Nyborg details the biophysical action of a single bubble vibrating under the influence of an ultrasonic field.

In Paper 27, by Hughes and Nyborg, it is shown that the violent, collapse-type cavitation is not a necessary condition for ultrasound to damage cells in suspension and macromolecules in solution. They speculate that such damage may "result from the shearing action associated with bubble-induced eddying and relative motions."

In Paper 28, under experimental conditions similar to that of the previous paper, a solution of the macromolecule DNA is subjected to "stable cavitation." By applying the theory of acoustic streaming, the velocity gradient in the neighborhood of the bubble is calculated and a relationship is obtained between the number of breaks in the DNA molecule and the calculated velocity gradient.

Whereas the techniques employed in the previous two papers used multiple bubbles, in Paper 29 Rooney developed a system to examine the interaction phenomenon in fluid suspensions by means of a single pulsating bubble, and in Paper 30 Williams et al. developed a similar system that employed a transversely oscillating wire. Both applied their techniques to determine the critical shear stress of hemolysis. Their results, within experimental error, agree numerically, and they conclude that the mechanism is a shearing process.

The pulsating bubble and oscillating wire produce both a first-order oscillatory stress and a second-order dc stress. In Paper 31, Rooney shows that, although the oscillatory stress is almost 10 times greater than the dc stress, the latter is the primary mechanism of hemolysis. He also suggests a synergistic effect between the stress and heat as mechanisms for hemolysis.

Understanding of the biophysical interaction between ultrasound and biological polymers in solution has centered on the question of whether or not degradation occurs in the absence of cavitation. Hawley et al. (1963) degraded DNA (molecular weight approximately 10^7) by backbone scission in aqueous solution in the absence of cavitation and suggested the mechanism of relative motion between the solvent and the more dense molecule. In Paper 32, by Macleod and Dunn, it is concluded that ultrasonic denaturation of enzymes (molecular weight approximately 60,000) in solution occurs only in the presence of cavitation. However, in Paper 33, by Coakley and Dunn, it is shown that transient cavitation will degrade DNA in solution but, at ultrasonic levels where transient cavitation is not detectable, greater denaturation of the molecule occurs.

In the cases where cells in suspension were exposed to ultrasound, cellular damage was generally attributed to cavitation. However, the site of action, either directly on the cell or indirectly by affecting the suspending medium, remained uncertain. Hrazdira examined this question in Paper 34 and showed that there was, indeed, an indirect effect on cell survival. In Paper 35, Clarke and Hill, using another cell system, confirmed the indirect effect but concluded that it was not the primary cause of cell death. The latter study also showed that both cell death and ultrasonically altered medium were caused by the same mechanism, stable cavitation.

Paper 36, by Coakley et al., demonstrates a strong correlation between ultrasonically produced cavitation events and cell survival, an amoeba in this case. The paper also points out that the data do not rule out a noncavitational contribution to cell death. In that regard, Williams (1972) subjected the same type of amoeba to the transversely oscillating wire and showed that intracellular streaming could easily be induced.

REFERENCES

Harvey, E. N., and A. L. Loomis. 1928. "High Frequency Sound Waves of Small Intensity and Their Biological Effects." *Nature,* **121,** 622–624.

Hawley, S. A., R. M. Macleod, and F. Dunn. 1963. "Degradation of DNA by Intense Noncavitating Ultrasound." *J. Acoust. Soc. Amer.,* **35,** 1285–1287.

Schmitt, F. O. 1929. "Ultrasonic Micromanipulation." *Protoplasma,* **7,** 332–340.

Williams, A. R. 1972. "Disorganization and Disruption of Mammalian and Amoeboid Cells by Acoustic Microstreaming." *J. Acoust. Soc. Amer.,* **52,** 688–693.

19
INVESTIGATION OF VIBRATORY CAVITATION IN LIQUIDS

R. Esche

This excerpt was translated expressly for this Benchmark volume by Floyd Dunn, University of Illinois at Urbana–Champaign, from "Untersuchung der Schwingungskavitation in Flüssigkeiten," Akust. Beih., No. 4, 217–218 (1952)

d) Cavitation in Blood and the Question of the Occurrence of Cavitation in Animal Tissues

From the results shown in Figure 13, the value of the sound pressure can be read for a specific frequency below which cavitation cannot occur in water, as long as boundary surface effects are not involved.

Since the damaging effect of intense therapeutic ultrasonic radiation cannot be explained completely by mechanical or thermal effects, the tendency has been to ascribe the tissue-destructive action totally or partially to cavitation. To contribute to the clarification of this question, the sound intensities that produce cavitation in blood and in animal tissues at 500 kHz were investigated.

1. Owing to gas saturation, sufficient nuclei are present in blood; in

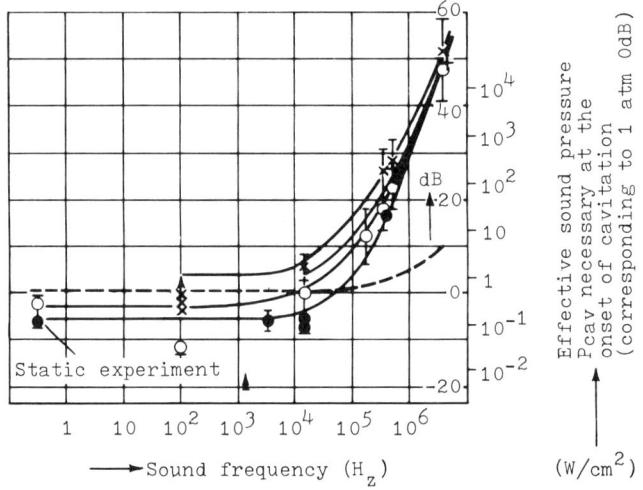

Figure 13 Frequency dependence of the onset of cavitation (parameter: condition of the liquid): x, distilled, filtered, and degassed water; +, distilled and filtered Mg_2SO_4 solution (0.017 mol/liter); o, fresh tap water; •, stale tap water, air supersaturated (20 percent) by heating. Reproduced from *Akust. Beih.*, **4**, AB217 (1952); copyright © 1952 by S. Hirzel Verlag.

particular cases these can be increased because of inadequate wetting of the erythrocytes. It has been observed, however, that cavitation in fresh cattle blood does not occur until the local peak intensity is 130–260 W/cm², or in other words at higher sound intensities than in tap water. Apparently the cavitation-inhibiting effect of the greater viscosity is more important than the cavitation-promoting effect of the high nuclei content. Qualitative investigations with gelatins of various viscosities confirm the results published by Briggs, Johnson, and Mason (2) that viscosity strongly reduces the ease with which cavitation can be produced.

2. In several experiments the demonstration of cavitation in animal tissues was attempted. Thick pieces of beef muscle tissue were placed in the experimental arrangement sketched in Figure 8* such that the region of greatest energy density was situated within the tissue. From measurements of the standing waves in front of the piece of tissue (in the plane sound beam, without the CCl_4 lens), it was confirmed that the reflection factor at the tissue surface was less than 10 percent.

In no case was cavitation demonstrated in the tissue for local peak intensities less than 1.7 kW/cm² at the focus. By taking into account the known absorption (3) and eventual scattering of the sound wave in the fibrous structure of the tissues, this maximum intensity value is reduced to about 400 W/cm². As a consequence of sound radiation pressure, small bubbles accumulated at the tissue surface, executed cavitation oscillations, and destroyed the tissue surface.

It appears, therefore, impossible that cavitation could appear in the blood or in the interior of tissue at the usual therapy frequency of 800 kHz and at the intensities employed, which are generally less than 5 W/cm². It is possible, of course, that degassing could occur, which could, in some cases, produce histological changes; however, the pressure pulses so emitted were so weak that the apparatus employed could not detect them.

LITERATURE

2. H. B. Briggs, I. B. Johnson, and W. P. Mason, Properties of liquids at high sound pressure. *J. Acoust. Soc. Amer.* **19,** 664 (1952).
3. R. Esche, Investigation of Ultrasonic Absorption in Animal Tissues and in Plastics, *Akust. Beih.*, **2,** 71 (1952).

* Figure 8 has been omitted, owing to limitations of space.

20

Copyright © 1959 by the Acoustical Society of America

Reprinted from *J. Acoust. Soc. Amer.*, **31**(1), 54–64 (1959)

Cavitation Microstreaming*

SAMUEL A. ELDER†
Brown University, Providence, Rhode Island
(Received July 21, 1958)

In the research reported here an attempt has been made to discover by experiment what physical assumptions and approximations are appropriate in the theory of cavitation microstreaming, especially for cavitation bubbles located near solid boundaries. A systematic investigation of the phenomenon has been made and its dependence on certain parameters (e.g., amplitude of sound) has been determined. The investigation has disclosed that as the sound amplitude is varied, other conditions remaining the same, the streaming changes discontinuously through several stable regimes. It appears that in order to account for the generation of vorticity one needs to assume different conditions at the boundaries for each regime. For at least one regime, a theoretical model due to Nyborg seems to be applicable; comparison is made with experimentally determined streaming velocities.

INTRODUCTION

IT is now generally accepted that gas and vapor bubbles play a major role in the physico-chemical action of high-amplitude sound on substances immersed in liquids.[1] There have been various attempts to account for the mechanism of the intermediary role of bubbles. Many have stressed the importance of the intense local pressures and temperatures which should be expected in and near a dynamically excited, highly compressible cavity.‡ Others[2–4] have given attention to effects such as shape-mode resonances and their associated violent mechanical deformations at the bubble surface. A recent Russian contribution[5] emphasizes the part played by surface energy phenomena when bubbles act to peel away surface films on solids.

Nyborg[6] has called attention to still another effect which may explain the action of bubbles in certain instances. This is the presence of small-scale acoustic streaming generated by bubble-scattered sound waves. An example of bubble streaming with practical interest was furnished by some experiments of Frings[7] and others at the Pennsylvania State University. They showed that bubble-associated turbulence can cause "sonic" destruction of paramecia (i.e., through simple mechanical buffeting). In quite another connection, Yeager[8] has noted that sonically treated electrodeposits show an improved quality when "gaseous cavitation bubbles" are present on the electrode surface, due to a depolarizing action produced by the bubbles. It would appear that the local motions of the electrolyte caused by streaming tend to dilute the ion cloud. It has been observed that this type of effect could be a mechanism for the acceleration of rate processes in general.[9]

---

* This work was supported in part by the Office of Naval Research and in part by the U. S. Air Force. The results were reported on at the 50th meeting of the Acoustical Society of America [J. Acoust. Soc. Am. **28**, 155(A) (1956)]; they are contained in a thesis submitted to the Graduate School of Brown University in partial fulfillment of requirements for the Ph.D. degree.
† Now at The Johns Hopkins University Applied Physics Laboratory, Silver Spring, Maryland.
[1] T. F. Hueter and R. H. Bolt, *Sonics* (John Wiley and Sons, Inc., New York, 1955), p. 242.
‡ E. Meyer, J. Acoust. Soc. Am. **29**, 4 (1957), has for example obtained Schlieren photographs of spherical shock waves produced by cavitation bubbles. V. Griffing, J. Chem. Phys. **20**, 939 (1952), has detected temperature-dependent chemical effects occurring in (gaseous) bubbles.
[2] M. Kornfeld and L. Suvorov, J. Appl. Phys. **15**, 495 (1944).
[3] E. Ackerman, Bull. Math. Biophys. **13**, 93 (1951).
[4] G. W. Willard, J. Acoust. Soc. Am. **26**, 933(A) (1954).

[5] Bebchok, Makarov, and Rosenberg, Soviet Phys.—Acoust. **2**, 111 (1956).
[6] Elder, Kolb, and Nyborg, Phys. Rev. **93**, 364(A) (1954).
[7] Ackerman, Reid, Kinsloe, and Frings, WADC Tech. Rept. 53–82.
[8] E. Yeager and F. Hovorka, J. Acoust. Soc. Am. **25**, 443 (1953).
[9] W. Nyborg, Amer. Inst. Bio. Sci. Symposium Proc. Ser., Vol. III (1955).

Earlier work by the author[10] and others has established, furthermore, that bubble-produced streaming is one of the important agents in sonic cleaning.

The present work is a continuation of earlier studies at Brown University by Kolb[11] and Nyborg. They found that (1) cavitation microstreaming is most pronounced for bubbles undergoing volume resonance, (2) that it is most readily observed for bubbles situated on solid boundaries, and (3) that it is an orderly process at low sound amplitudes. These facts led them to suggest that the streaming may be due to a viscous interaction of the bubble-scattered sound wave with nearby rigid boundaries. Nyborg[11,12] outlines a theoretical treatment of the problem (making use of Rayleigh's general perturbation approach) which he assumes might be applicable at low enough amplitude.

In the research about to be reported here an attempt has been made to discover what physical assumptions and approximations are appropriate in the theory of cavitation microstreaming. A systematic experimental investigation of the phenomenon has been made and its dependence on certain parameters (e.g., amplitude of sound) has been determined. The investigation has disclosed that as the sound amplitude is varied, other conditions remaining the same, the streaming changes discontinuously through several stable regimes. It appears that in order to account for the generation of vorticity one needs to assume different conditions at the boundaries for each regime. For at least one regime Nyborg's model seems to be applicable.

APPARATUS

Observations were made of the acoustic streaming which occurs near a vibrating air bubble in a viscous liquid. The bubble was situated at the bottom of a transparent-walled vessel which was filled with a liquid of known viscosity. The exposure vessel consisted of a rectangular open-end lucite box mounted firmly on a vibrating steel piston (Fig. 1). A magnetostrictive unit (a modified Raytheon Model DF-101 200 watt system) drove the piston at 10 kc. The exposure vessel was 2 in. wide and 1 in. deep. When filled with water it approximated a quarter-wavelength resonator for sound at 10 kc, having a pressure maximum at the bottom. (The vessel was actually operated somewhat less than full.) It was found that small bubbles could be made to adhere to the face of the piston. This seemed to be partly due to "radiation pressure" and partly due to the fact that the metal surface was sufficiently soiled§ to provide a large angle of contact. The bubbles are somewhat flattened due to this effect, often assuming a more nearly hemispherical than spherical shape.

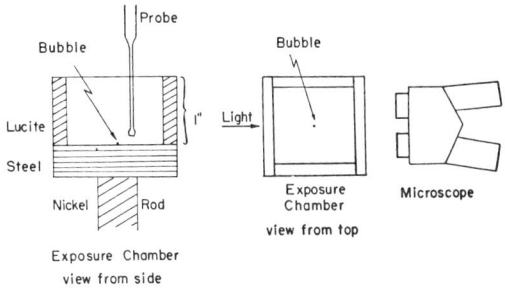

FIG. 1. Exposure chamber.

Water and glycol-water solutions of various proportions were the liquids used in these experiments. It was thus possible to vary the kinematic viscosity of the medium from 0.01 cm^2/sec to nearly 10 cm^2/sec at room temperature. For each solution, the procedure was to form a bubble at the center of the piston face, adjust the volume of air until the bubble was just smaller than resonant size, and then observe the streaming field as a function of driver amplitude.

In principle it should be possible to estimate how near a bubble is to resonance (at a given frequency) from its measured volume. Unfortunately the distorted shape of the bubble made the volume difficult to measure accurately. Alternatively another method was employed which makes use of the sudden phase changes and accompanying changes in local pressure field, which take place as either the volume or frequency vary about their resonant values.‖ (Phase proved a more reliable indication than amplitude, in this case, owing to the presence of a driving system resonance at 10 kc.)

Adjustments of the bubble size were made by means of an air-filled hypodermic needle (pressure control being provided by an eye-dropper bulb cemented to the base of the needle). Since the resonant diameter for air bubbles at 10 kc is about 0.06 cm, a microscope was necessary to observe the details of the field. A Bausch & Lomb SKW-5 stereomicroscope was used, observations being made from the side (Fig. 1).

A very dilute suspension of aluminum dust was employed to trace the fluid motion. With the field illuminated uniformly from the rear, aluminum particles show up as black specks on a white background.

The "first-order," or sound, field was estimated in two ways. Sound pressure measurements were made near the bubble by means of a small ($\frac{1}{16}$-in. diam) BaTiO$_3$ probe. Where possible, acoustic particle displacement amplitude at the bubble surface was obtained independently by measurement with a calibrated microscope reticule. Since acoustic pressure and particle displacement are theoretically related, a check on the experimental accuracy was possible for situations where both kinds of measurements could

---

[10] Elder, Kolb, and Nyborg, J. Acoust. Soc. Am. **26**, 933(A) (1954).
[11] J. Kolb and W. Nyborg, J. Acoust. Soc. Am. **28**, 1237 (1956).
[12] W. Nyborg, J. Acoust. Soc. Am. **30**, 329 (1958).
§ Although the surface was reasonably clean, no attempt was made to secure extreme chemical purity.

‖ Details of the method are given in the author's Ph.D. thesis [S. A. Elder, Dissertation Abstr. **17**, 159 (1957)].

be made. The discrepancy between the measurements was found to be of the order of 20% or less.

ASSUMPTIONS MADE IN ESTIMATING THE SOUND FIELD NEAR THE BUBBLE

Since the streaming motions were concentrated in the region very near the bubble, an attempt was made to estimate the strength of the sound field at the bubble surface. Where the bubble vibration amplitude was large enough to permit visual surface displacement measurements, the local acoustic particle velocity distribution could be estimated with fair accuracy (i.e., the surface velocity amplitude is given by the product of the displacement amplitude and the angular frequency).

Particle velocity estimates obtained from pressure probe data, on the other hand, were necessarily beclouded by difficulties of interpretation. This was because the relatively large size of the acoustic probe (1.59 mm diam as compared with the bubble diameter of 0.5 mm) made it necessary to gauge the local amplitude by inference from measured field values many bubble-diameters away. The following assumptions were made in extrapolating the sound pressure to the bubble surface:

(a) The "incident" wave radiated from the piston was assumed to be spatially constant in phase, with an amplitude varying only with height above the piston.

(b) The spherical wave scattered from the bubble was assumed to reflect from the piston boundary as from an infinite rigid plane in an infinite uniform medium.

(c) For bubbles slightly below resonant size (all the bubbles discussed below fall into this category) the scattered and incident waves were assumed to be in phase. (For an undamped bubble in a free field this would be very nearly true.)

Based on the above model the rms pressure amplitude sensed by the probe at a height h above the boundary and at a distance r from the bubble center would be

$$p(r,h) = p_0(h) + (a/r)p_s, \quad (1)$$

where p_0 represents the incident wave and p_s the scattered wave at the surface ($r=a$) of the bubble.¶ Thus if values of $p(r,h)$ were to be plotted as a function of $1/r$, holding h constant, one would expect to obtain a straight line of slope ap_s passing through the intercept $p(\infty,h) = p_0(h)$. The peak surface velocity, u_s, would then be equal to $\sqrt{2}(\omega\rho a)^{-1}p_s$ cm/sec, where ρ is the liquid density, and where p_s is given in microbars, rms.

In Fig. 2 where some actual data have been plotted this way, it may be seen that a fairly good straight-line portion does exist over the mid-range points. From the slope of the straight line portion we may estimate the rms scattered pressure amplitude at the bubble surface to be about 0.3 atmos, which corresponds to a surface particle velocity of about 190 cm/sec. This is in fairly good agreement with the value 160 cm/sec obtained by surface displacement measurements of the same bubble.

Extrapolation of the straight line portion in Fig. 2 to $1/r=0$ yields an amplitude for the exciting pressure wave of about 0.023 atmos.** This corresponds to a (plane wave) particle velocity of about 0.08 cm/sec. The contribution of the incident field to the particle velocity distribution near the bubble is therefore quite negligible. (This would still be the case even if the incident *pressure* amplitude were comparable to the scattered pressure.) Thus the strongly divergent character of the scattered field makes the piston appear rigid in comparison with the bubble surface.

Having seen that the extrapolation method actually "works," (i.e., gives values of acoustic particle velocity in fair agreement with those obtained by an independent method) let us examine the theoretical basis of Eq. (1) more closely. The approximate truth of assumption (a) has indeed been demonstrated by measurements made in the chamber with the bubble absent. Assumptions (b) and (c) depend for their justification on the circumstance that the acoustic wavelength involved was large compared with the scale of measurements and that the measurements were made close to the spherical source. In addition (c) supposes that there is an essential similarity between the behavior of a free bubble and one forced to vibrate in a confined space.

The dynamics of a free bubble are discussed in Appendix II, using formulae due to Rosenberg.[13] It appears that p_s and p_0 can usually be considered to be in phase if the bubble radius is less than 90% of resonant

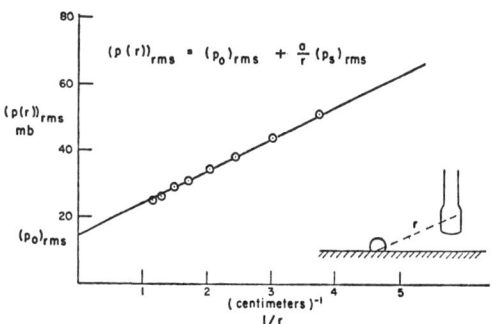

FIG. 2. Determination of p_s by extrapolation method. The scattered pressure amplitude at the bubble surface is proportional to the slope of the curve.

¶ A more exact expression, including the phase angle between p_0 and p_s, is given in Appendix I. The error incurred in neglecting this angle is there estimated.
** This figure has also been corrected for the variation of p_0 with h.
[13] M. Rosenberg, "Pulsations and growth of gas-filled bubbles in sound fields," Tech. Memo. No. 25, Div. of Appl. Sci., Harvard University (1952).

size. Furthermore, it is shown that for a bubble of given size, the phase error in Eq. (1) is always reduced when the damping constant is decreased. A similar law undoubtedly holds for bubbles confined on a rigid boundary. The exact relation between damping constant and phase for this case is difficult to derive on account of the geometry of the problem. From the measurements themselves, however, an upper bound on the damping constant can be obtained. The quantity (p_0/p_s) evaluated at the bubble surface has a minimum value for resonant-sized bubbles, the minimum value being equal to the damping constant (see Appendix III). The fact that reasonably small (below-resonance) values of this quantity were obtained for all but the most viscous cases is a good check on the internal consistency of the method.

RESULTS

Effects of Viscosity on the First-Order Field

Table I summarizes the data on the "damping ratio" p_0/p_s for various viscosities. The selected data in the second column are smallest measured values of the damping ratio at each viscosity. It should be emphasized

TABLE I. Damping ratio *versus* viscosity.

Kinematic viscosity of liquid	Minimum value of p_0/p_s	Free bubble damping constant
0.04 cm²/sec	0.032	0.062
0.07	0.044	0.064
0.16	0.054	0.068
0.40	0.095	0.097
1.34	0.190	0.160

that these values are for bubbles smaller than resonant size, and are therefore not equal to the damping constant itself. They represent, rather, an upper bound for the damping constant, since, as is shown in Appendix III, the resonance value of the damping ratio is its minimum value. In the third column the theoretical damping constant for a free bubble resonating at the same frequency is given for comparison. It will be noted immediately that the damping ratio for small viscosities is less than the damping constant of a free bubble. Physically this is due to the fact that a bubble located on a rigid boundary is actually a better scatterer (by a factor of two, approximately, assuming perfect reflection) than a free bubble of the same volume, compressed by the same amount. However it may be inferred from the trend of the data in Table I that the rate of increase of damping constant with viscosity for a bound bubble is greater than that for a free bubble. This excess damping is probably due to viscous drag in the so-called "acoustic boundary layer."

The acoustic boundary layer is the region near the boundary over which the tangential component of acoustic particle velocity drops to zero. Its thickness is of the order $(2\nu/\omega)^{\frac{1}{2}}$, where ν is the kinematic shear

TABLE II. Comparison of measured and theoretical boundary layer thickness.

Kinematic viscosity of liquid	Apparent boundary layer thickness	Theoretical boundary layer thickness
1.34 cm²/sec	0.0084 cm	0.0076 cm
7.48	0.0170	0.0155

viscosity coefficient.[14] At large viscosities the boundary layer is thick enough to have a visible effect upon the peak-to-peak displacement of the bubble surface around the base of the bubble. This effect makes it possible to measure the boundary layer thickness directly with a microscope reticule. Table II gives a comparison of the measured and theoretical values of boundary layer thickness for the liquids of largest viscosity. Considering the subjective problem involved in visually estimating the edge of a region which is not physically well defined, the agreement in Table II is surprisingly good.

Changes in the First-Order Velocity Field with Amplitude

Ordinarily the measured excess pressure and particle velocity of the sound near the bubble tend to have the same spatial symmetry. For certain conditions of amplitude and volume, however, it is possible for a vibrating bubble to lock into a surface or "capillary" mode of vibration in which the local velocity distribution at the bubble surface becomes comparatively complicated.[2–4,15,16] This is usually not accompanied by any noticeable changes in the pressure field, since the volume compressions associated with the surface instability are small compared with the over-all volume pulsation of the bubble and fall off very rapidly with distance away from the bubble surface.[17] As will be shown later, however, the local streaming field was found to be markedly affected by this first-order velocity distribution.

Another situation where differences in the symmetry of the observed velocity and pressure distributions are significant is the case of a bubble small enough to possess "surface viscosity."[18] A bubble with surface viscosity has a thin flexible skin which retards the free tangential flow of liquid past its surface. The skin is thought to be composed of certain organic materials.[19] It appears that many bubbles of diameter less than 0.02 cm tend to be naturally contaminated in this way.

---

[14] Reference 12, p. 1242.
[15] M. Strasberg and T. B. Benjamin, Paper U3, delivered at the 55th meeting of the Acoust. Soc. Am., May 10, 1958.
[16] T. B. Benjamin and M. Strasberg, Paper U4, delivered at the 55th meeting of the Acoust. Soc. Am., May 10, 1958.
[17] M. Strasberg, J. Acoust. Soc. Am. 28, 20 (1953).
[18] C. L. Pekeris, Columbia University DWR, OSRD 976, Sec. No. C4-sr20-326 (1942).
[19] K. F. Herzfeld and F. E. Fox, J. Acoust. Soc. Am. 26, 984 (1954).

For larger bubbles, similar effects may be obtained if certain organic substances are deposited on the bubble. This phenomenon has been mentioned in the literature as a possible explanation for the anomalous rate of rise[18] and resonant frequency[19,20] of small bubbles.

For a bubble oscillating to and fro, a surface skin would have the effect of introducing an additional acoustic boundary layer due to surface drag. As will be discussed later, characteristic streaming can result from such a velocity field. Some of the bubbles studied in the present work seemed to show this effect. It turned out to be a low-amplitude phenomenon; a critical threshold vibration amplitude was found at which the surface skin was broken up.

Streaming Fields

Introduction

Observations of microstreaming near excited bubbles were made in a number of different glycerin-and-water solutions, having kinematic viscosities varying from 0.01 cm^2/sec to 7.48 cm^2/sec. For purposes of discussion, the liquids will be considered in three groups: (1) low viscosity $\nu=0.01-0.05$ cm^2/sec, (2) medium viscosity $\nu=0.05-0.50$, (3) high viscosity $\nu=0.50-7.48$. For each group the streaming patterns are described as a function of sound amplitude. Where it is necessary to specify position in the field, (x,y) coordinates are used, the y axis being positive upward through the center of the bubble and the x axis being positive to the right.

Low-Viscosity Patterns

Figures 3(a) through 3(d) show a typical sequence of microstreaming patterns obtained as the driving amplitude is increased. The data used for this series are from a bubble in pure water ($\nu=0.01$ cm^2/sec). The maximum horizontal and vertical dimensions of the bubble were 0.048 cm and 0.040 cm, respectively.

At 4.4 millibars of incident sound pressure, the situation pictured in Fig. 3(a), occasional random motions of the tracer particles took place near the bubble, but nothing which could be classed as "steady-state" streaming occurred. The amplitude of the scattered wave pressure at the top of the bubble was estimated to be 19 mb, corresponding to a peak surface velocity of 11 cm/sec.

As the driving amplitude was increased conditions remained unchanged until suddenly, at an incident pressure of 8.6 mb, the bubble was seized with the surface mode vibration shown in Fig. 3(b). Simultaneously streaming motions became visible near the top of the bubble. The greatest acceleration of the streaming particles occurred very close to the two uppermost antinodes of the surface mode. The mean peak surface velocity was estimated to be about 22 cm/sec. Surface velocity is, of course, proportional to incident field amplitude for a given bubble, so long as the damping constant of the bubble does not vary (as, for example, by growth of the bubble).

When the driving amplitude had been raised until the surface velocity was 31 cm/sec, a new surface mode suddenly replaced the old one and the streaming direction reversed [Fig. 3(c)]. Comparison of Figs. 3(b) and 3(c) suggests that the direction of this type of

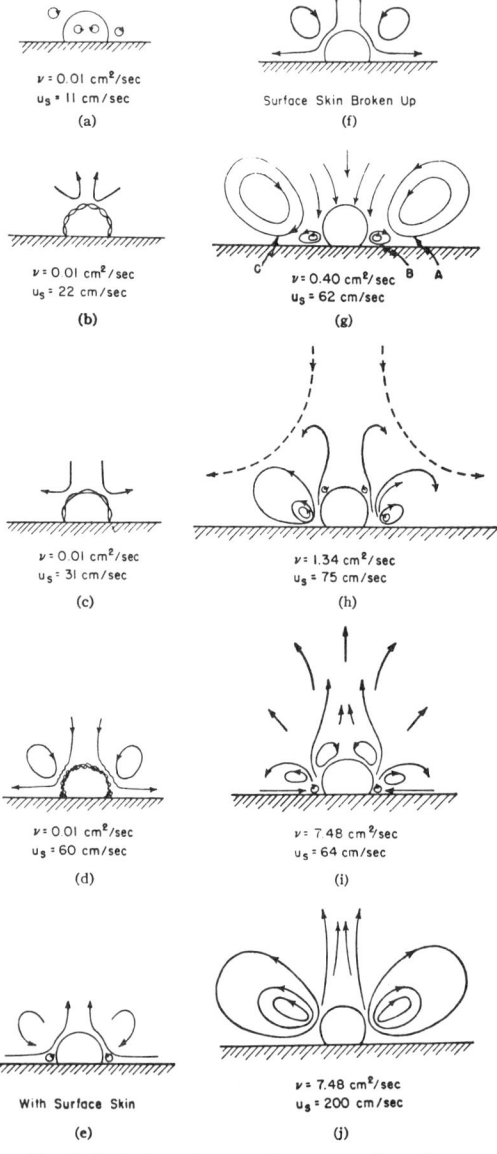

FIG. 3. Cavitation microstreaming patterns for various amplitudes and viscosities.

---

[20] M. Exner and W. Hampe, Acustica 3, 67 (1953).

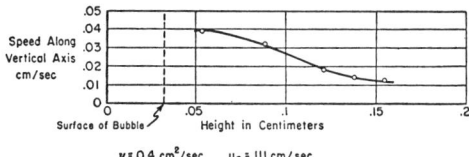

FIG. 4. Axial streaming velocity in Fig. 3(g).

streaming is such as to move liquid away from the nodes and toward the antinodes.

When the surface velocity had reached a value somewhat over 60 cm/sec, the stable surface mode dissolved into a chaotic surface agitation, the streaming pattern taking the form of a large vortex ring surrounding the bubble [Fig. 3(d)]. This mode and its associated streaming seems to represent the large amplitude limit for low-viscosity liquids. Further increases in driving amplitude caused no new effects other than to accelerate the rate of growth of the bubble by "rectified diffusion."[21] The increased surface area associated with the rough surface in the chaotic mode probably accounts for this.

In addition to the phenomena described above, which may be regarded as the "normal" low-viscosity streaming, an anomalous effect was found to occur both at low and medium viscosity whenever the bubble surface became contaminated. Ordinarily it was possible to keep the bubble surface reasonably clear of debris. When small amounts of common household detergent were added to the liquid, however, a thin surface film would form which caused the bubble to collect dust, tracer particles, etc., from the liquid. Under these circumstances the streaming at low amplitude consisted of a large and a small vortex ring surrounding the bubble as shown in Fig. 3(e). The upper ring rotated in such a way as to move liquid outward along the vertical axis. The lower ring rotated in the opposite sense, and was contracted into a narrow region near the base of the bubble. Although the tracer particles a short distance out from the surface were observed to be moving those actually on the bubble surface remained stationary, indicating the existence of a thin boundary layer region.

As driving amplitude was increased the streaming would speed up until, at some critical threshold amplitude, the surface skin would break up, destroying the boundary layer effect. At the same instant the streaming direction of the upper vortex would reverse while the lower vortex would apparently disappear altogether. The modified pattern is shown in Fig. 3(f). At still higher amplitudes the streaming would become more rapid, but the same pattern would be maintained. If the driving amplitude were now returned to its original value, neither the static surface condition nor the original streaming pattern would be restored for several minutes. In time, however, a new skin formed and the process could be repeated.

Medium-Viscosity Patterns

The predominant streaming regime for medium and high viscosities is shown in Fig. 3(g). The data from which this particular figure was obtained are for a bubble whose maximum horizontal and vertical dimensions were 0.051 cm and 0.034 cm, respectively. The liquid had a kinematic viscosity of 0.40 cm²/sec. This regime required somewhat larger driving amplitudes than those previously described, the threshold surface velocity for visible streaming motion being around 50 or 60 cm/sec. The pattern is somewhat similar to that shown in Fig. 3(e) except that the two vortex rings rotate in opposite senses.

At low driving amplitude a jet of upward streaming near the top of the bubble (not shown) occasionally disturbed the steady configuration. With increasing amplitude the pattern became more stable. Measurements were made of the streaming velocity along the vertical axis. A typical plot of streaming speed *versus* height is given in Fig. 4. In addition measurements of the horizontal streaming velocity at points A, B, and C [Fig. 3(g)] were made at various driving amplitudes. These results are summarized in Table III.

This regime, which we shall refer to later as regime II, proved stable over a wide range of amplitudes. The particular bubble shown in Fig. 3(g) maintained regime II streaming up to a surface velocity of 360 cm/sec. At that point a surface mode appeared, accompanied by the usual modifications in streaming pattern, the two vortex rings giving way to a vigorous outward circulation near the top of the bubble. At higher amplitude the bubble broke loose and disappeared. At the moment of its exit the bubble was generating a scattered sound pressure at its surface of about one atmosphere.

For the medium-viscosity liquids, in general, surface modes of vibration were not as readily excited as in

---

[21] F. G. Blake, Jr., ORN Tech. Memo., Nos. 9, 11, 12, Div. Appl. Sci., Harvard University (1949).

TABLE III. Measured horizontal streaming velocities.

Location	Streaming velocity	Surface velocity Estimated from pressure probe measurements	Estimated from surface displacement
$x=0.076$ cm $y=0.0084$	0.0035 cm/sec	62 cm/sec	
$x=0.042$ $y=0.003$	0.05	62	
$x=0.057$ $y=0.013$	0.0084	94	90 cm/sec
$x=-0.054$ $y=0.013$	0.024	102	90
$x=-0.071$ $y=0.013$	0.047	149	111

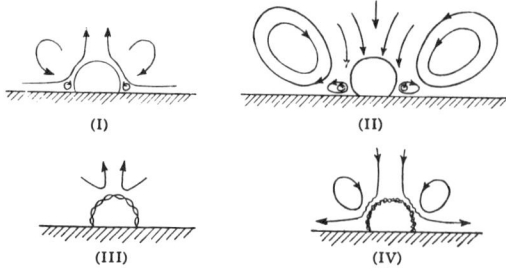

FIG. 5. Four regimes of streaming.

the low-viscosity case. The chaotic mode was not observed at all for kinematic viscosities greater than 0.07 cm²/sec.

High-Viscosity Patterns

The largest kinematic viscosity at which a surface mode was observed was $\nu = 1.34$ cm²/sec. The excitation of the mode required a surface velocity amplitude of 525 cm/sec in this case, corresponding to a scattered pressure of 1.2 atmos (rms). In general, surface modes were not excited at high viscosity, and the predominant streaming pattern was of the regime II type shown in Fig. 3(g). In fact regime II seems to be the limiting form for high amplitude and high viscosity.

Figures 3(h) and 3(i) show how this regime develops as the viscosity is increased. There are two noteworthy trends:

(1) The whole pattern expands, moving away from the boundary. In the extreme case of $\nu = 7.48$ cm²/sec [Fig. 3(i)] the lower ring has come to dominate the entire field, the upper ring having become so far removed from the vorticity sources that it is too weak to be observed.

(2) Additional vortex rings appear close to the bubble. These new rings also participate in the expansion of the pattern with increasing viscosity. The new rings tend to be squeezed out with increasing amplitude, however.

Figure 3(j) shows the limiting pattern obtained at high viscosity and amplitude. At first sight this pattern seems quite different from the pattern of 3(g), the predominant vortex rings in the two cases having opposite senses of rotation. Nevertheless, they may still be said to belong to the same family or regime in the sense that there are no discontinuous steps in the development from one to the other.

DISCUSSION AND CONCLUSIONS

Several Regimes of Streaming

The streaming fields consist, on the whole, of orderly patterns, often with symmetry about a vertical axis through the center of the bubble. The patterns depend, for a given frequency, on the viscosity of the liquid and on the amplitude of the bubble vibration.

Four different types of bubble-associated streaming may be distinguished in the data. These are represented by patterns I, II, III, and IV in Fig. 5. The first type or regime is that observed near a surface-contaminated bubble in a liquid of low viscosity. It is a low-amplitude effect. The second was observed over a wide range of amplitudes and viscosities. The third regime was observed most readily in liquids of low viscosity though it proved possible to excite it at higher viscosities with sufficiently large driving amplitudes. Its occurrence coincided with the onset of the first surface mode and the dissolution of the second regime at higher amplitude. Finally, the fourth regime, which occurred only in the least viscous solutions, seems to represent a return to the pattern of regime II, as the amplitude becomes too large to permit the existence of a single stable surface mode. The relationships between the regimes can perhaps best be understood by a sketch of the type given in Fig. 6 which shows the threshold lines plotted in the (ν, u_s) plane. (Regime I thresholds were not sufficiently well defined to be indicated here.)

That there should be more than one variety of bubble-associated streaming was not at all anticipated when the investigation was originally undertaken. At that time a single vortex ring rotating in the sense of the upper ring in regimes II and IV was expected from rough theoretical considerations. Consequently the first few observations gave a rather confusing picture of the phenomenon. Sometimes the streaming direction was one way and sometimes the other, with no rhyme or reason for the change. The underlying regularities of the patterns were finally disclosed by making a systematic study of the streaming as a function of amplitude and viscosity, i.e., the study reported in this paper. It then became apparent that the occurrence of

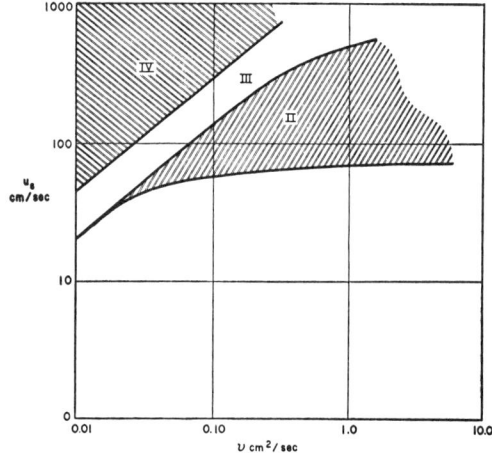

FIG. 6. Approximate threshold lines for regimes II, III, and IV.

"anomalous" types of streaming was associated with the presence of first-order effects not included in the simple theory. In particular it was found that in order to be able to predict the type of streaming, it is necessary to specify the surface condition of the bubble.

Source Mechanisms for Streaming

Regime I

We have already suggested that regime I streaming seems to be associated with the presence of an acoustic boundary layer on the bubble surface. Now it is interesting that this is the sort of situation which might be expected to give rise to streaming somewhat like that observed. Lane[22] and Thrasher[23] have studied the vortices generated by the presence of a small rigid sphere in a Kundt's tube. They find a pattern of four vortex rings symmetrical about a plane through the center of the sphere perpendicular to the axis of the tube. A typical configuration is sketched in Fig. 7. It may be seen that Lane's "outer vortex" ring is similar to the upper vortex ring in Fig. 5(a). The lower ring in Fig. 5(a) would seem to represent a distortion (due to the piston) of the lower outer vortex ring in Fig. 7.†† An "inner vortex" ring was not observed for the bubble. A rough calculation from Lane's data indicates that if a surface boundary layer was responsible for the bubble streaming, the width of the inner ring should have been large enough to detect; perhaps it escaped observation because tracer particles were not judiciously placed to bring it out. The possibility of the existence of a surface boundary layer is interesting since it ties in with the anomalous rate of rise[18] and resonant frequency[19,20] of small bubbles. As a matter of fact, the appearance of streaming under the application of a sound field might prove a useful test to determine whether or not a small bubble possesses a surface skin.

Regime II

Regime II received the greatest emphasis in the investigation because it was felt that this type of streaming held the most promise for theoretical treatment. For low enough amplitudes, the "vorticity" R_2 (i.e., the curl of the streaming velocity) should obey the usual second-order perturbation equation[24]

$$\nabla^2 \mathbf{R}_2 = \rho\mu^{-1} \nabla \times \nabla \cdot \langle \mathbf{u}_s \mathbf{u}_s \rangle, \quad (2)$$

---

[22] C. Lane, J. Acoust. Soc. Am. 27, 1082 (1955).
[23] R. Thrasher, J. Acoust. Soc. Am. 28, 155(A) (1956).
†† It might be argued that the presence of this piston wall invalidates the comparison between the situation described here and that of an oscillating sphere. However, an experiment performed by Dr. C. A. Lane and the writer, using a solid hemisphere oscillating near a plane boundary, confirms the intuitive feeling that the pattern on one side of the sphere is not greatly influenced by what is happening on the other side. Also a measurement of the width of the inner ring (or "dc boundary layer") showed satisfactory agreement with that to be expected for an isolated sphere under the same conditions of viscosity, frequency, liquid density, and sphere radius.
[24] P. J. Westervelt, J. Acoust. Soc. Am. 25, 63 (1953), Eq. (15).

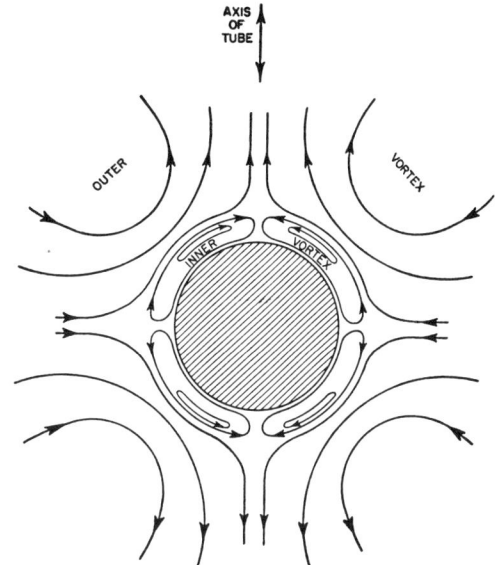

FIG. 7. Streaming due to a small rigid sphere in a Kundt's tube (sphere radius $= 17\delta_{ac}$).

where $\mathbf{u}_s$ represents the spherical wave first-order solution. One may assume for the first-order field‡‡ that

$$u_s(x,y,z) = \frac{A}{(x^2+y^2+z^2)} \times [\cos\omega t - \exp(-y/\delta_{ac})\cos(\omega t - y/\delta_{ac})]. \quad (3)$$

The flow u_s may be interpreted as the linear superposition of an ordinary inviscid longitudinal wave and a transverse "shear" wave generated by drag at the boundary. The shear wave is rapidly attenuated giving rise to a thin "boundary layer" of approximate width δ_{ac} over which the velocity falls from essentially inviscid amplitudes down to zero. When Eq. (3) is substituted into Eq. (2), using values of u_s in the $x-y$ plane, one obtains

$$\nabla^2 \mathbf{R}_2 = \mathbf{k} 2^{\frac{1}{2}} \rho\mu^{-1} \delta_{ac}^{-1} A^2 (x^2+y^2)^{-\frac{5}{2}} \times [\exp(-y/\delta_{ac})\sin(y/\delta_{ac}+\pi/4) \\ - 2^{-\frac{1}{2}} \exp(-2y/\delta_{ac})], \quad (4)$$

where $\mathbf{k} = \mathbf{i} \times \mathbf{j}$; $\mathbf{i}$, $\mathbf{j}$, and $\mathbf{k}$, being unit vectors in the direction of the x, y, and z axes, respectively. By analogy with the vector potential of electromagnetic theory, $\mathbf{R}_2$ is seen to be parallel to the negative of the "source" vector on the right in Eqs. (2) and (4). A plot of the source vector amplitude of Eq. (4) is given in Fig. 8.

Note (Fig. 8) that for positive x, the vorticity near the boundary is clockwise. This agrees with the sense

---

‡‡ Compare Eqs. (12) and (40), reference 12.

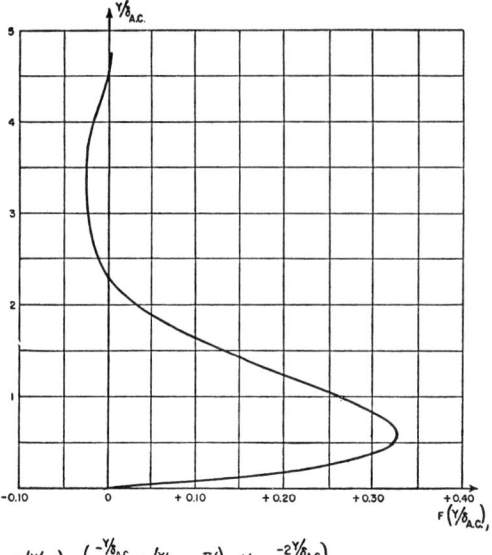

$$F(y/\delta_{A.C.}) = \left\{ e^{-y/\delta_{A.C.}} \sin(y/\delta_{A.C.} + \pi/4) - \frac{1}{\sqrt{2}} e^{-2y/\delta_{A.C.}} \right\}$$

FIG. 8. Plot of the theoretical vorticity source for regime II streaming [Eq. (4)].

of rotation of the so-called "lower ring" in Figs. 3(g) through 3(i). The crossover from clockwise to counterclockwise vorticity at $y=2.3\delta_{ac}$ evidently corresponds to the jet layer region between the lower and upper vortex rings. This checks roughly with data of Fig. 3(g) for which the mean height of the jet above the boundary was found to be about three times δ_{ac}. The possibility of an alternating sign for the vorticity at various heights above the boundary appears to arise from the presence of the phase term in Eq. (3) associated with the shear wave.

Nyborg[12] derives an approximate expression for the streaming velocity in the jet layer, taking advantage of the fact that the vorticity sources are mostly concentrated in the region within one or two boundary layer thicknesses from the plane (i.e., he ignores the "outer ring" sources). His order-of-magnitude formula for the limiting jet velocity, u_L, resulting from the first-order field produced by a hemispherical bubble of radius a_1 with surface velocity amplitude u_s is

$$u_L = \omega^{-1} u_s^2 a_1^4 x^{-5}. \quad (5)$$

Table IV summarizes the measured values of the jet velocity, calculated values of u_L§§ being given for comparison. The effective radius a_1 is taken to be one-half the maximum horizontal diameter of the bubble, while x is the mean horizontal coordinate of the timed path interval. On account of the finite length of the interval the apparent velocity should differ from the actual velocity at x by about a factor of two, assuming the velocity to vary as x^{-5}. This correction has been taken into account. As expected, there is order-of-magnitude agreement between theory and experiment.

Many situations in which microstreaming could be important as a mechanism for sonic effects occur at ultrasonic frequencies. The order-of-magnitude validity of Eq. (5) having been established, it is useful in giving some idea of the way in which streaming velocity varies with frequency. Taking the maximum value of u_L, namely $u_L = u_s^2/\omega a_1$, we find that the frequency dependency of u_L is due almost entirely to that of u_s, for the product (ωa_1) is essentially independent of frequency for bubbles near resonant size. At any given frequency u_s is limited by the maximum pressure $p_s = u_s/\rho\omega a$ which can be supported by the liquid without cavitation. At 10 kc (and at 1 atmos ambient pressure) the maximum value of p_s is about 1 atmos, corresponding to a particle velocity $u_s \sim 600$ cm/sec.|||| Thus the maximum streaming speed at 10 kc should be of the order $u_L \sim 200$ cm/sec. Since the cavitation threshold increases with frequency[25] it should be possible to obtain larger streaming speeds at higher frequency, independent of viscosity. (Correspondingly larger viscosities would be necessary, however, to maintain the streaming in regime II.)

Nevertheless, bubble-induced streaming would seem to be less important at higher frequencies. Since the damping constant is expected to increase with frequency, larger and larger incident fields should be required to obtain the same value of u_s as the frequency is increased. For very strong incident fields at very high frequencies, streaming due to objects other than bubbles should become competitive.

At the other extreme of very low frequencies or alternatively at very large viscosities, the boundary layer should become so diffuse that Nyborg's equations

TABLE IV. Comparison of experimental and theoretical values for streaming velocity.

Kinematic viscosity	Location	Observed streaming velocity	u_L (theoretical)
0.07 cm²/sec	$x=0.048$ cm $y=0.007$	0.061 cm/sec	0.30 cm/sec
0.40	$x=0.076$ $y=0.008$	0.0065	0.023
0.40	$x=0.057$ $y=0.013$	0.015	0.095
0.40	$x=-0.054$ $y=0.013$	0.043	0.14
0.40	$x=-0.071$ $y=0.013$	0.075	0.054

§§ Values of u_s calculated from pressure measurements were used for the first two points and from surface displacement measurements (since they were available and gave better agreement) for the last three.

|||| Round numbers are given here. The proportionality constant relating pressure and velocity (i.e., the bubble impedance) varies slightly from one bubble to another.

[25] R. Esche, Akust. Beih. 4, 208 (1952).

cease to apply. We have already noted the way in which the vorticity distribution spreads out with increasing viscosity [see Figs. 3(g) through 3(i)]. Over the region of applicability of Eq. (4) the height of the jet layer should be proportional to δ_{ac} (neglecting effects due to the boundary condition on the second-order field). In Table V are given values of D/δ_{ac} for various viscosities, the quantity D being the upper bound of the jet layer. It may be seen that the rough proportionality suddenly breaks down for viscosities somewhere between 1.34 cm²/sec and 7.48 cm²/sec.

The small counterclockwise (for positive x) vortices near the top of the bubble in Figs. 3(h) and 3(i) and near the bottom in Fig. 3(i) are as yet unaccounted for. This is not surprising since the region very close to the bubble is outside the range of validity of Eq. (4). [Eqs. (3) and (4) are derived on the assumption that $y \ll x$.]

Regimes III and IV

Strasberg[15] has independently observed the suddenness of the onset of surface modes. He reports that the phenomenon can be accounted for by a theory of surface mode-radial mode coupling. He finds that for low-amplitude volume pulsations, a bubble is dynamically stable against the excitation of shape oscillations. For any given size bubble at a given frequency, however, there is a critical amplitude at which a particular surface mode is violently excited, the amplitude of the excitation being limited only by nonlinear damping effects. The threshold for instability increases with the amount of (linear) damping present.

The source mechanism for regime III streaming is not completely clear at the present time. It would be interesting to explore regime III streaming in more detail, particularly to see whether or not these circulations are important in explaining the mobility of free cavitation bubbles.¶¶

Regime IV appears to be a return to regime II streaming. It is true that a lower vortex ring was not observed in regime IV but this can be explained: In the first place, the proximity of the turbulent bubble surface to the lower ring could have hindered the steady-state condition from being realized in addition to hindering observations of that region. Secondly, the speed of the tracer particles in the lower vortex was probably too rapid to be followed by eye in most cases. We may reason, then, that in the surface chaos the effects of surface modes are averaged out, so that the streaming not too close to the bubble surface is the same as though the bubble were smooth.

Regimes II and IV are probably the most important for sonic effects, since they seem to represent the limiting streaming configurations for large driving

¶¶ Earlier workers have commented on the "dancing" motions of cavitation bubbles, viz., N. Gaines, Physics 3, 209 (1932); also, reference 2.

TABLE V. D/δ_{ac} As a function of viscosity.

Kinematic viscosity	D/δ_{ac}
0.07 cm²/sec	4.5
0.16	4.0
0.40	3.6
1.34	5.3
7.48	∞

amplitudes at high and low viscosities, respectively.***
Jet layer streaming speeds for regime IV were too fast to be measured by eye. We have already estimated from Nyborg's formula, however, that the maximum jet layer speed should be about 200 cm/sec at 10 kc. For an air bubble in water the corresponding velocity gradient at the boundary would be $200/2.3\delta_{ac} = 1.5 \times 10^5$ cm sec⁻¹ per centimeter (i.e., a gradient equivalent to a Mach 1 change in velocity over one centimeter). It is easy to see why cavitation microstreaming is effective in surface cleaning.

ACKNOWLEDGMENTS

The author is grateful to Professor W. L. Nyborg for suggesting the problem and for the great quantity of time and energy which he has freely given to discussion. The author appreciates also the cooperation of the Johns Hopkins University Applied Physics Laboratory, which made its facilities available in the preparation of the final manuscript.

APPENDIX I. ESTIMATED ERROR IN SLOPE DUE TO PHASE APPROXIMATION

If p_0 and p_s differ in phase by an angle θ, the instantaneous total pressure in the near field of the bubble should be given by

$$p(r,h,t) = p_0 \cos\omega t + ar^{-1}p_s \cos(\omega t - \theta).$$

This may be written

$$p(r,h,t) = [(p_0 + ar^{-1}p_s \cos\theta)^2 + a^2r^{-2}p_s^2 \sin^2\theta]^{\frac{1}{2}} \cos(\omega t - \alpha),$$

where

$$\alpha = \arctan\left[\frac{ar^{-1}p_s \sin\theta}{p_0 + ar^{-1}p_s \cos\theta}\right].$$

Thus the rms amplitude of the total pressure is

$$p(r,h) = [(p_0 + ar^{-1}p_s \cos\theta)^2 + a^2r^{-2}p_s^2 \sin^2\theta]^{\frac{1}{2}}.$$

Differentiating,

$$\frac{dp(r,h)}{d(1/r)} = ap_s\left\{1 + \left[\frac{ar^{-1}p_s p_0}{2(p_0 + ar^{-1}p_s)^2} - \frac{p_0}{2(p_0 + ar^{-1}p_s)}\right]\theta^2 + \cdots \text{terms of order } \theta^4\right\}.$$

*** In this connection it might be useful to define a critical Reynold's number for the onset of the chaotic mode.

Now each of the terms in the square bracket on the right side of this equation necessarily has a value less than one. Therefore, the fractional error in taking $dp/dr^{-1} = ap_s$ cannot be greater than θ^2. Consequently, the error should be less than 20% if θ is less than 0.45 radians or about 27°.

APPENDIX II. DYNAMICS OF A FREE BUBBLE

For a free bubble the radius for resonant vibration a_r is given closely by[†††]

$$a_r/\lambda_0 = \tfrac{1}{2}\pi \cdot c_i/c_0 \cdot (3\rho_i/\rho_0)^{\frac{1}{2}} \approx 0.002$$

(for air bubble in water),

where ρ, c, and λ are density, phase velocity, and wavelength, respectively, and where the subscripts i and 0 refer to the media inside and outside of the bubble. It may be shown that for a lightly damped free bubble the phase relation between p_s and p_0 is a very sensitive function of a for values of a near a_r. By way of example, for a bubble damped only by radiation resistance the phase angle between these two waves is essentially 0° for values of a up to about $0.98a_r$. Thereafter it becomes suddenly 90°, switching again to 180° for values of a slightly greater than a_r. Clearly the phase approximation in Eq. (1) would be justified for this case. Studies by others have shown that actual bubbles are damped by irreversible heat conduction as well as by radiation resistance. Theory indicates that viscous shear should also contribute to damping, especially at high frequencies. The effect of additional damping is to make the phase transition more gradual. This may be seen in Table VI where the approximate phase angle at 10 kc has been calculated for several liquids, using formulae due to Rosenberg.[13] The damping constant δ

TABLE VI. Approximate phase angle between p_s and p_0 at 10 kc.

a/a_r	Ideal fluid (i.e., radiation resistance only) $\delta \simeq 0.014$	Water $\delta \simeq 0.060$	80% glycerin $\delta \simeq 0.088$	90% glycerin $\delta \simeq 0.16$	100% glycerin $\delta \simeq 0.59$
0.85	0°	10°	14°	22°	26°
0.90	0°	16°	22½°	34°	35½°
0.95	0°	28°	39°	53°	45½°
1.00	90°	89°	87½°	82½°	55½°

[†††] This formula, in a slightly different form, was originally given by M. Minnaert, Phil. Mag. 7th Ser. p. 240, (1933).

is defined as the ratio of the resistive to inertial impedance in the bubble at resonance.

In the first column of Table VI are given the values for an "ideal" fluid, i.e., one for which the damping is due to radiation only. It may be seen that water is far from ideal. However, a 10° phase error, such as occurs in water when $a/a_r = 0.85$, is not bad considering the general crudeness of the measurements. Analysis indicates that this would introduce no more than 4% error in estimating $p_s(a)$ (see Appendix I). With increasing values of δ or increasing nearness to resonance the phase angle error is expected to increase. An error of as much as 25° is perhaps tolerable if uncertainties in other measurements are as great as 20%.

APPENDIX III. PROOF THAT $(p_0/p_s)_r$ IS THE DAMPING CONSTANT

Let us imagine a hemispherical gas bubble of radius a vibrating on a rigid plane boundary under the excitation of a uniform incident pressure p_0. Considered as a lumped system, the fluid outside the bubble may be replaced by a thin shell of radius $r = a$ and of mass $M = 2\pi\rho_0 a^3$, [where M is defined in terms of the reactive part of the mechanical impedance of the scattered wave at $r = a$, i.e., $M = 2\pi a^2/i\omega \operatorname{Im}(p_s/u_s)$, it being further borne in mind that $ka \ll 1$ for resonant bubbles]. The scattered pressure p_s distributed uniformly over the area $2\pi a^2$ of the (upper) surface of the bubble has the effect of accelerating a shell of mass M to an instantaneous velocity u_s.

The equation of motion is therefore

$$2\pi a^2 p_s = M\dot{u}_s = i\omega M u_s.$$

Now by analogy with other single-degree-of-freedom systems, we may define the damping constants as

$$\delta = (Z/\omega M)_{\text{resonance}},$$

where the total impedance Z is defined as the ratio of the applied force, $2\pi a^2 p_0$, to the surface velocity $\dot{\xi}$.

Since the contribution of the incident wave to the velocity field is almost negligible, we have $\dot{\xi} \sim u_s$, and

$$u_s = 2\pi a^2 p_s/i\omega \approx 2\pi a^2 p_0/Z$$

or

$$|p_0/p_s| \approx Z/\omega M.$$

The ratio $|p_0/p_s|$ evidently has a minimum value when Z reduces to a pure resistance at resonance. The resonance value is, by definition, the damping constant.

21

Copyright © 1960 by the Institute of Electrical and Electronics Engineers, Inc.

Reprinted from IRE Trans. Med. Elec., ME-7, 163–165 (1960)

Ultrasonically-Induced Movements in Cells and Cell Models*

H. J. DYER† AND W. L. NYBORG‡

Summary—Normal- and high-speed cinemicrographs of events resulting from highly localized 25-kc vibration of small regions of an individual cell wall in Elodea leaf cells and in plastic cell models are discussed. In plant cells, complex patterns of ordered agitation are set up, similar to parts of the patterns observed in model cells. In models containing Newtonian fluids, steady circulation results with streamline positions and directions being functions of viscosity and the vibration pattern of the wall, the general features being accounted for by the theory of acoustic streaming. In models containing weak agar gels, a combination of plastic and fluid behavior is observed as setting ensues; immediately after vibration begins, suspended particles "flow" for a short distance following the streamlines of a viscous fluid, with displacement vectors distributed like the velocity vectors in acoustic streaming, but, as the sound ceases, return slowly almost to their original positions.

JUST over thirty years ago, Harvey [1], [2] and Schmitt [3], [4] demonstrated that vibrations of ultrasonic frequency caused physical dislocations of the cytoplasm of Elodea leaf cells. Their observation has been repeated by several subsequent workers, and has attracted our interest not only for its own sake but because of the possibilities that its study will provide additional knowledge about the effects of ultrasound on biological cells and that a useful tool for nonsurgical manipulation of cellular contents will result.

The apparatus used for applying ultrasonic vibrations directly to the walls of single cells while holding them under microscopic examination has already been described and illustrated [6]. It consists essentially of a Cavitron dental prophylaxis hand tool, modified by removal of the tip and replacement with a sharpened steel needle of 0.2-mm shaft diameter and 5-20-micron tip diameter. This magnetostrictive transducer is driven by a conventional audio-generator through a Heathkit 70-watt power amplifier (model W-6M), with a 1.5-ampere dc biasing current supplied to the jacket coils. The hand tool is touched to the cell wall or other test object by a micromanipulator mounted next to a compound microscope through which observations of tip diameter and amplitude of vibration and changes in the test objects are made and photographed. The tuned frequency of the transducer employed is about 25 kc; the system is tuned for each operation to its maximum amplitude for any given voltage input. Localized wall vibrations may be produced either by contacting the wall with the tip of a vibrating needle or by ultrasonic excitation of a small gas bubble resting on the wall.

Observations on sonically-excited plant cells, made by use of conventional microscopy and normal- and high-speed cinemicrography, reveal a complex pattern of intracellular movements which stops and starts promptly as the sound is turned on and off. These movements assume different forms, depending on the circumstances; possibilities include a) orderly motion such as might occur in simple fluids, b) displacement of intracellular particulates from equilibrium positions, to which they slowly return when the sound ceases, and c) violent and chaotic churning with destructive effects on the cell [6].

Moderate amplitudes and very low power inputs (less than 10 milliwatts) bring about decided displacements of cytoplasmic inclusions and cytoplasm as well as eddying motions of particles in the cell sap. Half-amplitudes from 1-2 microns are usually adequate to induce relatively gentle but positive motions quite different from normal, autogenous cyclosis; 3-5 microns usually results in orderly motions at very high speeds, while still higher levels of amplitude induce violent, chaotic motions which are obviously destructive. High-speed cinemicrography shows that motions below the destructive level are orderly, though not necessarily symmetrical (Fig. 1) Paths and directions of displacement of particles are, however, reproducible from cell to cell within limits possibly conditioned by the age of the cell, the concomitant thickness and rigidity of its wall, the shape of its interior, the shape of the particular plane of wall being directly excited, the sol-gel state of the cytoplasm, and possibly other factors and interactions of factors.

Studies using models were carried out for the purpose of reproducing phenomena observed in the plant cells under controlled conditions. The construction of a typical model is shown in Figs. 2 and 3. Representing the cell interior is a trough about 3 mm wide, 3 mm deep and 2 cm long milled in a ¼ inch thick piece of clear plastic. The "cell" is closed at one end by a plastic membrane M held firmly in position by means of clamping between pieces A and B. This "cell model" is filled with the desired fluid, and enclosed above by a microscope cover slip. An entry port in piece B permits the point of an exciting needle N to be brought into contact with the membrane at a point near

---

* Received by the PGBME, December 19, 1958. Presented at the Twelfth Annual Conf. on Electrical Techniques in Medicine and Biology, Philadelphia, Pa., November 10–12, 1959. This research was supported in part by the U. S. National Inst. of Health and in part by the U. S. Air Force Office of Scientific Res.
† Botany Dept., Brown University, Providence, R.I.
‡ Physics Dept., Brown University, Providence, R.I.

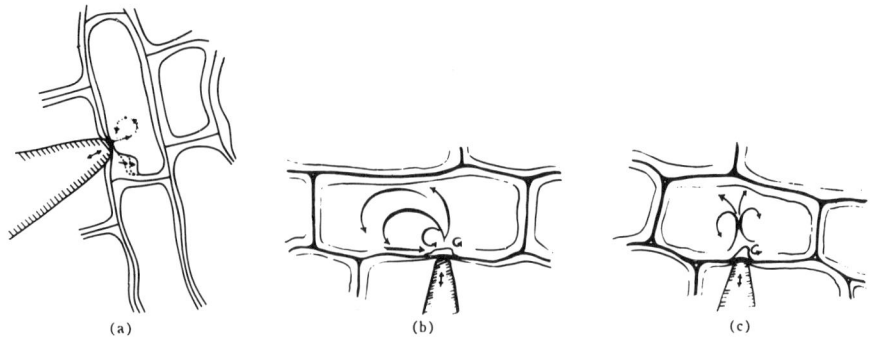

Fig. 1—Drawings taken from cinemicrograph frames showing paths of particles in Elodea leaf cells during ultrasound application through needle touching one wall; (a) shows initial motions as sound was turned on; (b) and (c) show paths maintained during sonation of two other cells. The mounds opposite the needle tips [(b) and (c)] are believed to be accumulations of gelated protoplasm; they build up in some cells during treatment.

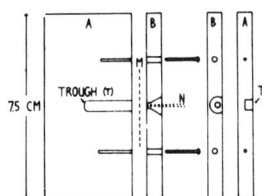

Fig. 2—Construction of plastic cell model. A membrane M of polyethylene or similar material is clamped between pieces A and B. The trough T is filled with fluid. A vibrating needle N contacts the membrane.

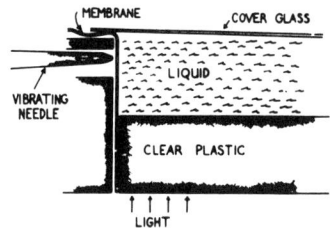

Fig. 3—Vertical section along plane of symmetry of cell model. Vibrating needle contacts membrane at area only a short distance below cover-glass; motions near this area may therefore be observed with high magnification.

the upper surface of the cell. An alternative to use of a vibrating needle is to place a small air bubble on the membrane, the bubble being set into oscillation by acoustic excitation of the surrounding fluid. With the cell mounted on the stage of a conventional research microscope, observations can be made in the fluid adjacent to the membrane in the vicinity of the small vibrating area, using magnifications up to 400 ×. Motions of the fluid are detected by viewing the paths of suspended particles, e.g., lycopodium spores, aluminum flakes.

When the model contains a Newtonian fluid, steady flow results, the nature of which depends on the fluid viscosity and the membrane thickness. For a highly viscous fluid [Figs. 4 and 5(b)] particles approach the membrane along the projected axis of the vibrating needle (or an analogous axis when the bubble is used). As a particle nears the membrane its speed increases suddenly, and it is projected outward along the membrane surface, away from the vibrating area. Regions of especially high speed and acceleration are indicated in Fig. 4.

Interesting and illuminating results involving viscoelastic behavior are obtained by filling the cell with a suitable dilute suspension of hot agar, which in time develops a weak gel structure. Observations are made at various stages in the gel formation (see Fig. 6). In very weak gels one finds that immediately after vibration commences, particles suspended in the gel "flow" for a short distance, following streamlines of a viscous fluid, as in Fig. 4. The flow velocity quickly decreases to zero, however, the particles coming to rest at positions such that the displacement vectors are distributed like the velocity vectors in acoustic streaming. The particles remain displaced as long as the sound continues, but return slowly to their original positions when the vibration source is turned off. When sound is applied in an on-off cycle, the particle displacement follows a time course similar to that indicated in the lower part of Fig. 6. As rigidity of the gel develops, the displacements occurring in such a cycle become smaller and finally may be too small to observe.

Motions and particle paths induced in ultrasonated cells of the Elodea leaf appear to be similar to motions of par-

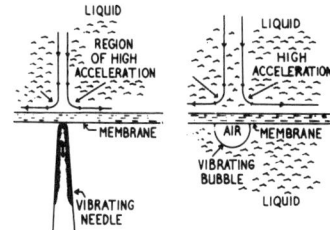

Fig. 4—Acoustic streaming in cell model containing liquid of high viscosity (0.1 poise or more). Most rapid motion may be localized in area no more than 0.1 mm in extent.

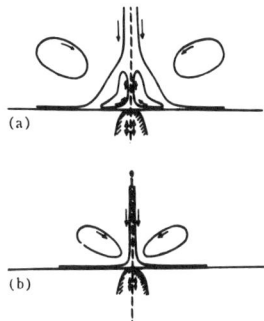

Fig. 5—Circulatory particle-paths near membrane; in (a) the liquid is water, and in (b) is a water-glycerine mixture with viscosity about 10 times that of water. Paths and directions taken by intracellular particles resemble those shown for the immediate neighborhood of the tip in case (a).

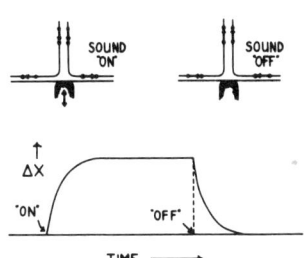

Fig. 6—Sonically-induced displacements in a weak gel. Displacement directions are similar to those in high-viscosity fluid (cf. Fig. 5(b). Magnitude Δx of displacement follows time course indicated, but decreases as gelation sets in.

ticles in the model systems described, particularly as found with low-viscosity fluids, as shown in the inner vortices of Fig. 5(a). It is not surprising, however, that deviations from the symmetrical flow patterns observed in models occur. These are not simply explained, but are believed to be conditioned primarily by asymmetries in cellular geometry and localized sol-gel discontinuities.

Bibliography

[1] E. N. Harvey and A. L. Loomis, "High frequency sound waves of small intensity, and their biological effects," *Nature*, vol. 121, pp. 622–624; April, 1928.

[2] E. N. Harvey, E. B. Harvey and A. L. Loomis, "Further observations on the effects of high frequency sound waves on living matter," *Biological Bull.*, vol. 55, pp. 459–469; December, 1928.

[3] F. O. Schmitt, "Ultrasonic micromanipulation," *Protoplasma*, vol. 7, pp. 332–340; 1929.

[4] F. O. Schmitt, A. R. Olson and C. H. Johnson, "Effects of high frequency sound waves on protoplasm," *Proc. Soc. Exptl. Biol. and Med.*, vol. 25, p. 718; April, 1928.

[5] D. E. Goldman and W. W. Lepeschkin, "Injury to living cells in standing sound waves," *J. Cell. and Comp. Physiol.*, vol. 40, pp. 255–268; October, 1952.

[6] W. L. Nyborg and H. J. Dyer, "Ultrasonically induced motions in single plant cells," *2nd Internatl. Conf. on Medical Electronics, Paris, France*, June 24-27, 1959, Iliffe and Sons, Ltd., London, Eng., publication probably 1960.

ULTRASONICALLY INDUCED MOTIONS IN SINGLE PLANT CELLS*

Wesley L. Nyborg, Ph.D.

Associate Professor, Physics Department, Brown University, Providence, R.I.

and Hubert J. Dyer, Ph.D.

Associate Professor, Botany Department, Brown University, Providence, R.I.

SUMMARY: An arrangement employing a longitudinally vibrating needle is used to set into vibration at 25 kc/s a selected small portion of plant cell wall. Results are viewed using conventional microscopy. Intracellular bodies are set into visible motion by the wall vibration. Under favourable low-amplitude conditions an orderly eddying motion results, similar to analogous motions which can be produced in simple physical models. At higher amplitudes complex and chaotic motions occur; cytoplasm may be torn from the cell wall and the wall itself ruptured.

SOMMAIRE: Emploi d'un appareil utilisant une aiguille à oscillation longitudinale amenant la vibration à 25 kc/s d'une petite portion choisie de la paroi d'une cellule végétale. Les résultats sont observés à l'aide d'un microscope conventionnel. Les corps intra-cellulaires sont rendus visibles par la vibration de la paroi. Dans des conditions favorables de faible amplitude, il en résulte un mouvement giratoire régulier, semblable aux mouvements analogues produits par des modèles de physique simple. À des amplitudes plus élevées, des mouvements complexes et chaotiques apparaissent; le cytoplasme peut être arraché de la paroi cellulaire et (rarement) la paroi elle-même peut se rompre.

THIS work represents efforts at a critical examination of the sonic effects first reported by Harvey,[1] Schmitt[2] and others.

A photograph of the microscope, the sonic applicator, and the supporting micromanipulator appears in Fig. 1. A plant leaf specimen under study is placed on the microscope stage, mounted centrally, in distilled water, between a small microscope slide and coverslip, the latter being separated about 300 μ by the use of glass wafers (Fig. 2). The applicator needle N is set into oscillation along its axis by means of a pencil-like magnetostrictive transducer. The latter is adapted from a unit manufactured for use in dentistry. The needle (N, Fig. 2c) is a 5 mm length of steel needle (0·23 mm dia) commonly used in mounting insects, its end sharpened by gentle honing in a jeweller's lathe. This needle is forced into a coaxial hole in a short length of brass rod B soldered to the squared-off end (T, Fig. 1c) of the transducer. Exercise of care in machining and mounting ensures that

* Work supported in part by the U.S. Public Health Service, National Institutes of Health.

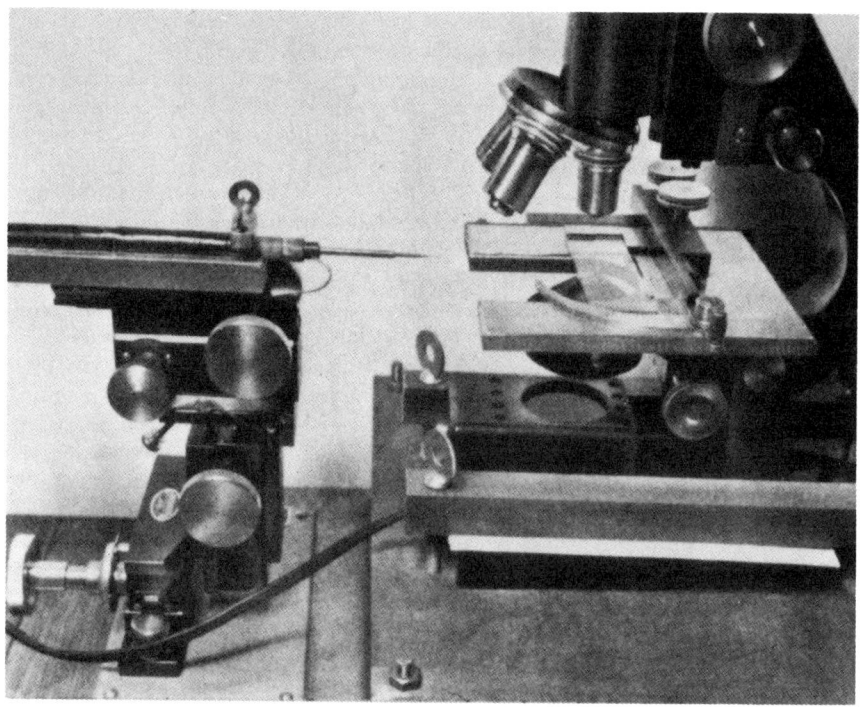

Fig. 1. Overall view of micromanipulator, transducer-needle unit, and microscope

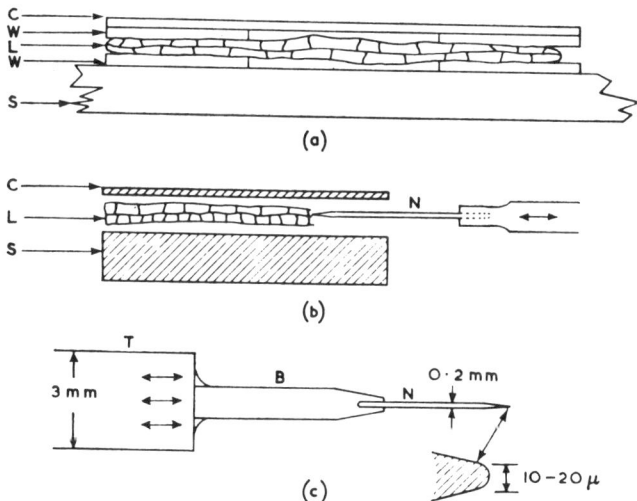

Fig. 2. Details of arrangements for application of oscillating needle to plant cells. (a) Arrangement for holding leaf, vertical section perpendicular to axis of needle; C—cover glass, 0·1 mm thickness; W—glass wafer, 0·1 mm thickness; L—leaf specimen; S—glass slide, 1 mm thickness. (b) Same arrangement, vertical section through axis of needle N. (c) Detail showing needle N and its mounting via brass piece B to end of transducer T

the vibration induced in the needle will be along its axis, with little lateral "whip".

Fig. 3 is a photomicrograph showing a portion of Elodea leaf, one cell being contacted by the ultrasonic applicator needle. This photograph comes from one frame of a 16 mm film taken at 16 frames/sec, the microscopic magnification being 440 diameters. The length and the width of the contacted cell in the plane of Fig. 3 are about 300 and 75 μ, respectively. The cell depth is probably 60 μ; the width of the needle tip is about 20 μ. One sees here the upper of the two cell layers commonly found in Elodea. (The cells in the lower layer are smaller, their dimensions being about one-half of those of the cells seen here.) In preparing a specimen for study, a thin leaf is cut along its length by means of a razor, thus exposing inner cell walls. It is commonly true that the exposed cells appear normal; e.g. they exhibit cyclosis as do other cells in the neighbourhood. The cells in the lower part of Fig. 3 are those exposed by this cut.

When a vibrating needle is brought near a cell wall, circulatory motion, i.e. eddying or whirling motion, is set up in the cell interior. If the vibrating amplitude is small, say, 0·1 μ or less, the motion will be highly localized; if the point of application is centrally located in a cell wall, it is commonly true that no disturbance will be seen in adjoining cells. In fact, the motion may then be confined to eddies only a few μ in extent, located near the

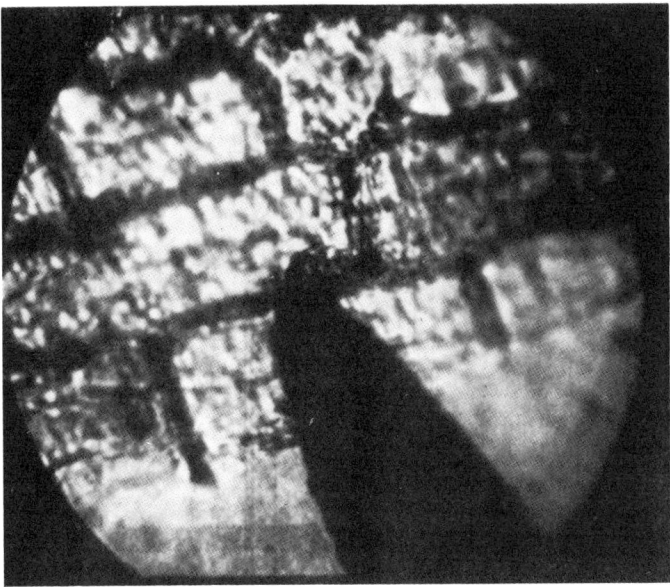

Fig. 3. *Photomicrograph of Elodea leaf, showing cell in foreground contacted by vibrating needle*

vibrating tip. The exact nature of these small eddies depends on accidental circumstances, especially on the location of irregularities in the cytoplasmic layer near the vibrating tip.

Examples of intracellular motion caused by the sound are indicated in Fig. 4; this figure is a sketch based on the photomicrograph in Fig. 3, and on movements revealed by the film from which Fig. 3 is taken. Normal cyclosis is not indicated on the figure. At low amplitudes the vibration produces no visible disturbance in any intact cell, except in that contacted by the needle. A variety of effects may occur in the latter cell. In one instance an eddy was observed in this cell just ahead of, and to the left of the vibrating tip; the motion is represented in Fig. 4 by showing successive positions of one of the chloroplasts at approximately 0·2 sec intervals. (Chloroplasts, such as the one whose motion is depicted, are apparently

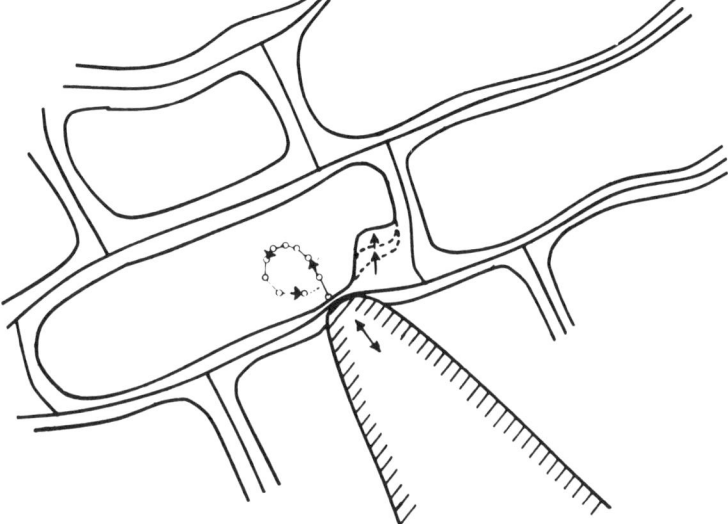

Fig. 4. Sketch based on Fig. 3, showing selected aspects of motion produced in cell by forced vibration of wall portion

forced from the cytoplasm into the vacuole by the sonic action. Whether they maintain thread-like attachments to the main cytoplasmic layer is a question as yet undecided.)

To the right of the tip (in Figs. 3 and 4) it was observed in one instance that the cytoplasm, initially in the lower right-hand corner of the contacted cell, was gradually displaced upwards, as shown in Fig. 4. Evidently the sound caused motion of some kind within the cytoplasm. At sufficiently high amplitude the sonic action is sufficient to tear the cytoplasm from the cell wall.

Eddies such as that shown very schematically in Fig. 4 are very frequently observed. They are apparently related to the circulations observed long ago by Faraday in the immediate vicinity of vibrating plates. Evidently the eddies result from the vibration of the cell wall, caused in turn by the vibrating needle. It is worth remarking that in producing such motions, it is *not* necessary for the vibrating tip of the needle to be in physical contact with the wall. The motions are most vigorous when such contact is established, but may be observable even when the separation is as much as 5 or 10 μ. Viewing the vibrating needle as a highly specialized kind of "stirrer" we thus note this unique property: the stirrer here is entirely outside (and not necessarily in contact with) the "vessel" (i.e. the cell) whose contents are being agitated.

One finds that considerable fluid motion takes place near the vibrating needle in the space where the needle is placed, outside the cell. However, the intracellular circulations are essentially independent of the external flow; we do not include a description of the latter motion in this discussion.

A considerable body of mathematical theory has become available in the last ten years, dealing with the phenomenon of *acoustic streaming*, i.e., with steady circulation produced by sound.[3] Especially pertinent to the phenomenon of intracellular eddying is the theory for streaming near a boundary. A consequence of this theory is that the near-boundary flow will be especially great over areas on a vibrating boundary where the vibration amplitude varies sharply with distance along the boundary, i.e. where surface gradients of the amplitude are great.[4] The detailed nature of the flow will depend on the amplitude distribution, on the location of boundaries, and on the liquid viscosity.

When, as in the experiments described here, a thin membrane is set into local vibration by a small needle, large surface gradients of the amplitude will

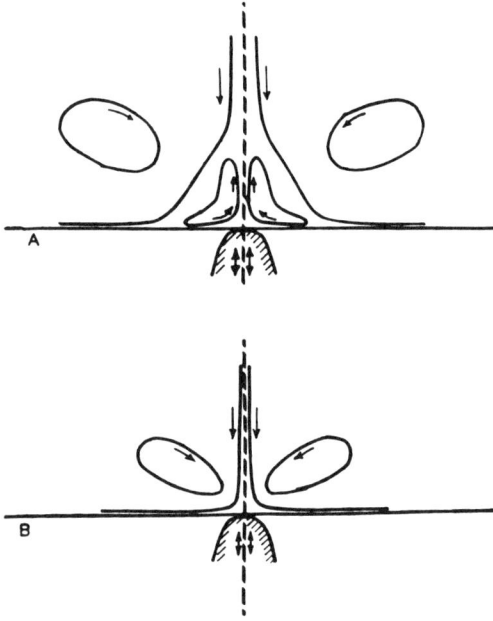

Fig. 5. *Circulatory motion observed near membrane in " model " situation using same vibrating needle as in plant cell experiments. Liquid in A is water, and in B is a water-glycerine mixture with viscosity ten times that of water. Motion is observed by visual microscopy, following small particles suspended in the liquid*

obviously exist near the driven area. Hence one expects especially vigorous eddying in the vicinity of this area. Actual flow phenomena in biological cells are necessarily complex because of inherent inhomogeneity. To obtain better understanding of the physical agents involved, experiments were performed in which the same vibrating needle used for plant experiments was used to set into local vibration an artificial membrane* about 15 μ thick. The latter was immersed in liquid and supported under moderate-to-weak tension in such a way that the surface free to vibrate (the " cell wall ") was several mm^2 in area. Eddying produced near the membrane on the side opposite the vibrating needle was observed under 400 × magnification. Typical patterns are shown in Fig. 5 for two liquids: (A) water and (B) a water-glycerine mixture with viscosity ten times that of water. In visual observations one is impressed by the rapid acceleration of particles as they approach the vibrating boundary, especially near the part immediately above the tip.

In visual observations we find the intracellular eddying produced in situations like that of Figs. 3 and 4 to possess features very similar to those of the streaming near the tip in the " model " situation, Fig. 5. Apparently the eddying in both cases is properly called *acoustic streaming*, and is ultimately explainable in terms of acoustic streaming theory. Using results of the latter theory one finds that intracellular eddying is to be expected whenever the cell walls are set into vibration, especially in a non-uniform manner. There are reasons, which space does not allow to be set out here, for believing that especially vigorous eddying is to be expected when gas bubbles, particularly of near-resonant size, rest on cell boundaries**.[4] Bubble-associated microstreaming may be an important mechanism for sonic and ultrasonic effects associated with " cavitation ".[5]

REFERENCES

1. Harvey, E. N., Harvey, E. B., and Loomis, A. L., " Further observations on the effect of high frequency sound waves on living matter ", *Biol. Bull.*, 55, 459 (1928).
2. Schmitt, F. O., Olson, A. R., and Johnson, C. H., " Effects of high frequency sound on protoplasm ", *Proc. Soc. Exptl. Biol. Med.*, 25, 718 (1928).
3. Raney, W. P., Corelli, J. C., and Westervelt, P. J., " Acoustical streaming in the vicinity of a cylinder ", *J. Acoust. Soc. Am.*, 26, 1006 (1954).
4. Jackson, F. J., and Nyborg, W. L., " Small scale acoustic streaming near a locally excited membrane ", *J. Acoust. Soc. Am.*, 30, 614 (1958).
5. Elder, S. A., " Cavitation microstreaming ", *J. Acoust. Soc. Am.*, 31, 54 (1959).

* " Saran ", i.e. vinylidene chloride polymer (Dow Chemical Co.).

** In the experiments reported on here it was ascertained that bubbles of this size were absent. In more typical ultrasonic irradiation set-ups, however, such bubbles are produced profusely, especially at surfaces.

Reprinted from *Proc. 3rd Intern. Conf. Med. Elec.*, London, 1960, pp. 445–449

Characteristics of Intracellular Motion Induced by Ultrasound

By H. J. Dyer and W. L. Nyborg

(*The paper was received 14th March, 1960. It was presented at the* Third International Conference on Medical Electronics *26th July*, 1960.)

SUMMARY

It was reported at the Second International Conference on Medical Electronics that localized intracellular movement of various kinds can be set up in plant cells via ultrasonic excitation of cell walls. Since that time more details have been obtained on the motions by employing high-speed cinematography. Events have been observed in cells of a variety of species, including ditchweed (Elodea sp.) leaves, onion (Allium cepa L.) epidermis, moss [Leptobryum pyriforme (L.) Schimper] protonemata, and Spirogyra sp. filaments. Excerpts of the moving-picture films obtained were shown at the present Conference. The paper lists a few of the phenomena observed and discusses physical principles which appear to be involved.

INTRODUCTION

In the experiments the excitation was provided by a needle set into oscillation[1] along its axis at either 25 or 90 kc/s. By micro-manipulation the vibrating needle tip is brought into contact or near-contact with a small portion of cell wall; this portion and a more-or-less restricted neighbouring area of the membrane are thus set into vibration. Depending on parameters, both of the physical agents (sonic amplitude and frequency, and shape and nature of tip) and of the biological system (size and shape of cell, and state of cytoplasm), one or more of the following may be observed:

(*a*) Regular eddying in the vacuolar sap.

(*b*) Forcing of cytoplasmic bodies, sometimes including the cell's nucleus, into vacuolar regions, with or without rupturing attachments of these bodies to the cytoplasm. If complete detachment occurs the bodies partake in the general vacuolar eddying; if the detachment is only partial, complicated motions result associated with time-dependent constraints (Fig. 1).

(*c*) Aggregation of intracellular bodies into a small region over the vibrating portion of the membrane, as if the bodies were attracted to this region (Fig. 1).

(*d*) Rotation of the entire protoplast about its axis, in the case of approximately cylindrical cells, e.g. Spirogyra cells which are partially plasmolyzed before ultrasonation.

(*e*) Visco-elastic behaviour, consisting of particle displacements which persist when the sound is 'on', and relax when it is 'off'.

Observations related to these were made much earlier by Harvey[2], Schmitt[3], and others.

In the following Section an attempt is made to present concepts of basic theory pertinent to describing the effects observed. The non-mathematician may prefer to proceed directly to Section 2, in which the phenomena are described in an essentially qualitative manner making use of the concepts developed in Section 1.

Mr. Dyer is, and Mr. Nyborg was formerly, at Brown University, Rhode Island, U.S.A.
Mr. Nyborg is in the Department of Physics, University of Vermont, U.S.A.
Supported in part by National Institute of Health Research Grants RG-4431 and A-3226.

(1) GENERAL THEORY

It is clear that localized vibratory excitation of a cell membrane leads to a complex variety of intracellular motions and structural changes. These are such as are hardly to be anticipated from the concepts emphasized in ordinary 'textbook acoustics'. The more familiar theory of sound treats the subject on a linearized basis, certain terms in the basic equations being neglected, as is permissible for the low amplitudes typically encountered in speech, music, etc. But under conditions of interest in connection with biological effects of sound, much higher amplitudes are typical and the linearizing approximation is no longer valid.

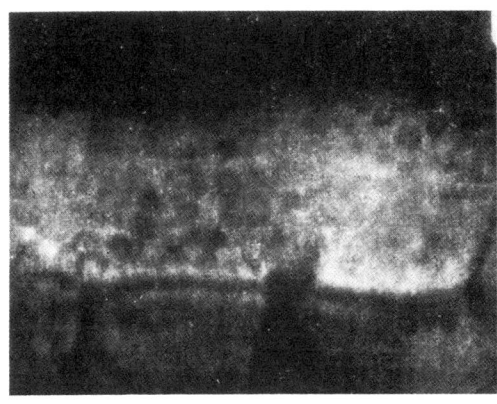

Fig. 1.—Some acoustic streaming effects in an Elodea leaf cell.

Reproduction of a frame from a 16 mm cinemicrograph taken at 400 diameters and 400 frames/sec; the plane of focus is the mid-plane of the cell, parallel to the leaf's surface. The tip of the longitudinally oscillating needle is seen to be in contact with a small area of one wall of an intact cell; on the inner side of the cell wall is an aggregation of coagulated cytoplasm, and the cloud of cytoplasmic droplets and chloroplasts in that cell, to the upper left of the needle, is rapidly cycling counter-clockwise in the vacuolar space.

Consider, for example, a fundamental principle of acoustics, embodied in the equation of motion (Newton's second law) for a particle of fluid. Let the motion be such as might exist near a small source of sound in liquid. We assume (i) that the motion is symmetrical and radially outward or inward relative to the source, and (ii) that the liquid is essentially incompressible. (The latter assumption is appropriate when, as here, our attention is confined to points whose distance from the source is much less than the wavelength λ for compressional sound waves in the liquid.) If r is the distance from the source, the equation of motion along r is

$$(\rho_0 \partial u/\partial t) + (\rho_0 u \partial u/\partial r) = f \quad \ldots \quad (1)$$

where $u(r, t)$ is the velocity of the particle (in the r-direction), ρ_0 is the density of the medium (assumed constant), and $f(x, t)$ is the force per unit volume exerted on the particle (in the

r-direction). The force f is usually entirely due to elastic and viscous stresses exerted on the surface of the particle by the surrounding fluid. The left-hand side of eqn. (1) is of the form $(\rho_0 a)$, where a is the acceleration of the particle and ρ_0 is the particle mass (per unit volume). In the linear acoustic approximation to eqn. (1), f and u are assumed to vary sinusoidally with time. Under the low-amplitude conditions where this assumption is valid, the second term on the left-hand side of eqn. (1) is negligible relative to the first term. The linearized equation, i.e. the first-order approximation to eqn. (1), is

$$\rho_0(\partial u_1/\partial t) = f_1 \quad \ldots \quad (2)$$

where (u_1, f_1) are the first-order approximations to (u, f), and vary sinusoidally with the time. Much of acoustics is based on equations similar to eqn. (2). When the second term on the left-hand side of eqn. (1) is not negligible, the equation becomes non-linear.

It has not proved fruitful so far to attempt exact solutions to basic (non-linear) equations of acoustics, such as eqn. (1). However, especially in the past 10 or 15 years, considerable progress has been made in the study of second-order phenomena, described mathematically by obtaining a second-order approximation to the equations. In obtaining this improved approximation to eqn. (1) a crucial step is to evaluate the second term on the left-hand side of that equation—by assuming that u is given for this purpose sufficiently well by u_1, the first-order approximation. In the field very near the point source we have an inverse square law, namely,

$$u_1 = Ar^{-2} \cos \omega t \quad \ldots \quad (3)$$

where A is a constant and ω is 2π times the frequency of the sound. Evaluating the non-linear term of eqn. (1) by making the anticipated approximation, we have

$$\rho_0 u_1 \partial u_1/\partial r = -2\rho_0 A^2 r^{-5} \cos^2 \omega t = -\rho_0 A^2 r^{-5}(1 + \cos 2\omega t) \, . \, (4)$$

Examining the expression on the extreme right-hand side of eqn. (4) we find that the non-linear term, as evaluated in the manner described, consists of two parts: a part which is independent of time, and a part which varies sinusoidally with time, but with angular frequency 2ω, twice that of the sound. Our concern here is chiefly with the time-independent or constant part of the non-linear term. Because this is not zero, it follows that the time average of the left-hand side of eqn. (1) over a large number of cycles is not zero; in fact, to the accuracy of the second-order approximation, the time average has just the value $(-\rho_0 A^2 r^{-5})$. Clearly the time average of the right-hand side of eqn. (1) should have the same value in the second-order approximation. Symbolically, using the bracket $\langle \, \rangle$ for time average over a suitably large number of cycles, this time average of eqn. (1) leads to

$$\langle \rho_0 u_1 \partial u_1/\partial r \rangle = \langle f_2 \rangle \quad \ldots \quad (5)$$

where $\langle f_2 \rangle$ is the time average value of f to the second approximation. We are thus led, by way of second-order acoustic theory, to the conclusion that when vibratory motion takes place in a medium, it may be accompanied by a field of steady, i.e. time-independent, stresses. Development of general theory for these steady stresses and their consequences, and applications to special problems, occupy a considerable part of the maturing subject of second-order theory.

In the special situation under consideration, that of the spherically symmetrical field near a point source, the steady stress will be entirely elastic. A slight outward movement of the fluid occurs when the sound source is first actuated, causing a rarefaction of the medium increasing in amount with decreasing r. A radial gradient of the time-independent pressure p_2 is thus developed, such that p_2 increases with distance r from the source.

In eqn. (5), the time-average force $\langle f_2 \rangle$ per unit volume in the r-direction is just $-\partial p_2/\partial r$. This is made equal to the left-hand side of eqn. (5), which can be rewritten so that the equation becomes

$$\frac{\partial}{\partial r}\langle \tfrac{1}{2}\rho_0 u^2_1 \rangle = -\partial p_2/\partial r \quad \ldots \quad (6)$$

Integrating and rearranging slightly, we have

$$\langle \tfrac{1}{2}\rho_0 u_1^2 \rangle + p_2 = \text{constant} \quad \ldots \quad (7)$$

Eqn. (7) has exactly the form of Bernoulli's principle in elementary physics. It is a statement of conservation of energy for particles partaking in steady fluid flow, valid when viscous effects are negligible. The bracketed term on the left-hand side of eqn. (7) gives the time-average kinetic energy per unit volume of the fluid due to the vibratory motion, while p_2 gives the potential energy per unit volume of the fluid due to steady compression. Eqn. (7) states that the sum of these is constant. At points where u_1, and hence the time-average kinetic energy, is large, the fluid will expand slightly, thus causing a decrease in p_2 and the time-average potential energy.

When heat is generated because of fluid viscosity, e.g. at boundaries, the Bernoulli energy-conservation principle is not valid. Somewhat related to this (although hardly obvious without examination of the equations) is the fact that when heat generation is important one commonly finds that the steady force $\langle f_2 \rangle$, in eqn. (5), can no longer be represented in terms of gradients of a pressure. By 'pressure' we mean, as usual, a stress exerted in a direction normal to the surface of any fluid particle in question. Under the circumstances considered, stresses with components tangential to the particle surface must be invoked, capable of exerting steady torque on the particle[4]. In a fluid these would be viscous stresses, associated with steady flow generated by the sound field. This sonically induced steady flow, sometimes in the form of large-scale circulations, and at other times in the form of microscopic eddies, has been given the name 'acoustic streaming'. Considerable attention has recently been given to this topic in second-order acoustics.

(2) APPLICATIONS

We now consider how these ideas may be relevant to the biological problems of interest. Let us first consider the matter of the eddying, which, as stated above, is found to occur in the vacuole of plant cells such as Elodea, when a portion of cell wall is set into vibration. It appears extremely likely that this is a special example of the phenomenon of acoustic streaming, discussed above. A detailed mathematical treatment of the flow employing the full theory would be very lengthy and quite impractical. Fortunately, some of the characteristic features of the streaming can be understood in fairly simple, essentially non-mathematical terms.

In the upper part of Fig. 2 is pictured schematically the oscillatory (first-order) flow field as it might be over a portion of

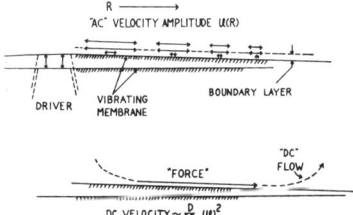

Fig. 2.—Sketch illustrating the build-up of acoustic streaming near a portion of a vibrating membrane.

vibrating membrane (the abbreviation 'a.c.', borrowed from electrical circuit terminology, is used to designate oscillatory or first-order quantities and 'd.c.' to designate time-independent second-order quantities) driven by a source at the left. The membrane motion is assumed normal to the membrane surface, but this normal motion produces a.c. flow in the adjacent fluid with components tangential to the surface. It is the tangential a.c. velocity which is particularly relevant to this discussion and whose amplitude is pictured with arrows above the membrane. As shown, the amplitude U is a decreasing function of R, where R is the distance measured from some fixed point at the left of the Figure, say, the centre of the driver. If the fluid were such as to slip at the boundary, the value of U at the membrane surface would not differ greatly from its value a short distance above this surface. In that hypothetical case, the only effect of the a.c. flow, in terms of earlier discussion [e.g. that accompanying eqn. (7)], is to cause a time-independent Bernoulli pressure gradient. However, the fluid certainly does not slip freely at the boundary; in fact, it is reasonable to assume that U is practically zero at the membrane-liquid interface. In consequence a very thin 'a.c. boundary layer' exists near the membrane, where the flow is relatively retarded[5]. Bernoulli's principle does not apply in this region. In line with earlier discussion [subsequent to eqn. (7)], the fluid is set into a steady circulatory flow in this situation. It turns out[6] that the flow is just what would be expected in the absence of a sound field if somehow, by means of an external agent, a force were exerted on the thin layer of liquid next to the membrane, as shown in the lower part of Fig. 2. Under suitably simple circumstances, the direction of this fictitious force is that of decreasing a.c. amplitude U, and the magnitude is proportional to $\partial(U^2)/\partial r$, i.e. to the associated Bernoulli gradient*. As one might expect, the 'force' gives rise rather directly to flow very near the membrane boundary, and the magnitude of flow is particularly great where the a.c. amplitude U is changing rapidly with distance R. More indirectly, the near-boundary flow induces circulatory motion at points distant from the membrane, i.e. in the main body of fluid. Thus the observed eddying motion is generated.

In Fig. 3 we examine more closely the motion we might expect near a membrane driven locally by a vibrating needle, as in the experiment with Elodea. The normal motion (upper two plots) is greatest over the vibrating tip, but it may extend out to some distance, depending on the membrane tension, etc. The

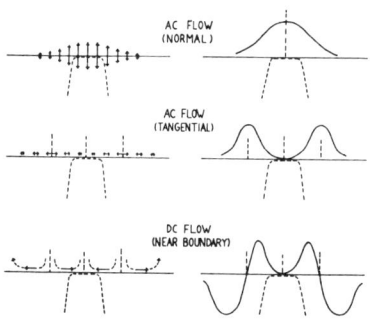

Fig. 3.—Velocity distributions, first and second order, near a portion of membrane excited locally by a vibrating tip.

*The statements made in this sentence are strictly true only when R is sufficiently large. At smaller values of R somewhat modified statements must be made with, however, a rather similar qualitative significance (see Reference 6 for details). For simplicity in this discussion we shall continue to ignore the modifications that should be made for small values of R.

tangential a.c. velocity amplitude (middle two plots) is zero at the centre (by symmetry), increasing with distance from the centre to a maximum at a distance which depends on tip diameter and membrane properties, and then decreasing again. The near-boundary acoustic streaming velocity is, according to results previously stated, in the direction of decreasing tangential a.c. amplitude, and is greatest approximately where the latter amplitude is changing most rapidly. We thus have streaming near the axis of the tip whose near-boundary flow is toward the centre of symmetry, and a streaming in the reverse sense at larger values of R. This near-boundary streaming induces eddies in the sense shown.

A rough idea of the nature of the induced eddies comes from adaptation of theory developed originally for elasticity[7]. It is not appropriate to give details here, but a special result is worth noting. Fig. 4 shows the shape of vortices which would be

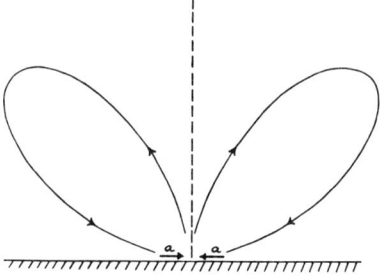

Fig. 4.—Vortex motion induced by a ring of radial flow (a–a) near a rigid boundary.

induced by a small ring of inward radial flow at the boundary. The equation of the streamline is

$$r = C \sin^2 2\theta \qquad \ldots \ldots \quad (8)$$

where C is a constant and the angle θ is measured from the axis of symmetry. Eqn. (8) is valid for points not too close to the origin $r = 0$. In obtaining a more detailed description of the eddying induced by the near-boundary flow, we would imagine the membrane surface (especially the small area near the vibrating tip) to be divided into a series of concentric rings, centred at the origin. In each of these rings the near-boundary flow is radial and has about the same value on all parts of the ring. However, the magnitude of the near-boundary flow (and, to some extent, the sign) will vary with the radius r of the ring. The induced flow resulting from each ring would be given by theory analogous to that leading to eqn. (8). The effect of all rings, and hence the effect of the near-boundary flow over the entire vibrating part of the membrane, would be obtained by superposition, i.e. by integration.

The general features of the predictions embodied in Figs. 3 and 4 are borne out in experiments with models[8, 9] and also seem reasonably applicable to the streaming in Elodea. However, streaming is very sensitive to surface roughness and to inhomogeneities of the medium. It is therefore not surprising that streaming patterns in Elodea cells are distorted when clumps of cytoplasm appear at the vibrating membrane.

At present, only rough estimates can be made of the magnitude of the streaming speeds to be expected within biological cells. Accurate calculations would require detailed knowledge of the vibration patterns of the membrane. An order-of-magnitude figure can be obtained by assuming that the sound field near the vibrating portion of the membrane is like that from a small source of sound. In that case[6] the streaming velocity u_2 near the boundary at a distance r from the centre of the source is given by

$$u_2 = Q_2(4\pi^2 \omega r^5)^{-1} \quad \ldots \quad (9)$$

where Q is the maximum during a cycle of the total flow from the source and ω is 2π times the frequency. Letting the area of the tip be A and its velocity amplitude be ωx_0, where x_0 is the displacement amplitude, we may take the product $A\omega x_0$ as a crude estimate of Q. To make a numerical calculation we assume the following: frequency, 25 kc/s; area A, 10^{-6} cm^2; distance r, 2×10^{-3} cm; displacement amplitude x_0, 10^{-4} cm. We obtain a speed u_2 of 12 microns/sec under these conditions. This result, expected to be only of order-of-magnitude accuracy, is not inconsistent with estimates of speed observed in Elodea cells. In weighing the significance of this result it is well to keep in mind that typical dimensions of the cell are 50–100 microns, that the speed increases with the square of the amplitude (so that increasing the amplitude to 10 microns would increase the speed by a factor of 100), and that the speed is very strongly dependent on r. One should realize also that the near-boundary flow referred to takes place at a distance no more than a few microns from the boundary, so that relatively high shearing rates may occur even for apparently low speeds.

In the above discussion we have emphasized the situation, used in our experiments, where the cell membrane is excited locally by a vibrating tip. This situation is advantageous for purposes of microscopic observation and cinematography. Also one can achieve especially high-amplitude gradients in this manner at whatever portion of the membrane one contacts with the vibrating tip. It follows from the preceding analysis that, within limits, one can produce localized eddying at any membrane site of interest.

It should be realized, however, that analogous phenomena are to be expected under more general conditions of ultrasonic application. Generation of the 'forces' which give rise to streaming depends on the existence of gradients in membrane-vibration amplitude. Passage of ultrasonics through tissues by whatever means will necessarily set membranes into vibration. Because of intersecting cell walls and other structures it is much more likely that the vibrations will take place non-uniformly than otherwise. A tendency for intracellular eddying is thus to be expected even in general irradiation of a mass of tissue, although not as predictably or as effectively as with localized excitation. Whether or not the eddying takes place will perhaps depend largely on whether or not the inner material of the cells which make up the tissue is fluid-like in nature.

As stated earlier, observations of intracellular motions caused by sound reveal that acoustic streaming is only one aspect of the total phenomenon. The relatively simple and orderly circulatory motion implied by acoustic streaming appears to dominate the picture either (a) at very low amplitudes where only small suspended particles in the vacuolar sap appear to move, or (b) at much higher amplitudes where chloroplasts and other bodies are torn from their normal connections with the cytoplasm and take part freely in the vacuolar eddying. At intermediate amplitudes the motions appear very complicated as cytoplasmic bodies 'attempt' to take place in orderly motion but are thwarted by 'collisions' with other bodies, or are hindered by connections to the general cytoplasmic continuum. It is not possible at present to give a satisfactory description of these motions, nor an account of just what is involved when bodies are forced from the cytoplasm.

It is possible, however, to advance comparatively simple arguments to explain the apparent attraction of cytoplasmic particles to the vibrating portion of a membrane. The phenomenon is apparently an example of another second-order acoustic phenomenon, called 'radiation pressure'. The mathematical theory of this subject is rather well developed[10]. The magnitude of the attractive force to be expected depends on properties, such as density and compressibility, of the intracellular particles. However, the prediction that such an attractive force should exist can be arrived at rather simply by choosing either of two extreme models.

First, in Fig. 5, we consider the case of a hard, incompressible particle, immersed in the alternating flow field near the tip. On

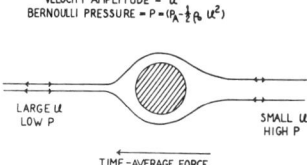

Fig. 5.—Model illustrating steady force acting on small rigid sphere in a sound field where the velocity amplitude varies in space.

the side of the particles near the tip (i.e. the left-hand side of Fig. 5) the velocity U is relatively large; while, on the other side, the velocity U is smaller. A difference of Bernoulli pressure thus exists between the near and far sides of the particle, such that the pressure is always greater on the far side. A steady force thus drives the particle toward the tip.

In the second and perhaps more plausible model (Fig. 6) the particle is assumed to be somewhat compressible relative to the fluid. In the sound field near the tip the alternating pressure

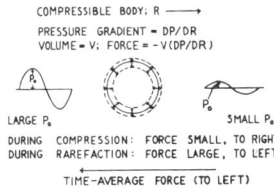

Fig. 6.—Model illustrating steady force acting on small compressible sphere in sound field where the pressure amplitude varies in space.

amplitude P_0 on the near side (represented by the left-hand side of Fig. 6) is greater than on the far side. During the compression part of the cycle the force on the particle due to sound pressure is away from the tip, toward the region of low P_0. The magnitude of the force is equal to the pressure gradient times the volume, the latter being relatively small during compression. In the rarefaction part of the cycle the force on the particle due to sound pressure is toward the region of high P_0, i.e. towards the tip. At the same time the volume, and hence also the force, is relatively large. Clearly the average effect is that of a net force towards the tip. The magnitude of the force depends on the compressibility of the particle. Conceivably, with refinements, acoustic means might ultimately be used in measuring the compressibility of particles.

(3) REFERENCES

(1) NYBORG, W. L., and DYER, H. J.. 'Ultrasonically Induced Motions in Single Plant Cells', *Proceedings of the Second International Conference on Medical Electronics*, 1960, p. 391.

(2) HARVEY, E. N., HARVEY, E. B., and LOOMIS, A. L.: 'Further Observations on the Effect of High Frequency Sound Waves on Living Matter', *Biological Bulletin*, 1928, **55**, p. 459.
(3) SCHMITT, F. O., OLSON, A. R., and JOHNSON, C. H.: 'Effects of High Frequency Sound on Protoplasm', *Proceedings of the Society for Experimental Biology and Medicine*, 1928, **25**, p. 718.
(4) NYBORG, W. L.: 'Acoustic Streaming Equations: Laws of Rotational Motion for Fluid Elements', *Journal of the Acoustical Society of America*, 1953, **25**, p. 938.
(5) LAMB, H.: 'Hydrodynamics. Ed. 6' (Dover Publications, 1945).
   *See also RANEY, W. P., CORELLI, J. C., and WESTERVELT, P. J.: 'Acoustical Streaming in the Vicinity of a Cylinder', *Journal of the Acoustical Society of America*, 1954, **26**, p. 1006.
(6) NYBORG, W. L.: 'Acoustic Streaming Near a Boundary', *Journal of the Acoustical Society of America*, 1958, **30**, p. 239.
(7) †LOVE, A. E. H.: 'A Treatise on the Mathematical Theory of Elasticity. Ed. 4' (Dover Publications, 1944).
(8) JACKSON, F. J., and NYBORG, W. L.: 'Small Scale Acoustic Streaming near a Locally Excited Membrane', *Journal of the Acoustical Society of America*, 1958, **30**, p. 614.
(9) DYER, H. J., and NYBORG, W. L.: 'Ultrasonically-Induced Movements in Cells and Cell Models', *I.R.E. Transactions on Medical Electronics*, 1960, **ME7**, p. 163.
(10) WESTERVELT, P. J.: 'Acoustic Radiation Pressure', *Journal of the Acoustical Society of America*, 1957, **29**, p. 26.

---

* The thickness of this boundary layer is of the order of $(2\eta/\omega)^{1/2}$, where η is the shear viscosity coefficient for the fluid and ω is 2π times the frequency. For liquids of interest here, and ultrasonic frequencies, this thickness is of the order of a few microns or less.

† One proceeds from eqn. (30), in Reference 7, interpreting displacements as velocities and the constant μ as the coefficient of shear viscosity; λ is assumed to be infinite.

24

Copyright © 1966 by the Acoustical Society of America

Reprinted from J. Acoust. Soc. Amer., 40(6), 1363–1370 (1966)

Deformation and Motion Produced in Isolated Living Cells by Localized Ultrasonic Vibration

WALTER L. WILSON,* FLOYD J. WIERCINSKI,† WESLEY L. NYBORG, R. M. SCHNITZLER, AND F. J. SICHEL

The University of Vermont, Burlington, Vermont 05401

and

Marine Biological Laboratory, Woods Hole, Massachusetts 02543

Naturally-isolated living cells were subjected to ultrasonic vibration by means of the tip of a steel needle applied directly to the cell surface or inserted into a drop of a suspension of cells in sea water. The needle was machined into the tip of a stainless-steel cone, and the base of the cone was glued to one end of an electroded and polarized barium titanate hollow cylinder. This composite transducer was driven at its resonant frequency of approximately 85 000 cps. Ultrasound applied to the surfaces of egg cells of marine invertebrates produces rotation, translation, deformation, and fragmentation of the nucleoli; rotation and deformation of the nuclei; acoustic streaming of nucleoplasm and cytoplasm; and deformation of the cellular surface and fragmentation of the cell. Cells and intracellular bodies are often attracted to the sound source as a result of acoustic-radiation pressure, and, especially for whole cells in suspension, this attractive tendency is typically superposed on a tendency of the body to take part in acoustic streaming of the surrounding medium. Some aspects of the motion can be accounted for in terms of presently available theory of nonlinear acoustics. Information on the physical properties of nucleolus, nucleoplasm, and cytoplasm is gained from use of this sonic technique.

INTRODUCTION

STUDIES have been reported by Dyer and Nyborg[1,2] on the effects of vibrations of ultrasonic frequency on selected individual cells of plants. In these cells, complex patterns of ordered motion of intracellular inclusions are produced by high-frequency vibration applied to selected regions of the cell wall, and the above authors[2] have discussed some of the physical principles probably involved. The present investigation is concerned with the effect on animal cells of similar localized high-frequency vibration. Specifically, the cells studied are eggs of marine invertebrates. Much information exists on these, since they have been extensively used as test objects in a variety of experiments.

I. MATERIALS AND METHODS

Cells were subjected to ultrasonic vibration by means of the tip of a steel needle applied directly to the cell surface or inserted into a droplet of a suspension of eggs in sea water. The apparatus described by Dyer and Nyborg[1] was used, with some modification. A stainless-steel needle was machined into the tip of a stainless-steel cone, the latter being glued at its base to one end of an electroded and polarized barium titanate hollow cylinder (Fig. 1). This cylinder and cone form a composite transducer similar to that described by Mason and Wick.[3] This transducer was mounted on a micromanipulator for positioning, and was driven at its resonant frequency of approximately 85 000 cps by means of an oscillator and a power amplifier. Needles of various sizes were used, a typical one having a shaft 0.2 mm diam, extending about 3 mm beyond the end of the cone and tapering to a tip whose radius of curvature in different needles varied over a range of about one to

---

* Present address: Biol. Dept., Oakland Univ., Rochester, Mich. 48063.
† Present address: Biol. Dept., Northeastern Ill. State College, Chicago, Ill. 60625.
[1] H. J. Dyer and W. L. Nyborg, IRE Trans. Med. Electron. 7, 163–165 (1960).
[2] H. J. Dyer and W. L. Nyborg, Proc. Intern. Conf. Med. Electron., 3rd, pp. 445–449 (1960).

[3] W. P. Mason and R. F. Wick, J. Acoust. Soc. Am. 23, 209–214 (1951).

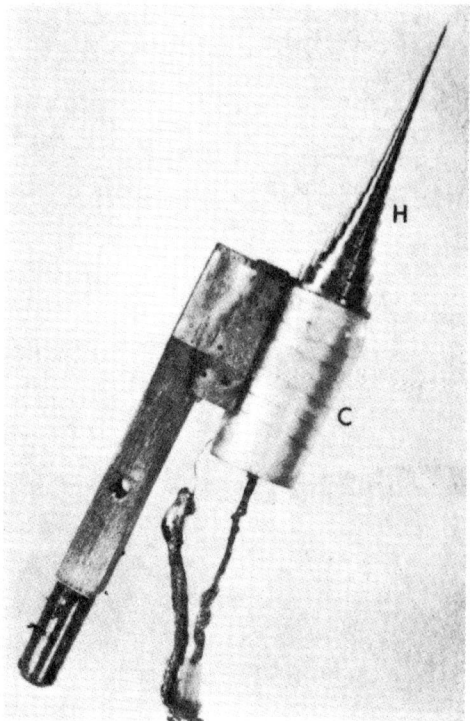

Fig. 1. Composite transducer consisting of stainless-steel stepped-exponential horn (H) cemented to ceramic cylinder (C). Resonant frequency about 85 kcps.

lasting more than a few minutes, the cells were enclosed in a moist chamber.

Observations were carried out either with an inverted microscope or with an upright microscope equipped with long-working-distance objectives, and both brightfield and phase microscopy were used. Motion pictures were taken at 18 frames per second on 16-mm film.

The eggs used were those of the sea urchin *Arbacia punctulata*, of the clam *Spisula solidissima*, and of the starfish *Asterias*. In the case of *Arbacia*, both mature and immature (germinal vesicle stage) eggs were used; the eggs of *Spisula* and *Asterias* were immature. The immature Arbacia egg has a diameter of about 74 μ. The nucleus (germinal vesicle) and nucleolus are clearly visible and have diameters of 38 and 11 μ, respectively. The diameter of the mature *Arbacia* egg is also about 74 μ, but the nucleus is smaller than that of the immature egg, about 12 μ in diameter, and is not as readily visible. The egg of *Asterias* is large, 110 μ in diameter, with a nucleus (germinal vesicle) and nucleolus 50 and 13 μ in diameter, respectively. Figure 2 shows a typical immature egg of *Asterias*. The *Spisula* egg is smaller than the egg of *Arbacia* or of *Asterias*. Its diameter is about 55 μ, and its nucleus (germinal

100 μ. Careful machining and mounting ensured that the vibration of the needle was in the long axis, with little lateral motion. The amplitude of the needle-tip motion can, of course, be varied continuously by adjusting the driving voltage; it can be determined by measuring the vibration blur with a microscope eyepiece micrometer. In the experiments reported here, the greatest value of the peak-to-peak amplitude was about 5 μ.

The vibrating needle was applied to the surface of an egg held stationary in sea water by either one of two methods. In one method (described by Tyler[4]), an egg was held on the fire-polished end of a thin-walled glass capillary tube with a pressure difference developed by a finely adjustable syringe. The syringe was connected by a steel tube to the glass capillary tube. Positioning of the capillary tube and egg was accomplished by a micromanipulator. In the second method, eggs were held stationary by distributing them over a thin layer of 1% agar gel in sea water on a glass slide.

In other experiments, cells were not held, but were allowed to move freely in a drop of sea water into which the vibrating needle was introduced. For observations

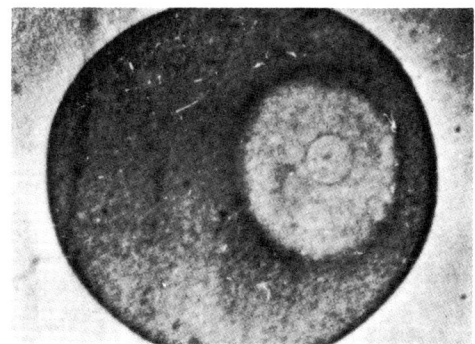

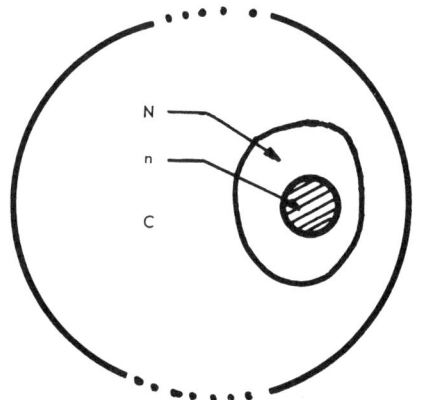

Fig. 2. Normal immature egg of Asterias. C: Cytoplasm. N: Nucleus. n: Nucleolus.

---

[4] A. Tyler, Biol. Bull. **109**, 369–370 (1955).

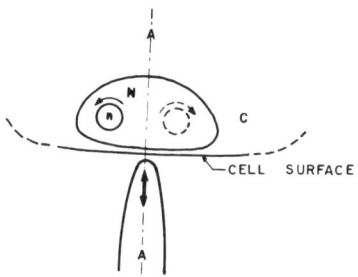

FIG. 3. Diagram showing sense of rotation of nucleolus (n) with respect to axis of symmetry (AA) of vibrating needle. N: Nucleus. C: Cytoplasm.

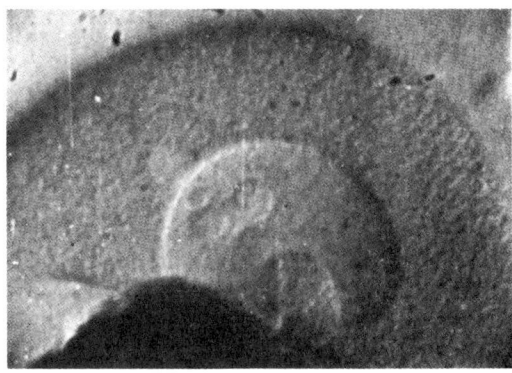

vesicle) and nucleolus have diameters of 30 and 13 μ, respectively. Harvey[5] has prepared a comprehensive review of the many researches that have been carried out on biological, physical, and chemical properties of *Arbacia* eggs.

II. RESULTS

With the needle tip applied directly to the surface of any of these eggs, effects on the nucleolus, nucleus, cytoplasm, and cell surface were observed. In the immature eggs of *Asterias*, *Spisula*, and *Arbacia*, the nucleolus rotates, turning more rapidly as the needle-tip amplitude is increased. Rotation tends to be in the sense of Fig. 3, in that the part of the nucleolus nearest the projected axis of the vibrating needle moves in the direction indicated by the needle. With low needle-tip amplitude, rotation of the nucleolus is the only motion seen in these cells. With higher amplitude, the nucleolus not only rotates but moves about within the nucleus. In the eggs of *Asterias* and *Spisula*, as needle-tip amplitude is increased, the nucleolus typically loses its spherical shape and becomes an irregularly shaped body from which one or more small bodies become separated (Fig. 4). Sometimes, the nucleolus disintegrates into a large number of small bodies. The nucleolus of the egg of *Spisula* often becomes a teardrop-shaped body and moves about the nucleus large-end first. In some cases, small bodies separate from the tail of the drop. The nucleolus of the *Spisula* egg sometimes elongates (Fig. 5) and then breaks into two bodies; in other cases, the nucleolus flattens, resembling a pancake, and undulates. In each type of egg, when sonation is terminated following deformation of the spherical nucleolus, the nucleolus reverts to its original spherical shape, and typically, following fragmentation of the nucleolus, the fragments slowly coalesce after termination of sonation.

At low needle-tip amplitudes, in the immature eggs of *Asterias* and *Spisula*, typically there is no visible movement of the cytoplasm. Previously it has been

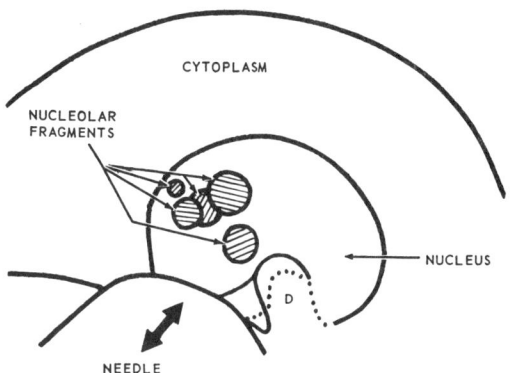

FIG. 4. Fragments (after sonation) of nucleolus of *Asterias* egg. Positive print from motion picture film. Line drawing for aid in identifying features of photograph. Vibrating needle is in contact with cell surface, which, in turn, is nearly in contact with the nuclear membrane. Surfaces of cell and nucleus have been deformed by the sound (see, especially, D) but have not been ruptured.

shown that the viscosity of immature eggs is rather high.[6] As needle-tip amplitude is progressively increased, rotation of the nucleus and streaming of the cytoplasm can be observed. Cytoplasmic streaming begins first in a small region close to the needle tip. However, in the mature egg of *Arbacia* (whether fertilized or not), cytoplasmic streaming or churning is a typical effect of sonation even at low needle-tip amplitudes; thus, the cytoplasm in these eggs appears to flow readily. According to Heilbrunn,[6] the viscosity of the protoplasm of the unfertilized egg of *Arbacia* is only 1.8 cP.

When the vibrating needle tip is in a drop of a suspension of eggs in sea water, the cells appear to be subject to two tendencies: (1) migration toward the source and (2) movement in a flow pattern typical of acoustic streaming near such a sound source (Fig. 6).

---

[5] F. B. Harvey, *The American Arbacia and Other Sea Urchins* (Princeton University Press, Princeton, N.J., 1956).

[6] L. V. Heilbrunn, *The Viscosity of Protoplasm*: "Protoplasmatologia II, C, 1" (Springer-Verlag, Wien, 1958).

When some cells approach or are held stationary near the tip of the vibrating needle, the part of the cell closest to the needle forms a conelike protuberance (Figs. 7 and 8). Upon increase of the sonic amplitude, this protuberance will commonly appear to grow in length, tending to extend itself toward the vibrating needle or along its surface. When the protuberance reaches a sufficient length, it apparently becomes unstable and one or more droplets are separated from it (Fig. 8). Sometimes, the fragmentation into droplets continues until the cell is reduced to a small fraction of its original volume. In Arbacia eggs, the growth of a protuberance, and its disintegration into droplets, occurs at much smaller needle amplitudes for fertilized than for unfertilized eggs.

The protuberances also occur on many cells lying along the shank of the needle, and in these cells several protuberances generally occur, forming and regressing, giving a wavelike or rippling appearance to the cell surface. This attraction between cell and needle also occurs in cells that are held stationary; in these, in addition to a protuberance on the cell surface, a protuberance often occurs on the side of the nucleus closest to the needle (Fig. 9). Protuberances are also formed on the nucleolus.

Protuberances on cell and nucleolus do not occur invariably. In some eggs, a flattening or a concavity of the cell surface and of the nuclear surface closest to the needle is produced by the sound (Fig. 10). When two cells (A, B, Fig. 11) are swept by the acoustic streaming toward the needle, a concavity typically develops on the more distant cell (B) in the region bordering the cell (A) closer to the needle. It is possible that the closer cell is acting as a secondary source.

III. DISCUSSION

The present work confirms Schmitt's observation[7] of rotation and fragmentation of the nucleolus of Asterias eggs produced by ultrasound. His method was similar

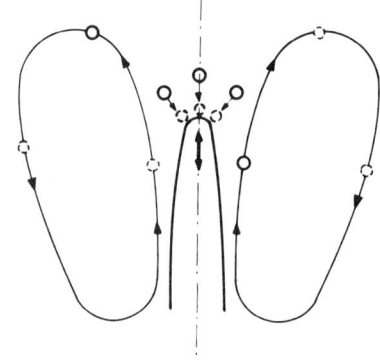

FIG. 6. Motion of suspended eggs or particles in vicinity of longitudinally vibrating needle. The eggs are subject to two tendencies: (1) to move toward the needle source; (2) to follow a circulatory pattern. The first tendency predominates near the needle tip and results in agglomeration of eggs on the surface there; the second tendency predominates at greater distances from the source and, especially, at the sides.

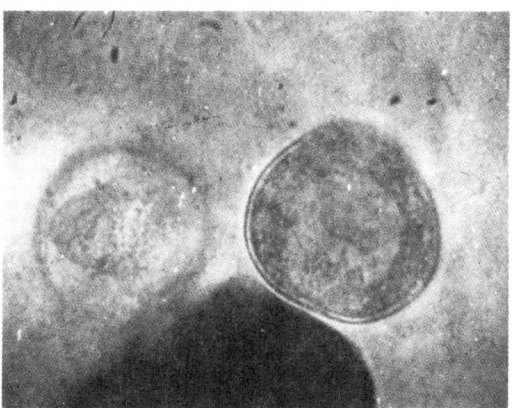

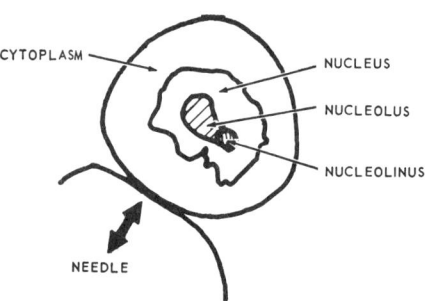

FIG. 5. Elongated nucleolus of Spisula egg prior to breaking into two parts. Positive print from motion picture film. Contortions of nuclear membrane seen here are natural in this egg (and are not produced sonically). Migration of nucleolinus (a small body in the nucleolus) to one end of nucleolus is a typical result of sonation.

to that used in the present investigation in that a "point" source of ultrasound was employed. However, Schmitt worked with a frequency much higher than that used by us, approximately 750 000 cps. Also, in his work, the rotational and whirling effects were attributed to diagonal and transverse movement of the vibrating needle. Apparently this is not the proper interpretation of the present effects, for in the system used by us the motion of the needle is longitudinal rather than diagonal and transverse.

At low needle-tip amplitudes the rotating nucleolus of Asterias and Spisula eggs is confined to a fixed position in the nucleus. As needle-tip amplitude is progressively increased, a level is reached at which the nucleolus begins to move about within the nucleus.

---

[7] F. O. Schmitt, Protoplasma 7, 332–340 (1929).

Presumably, this movement is the result of acoustic streaming of the nucleoplasm produced by vibration of the nuclear membrane. Some of the characteristics of acoustic streaming near a vibrating membrane are discussed elsewhere.[2,8] At present, an adequate physical explanation of nucleolar rotation in a fixed position within the nucleus is not available. Apparently, the nucleolus is not confined by structure within the nucleus. According to Harding,[9] the nucleoplasm of

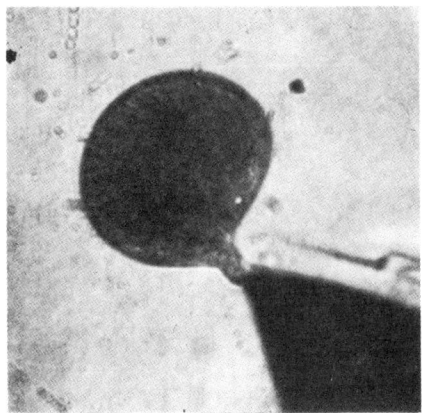

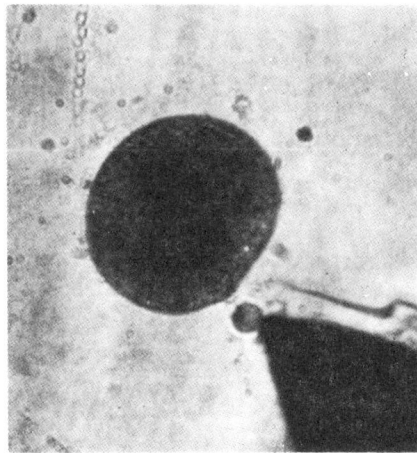

FIG. 8. *Upper*: Formation of protuberance on a fertilized *Arbacia* egg held in position by agar gel. *Lower*: Subsequent droplet formation from the protuberance shown above. As is apparent, the vibrating needle used here has a tip that is sharp as compared to that used in obtaining Fig. 7.

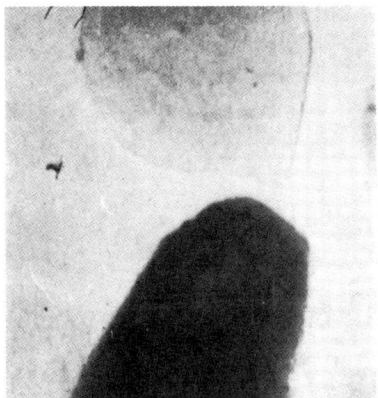

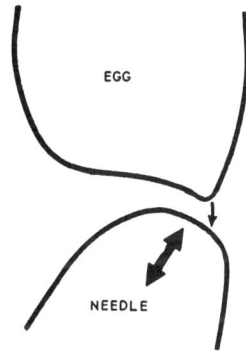

FIG. 7. Conelike protuberance on *Arbacia* egg suspended in sea water. Positive print from motion picture film. Arrow shows direction in which protuberance "grows."

Asterias egg is liquid with a viscosity of about 7 cP (23°C), through which the nucleolus falls freely under the influence of gravity.

These results give information about the physical characteristics of the nucleolus. The fact that the original spherical shape of the nucleolus can be changed reversibly means that, in a normal cell, either (1) the nucleolus is a nonrigid body, or (2) treatment of the nucleolus with ultrasound has changed a rigid body to a nonrigid one. However, the latter possibility seems unlikely. It is of interest that nucleoli isolated from cells and kindly supplied to us in suspension by Dr. W. S. Vincent did not exhibit deformation when subjected to ultrasound; apparently the isolation technique alters the physical properties of the nucleolus. Various workers have presented evidence (for example, from electron microscopy) that shows that the nucleolus is a body devoid of an enveloping membrane,[10] and that the nucleoplasm of the egg of *Asterias* is a liquid. Therefore, the nucleolus must be either a gel or a liquid immiscible with the liquid nucleoplasm. Because of the tendency, noted in the present work, for nucleolar fragments to

---

[8] F. J. Jackson and W. L. Nyborg, J. Acoust. Soc. Am. **30**, 614–619 (1958).

[9] C. V. Harding, Proc. Soc. Exptl. Biol. Med. **70**, 705–708 (1949).

[10] E. Borysko and F. B. Bang, Bull. Johns Hopkins Hosp. **89**, 468–473 (1951).

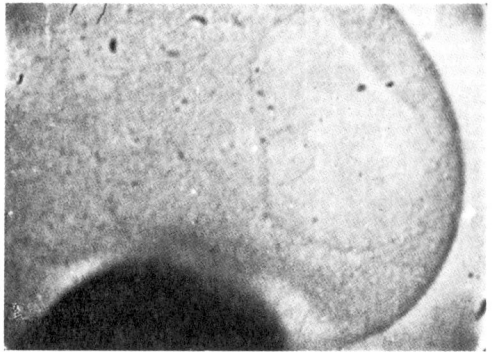
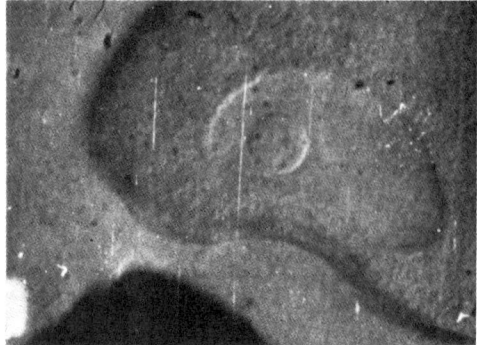

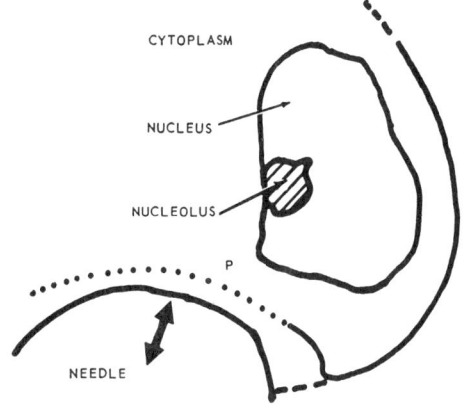
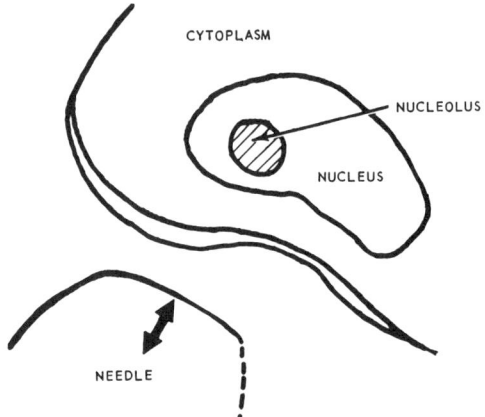

FIG. 9. Protuberance (P) on nucleus of *Asterias* egg held in position by agar gel. Positive print from motion picture film. Note distortion of nucleolus.

FIG. 10. Concavities on nuclear and egg surfaces of *Asterias* egg held in position by agar gel. Positive print from motion picture film.

coalesce in the absence of sound, we conclude that the sonated nucleolus must be regarded as a liquid.

Information is also gained about the physical state of the cytoplasm. It is an interesting fact that, in the unfertilized eggs of Asterias and Spisula, low needle-tip amplitude causes movement of the nucleolus and streaming of the nucleoplasm without apparent motion and streaming of the intervening cytoplasm. As needle-tip amplitude is progressively increased, streaming in the nucleus becomes more rapid and streaming of the cytoplasm begins, first close to the needle tip. According to acoustical theory, this is evidence that the cytoplasmic suspension behaves as a non-Newtonian substance.[11] In studying the viscosity of the cytoplasmic suspension of the egg of Spisula by a centrifuge method, Wilson and Heilbrunn[12] used forces resulting from centripetal accelerations between 2250 and 9000 g and found that the cytoplasm exhibited Newtonian behavior. The forces involved in the ultrasonic method are typically much lower than those used by Wilson and Heilbrunn in the centrifuge method, and it may be that the cytoplasmic suspension of the egg of *Spisula* exhibits different properties as the shearing force is varied. Shear-dependent properties have also been reported by Larson[13] for the cytoplasm of two species of paramecium, *P. aurelia* and *P. trichium*. In the case of *P. trichium*, with centripetal accelerations up to about 800 g, the cytoplasm is dilatant; with accelerations between 800 and 1950 g, the cytoplasm is thixotropic; and for accelerations between 1950 and 2550 g, according to Larson's graph, the cytoplasmic viscosity remains constant—that is to say, it exhibits Newtonian behavior. However, in this latter range the results would be more convincing if more samples had been taken.

---

[11] According to second-order acoustic-streaming theory for Newtonian liquids, the flow pattern is *independent* of amplitude, and the speed of any point is proportional to the square of the amplitude. See Ref. 14 for a bibliography on this subject.

[12] W. L. Wilson and L. V. Heilbrunn, Quart. J. Microscopical Sci. **101**, 95–103 (1950).

[13] E. Larson, Doctoral thesis, Univ. Pa., Philadelphia, Pa. (1955); cited in Ref. 6.

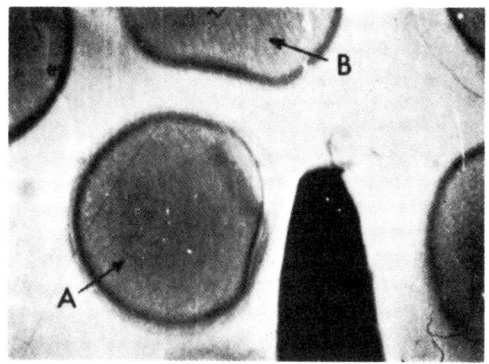

FIG. 11. Concavity of distant egg B, apparently as a result of its proximity to egg A, the latter being acted on directly by the sound source. Eggs in sea water suspension. Positive print from motion picture film.

IV. PHYSICAL ANALYSIS OF THE MOTIONS

Basic physical principles underlying the acoustic streaming are described in the literature of nonlinear acoustics; a review of the subject has been given recently by Nyborg.[14] It should be realized that acoustic-streaming theory has so far been applied only to Newtonian fluids; further development is needed in using such a theory to describe sonically-induced motions in the cytoplasm.

The apparent attraction of cells and intracellular particles to the source is evidently an example of acoustic radiation pressure, another topic treated in the literature on nonlinear acoustics. Particularly relevant here is a mathematical treatment by Embleton[15] of the apparent force on a spherical rigid body of radius a at a distance r from a point source of sound. His general result simplifies considerably when $a \ll r \ll \lambda$, where λ is the wavelength for compressional waves. The reduction of Embleton's result is discussed in Appendix A; it is shown there that the validity of the approximation depends also on the difference between the densities ρ_s and ρ, of the sphere and fluid, respectively. A particularly simple expression applies when ρ_s differs from ρ by a few percent, as is true in our situation. The apparent force between body and source is then found to be attractive when $\rho_s > \rho$ and repulsive when $\rho_s < \rho$.

A particle moving with constant velocity dr/dt under the influence of this apparent force would, in analogy to the motion in a centrifuge, be acted on by an equal and opposite viscous force proportional to dr/dt. Using this condition, together with the reduced Embleton result, one finds an expression that can be written as

$$dr/dt = -s(u_0^2/r), \quad (1)$$

in terms of symbols defined here. The velocity dr/dt is that achieved by the particle at a distance r from the source; s is the sedimentation constant for the particle, defined as the velocity achieved per unit "applied acceleration" in, for example, a gravitational or centrifugal field. An explicit expression for s for the sphere is given in Eq. (A11). The quantity u_0 is the velocity amplitude at distance r from the source; if u_0 is assumed to be proportional to r^{-2}, one finds that dr/dt is proportional to r^{-5}. Equation 1 would be the equation for sedimentation in a centrifuge of angular velocity ω if (u_0^2/r) were replaced by $(-\omega^2 r)$; one should note the difference in direction.

An order-of-magnitude calculation for the attraction may be of interest. In the liquid just outside the surface of a spherical sound source of radius 10μ, vibrating at 10^5 cps with surface radial amplitude of 1μ, a small particle would "sediment" at a rate equal to that in a "4000 g" centrifuge. For example, a spherical particle (in sea water near the source) of relative density 1.05 and radius 10μ will migrate under these conditions with a velocity of about 4 cm/sec. This agrees within an order of magnitude with observed speeds. The apparent force is highly localized; at a distance from the vibrating sphere equal to its radius, the "sedimentation" rate is less by a factor of 2^5 or 32.

Admittedly, the living cell is not a rigid sphere, nor is our vibrating needle a symmetrical point source; however, it is believed that the model of such a sphere in such a field is a useful one for giving insight into the actual problem.

ACKNOWLEDGMENT

This work was supported by the National Institutes of Health, Public Health Service, U. S. Department of Health, Education, and Welfare.

---

[14] W. L. Nyborg, "Acoustic Streaming," in *Physical Acoustics*, W. P. Mason, Ed. (Academic Press Inc., New York, 1964), Vol. 2, Chap. 11.
[15] T. F. W. Embleton, J. Acoust. Soc. Am. 26, 40–45 (1954).

Appendix A

We give here an outline of steps involved in obtaining Eq. 1, starting from the theory given by Embleton in Ref. 14. Equations cited from that reference are identified by the letter **E**; for example, Eq. 9 in Ref. 14 is designated here as Eq. 9(**E**).

The sound field in the fluid is an outgoing spherical traveling wave in which the velocity potential φ at a distance r from the origin is [see Eq. 27 (**E**)]

$$\varphi = (A/kr)e^{i(\omega t - kr)}, \quad (A1)$$

in which the constant A is chosen here to be real. A rigid sphere of radius a is immersed in the fluid, with its center a mean distance r from the origin. We define β as the ratio (ρ/ρ_s), where ρ_s is the density of the sphere and ρ that of the surrounding fluid.

By use of radiation-pressure theory, Embleton calculated the time-averaged force $\bar{P}$ on the sphere in the direction of the origin of the sound field. For a solid sphere in air, he found that the force would be either positive (attractive) or negative (repulsive) depending on the distance r, the radius a, and the angular frequency ω. In this paper, we consider a general result of Embleton's, applied for simplicity to a very special case. We take all dimensions of interest to be small as compared to the sonic wavelength λ. Thus, we assume that

$$ka \ll 1, \quad kr \ll 1, \tag{A2}$$

where the propagation constant k is equal to $(2\pi/\lambda)$ and also to (ω/c), where c is the velocity of sound in the fluid. It is also convenient to make use of the Conditions

$$(ka)^2 \ll 3|1-\beta|; \quad (kr)^2 \ll 3|1-\beta|. \tag{A3}$$

In justification of these conditions, we note that, in the experiments reported in this paper, k is of the order of 3 cm^{-1} and a is less than 0.005 cm; we consider distances r of the order of, say, 5 a. Thus 0.015 and 0.075 are typical magnitudes for ka and kr, respectively, and Eqs. A2 are satisfied. For a marine egg in sea water, a typical value for β is about 0.95; hence $(1-\beta)$ is about 0.05 and Eqs. A3 are satisfied.

A series expansion for $\bar{P}$ is given in Eq. 33 (E) of Ref. 14; when Eqs. A2 and A3 are satisfied, and also $a \ll r$, the primary contribution to $\bar{P}$ comes from the term characterized by $n=1$. We introduce the symbols σ, B, and Q where

$$\sigma = [\alpha^2 - 3(1-\beta)], \quad \alpha = ka;$$
$$B = (C_1 C_2 + D_1 D_2); \tag{A4}$$
$$Q = H_1 H_2 \alpha^5.$$

Expressions for the C_i, D_i, and H_i are given on p. 43 of Ref. 14; from these expressions, in view of Eqs. A2 and A3, we find that

$$\sigma \cong -3(1-\beta); \quad B \cong 3; \quad Q \cong 18\alpha^{-3}(1+\tfrac{1}{2}\beta). \tag{A5}$$

From Embleton's Eq. 33(E), keeping only the term $n=1$, we obtain

$$\bar{P} = -(2\pi\rho A^2/k^2 r^2)(2\sigma^2/Q^2)$$
$$\times [B/(kr)^3][\sigma^{-1}(Q^2-\sigma^2)^{\frac{1}{2}}+1]. \tag{A6}$$

In view of Eqs. A2 and A5, we see that $\sigma^2 \ll Q^2$ and that unity is negligible on the right of Eq. A6. The latter equation then becomes, making use of Eqs. A5, and letting V be the volume of the sphere,

$$\bar{P} = \rho V A^2 (1-\beta)/bk^2 r^5, \quad b=(2/3)(1+\tfrac{1}{2}\beta). \tag{A7}$$

When β is 0.95, b is about 0.98; for present purposes, we may set $b \cong 1$.

The velocity amplitude u_0 at a distance r_0 from the source is, by Eq. A1, for $kr \ll 1$

$$u_0 = |\partial \varphi/\partial r| \cong A/kr^2. \tag{A8}$$

Eliminating A from Eq. A7 by use of Eq. A8 and letting $b=1$, we obtain

$$\bar{P} = \rho V (1-\beta) u_0^2 / r. \tag{A9}$$

Remembering the sign convention, we see that $\bar{P}$ is inward toward the origin when $\beta < 1$ and outward when $\beta > 1$. Thus the body is attracted to the source when $\rho_s > \rho$ and is repelled when $\rho_s < \rho$.

We now assume that the force causes the spherical particle to move with constant velocity dr/dt against a resisting force $f\, dr/dt$, which is equal and opposite to $\bar{P}$. Setting $\bar{P}$ equal to $(-f\, dr/dt)$, we obtain, from Eq. (A9),

$$dr/dt = -s u_0^2 / r. \tag{A10}$$

This is just Eq. 1, the sedimentation constant s being given by

$$s = \rho_s V_0 (1-\beta)/f. \tag{A11}$$

25

Copyright © 1970 by Academic Press, Inc.

Reprinted from *Exptl. Cell Res.*, **60**, 78–85 (1970)

MORPHOLOGICAL CHANGES INDUCED IN THE FROG SEMITENDINOSUS MUSCLE FIBER BY LOCALIZED ULTRASOUND

M. J. RAVITZ and R. M. SCHNITZLER

Department of Anatomy, Albert Einstein College of Medicine, New York, N.Y. 10461, and Department of Physiology and Biophysics, University of Vermont, Burlington, Vt 05401, USA

SUMMARY

The effects of highly localized ultrasonic vibration on frog semitendinosus muscle fibers were examined by electron microscopy. A barium titanate transducer, resonant at 85 kHz, to which a stainless steel acoustic horn was cemented, served as the sound source. Sound displacement amplitudes of one to five micra were produced at the horn tip. Thermal effects and cavitation were absent.

A spectrum of structural changes was observed which depended on amplitude and duration of sonation. The mitochondrial cristae and components of the sarcotubular system appear most sensitive to ultrasound. With increasing amplitudes and treatment durations, decrease in glycogen content, Z and M line disruption, and misalignment of the filaments within the myofibrils occur. In the most severely treated fibers, there is a complete breakdown of band structure. The generation of steady intracellular stresses produced by the sound field, predicted by non-linear acoustic theory, is postulated to explain these results. The results provide evidence that many effects of sound on muscle reported in the literature and ascribed to heating or cavitation can be produced in the absence of these factors.

Since the pioneering efforts of Chambers & Harvey [6], numerous reports have appeared concerning the effects of ultrasonic waves on skeletal muscle morphology [1, 3, 13–18, 20, 23, 24, 27, 28]. The reported effects have ranged from subtle modifications in the striation pattern to severe disruption of the fibers. In more recent work electron microscopy has indicated that components of muscle cells may be differentially sensitive to ultrasonic treatment [4, 10, 25]. Although volume and interface heating as well as cavitation have been implicated as mechanisms, it is not clear what role these play in producing structural changes. Recently, Wilson et al. [32, 33] have shown that injury can be elicited in single muscle fibers using a sound source in which thermal and cavitational effects are absent. The results of an electron microscope study of muscle cells treated with this sound source is reported here.

METHODS AND MATERIALS

All experiments were done on the semitendinosus muscle of *Rana pipiens*. Fibers were subjected to ultrasound by means of a vibrating needle machined into the tip of a stainless steel acoustic horn. The horn was driven by an electroded and polarized barium titanate hollow cylinder with a resonant frequency of 85 kHz. The tip of the needle had a 20 μ radius of curvature. The amplitude of the needle tip motion varied over a range of from 1 to 5 μ (maximum pressure amplitude of 0.125 atm) so that transient cavitation was avoided. Details and an illustration of this sound generator are published elsewhere [31]. Experiments by the authors with temperature crayons indicated a negligibly small temperature rise at the needle tip. Calculations by Brown [5] indicate an upper limit value of temperature rise of 0.4°C. He assumed that viscous heating of the Ringer solution near the surface of the horn tip would be the most plausible cause of any temperature rise. In order to simplify the analysis, Brown made the approximation of representing the vibrating horn tip by a pulsating sphere whose radius equals that of the horn tip. In this situation any heat generation would occur in a thin layer near the sphere surface as a result of viscous drag exerted by the pulsating surface on the Ringer solution.

Single fibers or a small bundle of fibers were mounted in a Lucite chamber on the bottom of which was inscribed a scratch line. The vibrating needle was then carefully positioned over the scratch line by means of a micromanipulator; in this way the location of the sonated area of the fibers was determined. The duration of sound treatment varied from one to 30 min.

The fibers were stretched to about 130 % of rest length, treated with sound, and fixed for 1 h in a 0.5 % solution of glutaraldehyde (Ringer solution, pH 7.1) followed by a 3 h rinsing in a 10 % sucrose in Ringer solution. A 2 mm long segment of a sonated fiber was cut (1 mm on each side of the scratch line). This segment was post-fixed for 1.5 h in 1 % phosphate buffered osmium tetroxide solution at 4°C. Dehydration was done in a graded series of ethanol–water solutions. The fiber segments were embedded according to the method of Luft [21]. Sections were cut on a Sorvall MT-2 Porter-Blum Ultramicrotome using diamond and glass knives and were mounted on 200-mesh screens. The sections were stained with uranyl acetate for 10 to 15 min and double stained with lead citrate for 5 min. After examination in a Philips 200 EM operated at 60 kV, photographs were taken of areas of the muscle fiber showing the greatest structural change for a particular sound treatment.

RESULTS

The results of this study indicate that ultrasonically induced changes in the structure of muscle are dependent on sound amplitude and duration of treatment. The general architecture of a control fiber is shown in fig. 1 where the orderly arrangement of actin and myosin filaments within the myofibrils can be seen. Glycogen granules are in abundance and the components of the transverse tubule system and sarcoplasmic reticulum can be distinguished easily. The insert of fig. 1 shows mitochondria from an unsonated fiber.

With mild sonic treatment, such as a 2μ needle tip displacement amplitude for a period of 1 min, disruptive changes can be seen mainly in the mitochondria (figs 2, 3). Some cristae appear torn away from the outer mitochondrial membrane. With this amplitude and duration of sonation some mitochondria become vacuolar in appearance and in some the destruction is so extreme that the contents disappear altogether. The band structure, sarcotubular system, and glycogen content of these fibers remain normal, although some micrographs show small displacements in the M line. Fig. 3 clearly indicates that the components of the T-tubule system and sarcoplasmic reticulum remain intact with this level of treatment. With longer duration (5 min) of sonation at the same amplitude, some disorganization of the sarcotubular system can be detected. In some areas the triad geometry appears altered, although individual components of the T-tubule system and sarcoplasmic reticulum can be identified. The glycogen granules are still copious, but they are less distinct, giving the impression that they are degraded by the ultrasonic waves.

If the sound amplitude is increased to 5 μ, changes in structure as shown in fig. 4 results after a 1 min treatment. Individual filaments within the myofibrils are displaced with respect to one another. There is a range of sarcomere lengths from about 2.5 to 3.5 μ so that some myofibrils have been considerably stretched. The Z line is tortuous; in some places it is ruptured. In those sarcomeres showing the greatest Z line damage, the M line appears split apart. Despite these alterations in the band structure the actin and myosin filaments remain intact. Glycogen is sparse and the sarcotubular system is now completely disorganized. Large empty spaces appear where the terminal and intermediate cisternae of the sarcoplasmic reticulum and the glycogen were located previous to sound treatment. Fig. 5 (same treatment as fiber in fig. 4) shows the accumulation of spherically shaped bodies at the cell surface. The accumulation at the locus of the sonation of

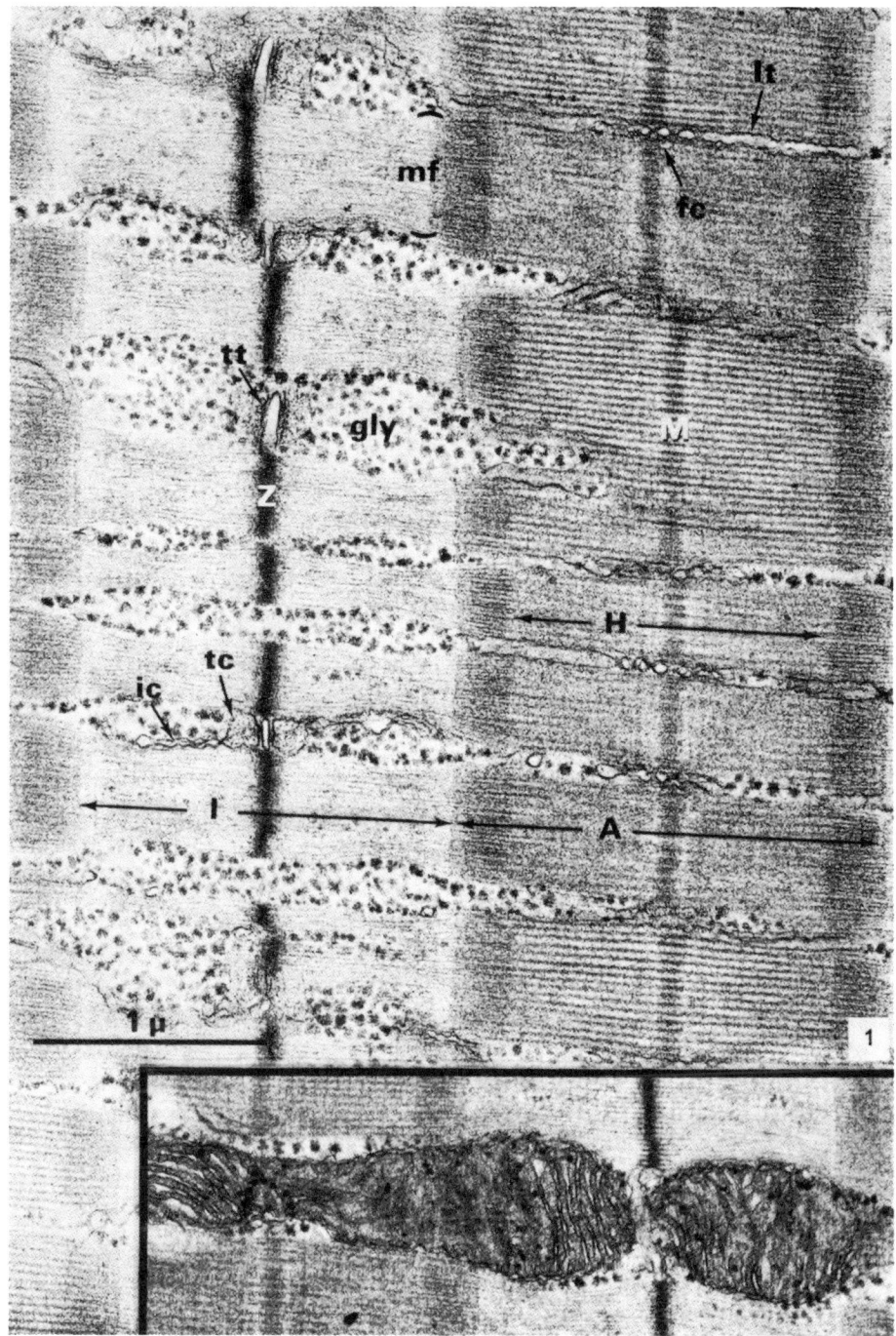

Fig. 1. Control muscle. *mf*, myofibril; *A*, A band; *I*, I band; *H*, H zone; *Z*, Z line; *M*, M line; *tt*, transverse tubule; *tc*, terminal cisterna; *ic*, intermediate cisterna; *lt*, longitudinal tubule; *fc*, fenestrated collar; *gly*, glycogen granules. ×36,000. *Insert* shows mitochondria from a control fiber. ×36,000.

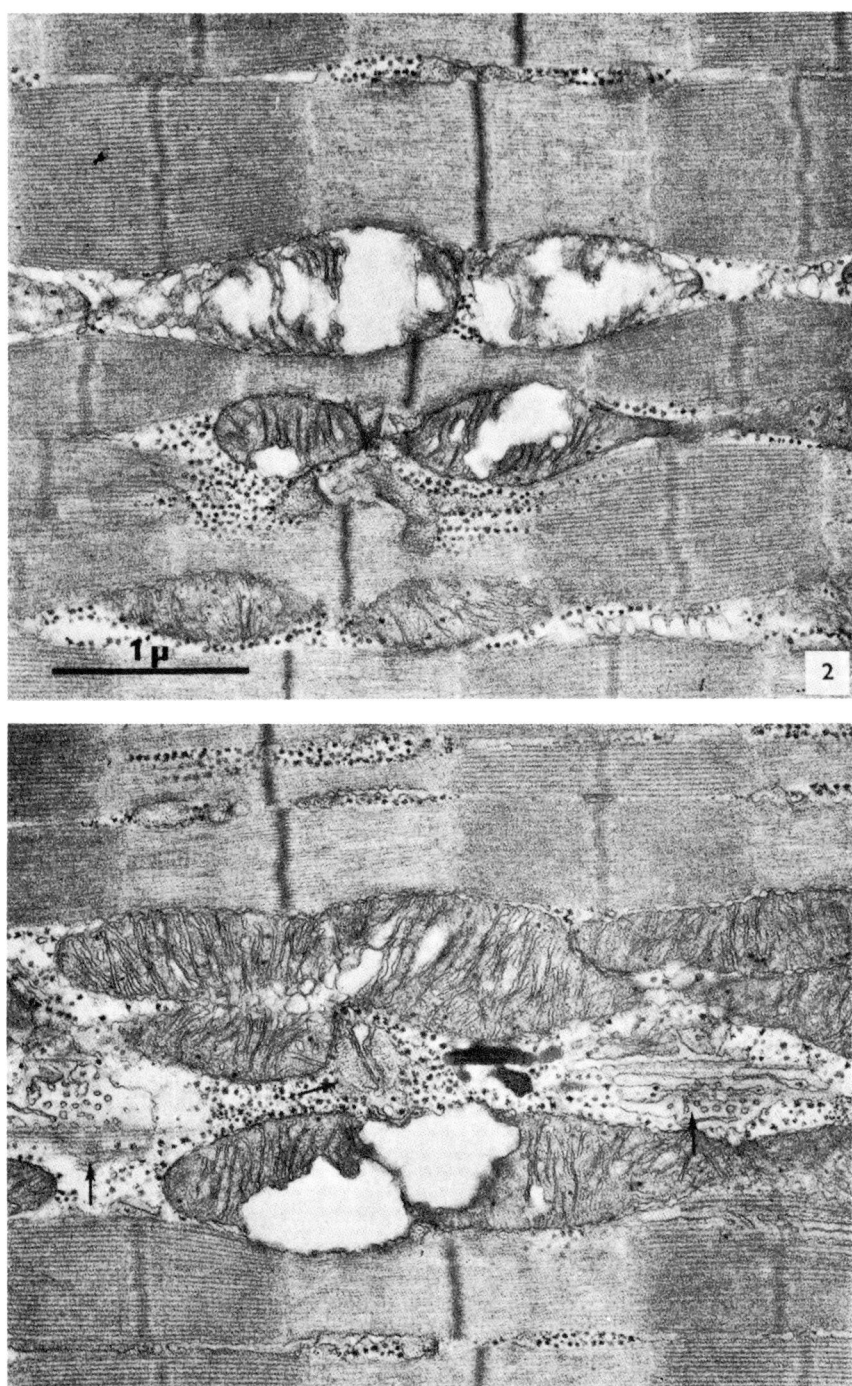

Fig. 2. A fiber showing disruption of the mitochondrial cristae. ×30,350.
Fig. 3. A fiber showing mitochondrial disruption but with components of the sarcotubular system intact (*arrows*). ×30,350.

Fig. 4. A more severely treated muscle fiber showing changes in the band pattern, the variation of sarcomere length, the misalignment of filaments within the myofibrils, disruption of the Z line (*Z*), splitting of the M line (*M*), destruction of the transverse tubules and sarcoplasmic reticulum (*arrows*) and decrease of glycogen content. × 24,650.

Fig. 5. The fiber surface of a muscle cell treated in the same way as that in fig 4. Note the accumulation of spherically shaped bodies (*arrows*). The rupture of the Z line (Z), splitting of the M line (M), misalignment of filaments, and variation of sarcomere length can be seen. ×36,000.

these unidentified bodies is characteristic of treatment using this sound source. In some cases these spherical bodies apparently invade the cell interior. The higher magnification of this micrograph shows the Z line and M line alterations more clearly.

With prolonged ultrasonic treatment (5 μ amplitude, 10 min), the Z line disruption becomes progressively more severe, although the A and I bands can still be distinguished. Electron-dense material appears between myofibrils and is probably debris resulting from the destruction of the mitochondria and sarcotubular system. With very prolonged sonation (5 μ amplitude, 30 min) complete disorganization of the band structure accrues. Despite this extreme perturbation it is often possible to find a few intact myofilaments.

DISCUSSION

Because of the diversity of sound field parameters and conditions used in different laboratories, comparisons are difficult. In general this report confirms previous work of the sensitivity of muscle mitochondria [4] and the Z line [25] to ultrasound. Ultrasonically induced decreases of glycogen content have been reported in liver by Bell [2] and Curtis [7]. The decrease of glycogen reported here for muscle is in disagreement with the biochemical work of Zimny & Head [34]. Other differences between this study and others are probably due to the presence of thermal or cavitational effects in previous work.

There is a considerable literature now accumulating which indicates that if a source of sound similar or identical to the one used here is brought near a cell, the intracellular contents are subjected to two tendencies [8, 9, 11, 12, 22, 26, 29–31]. There is a tendency for both intracellular and extracellular particles to be attracted to the acoustic horn tip. This phenomenon is satisfactorily explained by assuming that these particles are subjected to forces as a result of radiation pressure induced in the liquid by the vibrating source. Radiation force could conceivably be the explanation for the accumulation of the spherically shaped bodies at the site of the sonation seen in fig. 5. The other tendency is for intracellular and extracellular elements to partake in acoustic streaming—a time independent flow in the body of a sound irradiated fluid. For example, Hughes & Nyborg [19] introduced a vibrating needle into a suspension of erythrocytes and found that cell breakage was due to the eddying motion produced in the liquid by the needle vibration. Acoustic streaming and movement due to radiation force will occur intracellularly if the organelles and inclusions suspended in the cytoplasm are free to move, i.e., free from constraints imposed on the cell components by the physico-chemical properties of the cell. In the highly ordered structure of the muscle cell, the protein filaments and membranous systems are not free to assume any random geometrical configuration within the cell. However, the non-uniform nature of the sound field used there generates steady intracellular stresses in these cells; the structures nearest the source, i.e., near the sarcolemma, are subjected to particularly high viscous stress. Such stresses set up in the sarcoplasm may cause stretching and twisting of the internal membranes and filamentous structures.

At low sound amplitudes the stress is not great enough to induce any permanent change in the configuration of the contractile proteins. The mitochondrial cristae, however, may not be as rigidly fixed as the contractile proteins, and so any mixing of the mitochondrial contents disrupts the cristae. The contractile proteins are also subjected to stress, and there may be an associated strain but at low sound amplitudes the strain is not great enough to cause a permanent set, and so the filaments resume their former equilibrium positions when the sound is terminated. At higher amplitudes (figs 4, 5) it appears that the elastic limit has been exceeded and that a permanent set has been imposed on the filaments. The variation of sarcomere length is a characteristic feature of muscle fibers subjected to this particular dose. The micrographs give the impression that the yielding structure is the Z line. Particularly injurious deformations occur in those myofibrils where the Z line is ruptured. This result indicates that the Z line may have a more fragile structure compared with the remainder of the contractile complex. Higher sound amplitudes result in breaking of the protein strands as the ultimate strength of these filaments is reached.

This study indicates that many of the effects of ultrasonic waves reported in the literature and attributed to heat or cavitation can be produced in the absence of these factors. The results suggest the possibility of subjecting muscle cells to controlled doses of ultrasound and mechanically altering their components in a selective manner.

We would like to thank Dr Wesley L. Nyborg, Department of Physics, and Dr Rodney L. Parsons, Department of Physiology and Biophysics, University of Vermont, and Dr George D. Pappas, Department of Anatomy, Albert Einstein College of Medicine, for helpful discussions during the preparation of this manuscript.

This research was supported by the United States Public Health Service, National Institutes of Health, research grants GM 08775 and GM 00102.

REFERENCES

1. Bejdl, W, Nuovo cimento 7 (1950) 461.
2. Bell, E, J cell comp physiol 50 (1957) 83.
3. Biancani, E, Biancani, H & Dognon, A, La presse med 42 (1934) 1503.
4. Borovyagin, V L & El'piner, I E, Biophysics 9 (1964) 335.
5. Brown, R L, M S Thesis, University of Vermont (1967).
6. Chambers, L A & Harvey, E N, J morph physiol 52 (1931) 155.
7. Curtis, J C, Ultrasonic energy (ed E Kelly) p. 85. Univ. Illinois Press, Urbana (1965).
8. Dyer, H J & Nyborg, W L, IRE trans med electron 7 (1960) 163.
9. — Proc 3rd intern conf med electron (1960) 445.
10. Eggleton, R C, Kelly, E, Fry, F J, Chalmers R & Fry, W J, Ultrasonic energy (ed E Kelly) p. 117. Univ. Illinois Press, Urbana (1965).
11. El'piner, I E, Faikin I M & Basurmanova, O K, Biophysics 10 (1965) 889.
12. El'piner, I E, Faikin, I M & Basurmanova, O K, Fed proc 25 (1966) T716.
13. Garay, K & Gerendas, M, Experientia 5 (1949) 410.
14. Gessler, U & Schmitz, W, Strahlentherapie 93 (1954) 617.
15. Gloggengiesser, W, Beitr path anat 11 (1951) 457.
16. — Ultraschall in Med 6 (1953) 139.
17. Haefely, W, Acta anat 29 (1957) 344.
18. Horatz, K, Ultraschall in Med 1 (1949) 249.
19. Hughes, D E & Nyborg, W L, Science 138 (1962) 108.
20. Kelly, E, Ann prog rep contract NONR 1834 (20), NR 101-075, Univ. Illinois (1961).
21. Luft, J H, J biophys biochem cytol 9 (1961) 409.
22. Nyborg, W L & Dyer, H J, Proc 2nd intern conf med electron (1959) 391.
23. Ponzio, M, Minerva med 41 (1951) 85.
24. Rebello, M A, Compt rend soc biol 150 (1956) 1657.
25. Samosudova, N V & El'piner, I E, Biophysics 11 (1966) 821.
26. Schmitt, F O, Protoplasma 7 (1929) 332.
27. Schmitz, W & Gessler, U, Verh deutsch Ges inn Med 58 (1952) 276.
28. — Strahlentherapie 93 (1954) 626.
29. Wilson, W L & Schnitzler, R M, Biol bull 125 (1963) 397.
30. Wilson, W L, Wiercinski, F J, Nyborg W. L. & Sichel, F J, Biol bull 123 (1962) 518.
31. Wilson, W L, Wiercinski, F J, Nyborg, W L, Schnitzler, R M & Sichel, F, J acoust soc Am 40 (1966) 1363.
32. Wilson, W L, Wiercinski, F J & Schnitzler, R M, Biol bull 127 (1964) 396.
33. — J acoust soc Am 37 (1965) 1217.
34. Zimny, M L & Head, L H Am j physiol 200 (1961) 672.

Received June 18, 1969
Revised version received November 11, 1969

Mechanisms for Nonthermal Effects of Sound

Wesley L. Nyborg

Physics Department, University of Vermont, Burlington Vt. 05401

Research on cell suspensions, single cells, macromolecular suspensions, and model (nonliving) systems reveal a variety of changes brought about by sound in the absence of gross-heating or transient ("collapse"-type) cavitation. Many of these phenomena are explainable in terms of acoustic streaming, radiation pressure, and other characteristics of ultrasound that arise from nonlinearity. Commonly, the sonic effect depends on nonuniformity in the sound field. Interestig results are obtained with techniques by which one can set up nonuniform vibration in the walls or membranes of individual cells. In ultrasonic beams acting on tissue, it is possible that variations may arise from gradients in the incident field, or from inhomogeneities in the tissue, which scatter sound. Gaseous pockets or bubbles would be especially effective as such inhomogeneities.

VARIOUS investigators have shown that ultrasound can cause heating in tissue and that the temperature elevation can have important consequences.[1] It has also been shown that important biological changes can result from sonication even when measures are taken to control the temperature.[2-4] In this paper, specific aspects of nonthermal sonic action are discussed, especially those involving steady stresses, displacements, and circulatory movements.

I. STUDIES WITH INDIVIDUAL CELLS AND CELL MODELS

It was reported by Harvey[5] and Schmitt[6] that sound produces movements on the interior of certain plant and animal cells. Their results were confirmed and extended in experiments by H. J. Dyer and the writer[7,8] involving a variety of plant cells. Eddying motions are readily set up in the large vacuoles of plant cells, in which the medium is essentially a waterlike fluid. In the protoplasm, steady stress fields are set up, which cause stretching, twisting, and rupture of delicate membranous and fibrous structures. Also, breaks will occur in the cell wall, if amplitudes are sufficiently high.

Such movements and deformations are especially likely in a cell if its wall is contacted by a small sound source, with dimensions no greater than the (smallest) radius of curvature of the cell surface being contacted. Possibilities are suggested schematically in Fig. 1. A number of experiments have been done in which the cell is contacted with the rounded tip of a needlike object set into oscillation at ultrasonic frequencies. So far as the cell is concerned, the situation is as in Fig. 1 (a). If the cell wall is fairly flexible, it will be set into localized vibration by the oscillating source.

The same general kind of situation is achieved if a volume oscillator exists near the cell as suggested in Fig. 1(b). This condition is realized for a cell surrounded by liquid, with a vibrating gas bubble at its surface. The bubble can be set into vigorous radial oscillation by a general sound field generated, let us say, by a distant source. As the bubble expands and contracts cyclically, it causes the adjoining small area of cell surface to be

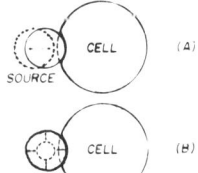

Fig. 1. Methods for setting up vibration in selected parts of cellular membrane. (A) translating source. (B) Pulsating source.

---

[1] See, for examples, P. P. Lele, Exp. Neurol. **8**, 47–83 (1963).
[2] W. J. Fry, V. J. Wulff, D. Tucker, and F. J. Fry, J Acoust. Soc. Amer. **22**, 867–876 (1950).
[3] F. Dunn, Amer. J. Phys. Med. **37**, 148–151 (1958).
[4] J. C. Curtis, in "Ultrasonic Energy," E. Kelly, Ed. (University of Illinois Press, Urbana, 1965), pp. 85–116.
[5] E. N. Harvey and A. L. Loomis, Nature **121**, 622–624 (1928).
[6] F. O. Schmitt, Protoplasma **7**, 332–340 (1929).
[7] W. L. Nyborg and H. J. Dyer, Int. Conf. Med. Electron., 2nd, Paris, France, 24–27, June 1959, 391–396 (1960).
[8] H. J. Dyer and W. L. Nyborg, IRE Trans. Med. Electron., **ME-7**, 163–165 (1960).

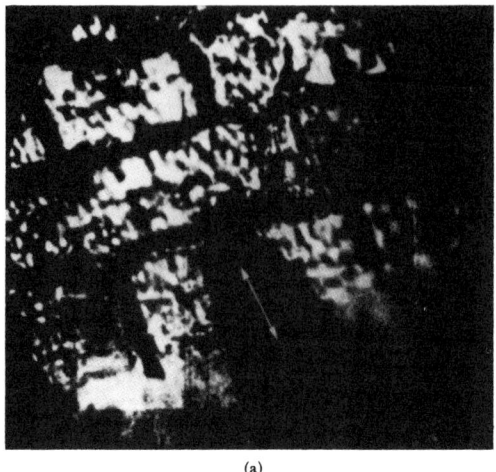
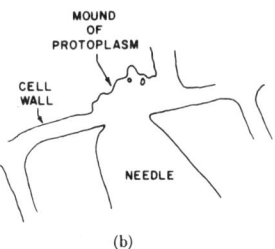

FIG. 2. (a) Frame from motion picture in which longitudinally oscillating needle is shown contacting cell of Elodea (from Ref. 7). Direction of needle oscillation shown by arrow, (b) Identification of structures in Fig. 2(a).

subject to sound pressure of relatively high amplitude. As a result, one expects localized vibration of the surface of the same nature as that from directly applied vibration.

A. Rôle of Radiation Pressure

When a membrane is thus set into vibration in a given small area, a steady stress field is set up in the medium nearby. In terms of theory for physical acoustics, this stress field arises from nonlinearity of the basic equations for sound waves. As one aspect of this field, the sound causes an alteration of the static pressure by an acoustic form of the Bernoulli effect. According to this, the static pressure at a given point is increased by an amount p_2 given by[9]

$$p_2 = C + V - T, \quad (1)$$

where C is a constant, while V and T are time-averaged volume densities of potential energy and kinetic energy, respectively. In terms of the instantaneous particle velocity u_1 and excess pressure p_1, one can write for V and T:

$$V = \langle p_1^2 \rangle / 2\rho_0 c_0^2; \quad T = \tfrac{1}{2}\rho_0 \langle u_1^2 \rangle, \quad (2)$$

where ρ_0 and c_0 are the density and velocity of sound, respectively, in the medium, and where the bracket $\langle \ \rangle$ signifies a time average.

Near a small source such as either of those suggested in Fig. 1, the kinetic energy term T involving u_1^2 is much more important than the potential energy term V involving p_1^2. One, then, has at any point in the field a deficit in pressure proportional to the value of T at that point. Since T varies in space (tending, e.g., to decrease with distance from the source), the pressure p_2 also varies. A particle (such as a chloroplast) suspended in such a field is subjected to unequal forces on opposite sides; since it is thus acted on by a net force, the particle will migrate if not restrained by the surrounding cytoplasm. Such migrations do occur in a cell when part of its wall is caused to vibrate, as exemplified by Fig. 2. Here we see a photograph in which the subject is a plant cell contacted by the tip of a small steel needle, set into oscillation along its axis at 25 kHz. The oscillation had been maintained for a few seconds before the photograph was taken. It can be seen that a mound of protoplasm has accumulated near that portion of the cell wall that the source has caused to vibrate. This accumulation is believed to be a result of sonic attraction and essentially due to gradients of the Bernoulli pressure mentioned earlier. Put in other terms, such a force is a result of acoustic radiation pressure of a particular kind.

Confidence in this interpretation was gained by doing experiments with a model cell, one of whose walls was a thin plastic membrane. The cell was filled with a suspension of lycopodium spores in water, and a small area of the membrane was caused to vibrate ultra-

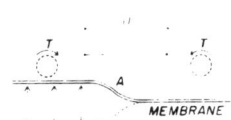

FIG. 3. Portion of membrane near the edge of a vibrating area.

[9] L. V. King, Proc. Roy. Soc. (London) **A147**, 212–240 (1934).

FIG. 4. Circulations near membrane, part of which is caused to vibrate (from Ref. 15).

FIG. 5. Circulation in plant cell part of whose wall is vibrated. Based on results described in Refs. 7 and 8.

sonically, as in the plant-cell situation. Motions of the spores were immediately evident; within a few seconds, a mound (10–20) of spores had accumulated over the vibrating area, analogous to the accumulation seen in the plant cell.

It may be surprising that the acoustic radiation force should be *toward* the source, particles being attracted to it; in plane-wave situations, we are familiar with objects being repelled, the radiation force then being *away* from the source. However, it was shown by Embleton[10] that in the spherical sound field generated by a small source, an object may be attracted to the source. Special results for small particles have been developed by Gor'kov[11] and by the writer.[12] In terms of T and V given in Eqs. 2, the force F in the x direction on a particle of volume v is[12]

$$F = v(B\partial T/\partial x - \partial V/\partial x) + \Delta, \quad B = 3(1-\beta)/(2+\beta), \quad (3)$$

where β is the ratio of the density ρ_0 of the medium to that ρ of the particle; Δ is a term (requiring a rather complicated expression), which is not important for the divergent fields considered here. Near a small source, the term involving T predominates; when β is nearly unity, as is true for biological particles in typical media, we have simply

$$F = v(1-\beta)\partial T/\partial x. \quad (4)$$

We see that the force is in the direction of increasing T (i.e., toward a small source), if $\beta < 1$ (particle more dense than the medium), and the opposite is true if $\beta > 1$. Tests have been made that confirm various aspects of this result.[13,14]

B. Rôle of Sonic Torque and Acoustic Streaming

We return to the general subject of stress fields set up by sound in a medium near a vibrating membrane. Because of radiation forces, as just discussed, particles suspended in the medium tend to move toward definite points or regions of the sound field. A radiation force field actually exists on each volume element of the medium; this force $\mathbf{F}_i$ is of the irrotational type, like an electrostatic field described by Coulomb's law. But the medium is also subject to another kind of force field $\mathbf{F}_s$ of the solenoidal type. Because of $\mathbf{F}_s$, each volume element of the medium is subject to *torque*, causing it to twist. Suspended particles may be

---

[10] T. F. W. Embleton, J. Acoust. Soc. Amer. **26**, 40–45 (1954).
[11] L. P. Gor'kov, Soviet Phys. Doklady **6**, 773–775 (1962).
[12] W. L. Nyborg, J. Acoust. Soc. Amer. **42**, 947–952 (1967).
[13] J. A. Rooney, MS Thesis, Univ. of Vermont, 1967.
[14] W. L. Nyborg and J. A. Rooney, J. Acoust. Soc. Amer. **41**, 1583(A) (1967).

set into steady rotation or may tend to move in circulatory paths.

In studying the acoustic solenoidal force field $\mathbf{F}_s$, one finds that this arises only when the acoustic Bernoulli equation fails. The latter equation describes the situation completely only when there are no absorption mechanisms or boundary effects that convert sound energy into heat. In the vibrating-membrane situation, energy dissipation (i.e., heat production) occurs primarily along the surface in a rather thin layer called the *boundary layer*. Here, the heat generation is essentially the result of viscous drag exerted by the membrane on the tangential component of oscillatory flow. The velocity magnitude of this component tends to be large in regions where there is a large gradient of (normal) vibration amplitude of the membrane. This point is illustrated in Fig. 3.

Here a membrane is shown at a moment when part of it, on the left of "A," is moving upward while the other part is stationary. Part of the fluid above the membrane in the vicinity of A is then forced to the right, causing a momentary velocity U in that direction. A half-cycle later, in the oscillation the motions are reversed. It is therefore plausible that a relatively high tangential component U of oscillatory velocity is set up in the fluid near Point A. In general, large values of U tend to occur when there is a large gradient of the membrane amplitude, as at the edge of a vibrating region; it is such an edge that characterizes the neighborhood of A in Fig. 3.

Returning to the acoustic solenoidal force $\mathbf{F}_s$, its magnitude depends on nonuniformity even more than does that of U. In fact, $\mathbf{F}_s$ tends to be large where there are large gradients of U itself. In Fig. 3, it is clear that U decreases both to the left and to the right of the edge of region A. Hence, large gradients of U and corresponding large values of $\mathbf{F}_s$ occur in neighboring regions on either side of A. Because of $\mathbf{F}_s$, volume elements of medium near the membrane are subjected to steady torque, causing them to twist or rotate. Near an edge of a vibrating region such as A in Fig. 3, the sense of the torque on a given volume element to the left of A is opposite to that on a corresponding element to the right. The situation is suggested schematically for a pair of elements by the arrows "T" in Fig. 3.

Torque distributions near the membrane give rise to circulation or eddying in the adjoining medium if the latter is fluidlike. Studies of these circulations were

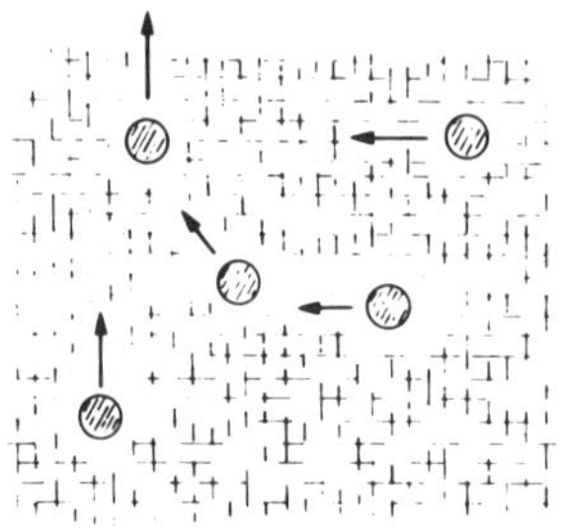

FIG. 6. Sketch suggesting in a general manner the relationship of chloroplasts (circles) to their cytoplasmic environment, consisting of folded fibers and membranes in a waterlike fluid. Arrows show sonic forces on different particles in an inhomogeneous sound field.

FIG. 7. Chloroplast sonically separated from general cytoplasmic region, but still somewhat attached through a filamentary connection. Increase of sonic amplitude causes the connection to be ruptured.

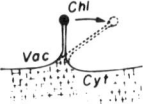

made by Jackson and the writer,[15] with some results shown in Fig. 4. Here, a thin plastic membrane (thickness about 25 μ) formed the bottom of a vessel that contained a glycerine–water mixture in which were suspended some small polystyrene spheres. The latter acted as indicators for liquid movement, and were observed after successive intervals of time when a small circular area of the membrane was set into vibration at a frequency of 40 kHz. Eddies are shown schematically. Motions are particularly vigorous near the membrane but also extend into the medium. A pair of counter-rotating vortex rings are shown, one inside the edge of the vibrating area, the other outside. For one particle, the path is shown quantitatively, though only roughly; successive positions occur at equal time intervals, so that the distance between points is proportional to the speed. Clearly, the highest speeds occur near and tangentially to the membrane, and near the edge of the vibrating area. (Standing waves are sometimes set up in the membrane, causing interesting variations of the flow.)

Acoustic streaming theory was applied to certain aspects of the motion, and found satisfactory. However, a complete description of the circulations would need to be very elaborate, and has not been carried out.

In normal cells, such symmetrical eddying or circulation probably occurs only in the vacuoles, i.e., in spaces where the medium is essentially a simple liquid. Simple eddying also sometimes occurs in damaged cells when fibrous and membranous components of the cytoplasm have coagulated; then relatively large regions of fluid may appear where the viscosity is similar to that of water.

One readily observes acoustic streaming in a variety of healthy cells, but the circulation pattern is less simple and symmetrical than in model situations. Figure 5 (from a film by H. J. Dyer and the writer) shows the orbit of a detached chloroplast in a plant cell part of whose wall is vibrated ultrasonically; conditions are similar to those for Fig. 2. Plotted points are positions at regular time intervals. Speeds vary from a fraction of a circulation per second to hundreds of circulations per second, for wall amplitudes ranging from tenths to tens of microns. Such circulations continue more or less indefinitely as long as the vibration of the cell wall is maintained (if the amplitude is not too great). The circulation shown was the only one observed in this instance; no counter eddy was seen. Such asymmetry is common; it is probably associated with irregularity in the wall itself or in its vibration.

Analogous acoustic streaming motions have been observed in other plant cells and in a number of different animal cells; thus eddying is readily induced in the cytoplasm of mature eggs of Arbacia.[16]

Usually the motions are more complicated than those of simple eddying, as might be expected considering the nature of intracellular media. In plant cells, the most visible indicators of motion are the chloroplasts. These are normally not free, but instead are somehow bound to structures in the cell. The situation is suggested diagrammatically in Fig. 6. In a nonuniform sound field (as exists when part of the cell wall vibrates), a radiation force may act on each chloroplast, but on some more than others. Hence, stresses are set up in the cytoplasmic structure.

For moderate values of the sonic amplitude, chloroplasts appear to be partially but not completely forced from the gel-like cytoplasmic part of the cell into the fluidlike vacuolar parts (see Fig. 7). At such amplitudes, they are observed to take part in vigorous but rather chaotic motion. It is possible that the chloroplasts tend to move in simple circulations, but are hindered by fibrous or membranous connections (as indicated) and other obstacles.

When the amplitude is increased further, the chloroplasts are apparently freed completely and circulate in the vacuole in the manner of a particle suspended in an ordinary liquid.

In animal cells sometimes vibration of the surface also gives rise to motions that are related to, but different from simple acoustic streaming, presumably because of intracellular constraints. In the small eggs of such marine organisms as the starfish, clam, and sea urchin, interesting effects are readily seen in the nucleus.[16] At low amplitudes, the nucleolus spins about its own axis with speed and direction that depend on the amplitude and position of the source. At somewhat higher amplitudes, it circulates about the nucleus and acquires a distorted shape, which returns to normal when the sound is turned off. At still higher amplitudes, it frequently divides into two or more separate bodies, which may recombine after cessation of sound. Perhaps there is some relationship between this fragmentation,

---

[15] F. J. Jackson and W. L. Nyborg, J. Acoust. Soc. Amer. **30**, 614–619 (1958).

[16] W. L. Wilson, F. J. Wiercinski, W. L. Nyborg, R. M. Schnitzler, and F. J. Sichel, J. Acoust. Soc. Amer. **40**, 1363–1370 (1966).

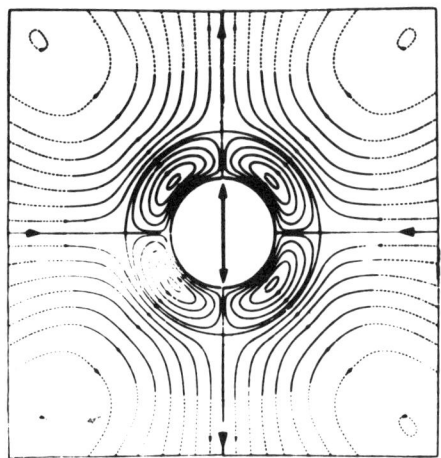

FIG. 8. Acoustic streaming near a vibrating cylinder (from Ref. 17).

and that reported by Curtis[4] for mitochondria in liver cells.

II. STUDIES WITH BIOLOGICAL SUSPENSIONS

In the preceding discussion we have referred to experiments with individual biological cells that are held fixed, then contacted with a small sound source. We now turn to other situations, involving suspensions of cells and macromolecules. Here, the biological objects are subject to the action of small ultrasonic sources immersed in the suspension. In these situations, we must again consider the consequences of "second-order" acoustic quantities, especially acoustic streaming and radiation pressure.

A. Acoustic Streaming and Radiation Pressure near a Small Source

Basic aspects of the acoustic streaming near an oscillating solid source are illustrated by the results for a cylinder, for which the analysis has been carried out in detail. Figure 8 shows the symmetrical circulations that occur near a cylinder set into translational oscillation along a line perpendicular to its axis, in a Newtonian fluid. The Figure is from theory by Holtzmark et al.,[17] who also made experimental tests, finding good agreement with theory under their conditions. Further important information on the streaming near a cylinder has been given by Raney, Corelli, and Westervelt.[18] One sees an inner set of four vortices, defining an *acoustic-streaming boundary layer* as well as an outer set. At

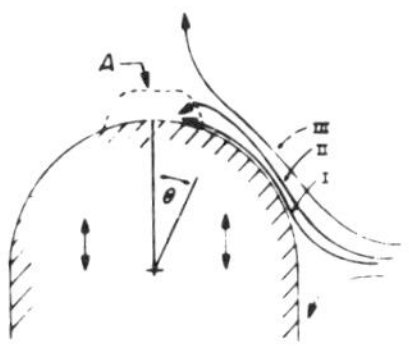

[17] J. Holtzmark, J. Johnsen, T. Sikkeland, and S. Skavlem, J. Acoust. Soc. Amer. 26, 26–39 (1954).
[18] W. P. Raney, J. C. Correli, and P. J. Westervelt, J. Acoust. Soc. Amer. 26, 1006–1014 (1954).

FIG. 9. Trajectories of suspended particles near a vibrating object. Particles more dense than the suspending liquid are attracted to Region A (from Ref. 13).

low and moderate amplitudes, the thickness of the acoustic-streaming boundary layer is of the order of δ, where

$$\delta = (\eta/\pi\rho f)^{\frac{1}{2}}; \quad (5)$$

here η and ρ are, respectively, the viscosity coefficient and density of the fluid, and f is the frequency. At 100 kHz in water, δ is equal to 1.8 μ. At higher amplitudes, it has been shown that this boundary layer shrinks.[18] Very near the boundary, one thus has a region of relatively high velocity gradient. We can expect this region to be important when a vibrating object exists in a suspension of such particles as macromolecules, droplets, or cells. When a particle is swept into the boundary layer, it is subjected to the relatively high viscous stress that exists there.

Figure 9 shows observed particle trajectories in an analogous situation.[13,14] In this experiment, the vibrating object was essentially a needle with rounded tip 750 μ in diameter; the frequency was 85 kHz. The trajectories are of polystyrene spheres (diameter range 7–14 μ) suspended in liquids of varying density. When the densities of particle and liquid are equal, the paths near the rounded tip are similar to those of the cylinder in Fig. 8 (when the former are viewed in a plane through the axis of symmetry). The path labeled III is of this kind. But in the present experiment, the density of the particles was about 7% greater than that of the liquid. Under these conditions, radiation pressure causes deviation from the streaming for particles whose paths lie close to the source; see paths I and II. The direction of the radiation force is such as to attract particles to the Region A at the tip; an aggregate forms here rather quickly. Measurements of actual speeds were made and found to be in reasonable agreement with expectations from theory for acoustic streaming and radiation pressure.

As small particles flow into the near-boundary region of such a source as this, they are subject to boundary-layer shear; estimates from theory give values for velocity gradients from 10^4 sec^{-1} to 10^6 sec^{-1} under reasonable conditions. In experiments by Hughes and the writer,[19] human erythrocytes were suspended in saline and subjected to action of this kind at 85 kHz. Haemolysis was readily produced, evidenced by release of haemoglobin into solution.

More detailed studies were made by Wilson et al.,[16] with suspensions of marine eggs. Motion pictures were

[19] D. E. Hughes and W. L. Nyborg, Science 138, 108–114 (1962).

FIG. 10. Distortion of a suspended droplet in flow with a velocity gradient. Velocity horizontal, increasing vertically. Droplet initially spherical, becomes elongate and may subdivide (from Ref. 20).

taken, revealing considerable distortion of the eggs as they flow into the near-boundary region. At the higher amplitudes (up to about 10 μ peak to peak), cells are disintegrated. At more moderate amplitudes, a protuberance of the cell may form in the direction of the sound source; this protuberance frequently extends outward from the cell until a small body is formed that separates from the rest of the cell.

B. Effects of Hydrodynamic Shear on Cells and Macromolecules

For insight into such phenomena, it is natural to look toward purely hydrodynamical studies. Rumscheidt and Mason[20] have made detailed studies of the effect of well-defined hydrodynamic shear on suspended droplets. As shown earlier by Taylor[21] and illustrated in Fig. 10, shear causes a droplet to assume an elongate shape. Because of surface tension, such a shape becomes unstable, resulting in bifurcation, if the velocity gradient G exceeds a critical value G_c. For a droplet of radius a suspended in a liquid of viscosity coefficients η, the condition is approximately

$$\eta G_c = T/2a, \qquad (6)$$

where T is the interfacial tension. (More general expressions are given in Refs. 20 and 21.) For an oil droplet of radius 10^{-2} cm and interfacial tension 10 dyn/cm^2, suspended in a liquid of viscosity coefficient 10^{-2} P, we obtain 5×10^4 sec^{-1} for G_c. For biological cells, the interfacial tension quoted in the literature is frequently smaller than the above, leading to a correspondingly small value for G_c. Values of G large enough to cause droplet fragmentation readily occur in the boundary layer of ultrasonically vibrating objects. However, no direct test has yet been made of the applicability of this theory to ultrasonic emulsification or cell disintegration on the one hand, or to sonic disruption of intracellular bodies on the other.

In connection with some experiments on DNA, Levinthal[22] was led to analysis of effects of hydrodynamic shear on macromolecules. The molecule is represented by a rod, which, when suspended in a velocity gradient G, tends to rotate, though at an uneven rate. Viscous flow relative to the rod tends to cause tension, which varies with the angle of inclination θ between the rod and the flow. When θ is 45°, the tension

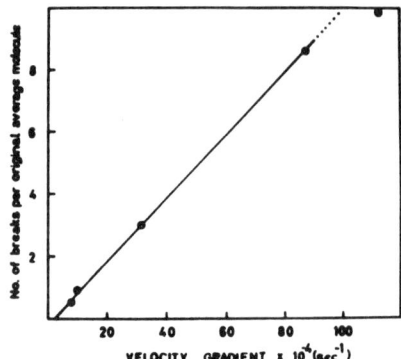

FIG. 11. Breaks produced in DNA by vibrating bubbles. The velocity gradient is that for the bubble-associated acoustic streaming (from Ref. 23).

F is maximum and given by

$$F = fGL^2/16, \quad f = 3\pi\eta/\ln(L/a), \qquad (7)$$

where L is the length and a the radius of the rod. If $L = 0.003$ cm, $(L/a) = 200$, and $\eta = 10^{-2}$ P, one obtains $(10^{-8} G$ dyn$)$ for F. Hence, if the breaking value for F is of the order of 10^{-4} dyn, the critical value for G is 10^4 sec^{-1}. Experiments show that degradation of DNA occurs when G is about 4×10^4 sec^{-1}.

Such values of G occur readily in the boundary layer of ultrasonically vibrating objects. A remarkable test of the theory has been carried out by Pritchard, Hughes, and Peacocke.[23] In their work, DNA suspensions were subjected to the effects of vibrating gas bubbles at a frequency of 20 kHz at amplitudes sufficiently low to prevent cavitation in the "collapse" or "transient" sense. Acoustic streaming is generated by these bubbles in a manner that was studied in detail by Elder.[24] Figure 11 shows a measure of molecular degradation plotted as a function of velocity gradient G for the acoustic streaming. The ordinate is given as "number of breaks," a quantity inferred from the measured viscosity of DNA suspension. A straight-line relationship is expected when the breaks are indeed a result of acoustic streaming, and it fits the data very well. From the slope of this curve, they deduced a value for the breaking strength of a DNA macromolecule, obtaining results that are reasonable for the chemical bonds involved.

III. DISCUSSION

A. Other Studies on Individual Cells with Localized Ultrasound

In the foregoing, we have emphasized observations that are explainable in terms of radiation pressure and

---

[20] F. D. Rumscheidt and S. G. Mason, J. Colloid Sci. 16, 238–261 (1961).
[21] G. I. Taylor, Proc. Roy. Soc. (London) A146, 501–523 (1934).
[22] C. Levinthal and P. Davison, J. Mol. Biol. 3, 674–683 (1961).
[23] N. J. Pritchard, D. E. Hughes, and A. R. Peacocke, Biopolymers 4, 259–273 (1966).
[24] S. A. Elder, J. Acoust. Soc. Amer. 31, 54–64 (1959).

acoustic streaming. In other observations, the mechanisms have not yet been established. Some of the latter observations come from experiments with individual cells contacted by a small source as in experiments already described. Thus, Wilson and Schnitzler[25] have found controllable effects on cell division in Arbacia eggs subjected to localized excitation at 85 kHz. Dyer[26] has seen marked morphological and biochemical changes in progeny of single moss cells which had been subjected to similar excitation. Wilson, Wiercinski, and Schnitzler[27] have reported changes in the calcium permeability of muscle cells as a result of ultrasound applied to part of the cell surface. Sichel and Brown[28] have found changes in the thresholds for electrical excitation of Nitella, when vibration at 85 kHz is applied to part of its wall.

These results demonstrate convincingly that important structures and processes in a cell can be significantly altered by ultrasonic vibrations applied to it *via* only a small part of its membrane. While specific mechanisms for these alterations have not been identified, we can rule out the possibilities of gross heating or cavitation. It is probable that stresses and motions are involved, related to those of acoustic-radiation pressure and acoustic streaming. While physical theory for the latter has been developed rather extensively, it has so far been applied only to Newtonian fluids. To be biologically relevant, the theory must obviously be extended to viscoelastic media.

Model experiments were carried out by H. J. Dyer and the writer[8] from which the following conclusions were reached. In a gel-like media, a sound field has associated with it forces and torques, just as in a Newtonian fluid. However, the visible particle movements that result are quite different. For example, in a Newtonian liquid, solenoidal force fields set up when the sound field is established tend to cause fluid circulations or eddies to appear quite promptly and to continue indefinitely without change (as long as the sound field is unaltered). In a weak gel under the same conditions, particles tend to proceed initially (just after the sound field is set up) along the same streamlines as in the fluid; but constraints soon develop that cause a given particle to travel more and more slowly until it comes essentially to rest. If the sound field is then turned off, each particle tends to return to its initial position; thus the constraints are evidently somewhat elastic as if the particles were attached to springs that become extended or compressed during the motion. These "springs" must be fibers or membranes formed by aggregation of the basic macromolecules of the gel.

If sound can cause distortion of fibers or membranes in a gel, it evidently can also do so in cytoplasm. Perhaps the above-mentioned changes in cell permeability and electrical activity are a result of membrane disturbance, and the changes in cell division and inherited characteristics are the result of distortions in mitotic apparatus and hereditary structures.

It is characteristic of the experiments described so far that vibration is applied to only part of a cell, so that its membrane is excited nonuniformly. This is done deliberately, since theory for acoustic streaming and radiation pressure show that these phenomena require nonuniformity for their existence. For example, it is seen in Eq. 4 that the force F depends on the *variation* of T with x. Similarly, it can be shown that no acoustic streaming at all is to be expected if an entire cell is translated uniformly, i.e., if it is caused to vibrate to and fro as a whole.

B. Possible Implications for Irradiation with Ultrasonic Beams

Questions arise as to the bearing the above considerations have on the important situation where a beam of ultrasound passes through tissue or through a suspension. From careful experimentation and analysis, it has been shown, for example, that motor paralysis in frogs[29] and degradation of DNA in solution[30] are produced by ultrasonic beams without manifestations of gross heating or violent cavitation.

Considerations of mechanisms for these effects obtained with beams are necessarily speculative. It is possible that intracellular stresses, movements, and deformations occur in the tissue somewhat as were observed in the single-cell experiments discussed above. If so, it appears that small-scale nonuniformities must exist in the sound field in the vicinity of irradiated cells. In transmission of a beam through tissue, such nonuniformities may arise from inhomogeneities in the medium. Especially effective sonic excitation would be achieved if the sound were to act on small gas bubbles, possibly after causing growth or aggregation of smaller ones. In tissues, these might appear in the vicinity of cells as suggested in Fig. 1(b). Such bubbles need not be large to be effective, nor need the vibration amplitude be so high as to lead to the noisy and violent effects commonly associated with cavitation. Furthermore, the bubbles or "microbubbles" might be small enough to escape notice by methods that have so far been used to detect them.

Although a detailed analysis of the various ways in which microbubbles might be effective is inappropriate here, a representative calculation is worthwhile. For purposes of illustration, we make order-of-magnitude estimates of acoustic streaming fields that can be

---

[25] W. L. Wilson and R. M. Schnitzler, Biological Bull. **125**, 397(A) (1963).
[26] H. J. Dyer, J. Acoust. Soc. Amer. **37**, 1195(A) (1965).
[27] W. L. Wilson, F. J. Wiercinski, and R. M. Schnitzler, J. Acoust. Soc. Amer. **37**, 1217(A) (1965).
[28] F. J. M. Sichel and R. L. Brown, in MS thesis by R. Brown, Univ. of Vermont, 1967.

[29] W. J. Fry, "Intense Ultrasound in Investigations of the Central Nervous System," in *Advances in Biological and Medical Physics*, C. A. Tobias and J. H. Lawrence, Eds. (Academic Press Inc., New York, 1958), Vol. VI, pp. 341–343
[30] S. A. Hawley, R. M. Macleod, and F. Dunn, J. Acoust. Soc. Amer. **35**, 1285–1287 (1963).

expected. A simple formula for a representative streaming velocity U_L near a pulsating hemispherical bubble on a solid boundary is[31]

$$U_L = u_s^2/2\pi f a, \quad (8)$$

where f is the frequency, a the bubble radius, and u_s the velocity amplitude of the bubble surface. This result was tested by Elder,[24] who found experimental values to be lower than those predicted by Eq. 8, but nevertheless in order-of-magnitude agreement with that equation. (Statnikov[32] has recently derived an alternative expression for the acoustic streaming near a vibrating bubble; while he shows better agreement with Elder's results, his boundary conditions require justification, and it is not clear how he made his numerical calculations.)

In deriving Eq. 8, it was assumed that the bubble rests on a surface that presents a nonslip boundary condition to the tangential flow. There are various possible ways in which this condition might apply approximately to the situations under consideration. For example, a bubble might rest on a membrane or other structure. But particularly interesting is the possibility, as first hypothesized by Harvey,[33] that small solid particles or "cavitation nuclei" exist in solution; such a particle may contain small cracks or crevices that trap gas at its surface. There is much evidence for the existence of these "nuclei."[34] A trapped gas pocket behaves like a bubble in the sense required here, although the (hemispherical) model assumed for Eq. 8 does not literally apply.

When the bubble exists near a surface that presents a nonslip or ("reduced slip") condition, a boundary layer exists near this surface (similar to that near the vibrating cylinder of Fig. 8), in which there exist relatively high gradients of the acoustic streaming velocity. To estimate the possible magnitude of these velocity gradients, we return to Eq. 8; if the displacement amplitude of the bubble surface is ξ, we may replace u_s in Eq. 8 by $(2\pi f \xi)$ and obtain

$$U_L = (2\pi f a)(\xi^2/a^2). \quad (9)$$

Let us now assume that the frequency is 10^6 Hz and that the bubble is of radius 10^{-4} cm, much smaller than the resonant value. Assume relatively weak excitation of the bubble such that the maximum fractional change in its radius (ξ/a) is just 0.05; to achieve this, one requires a pressure amplitude in the vicinity of 0.2 atm, a very modest amplitude indeed. One then obtains from Eq. 9 a value of about 1.6 cm/sec for U_L. This quantity U_L is typical of the velocity a short distance δ from the nonslip surface, where δ is given by Eq. 5. In the boundary layer near this surface, the velocity gradient G is of the order of U_L/δ. For a waterlike medium, we let η and ρ be 10^{-2} P and 1.0 g/cm^3, respectively. Then δ has the approximate value 3×10^{-5} cm and hence

$$G \cong U_L/\delta \cong 5 \times 10^4 \text{ sec}^{-1}. \quad (10)$$

Remembering earlier discussion, we note that this value of G is high enough to cause degradation of DNA and fragmentation of small droplets. Calculations leading to Eq. 10 were made under the assumption of a pressure amplitude of only 0.2 atm; approximations made in the theory would be less valid at higher amplitudes. In the absence of fully adequate theory, it is nevertheless to be expected that values of G much greater than that in Eq. 10 are achieved at the relatively high amplitudes typical of ultrasonic irradiation experiments.

ACKNOWLEDGMENT

Reference is made in this paper to results of research sponsored by two grants (GM-08209 and GM-08775) from the National Institutes of Health to the University of Vermont.

---

[31] W. L. Nyborg, "Acoustic Streaming," in *Physical Acoustics*, W. P. Mason, Ed. (Academic Press Inc., New York, 1965), Vol. II, Part B, pp. 320–322.

[32] Yu. G. Statnikov, Soviet Phys.—Acoust. **13**, 398–9 (1968), [transl. from Akust. Zh. **13**, 464–466 (1967)]. It is assumed here that the bubble surface presents a nonslip boundary condition to the first-order, but not the second-order, velocity field.

[33] E. N. Harvey, D. K. Barnes, W. D. McElroy, A. H. Whiteley, D. C. Pease, and K. W. Cooper, J. Cellular Comp. Physiol. **24**, 1–34 (1944).

[34] H. G. Flynn, "Physics of Acoustics Cavitation in Liquids," in *Physical Acoustics*, W. P. Mason, Ed. (Academic Press Inc., New York, 1964), Vol. I, Part B, pp. 116–127.

Cell Disruption by Ultrasound

Streaming and other activity around sonically induced bubbles is a cause of damage to living cells.

D. E. Hughes and W. L. Nyborg

Ultrasonic methods for injuring or disrupting microorganisms or animal cells in vivo are now widely used (*1, 2*), but there is as yet no generally accepted theory to explain their disruptive effects (*3*). In connection with fragmentation of microorganisms in suspension, it is an accepted view that a major role is played by a sonically maintained activity known as "gaseous cavitation."

This activity occurs in liquids containing dissolved gases and takes place especially readily in the presence of solid or quasi-solid surfaces containing tiny cracks or crevices. These surfaces may be presented by the vessel walls, or by tiny particles suspended in the liquid. Alternations of pressure in the sound field cause bubbles (containing gas or vapor, or both) to grow and take part in a complex and extremely energetic motion. Noltingk and Neppiras (*4*) have shown that under suitable conditions a tiny pocket of air of, say, a few microns in radius expands to a size thousands of times greater than the original volume, then violently collapses to a fraction of the original size, all within a time less than that for one cycle of the sound. In the collapse phase Noltingk and Neppiras predict instantaneous temperatures of the order of 10^{4} °K and pressures of the order of 10^{6} atmospheres. The equations for such nonlinear behavior of the bubbles have recently been developed further by Flynn (*5*). Among the effects attributed to cavity collapse are local heating and electrical discharges, sonoluminescence, chemoluminescence, and free radical formation (*6, 7*).

Dr. Hughes, a member of the Medical Research Council, Unit for Cell Metabolism Research, department of biochemistry, Oxford, England, is at present visiting professor in the department of microbiology, Dartmouth Medical School, Hanover, N.H. Dr. Nyborg is professor of physics at the University of Vermont, Burlington.

Cell Disruption by Ultrasound

In addition to such possible action, gas bubbles are capable of resonance vibration. The resonance radius a of an air bubble in water at atmospheric pressure is given rather accurately (at frequencies of less than 10^6 cycles per second) by the equation

$$a = 3.0/f,$$

where a is in millimeters and the frequency f is in kilocycles per second. Bubbles of less than resonance size attract each other; in a representative high-amplitude sound field such as is used in cell disintegration they collide at high speed and coalesce. By such coalescence and by a "rectified diffusion" process (8), bubbles grow to resonance size (9). Violent volume and surface vibrations then occur (10). As a consequence of this vibration, probably by way of a complex surface motion, the resonantly vibrating bubble surrounds itself with a cloud of very tiny bubbles, or microbubbles, of its own creation (11). Also a vigorous eddying occurs very near the bubbles (9, 12). This eddying or microstreaming is of itself sufficient to accelerate certain types of reactions (13), to induce surface modes of the exterior of cells (14), or to break cells (15).

Activities such as bubble growth, coalescence, surface excitation, microbubble production, and bubble-associated eddying occur, under favorable conditions, at moderate pressures; typical pressures are in the range of 0.01 to 1.0 atmosphere. The threshold for "collapse" type cavitation is usually somewhat greater than 1 atmosphere and may be much greater. This article describes part of a study undertaken to follow by high-speed cinematography the development of sonically induced bubble growth and cavitation and to correlate the various stages seen by this means with the effect on cells and other biological material.

It has been previously shown that when a needle (end diameters 10 to 15 μ) vibrating at 85 kilocycles per second at amplitudes from 0.05 to 5 microns, is placed next to a cell wall, it causes the contents of plant cells to become violently agitated (16). The motions agree with those predicted from an approximate theory of microeddying in that particles in the cell at first tend to move along the wall toward the region of contact with the needle. The pressures were too low for the formation of bubbles and for cavity collapse. In the work reported here a similar vibrating needle was tested with suspensions of microorganisms and other cells to see whether eddying in the liquid phase would develop sufficient shear to injure the cells.

Suspensions of fresh erythrocytes in 0.9-percent saline were treated in the apparatus shown in Fig. 1, which consists of a small plastic container (0.2-ml volume), mounted on the stage of a microscope in such a position that the vibrating needle can be dipped into it with the aid of a micromanipulator. Cell breakage was estimated by measuring the amount of hemoglobin released from the cells into the liquid after removal of cells by centrifugation at 2000g for 5 minutes. Under a variety of conditions it was found that erythrocytes were readily damaged; the number of cells damaged increased as amplitude of vibration increased (Fig. 1). Neither bubble formation nor cavitation was detected under these conditions, and it is clear that the cells were broken by shearing due to the eddying motions (indicated by arrows in Fig. 1) which are induced by the vibrating needle. These may be of the order of 10 meters per second near the tip of the needle. Suspensions of the protozoan *Tetrahymena pyriformis* were treated under the same conditions. Various degrees of injury, from temporary inhibition of motility to complete disruption, were found, depending mainly upon the amplitude. A film was taken at 3000 frames per second of *Tetrahymena* treated under relatively mild conditions which produced no visible disruption. It was seen that not only were the cells violently distorted as they entered the region of highest streaming speeds near the needle tip but that the contents of the cells tended to move in a circular manner relative to the cell motion. Both erythrocytes and *Tetrahymena* are relatively easily disrupted by low rates of shear, produced for instance by filtration through sintered glass filters. It was of interest, therefore, to test the disruptive effect of the vibrating needle on bacteria, which need much higher shear rates for disruption (17). Suspensions of *Escherichia coli* in water or in 0.9-percent sodium chloride [50 mg (dry weight) of cells per milliliter] were treated for various lengths of time and at different amplitudes. Injury was estimated by measuring the amount of protein re-

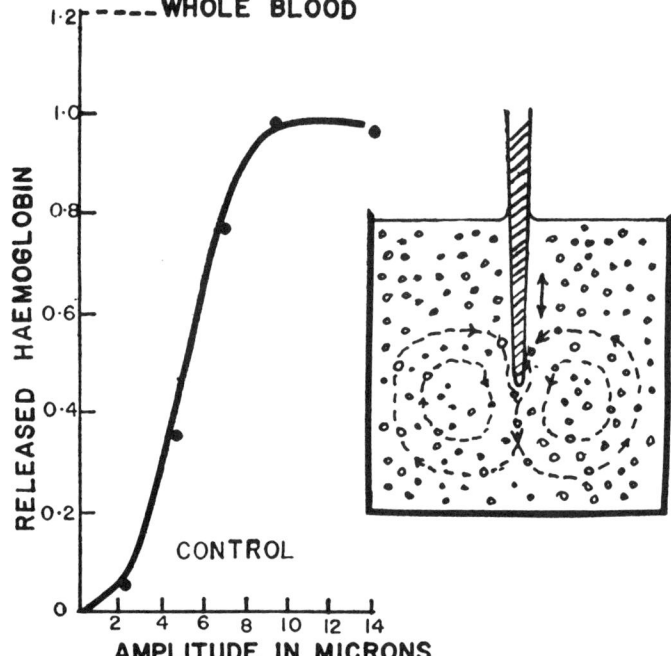

Fig. 1. (Left) The effect of amplitude of vibration on human erythrocytes as measured by hemolysis. Released hemoglobin is expressed as optical density at 490 millimicrons. (Right) Diagram of vibrating needle and cell. The double arrow indicates the direction of vibrations; the dashed line indicates the streaming motions.

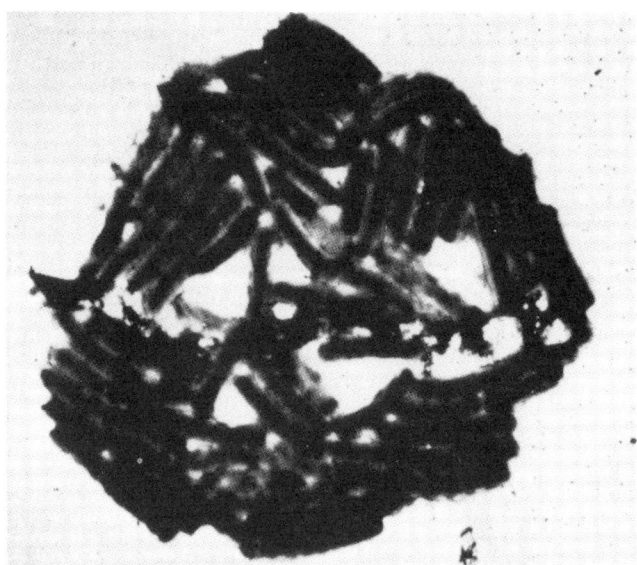

Fig. 2. Electron micrograph showing empty and partly empty hulls of *Escherichia coli*, which accumulated on the needle tip during treatment for 30 minutes.

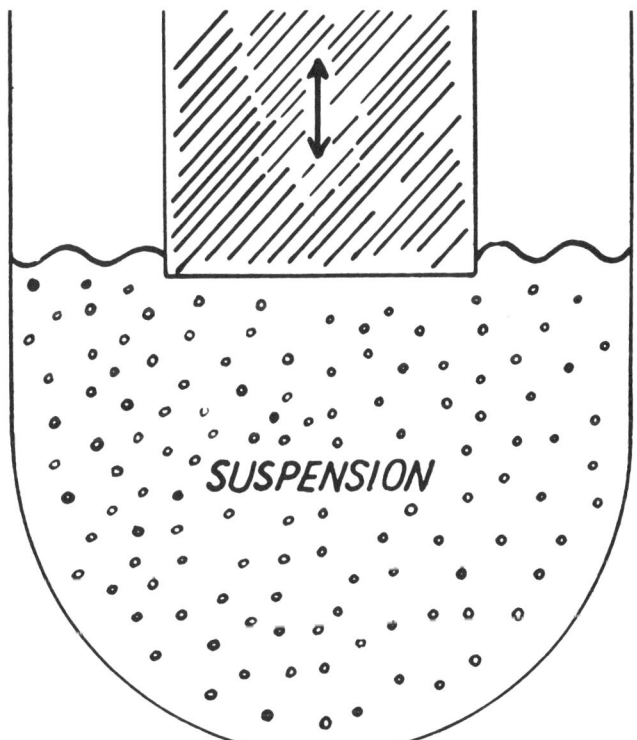

Fig. 3. Diagram showing the arrangement of a vibrating probe dipped into a cell suspension, such as is used in the M.S.E. 20-kilocycle-per-second cell disintegrators.

leased from the cells and remaining in the supernatant after centrifugation at 6000 to 8000g for 10 minutes ([18]). Significant amounts of protein were released from the cells upon treatment for relatively long periods (up to 1 hour), and the amount of protein depended approximately linearly on the amplitude of the needle vibration. Examination by light and electron microscopy showed many empty cells in the treated suspensions. In particular, clusters of empty cells were found in a small sheath of material which often accumulated on the needle just below the tip (Fig. 2).

Cavitation Streamers

It is evident from these preliminary results that streaming motions induced by the vibrating needle produce sufficient shear to injure red blood cells, *Tetrahymena*, and bacteria. In the latter case, however, the damage is small compared with the usual effects of ultrasound, despite the fact that treatment periods were very long (up to 1 hour) compared with the time (3 to 5 minutes) required for complete disruption of *Escherichia coli* under the usual conditions ([18]).

The question arises whether the mode of action of ultrasound in commercial disintegration devices is like that of the vibrating-needle arrangement. One's first impression is that the situations are quite unrelated. The M.S.E. disintegrator ([19]), in which vibrations of 10- to 20-micron amplitude are generated by a vibrating bar (19 to 20 kcy/sec), is in fairly common use for disintegrating bacteria (Fig. 3) ([18, 20]). This instrument may be tuned by ear for maximum hissing noise or with the aid of signals from a strain gauge or accelerometer. The hissing noise is associated with so-called "cavitation streamers." This term is generally used to describe transient clouds of bubbles formed in a liquid by sound intensities above the cavitation threshold. These give the appearance of cloudy lines, often originating at the point of highest pressure amplitude, in this case on the vibrating bar or probe. Under suitable conditions, sound-induced chemoluminescence and sonoluminescence can be seen to originate in such streamers.

We have made high-speed motion pictures of the development of cavitation streamers on the surface of a bar probe. The streamers appear to be formed by the continued appearance, growth, and coalescence of small bubbles attracted to each other and to larger ones which grow to approximately resonance size. An unexpected finding was that as these large bubbles reach a given size, near that for resonance, they suddenly vanish, leaving in their place a cloud of tiny bubbles (Fig. 4). It is possible that this kind of bubble collapse is a catastrophic form (occurring at sufficiently high amplitude) of the Willard phenomenon, referred to earlier, of microbubble production associated with surface vibrations. Its relationship to the simple, radially symmetrical collapse considered by Noltingk and Neppiras (4) is of course not known. The small bubbles formed at the time of collapse hasten to coalesce and join larger bubbles, thus maintaining a continuous coalescence-growth collapse cycle.

The formation of cavitation streamers is often associated with the formation of free radicals and sonoluminescence (21), which are thought to occur at the collapse of cavities (7). Cell breakage, however, is independent of free radical formation (22). One is led to suggest that cell breakage is not dependent on violent collapse but may well be the result of shearing action associated with bubble-induced eddying and related motions.

Low-Amplitude Bubble Activity

To test this suggestion, means were considered for obtaining bubble activity at relatively low amplitudes at which collapse would not occur. From the experiments with the vibrating needle it seemed likely that at least a slow rupture of cells could take place under these conditions. However, a valid test of the effectiveness of low-amplitude bubble activity requires a special arrangement. Commonly the amplitude must be relatively high in order to produce the bubbles themselves. A method described below proved satisfactory for causing bubbles of suitable size and numbers to appear at low amplitude. Using this, we were able to study the effects of vibrating bubbles over a range of amplitudes; at very low values the bubbles appear to act as simple (secondary) sound sources, while at high values streamer formation and collapse events are observed.

The face of a brass probe (2 cm in diameter) was drilled with a series of about 50 holes, each slightly smaller than a resonant bubble at 20 kilocycles

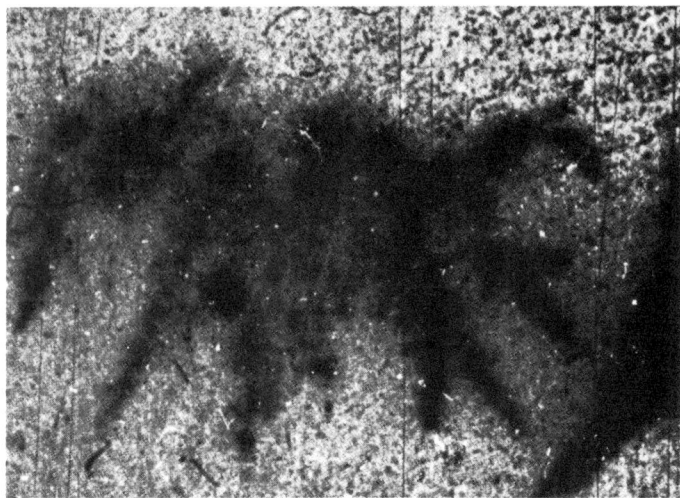

Fig. 4. A single frame from a film of cavitation streamers taken at about 6000 frames per second. The large bubbles appear as dark spots, and the streamers, consisting of masses of smaller bubbles, appear as the branching, less dense background. The edge of the vibrating bar is at lower right.

per second, 200 microns in diameter and 200 microns deep (Fig. 5, bottom right). At low amplitudes (about 3 μ), bright and shining bubbles appeared to grow from the trapped air in the holes and extruded from the surface of the probe into the liquid but did not leave the holes. With increasing amplitude, the bubbles became slightly opaque and the surfaces appeared to be actively vibrating. Occasionally at this stage a bubble would leave the hole and move either toward another bubble or to the center of the probe. Slight hissing was generally heard at this stage. A further increase in amplitude caused most bubbles to leave the holes and caused small opaque bubbles to move about the surface of the probe. Marked hissing could then be heard. Increasing the amplitude still further caused marked movement in the surface of the liquid next to the probe. Small bubbles, both from the holes and from the surface, then moved toward the larger bubbles, and streamer activity (in terms of growth by coalescence and collapse, with small-bubble formation) was established. The threshold for streamer formation, 1.3 microns, is indicated by the vertical dashed line in the graphs shown in Fig. 5. By carefully adjusting the amplitude, each of the described stages of bubble activity could be held for periods up to 60 minutes at the lower amplitudes and for shorter periods (because of heating) at the higher amplitudes. A highly polished probe without holes drilled in the surface did not show streamer formation until it was driven at much greater amplitudes than the probe with holes in the surface. The efficiency of these probes for disrupting *Escherichia coli* is shown in Fig. 5 (bottom left), where it may be seen that cell breakage occurs at much lower amplitudes with the drilled probe than with the polished probe. With the former, significant breakage occurred at the first stage of visible bubble formation and increased almost linearly as the amplitude increased.

As previously described, no phenomenon resembling the collapse stage of cavitation was evident at low amplitudes. Cavitation streamers occurred only when the amplitude was great enough to result in coalescence and collapse—that is, about 1.3 microns (Fig. 5, top left and top right). Free radical formation, as indicated by the formation of iodine from potassium iodide in the presence of carbon tetrachloride, was measured (18). No free radical formation was detected at amplitudes below the streamer threshold during periods of up to 40 minutes, whereas rapid iodine release was detected in a significant amount after a few seconds of marked streamer formation. The onset of free radical formation occurred suddenly when the amplitude reached 1.3 microns (Fig. 5, middle right).

Breakdown of Large Polymers

The breakdown of a large polymer such as deoxyribonucleic acid (DNA) was of interest under these conditions, as this molecule may be broken by liquid shear (23) and also by free radical attack. Shearing in the absence of free radicals is thought to reduce the molecular size by producing breaks across the phosphate-sugar backbone, whereas free radical attack breaks the hydrogen bonding between bases and reduces hyperchromicity. It was found that rapid breakdown of DNA, as judged by reduction in viscosity, occurred at the lower amplitudes and, in the absence of free radicals, at amplitudes below the threshold for streamers; there was no reduction in hyperchromicity (Fig. 5, top right and middle right). Although the breakdown of DNA occurred more rapidly at the higher amplitudes, denaturation, as judged by an increase in optical density, also occurred (Fig. 5, top right). Thus the chemical effects of free radical formation were clearly separated from the chemical effects resulting from shear due to violent eddying of the liquid around bubbles.

The finding that cells such as bacteria may be disrupted by ultrasound in the absence of free radical formation is particularly important for the preparation of biologically active material such as enzymes from the cells. It has been shown, for instance, that in certain cases enzyme inactivation is entirely due to oxidation by free radicals (18) and may be partly prevented by the addition of a scavenger such as cysteine. In other instances, the passage of hydrogen may also reduce enzyme inactivation by reducing free radical formation (3). It has also been suggested that shearing alone may inactivate certain enzymes—for example, polymetaphosphatase (18). It would be of interest to see whether the lower shear produced by bubbles would also inactivate this enzyme.

Implications

The apparatus used in these studies cannot be regarded as having many practical applications since the volumes treated are relatively small and the times of treatment are long. However, the results suggest that modification of the M.S.E. disintegrators may improve their overall performance as well as give more control over the disintegrative effects. For instance, the effects of the small bubbles formed in the holes drilled in the face of the probe appear to be related to the effects of adding powders on disintegration by the higher powered M.S.E. instruments. It was found that the addition of diatomaceous earth (Embacel) could, under certain conditions, triple the rate of cell breakage (18). Embacel traps a considerable amount of air in the form of small bubbles of somewhat less than resonance size. Nonwettable solid particles of the same size, which were suggested as a source of nuclei, were without effect (24). Powdered glass and other materials that trap less air than Embacel were also less effective. At first it was thought that these results might be explained by assuming that the Embacel and other powders served as a source of nuclei, particularly as their effect was most marked under conditions where air exchange between liquid and air was difficult. However, an alternative possibility, suggested by the experiments described, is that the increase in the number of air bubbles throughout the liquid provides a greatly increased

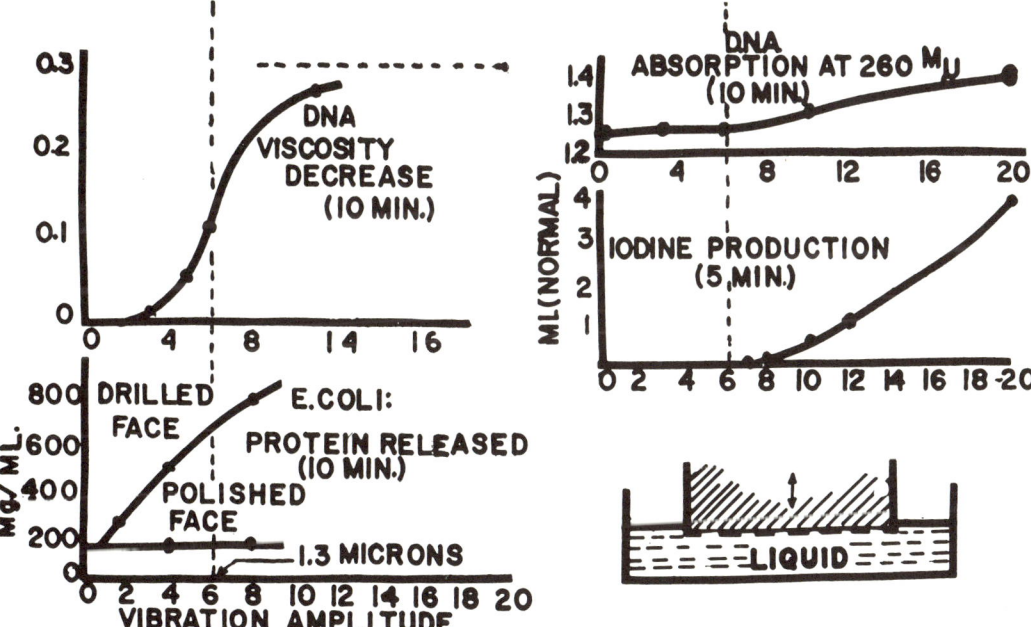

Fig. 5. (Bottom right) Drilled probe. (Bottom left) Disruption of *Escherichia coli* by drilled and polished probe. The other graphs show the effect of amplitude on DNA breakdown and on free radical formation. In each case the amplitude is expressed in microns. The decrease in viscosity of DNA (top left) is expressed as a fractional decrease. The absorption of DNA (top right) is expressed as the optical density with a 1-centimeter light path. The release of protein was estimated by means of the Folin reagent.

number of rapidly moving objects as well as of centers of eddying activity, each of which contributes toward the total disruptive effects. The results also suggest that cavitation streamers per se may be unnecessary for cell damage and that the improved rate of disruption generally observed when these are present is due mainly to the increased number and activity of bubbles, and not to the effects of the collapse of bubbles, as is generally assumed to be the case.

Ultrasonic methods of disrupting microorganisms are generally found to yield extracts containing finer fragments than are yielded by other methods—for example, use of a Hughes press or Milner press (21). However, recent studies on time constants for the release of various cell components of yeast, together with electron microscope studies, suggest that the initial cell rupture may not differ radically in the various methods and that the fine fragments produced by ultrasonics result from further comminution after release from the broken cells. In addition, there is a suggestion that some components may be released by ultrasound even before major damage has been caused to the cell wall. This is illustrated in Fig. 6, which compares the rate of release of total protein with the rates of release of two enzymes—alcohol dehydrogenase, which is regarded as a "soluble" enzyme, and succinic dehydrogenase, which is bound to particles resembling mitochondria (25). Such experiments strongly suggest that examination of the actual process of disruption may give valuable information about the structure of the cells and the location of enzymes within them (26).

With regard to the surgical use of ultrasound for producing brain lesions and destroying the labyrinth (27), there is still no agreement about the mechanism of the disruptive effects which result in the death of the cells, generally after some delay following application (2). There is evidence suggesting that the effects are not due to cavitation, but whether they are due to local heating is in some doubt (2, 28). In experimental animals no immediate gross histological changes are found, but the rapidity of loss of motor or sensory activity suggests that some immediate biochemical lesion is caused. It is likely that such immediate effects may be produced by causing motions of cell contents similar to those caused by the vibrating needle in plant cells or *Tetrahymena*. In animal cells such violent motions might destroy cell function by injuring membrane systems such as the endoplasmic reticulum or other parts of cells not normally detected by the histological method used. A mechanism by which such motions may be brought about by focused sound fields is suggested by the experiments described, which have shown that a vibrating bubble as well as a vibrating needle may act as a sound source in setting up eddying systems of its own within a larger sound field. Such point sources might occur particularly at cell junctions or at membrane interfaces within a nonuniform sound field, such as is used for labyrinth destruction. The presence of small bubbles cannot be altogether dismissed on the basis of histological examination because it is not likely that they would persist after the sound source has been turned off (29).

In addition to giving a description of the events leading to ultrasonically induced cavitation, the experiments briefly reported here are of interest in showing that cell rupture, normally associated with "gaseous cavitation," can occur in the absence of the collapse phenomenon and at surprisingly low power. This finding is important for its practical application in preparing extracts of microorganisms and other cells with ultrasonic vibrations, and important for the further insight it gives into the disruptive action of sound on free cell suspensions and on cells in tissue aggregates (30).

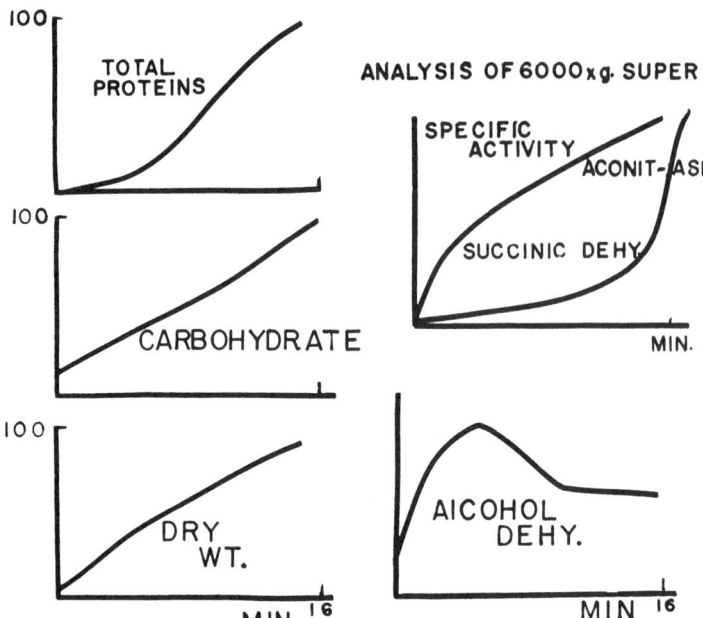

Fig. 6. Time constants for the release of cell components by treating yeast suspension with ultrasound. Yeast (30 ml of a 1.5 wt/vol suspension in water) was treated with the small titanium probe at power setting 4 in the 500-watt M.S.E. disintegrator. Samples were centrifuged at 6000g for 10 minutes, and the components in the supernatants were estimated. Protein, carbohydrates, and dry weight are expressed as percentages of the totals released when all the cells are broken. The enzymes are expressed as specific activities—that is, units of enzyme released per unit of protein released.

References and Notes

1. W. B. Hugo, *Bacteriol. Rev.* **18**, 87 (1954).
2. W. G. Fry, *Advances Biol. Med. Phys.* **4**, 282 (1958).
3. P. Grabar, *ibid.* **3**, 191 (1958).
4. B. E. Noltingk, and E. A. Neppiras, *Proc. Phys. Soc. London* **63B**, 6 (1950); *ibid.* **64B**, 1032 (1951).
5. H. Flynn, in preparation.
6. O. Lindstrom, *J. Acoust. Soc. Am.* **27**, 68 (1956).
7. P. Jarman, *ibid.* **32**, 1459 (1960).
8. F. G. Blake, *Office Naval Res. Tech. Rept. No. NR 014-903* (1949); Rosenberg, *Office Naval Res. Tech. Rept. No. NR 384-903* (1953).
9. J. Kolb and W. L. Nyborg, *J. Acoust. Soc. Am.* **28**, 1237 (1956).
10. M. Kornfeld, and L. Surorov, *J. Appl. Phys.* **15**, 495 (1944).
11. G. W. Willard, *J. Acoust. Soc. Am.* **26**, 933A (1954).
12. S. A. Elder, *ibid.* **31**, 54 (1959); R. K. Gould, thesis, Brown University (1961).
13. W. L. Nyborg, R. K. Gould, F. I. Jackson, C. E. Adams, *J. Acoust. Soc. Am.* **31**, 706 (1959).
14. E. Ackerman, *Proc. Intern. Congr. Med. Electron. 3rd, London* (1960).
15. R. E. P. Smith, thesis, Brown University (1961).

16. H. J. Dyer and W. L. Nyborg, *IRE (Inst. Radio Engrs.) Trans. Med. Electron.* **7**, 163 (1960).
17. D. E. Hughes, *Brit. J. Exptl. Pathol.* **32**, 97 (1951); H. W. Milner, N. S. Lawrence, C. S. French, *Science* **111**, 633 (1950).
18. D. E. Hughes, *J. Microbiol. Biochem. Eng. and Technol.* **4**, 405 (1962).
19. The M.S.E. disintegrator is manufactured by Measuring and Scientific Equipment, Ltd., Westminster, England.
20. E. A. Neppiras and D. E. Hughes, in preparation.
21. E. N. Harvey, *J. Am. Chem. Soc.* **61**, 2392 (1939).
22. D. E. Hughes and A. Rogers, *Medical Electronics* (Illiffe, London, 1960), p. 397.
23. P. Doty, B. B. McGill, S. A. Rice, *Proc. Natl. Acad. Sci. U.S.* **44**, 432 (1958).
24. G. Bradfield, *Symposium on Cavitation in Hydrodynamics* (Her Majesty's Stationery Office, London, 1953).
25. M. F. Utter, D. B. Keech, P. M. Nossal, *Biochem. J.* **68**, 431 (1958).
26. A. G. Marr and E. H. Cota-Robles, *J. Bacteriol.* **74**, 79 (1957); M. M. Matthews and W. R. Sistrom, *ibid.* **78**, 778 (1959).
27. A. R. James, G. A. Dalton, M. A. Bullen, M. F. Freundlich, J. C. Hopkins, *J. Laryngol. Otol.* **54**, 730 (1960).
28. H. T. Ballantine, E. Bell, J. Manaplatz, *J Neurosurg.* **17**, 858 (1960).
29. E. N. Harvey, A. H. Whiteley, W. D. McElroy, D. C. Pease, D. K. Barnes, *J. Cellular Comp. Physiol.* **24**, 23 (1944).
30. We are indebted to Sir Hans Krebs and to E. Neppiras for advice and criticism. The experiments were aided by a loan of equipment from Measuring and Scientific Equipment, Ltd., and by grants from the U.S. National Institutes of Health and the Rockefeller Foundation. The investigations were carried out while one of us (W.L.N.) was on leave of absence at Oxford in 1960–61.

28

Copyright © 1966 by John Wiley & Sons, Inc.

Reprinted from *Biopolymers*, 4(3), 259–273 (1966)

The Ultrasonic Degradation of Biological Macromolecules under Conditions of Stable Cavitation. I. Theory, Methods, and Application to Deoxyribonucleic Acid

N. J. PRITCHARD, D. E. HUGHES,* and A. R. PEACOCKE,
Departments of Biochemistry and Clinical Biochemistry, University of Oxford, England

Synopsis

Solutions of calf thymus DNA have been degraded in the presence of vibrating air bubbles in ultrasonic fields of low power which would not normally induce ultrasonic cavitation. The DNA was degraded to a limiting intrinsic viscosity, after which further irradiation by ultrasound had little or no effect. This limiting intrinsic viscosity decreased with increase in the ultrasonic intensity. Previously developed theories have been adapted to calculate the maximum velocity gradient associated with the streaming of the solution around such vibrating air bubbles. The tensile force which is induced and which acts on the DNA has been calculated on the basis of current theories of degradation by hydrodynamic shear. These calculations indicate that the degradation of the DNA by ultrasound under conditions of "stable cavitation" is mainly the result of the shearing forces engendered in the solution around the oscillating bubbles.

INTRODUCTION

Irradiation by ultrasound has found wide biochemical application as a method for the disruption of cells and extraction of enzymes, and in clinical diagnosis and therapy. That biological macromolecules may lose some degree of organization or be degraded by ultrasound is often overlooked, although it was demonstrated in 1933 that high frequency sound reduces their viscosity.[1,2] It has recently been shown that protein synthesis in yeast cells ceases in the presence of a critical sonic field of low power, but that the cells remain viable after treatment.[3] This effect is ascribed to "supramacromolecular disorganization" within the cells when the sound field is applied. Mikhailov and Fedorova have suggested that reversible structural changes produced in synthetic polymers by ultrasound are related to the number of short-lived "nodes" in the polymeric network at which van der Waals' forces balance, and that these changes are measurable by the higher absorption coefficient of the ultrasound during irradiation.[4]

* Present address: Department of Microbiology, University College, Cathays Park, Cardiff, Wales.

Hence reversible and irreversible changes can be distinguished in the effects of ultrasound on macromolecules. While it may well be that the harmful effects of ultrasound on living tissues and tissue extracts are partly associated with reversible changes caused by the sound field, the present work is restricted to the irreversible changes.

Although the fact of irreversible degradation of high polymers by ultrasound is well established, the mechanism of the process is not well understood. Degradation could occur: (*1*) by unequal distribution, caused by localized heating effects in the sonic field, of the vibrational energy associated with monomers along the polymer chain; (*2*) by attack of free radicals produced in the field; (*3*) by a mechanicochemical or "shearing" mechanism, although this needs further clarification.

Frenkel first pointed out that purely hydrodynamic forces on long polymers could be great enough to break covalent bonds. Subsequent work on the degradation of both synthetic polymers and biological macromolecules has described the process more quantitatively.[6-11] According to Gooberman the sound field produces cavitations from which shock waves radiate into the liquid; the shearing stresses associated with the flow of solvent from these shock waves are thought to cause degradation and the bond most likely to be broken is close to the center of the polymer.[12] Freifelder and Davison confirmed that the ultrasonic degradation of virtually monodisperse T7 phage DNA was a nonrandom process which resulted in the preferential halving of molecules.[13]

No attempt has yet been made to correlate the degradation of polymers by hydrodynamic shear with degradation by ultrasound, principally because it is difficult to design experiments which clearly distinguish the different mechanisms of ultrasonic degradation. Nevertheless, most workers favor the shearing mechanism. A common feature of all published theories of ultrasonic degradation is the postulate of frictional forces between molecules of solvent and the macromolecules,[14] which are thought to be increased by cavitation.[15]

In most of the previous work, gaseous cavitation has been induced by treating liquids at sound pressures above the cavitation threshold. The cavitation threshold depends upon frequency, on the presence and nature of dissolved gases, and on the number of cavitation nuclei of a suitable size which are present; this threshold is generally of the order of 0.7 to 1.0 atm. cm.$^{-2}$ at 20 kc./sec. under the conditions used in these experiments. At pressure amplitudes above this threshold, collapse cavitation is present with its concomitant heat, shock-wave, and chemical effects.[16-18] Hughes and Nyborg showed that some features of the degradation of DNA are not dependent on collapse cavitation at low frequencies (20–25 kc./sec.) when their system was well below the cavitation threshold.[19] It appears that at frequencies greater than about 200 kc./sec. collapse cavitation is less likely to occur and is probably not responsible for macromolecular degradation.[4,20] It is therefore reasonable to suggest that an effect, responsible for degradation and associated with bubble activity in lower frequency

ULTRASONIC DEGRADATION OF BIOLOGICAL MACROMOLECULES 261

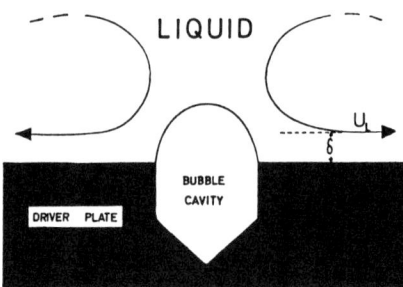

Fig. 1. Acoustic microstreaming around a pulsating bubble at low pressure amplitudes of the driver plate.

systems at pressure amplitudes below the collapse cavitation threshold, persists also at higher frequencies in the absence of cavitation. Effects which are associated only with collapse cavitation (e.g., sonoluminescence) can be disregarded, since degradation begins in cavitating systems long before their onset.

In this work, an apparatus was used similar to that of Hughes and Nyborg in their experiments on cell disruption.[19] The solution to be treated was seeded with air bubbles at atmospheric pressure and these enhanced the magnitude of the time-independent flow associated with an applied sonic field of low power. The solution is then said to be irradiated under conditions of "stable cavitation." A more important aspect of this system is that the magnitude of this time-independent flow (acoustic microstreaming) is calculable, as is the associated velocity gradient, when the bubble radius is pulsating sinusoidally. The system of vibrating bubbles, with the characteristic pattern of eddying liquid surrounding it, constitutes a "regime."[21] The situation shown in Figure 1 is attainable at low pressure amplitudes, and has been called regime I[22] (this is not the same as Elder's regime I). At higher pressure amplitudes, other modes of vibration of the bubble are observed and the shear stresses in the liquid cannot be evaluated. In the experiments described here, bubble vibration was largely restricted to regime I.

ULTRASONIC IRRADIATION UNDER CONDITIONS OF STABLE CAVITATION

Apparatus

A Mullard 500-W. oscillator and amplifier fed a 20 kc./sec. signal to a water-cooled magnetostrictive transducer (Fig. 2). This signal was monitored by a Phillips GM 6000 valve voltmeter and also by one trace of a dual-trace oscilloscope. The steel driver was circular in cross section, had a length equal to half the wavelength of sound in the metal, and was machined to form a shoulder at its midpoint. An annular ring of Perspex, $1/2$ in. thick, was secured to the driver on this shoulder, with four steel screws $1^1/_2$ in. in length. The Perspex ring supported a cylindrical brass

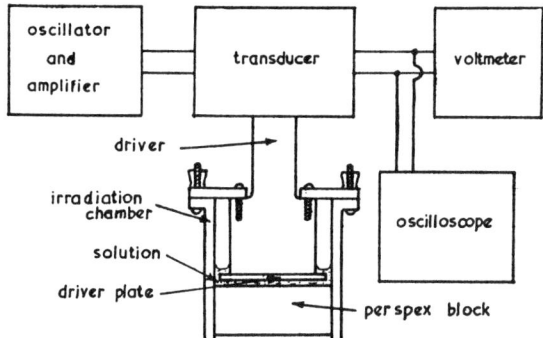

Fig. 2. Apparatus used for the ultrasonic irradiation of solutions under conditions of stable cavitation.

chamber with a Perspex base 3-in. thick. The underside of the driver was formed by a circular hard brass plate, radius 3.42$_5$ cm., drilled with 80 circular cavities 200 μ in diameter and in depth, in a square graticule pattern with no central hole. When the solution to be irradiated (15 ml.) was raised in the irradiation chamber on to the driving plate, bubbles of air were trapped in the drilled cavities. All solutions were degassed for 20 min. at a reduced pressure of approximately 3 cm. of mercury immediately before irradiation. At any power setting of the ultrasonic generator, the apparatus was tuned to give the lowest voltage across the transducer.

Theory

A hemispherical bubble resting on a plate boundary is assumed to pulsate radially. Let ξ be the displacement amplitude of the bubble surface, and a be the radius of the hemisphere. The following theory depends on the assumption that ξ/a is less than 0.1; approximate calculation shows $\xi/a \leqslant 0.075$ for the system used here. Nyborg[23] has derived an expression for the maximum limiting flow, u_L, to the plane boundary of the driver plate:

$$u_L = U^2/\omega a \tag{1}$$

where ω is the angular frequency of pulsation and U is the bubble wall radial velocity amplitude. Connolly,[24] by using an equation derived from the spherical wave equation of acoustics,[25] has related U to the pressure amplitude, p_s, scattered by the bubble:

$$U = p_s/\rho\omega a \tag{2}$$

where ρ is the density of the medium. p_s can be related to the pressure amplitude, $p_{a,x}$, which would result in the absence of the bubble. The variation of pressure amplitude over the surface of an undrilled driver plate is taken from an expression due to Jackson and Nyborg[21] and may be expressed as

$$p_{a,x} = \rho\omega^2 W(r^2-x^2)/4h \tag{3}$$

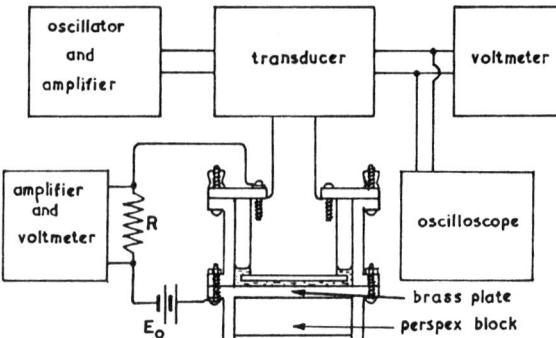

Fig. 3. Apparatus used for investigation of the relationship between $p_{a,0}$ and E_t.

where W is the displacement amplitude of the driver plate, ρ is the density of the medium (assumed incompressible) which separates the driver plate–base plate assembly, h is the distance between driver plate and base plate, r is the radius of the driver plate, and x is the distance from the center of the plate to the position where the value of $p_{a,x}$ is required.

The relationship between p_s and $p_{a,x}$ has been examined by Devin,[27] who showed that the pressure amplification factor,

$$A = p_s/p_a \qquad (4)$$

is 15 at a frequency of 20 kc./sec. Elder suggests a value of 17 for an exciting frequency of 10 kc./sec. Throughout this present work the value of A will be taken as 15. Equation (1) may be now rewritten as

$$u_L = A^2 p_{a,x}{}^2/\rho^2\omega^3 a^3 \qquad (5)$$

For the determination of the variation of $p_{a,x}$ with the voltage E_t, applied to the transducer, it was necessary to measure the corresponding variation of W. For this purpose a driver was used which was similar in all respects to the drilled one, but without bubble cavities. The base of the irradiation chamber was made of metal so that the base plate assembly constituted a vibrating plate capacitor. The plate separation was less than 2 mm. and the intervening gap was filled with degassed ethylene glycol so that the driver was working under almost the same load as when aqueous DNA solutions were irradiated.

The displacement amplitude W of the driver plate was determined by the relation between the capacitance of the driver plate–base plate assembly and the voltage, E_t, applied to the transducer. The 20 kc./sec. signal across a standard 100 kΩ resistance R was measured in the circuit of Figure 3 with the aid of a high impedance amplifier and valve voltmeter. The measured valve voltmeter reading was the root mean square valve. Let the actual voltage amplitude be E_R. It can be shown that

$$|W| = (hE_R/\omega RC_0 E_0)\,[1 + (\omega RC_0)^2]^{1/2} \qquad (6)$$

where E_0 is the applied EMF across the capacitance and standard resistance and C_0 is the static capacitance of the plates.[25] For large values of ωRC_0, eq. (6) reduces to

$$|W| = hE_R/E_0 \qquad (7)$$

$p_{a,0}$, the pressure amplitude at the driver plate center, is given by eq. (3) with $x = 0$. Hence

$$|p_{a,0}| = \rho\omega^2 W r^2/4h \qquad (8)$$

and, substituting for W,

$$p_{a,0} = \rho\omega^2 r^2 E_R/4E_0 \qquad (9)$$

It was possible in this way to relate the calculated pressure amplitude associated with a given vibrating bubble with variation of the voltage applied to the transducer. The variation of $p_{a,x}$ with x shows that p_s was a maximum for the bubble nearest the center of the driver. By eq. (5) the magnitude of the corresponding microstreaming, u_L, will also be a maximum at this point. The boundary layer thickness, δ, through which u_L is reduced to zero is given by Nyborg as

$$\delta = (2\eta/\omega\rho)^{1/2} \qquad (10)$$

where η is the viscosity of the liquid or solution.[23]

Thus, the assumed linear maximum velocity gradient G, generated by microstreaming near the driver plate boundary, is, from eqs. (5) and (10),

$$G = u_L/\delta = (A^2 p_{a,x}^2/a^3)(1/2\eta\rho^3\omega^5)^{1/2} \qquad (11)$$

The value of a in this expression is the equilibrium resonant radius of the bubble. At 20 kc./sec. the bubble, originally of radius 100 μ, grows by a process of rectified diffusion to a resonant equilibrium radius which may be derived from an equation of Minnaert.[28] The equilibrium resonant radius is given by

$$a = 3.0/f \qquad (12)$$

where a is in millimeters and f, the frequency, is in kilocycles per second. Thus at 20 kc./sec., a is 150 μ.

For a non-Newtonian solution, η and G are interdependent, but the shear dependence of viscosity for DNA solutions is insignificant at the high shear rates which prevail in the microstreaming around the resonant bubble, so that the value of η should be that corresponding to a velocity gradient of the order of 10^2 sec.$^{-1}$.

MATERIALS AND METHODS

DNA

Calf thymus gland was stored at $-15°C$. for several days after excision and extracted by the method of Kay et al.[29] After the final stage of purification the product was freeze-dried from a solution in $0.001M$ sodium chloride. The extinction per mole of phosphorus at 259 mμ was 6720. The total protein content by the Sakaguchi[30] method was 1%; the RNA content determined by the method of Ceriotti[31] was 3–4 wt.-%. DNA (0.35 mg./ml.) was irradiated in 0.015M sodium chloride together with 0.0015M sodium citrate.

Human Erythrocytes

A mixture of whole blood and physiological saline solution (0.9% sodium chloride solution) in the ratio 10 parts of whole blood to 90 parts saline was used. The mixture was degassed for 20 min. before treatment, and afterwards centrifuged at 1000g for 5 min. The optical density of the cell-free supernatant was measured at 540 mμ.

Viscosity

Measurements were made in a rotating inner cylinder viscometer of the type designed by Zimm and Crothers.[32] Initial experiments were conducted in order to determine at what shear stress solutions of the DNA used became effectively Newtonian. The results of this aspect of the work are being presented elsewhere.[33] At the shear stress at which viscosity measurements were made in this work (0.0011 dyne/cm.^{2}) the determined viscosity coefficients were equal to those prevailing at zero-shear stress, without need for extrapolation.

Ultraviolet Absorption

A Unicam SP 500 spectrophotometer, adapted by K. J. Thrower, was used. A thermostatted cell holder (A. Adkins Ltd., Leicester) was used for measurement of DNA melting profiles. The controlled temperature in the sample cell was measured with a calibrated thermistor. The optical density at 259 mμ of the solution was measured at each temperature; the temperature of holder and solution was then raised to a new level and the optical density determined again; this was repeated to obtain the full melting curve. Heating–cooling melting profiles were measured with the same spectrophotometer. Samples were heated for 9 min., cooled in melting ice for 4 min., and their optical density measured at 259 mμ.

Sedimentation

A Beckman Model E ultracentrifuge using ultraviolet optics was used for the measurement of integral s distribution curves at one concentration of DNA (rotor speed 44,770 r.p.m.; temperature 20°C.).

RESULTS

Determination of Maximum Velocity Gradient due to Microstreaming

The linear relationship between the pressure amplitude at the center of the nondrilled driver plate and the voltage applied to the transducer is shown in Figure 4.

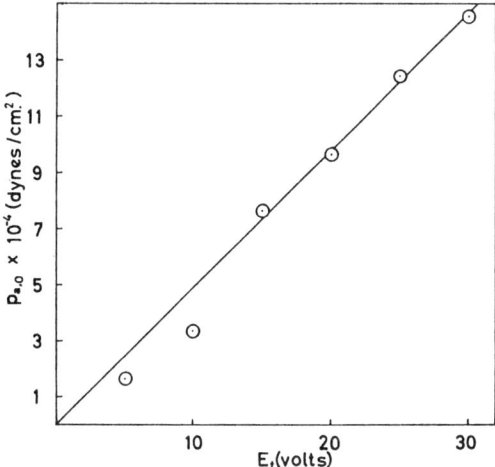

Fig. 4. Variation of $p_{a,0}$ with E_t.

The maximum pressure amplitude scattered from a resonant bubble was related to $p_{a,0}$ as measured, by combining eq. (3) and (4). With a 0.5 cm. square graticule pattern of holes on the driver plate, radius 3.42_5 cm., the maximum p_s is given for the hole nearest the center of the driver plate by $Ap_{a,0,s}$, or $15p_{a,0} \times (11.481/11.731)$, the last term correcting for its position on the plate. If the viscosity of the solution did not change with irradiation, then there should exist a relationship between E_t and the maximum velocity gradient in the solution resulting from microstreaming. This relationship is more complex for macromolecular solutions which undergo degradation, since the decreased viscosity of the solution enhances the velocity gradient by reducing the boundary layer thickness. The maximum velocity gradient in the solution of DNA, irradiated until there was no further decrease in viscosity, varied with E_t as shown in Figure 5.

Irradiation of Human Erythrocytes

Connolly has tried to correlate the leakage of hemoglobin from erythrocytes with bubble activity in an ultrasonic field.[24] He obtained results similar to those summarized in Figure 6. He ascribed the inflection at $E_t = 25$ V. to a change in the mode of bubble oscillation. Figure 7 also shows the reduction of intrinsic viscosity of DNA under the same conditions. Since the index in the Staudinger equation is unity for DNA solutions in this molecular weight range,[9,34] the ratio of intrinsic viscosity after a period of irradiation to that before irradiation (i.e., $[\eta_t]/[\eta_0]$) will be a direct measure of the resultant reduction in the weight-average molecular weight of the DNA. It is seen that the degradation of DNA follows a different course from that of the leakage from blood cells, and that the DNA is degraded at low powers when the permeability of the red cell membrane is virtually unaffected.

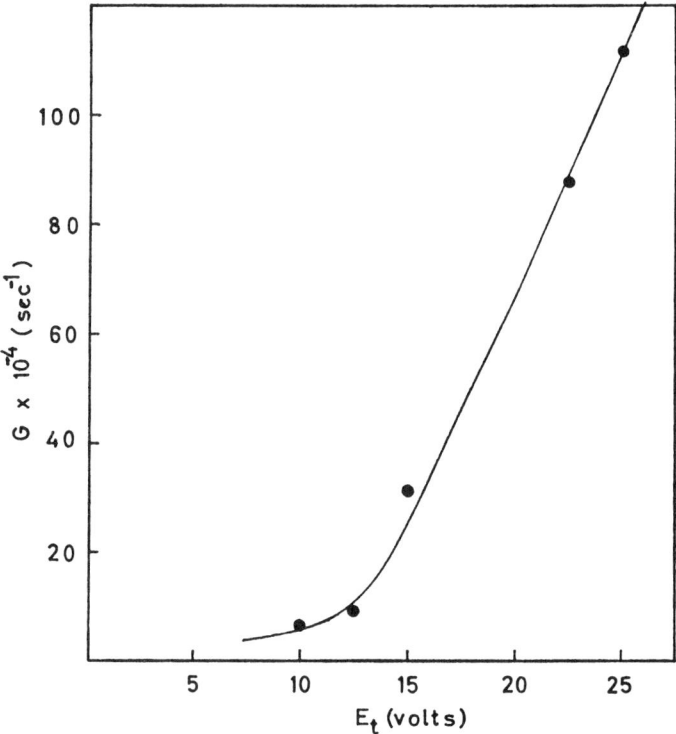

Fig. 5. The maximum calculated velocity gradient in solutions of DNA ultrasonically irradiated at varying E_t.

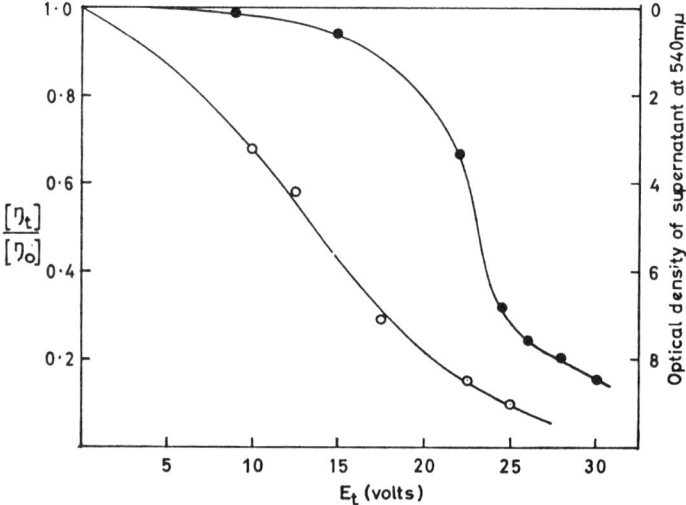

Fig. 6. The degradation of DNA (time of irradiation 4 min.; left-hand ordinate, ○) compared with the release of hemoglobin from human erythrocytes (time of irradiation 2 min.; right-hand ordinate, ●) at varying E_t.

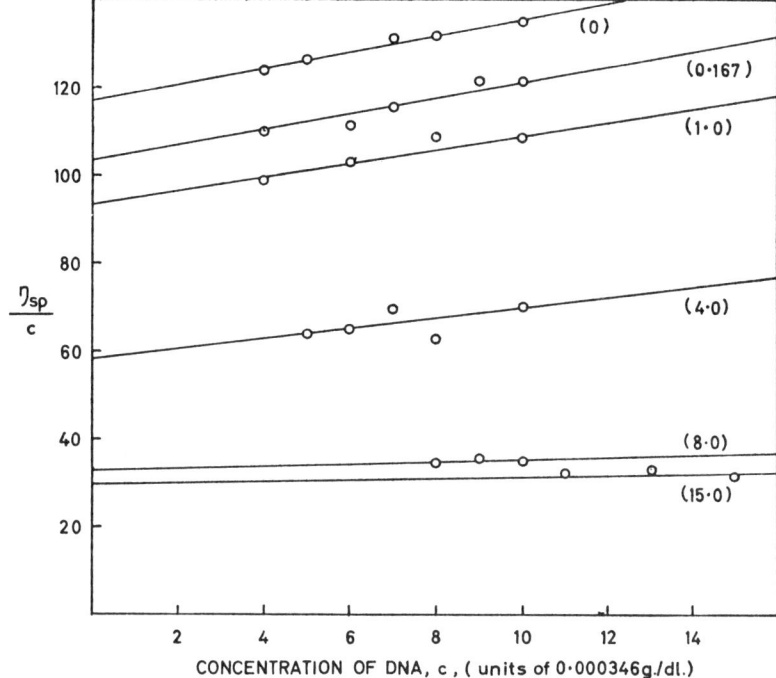

Fig. 7. Typical viscosity data for the investigation of the time-dependent degradation of DNA at fixed E_t. Figures in parentheses adjacent to each Huggins' plot indicate the time of irradiation. $E_t = 15$ V.; buffer, 0.015M sodium chloride + 0.0015M sodium citrate.

Ultrasonic Irradiation of DNA under Conditions of Stable Cavitation

In the following it is demonstrated that, even at very small pressure amplitudes when degradation of DNA is not ordinarily possible, considerable degradation takes place in the presence of stable cavitation. At these low powers there was no free-radical formation as measured by the formation of iodine from potassium iodide in the presence of carbon tetrachloride. The intrinsic viscosity of the DNA was reduced to a limiting value, and further irradiation had little effect. This behavior is shown in Figure 8. It is seen that the initial rate of degradation increases with increasing sound intensity, as does the time taken to reach a stable intrinsic viscosity. The relationship, at a given power, between the shear rate at this limiting viscosity and the number of breaks produced per original average molecule was calculated. If the intrinsic viscosity of the undergraded DNA is given by $[\eta_0]$ and that of DNA degraded for a time of t minutes is $[\eta_t]$, then the number of breaks per original average molecule produced in the degraded solution after t minutes will be $([\eta_0]/[\eta_t]) - 1$. It can be seen from the results summarized in Figure 9 that the number of breaks produced in a macromolecule of mean molecular weight when the solution had reached a stable intrinsic viscosity at fixed ultrasonic intensity was a linear function of the maximum velocity gradient.

ULTRASONIC DEGRADATION OF BIOLOGICAL MACROMOLECULES 269

Examination of the melting profiles of sonically treated native DNA showed that profiles measured at the temperature of heating were little effected by treatment of the DNA with ultrasound. However, profiles measured by the heating–cooling technique showed that T_m was reduced by the ultrasonic treatment, with no loss of hyperchromicity. A typical set of thermal denaturation curves of DNA solutions, sonically irradiated at 18 V. applied to the transducer, is shown in Figure 10.

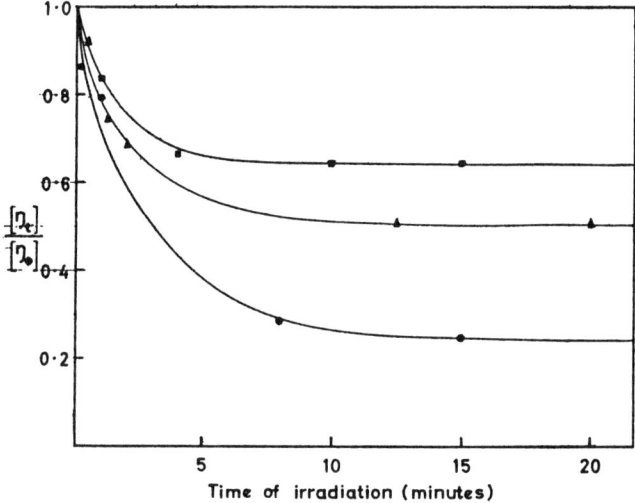

Fig. 8. Time dependence of degradation of DNA by ultrasonic irradiation under conditions of stable cavitation: (■) 10 V.; (▲) 12.5 V.; (●) 15 V.

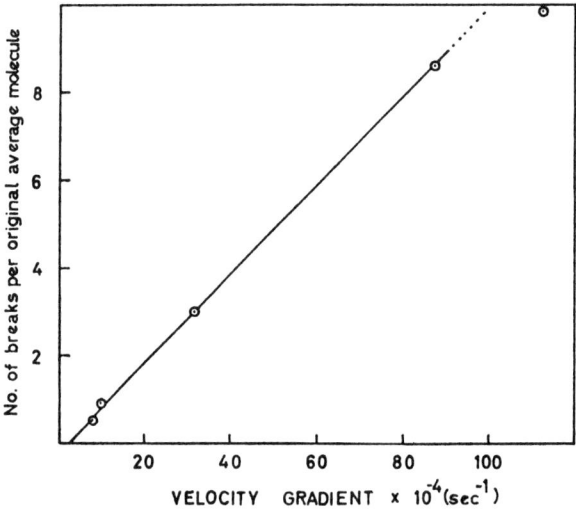

Fig. 9. Maximum calculated velocity gradient, in solutions of DNA degraded to a stable intrinsic viscosity, related to the number of breaks per original average DNA molecule.

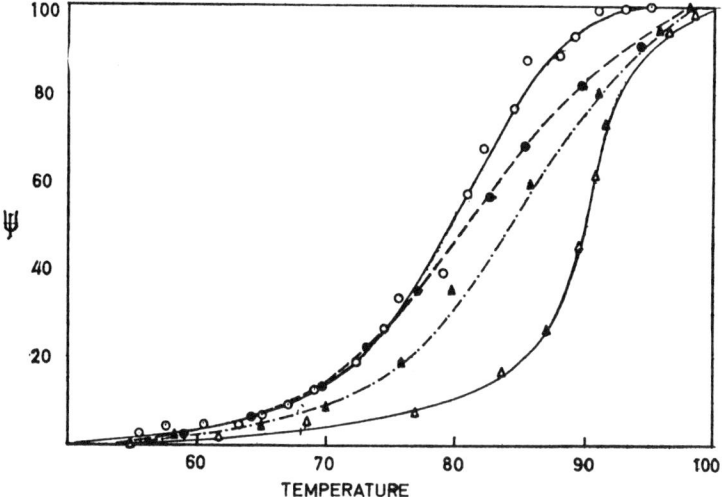

Fig. 10. Melting profiles of DNA in 0.1M sodium chloride: (○) native DNA, extinction measured at temperature of heating; (△) native DNA, heating–cooling profile. DNA treated under conditions of stable cavitation by ultrasound at $E_t = 18$ V. for 5 min.: (●) extinction measured at temperature of heating; (▲) heating–cooling profile. ψ denotes the percentage of total change of optical density at 259 mμ.

Sedimentation analysis showed a uniform shift of the integral distribution curve towards lower values of sedimentation constants for solutions treated with ultrasound. No change in the distribution at one concentration was observed as resulting from the irradiation.

DISCUSSION

The degradation caused by stable cavitation in an ultrasonic field is comparable with that produced in the absence of resonant bubbles by ultrasonic irradiation at a greater intensity. Even at the lowest power at which the transducer could be excited and tuned reproducibly, degradation of DNA was observed. There was no denaturation of the DNA, although the difference between the heating–cooling melting profile for native and for sonically treated DNA means that the irradiated DNA renatured less readily than did the original DNA.

The velocity gradients which have been shown to be necessary to degrade DNA solely by hydrodynamic shear may be compared with those occurring during stable cavitation. Levinthal and Davison[10] showed that T2 phage DNA (molecular weight, 90×10^6 to 130×10^6) degraded in a velocity gradient of 2.25×10^4 sec.$^{-1}$, but not at 1.87×10^4 sec.$^{-1}$. They also derived an expression for the maximum stretching force, τ_{max}, which a rigid-rod model molecule would experience in a uniform velocity gradient, namely,

$$\tau_{max} = 3\pi\eta_0 GL^2/16 \log(L/r) \tag{13}$$

where L is the length, r the radius of the rod, and η_0 is the viscosity of the solvent. The value of τ_{max} which was required to rupture the phage DNA by capillary shearing was calculated from eq. (13) and was in good agreement with that derived from known bond strengths, namely, 9–18 × 10^{-4} dyne. The maximum velocity gradient which was induced in the present experiments by ultrasound and which was required to produce one break per original average molecule was calculated from eq. (11) and Figure 9 to be 12 × 10^4 sec.$^{-1}$. If this value for G is substituted in eq. (13) with $r = 9$ A. and $L = 3\ \mu$ (assuming the molecular weight is 6 × 10^6), the maximum stretching force to cause one break per molecule is calculated as 23 × 10^{-4} dyne. Considering the approximations involved, this may be regarded as near to the value of 9–18 × 10^{-4} dyne calculated by Levinthal and Davison to be necessary to rupture the DNA double helix. This would mean that the hydrodynamic forces generated by microstreaming around the resonant bubbles are sufficient to cause degradation under conditions of stable cavitation during ultrasonic irradiation.

Recently Zimm and Harrington have examined hydrodynamic degradation by a variety of different shearing devices.[35] They derived a more general expression than Levinthal and Davison for the average force, $\langle f \rangle_{av}$, sustained by a long-chain solute molecule in a velocity gradient G, namely,

$$\langle f \rangle_{av} = G(\eta - \eta_0)/nz \tag{14}$$

where η is the viscosity of the solution and η_0 is the viscosity of the solvent; the quantity nz is the number of macromolecules crossing a unit area which is in the direction of the velocity gradient; n is the number of macromolecules per unit volume, and z is the average of the reciprocals of the projections of the end-to-end lengths of the macromolecules on the normal to the defined unit area. This treatment has the advantage that it makes no assumptions concerning extensibility at high shear rates. Taking z to be $3^{-1/2}$ times the root-mean-square end-to-end length as measured by light scattering, Zimm and Harrington calculated tensile forces required to break T2 phage DNA which were 30 times smaller than those calculated by Levinthal and Davison.

When the theory of Zimm and Harrington is applied to the present work, the calculated value of $\langle f \rangle_{av}$ required to cause one break in each original molecule is 2 × 10^{-5} dyne, which is again smaller, by a factor of 10, than the value calculated by eq. (13) from the same results.

The velocity gradient generated in microstreaming around a vibrating bubble is inversely proportional to the square root of the viscosity of the solution, according to eq. (11). Since the macromolecules degrade at a fixed ultrasonic power and the viscosity decreases, the maximum velocity gradient G should increase, reaching a limit when the viscosity has fallen to that of the solvent. According to eq. (14), such an increase in G would, by itself, lead to an ever-increasing value of the tensile force, $\langle f \rangle_{av}$. However, this effect is counteracted, in eq. (14), by the actual decrease in the viscosity η of the solution. The quantity nz is also not constant for different stages of the degradation. Degradation was, in fact, observed to reach a limit, characteristic of each value of E_t and did not proceed to complete disruption into monomers. It must be presumed that as the solution is degraded, the average tensile force experienced by the macromolecules is reduced until it is not great enough to rupture any molecules present, and the solution reaches an intrinsic viscosity which is not affected by further irradiation. Furthermore, at all stages during irradiation the value of the maximum velocity gradient in the solution, as calculated from eq. (11), is proportional to the square of the pressure amplitude of the driver plate, and thus, by experiment (Fig. 4), to the square of the voltage applied to the transducer. The limiting value of the intrinsic viscosity of the solution should therefore decrease as the ultrasonic intensity is increased.

An analysis of the kinetics of degradation using the theory of ultrasonic stable cavitation applied to this apparatus is complicated since the measured viscosity is that of the bulk solution, whereas the degradation occurs in only special limited regions. A probability factor is involved in the entry of a macromolecule into volume elements of the solution where the velocity gradient is maximal; all the macromolecules must pass through these regions before the solution can reach a limiting intrinsic viscosity. The orientation of the macromolecules as they pass into such regions may also be an important factor upon which the extent of degradation depends.

In this work, as in other work on hydrodynamic shear degradation, the part played by solute–solvent interaction has not been considered: in any complete treatment of the hydrodynamic degradation of macromolecules this factor must surely be of importance.

The authors acknowledge the help they have received from discussions with Mr. C. C. Connolly and his making available the thesis which he submitted to the University of Vermont. They have had valuable advice from Professor W. L. Nyborg, of the University of Vermont, and are grateful for the facilities afforded by Professor Sir Hans Krebs, F.R.S. and Mr. J. R. P. O'Brien in their respective departments. One of them (N. J. P.) is indebted to the Nuffield Committee for the Advancement of Medicine of the University of Oxford for support during this work.

References

1. Szent-Gyorgi, A. G., *Nature*, **131**, 278 (1933).
2. Flosdorf, E. A., and L. A. Chambers, *J. Am. Chem. Soc.*, **55**, 3051 (1933).
3. Burns, V. W., *Science*, **146**, 1056 (1964).
4. Mikhailov, I. G., and N. M. Fedorova, *Akust. Zh.*, **9**, 1 (1963).
5. Frenkel, J., *Acta. Physicochim. URSS*, **14**, 51 (1944).
6. Johnson, W. R., and C. C. Price, *J. Polymer Sci.*, **45**, 217 (1960).
7. Overend, W. G., and A. R. Peacocke, *Trans. Faraday Soc.*, **46**, 794 (1950).
8. Lee, W. A., and A. R. Peacocke, *J. Chem. Soc.*, **1951**, 3374.
9. Doty, P., B. B. McGill, and S. A. Rice, *Proc. Natl. Acad. Sci. U.S.*, **44**, 432 (1958).
10. Levinthal, C., and P. F. Davison, *J. Mol. Biol.*, **3**, 674 (1961).
11. Burgi, E., and A. D. Hershey, *J. Mol. Biol.*, **3**, 458 (1961).
12. Gooberman, G., *J. Polymer Sci.*, **42**, 26 (1960).

13. Freifelder, D., and P. F. Davison, *Biophys. J.*, **2**, 235 (1962).
14. Okuyama, M., *Z. Elektrochem.*, **121**, 46 (1954).
15. Webster, E., *Ultrasonics*, **1**, 39 (1963).
16. Noltingk, B. E., and E. A. Neppiras, *Proc. Phys. Soc. London*, **63B**, 6 (1960).
17. Noltingk, B. E., and E. A. Neppiras, *Proc. Phys. Soc. London*, **64B**, 1032 (1951).
18. Hughes, D. E., *J. Biochem. Microbiol. Tech. Eng.*, **3**, 405 (1961).
19. Hughes, D. E., and W. L. Nyborg, *Science*, **138**, 108 (1962).
20. Hawley, S. A., R. M. Macleod, and F. Dunn. *J. Acoust. Soc. Am.*, **35**, 1285 (1963).
21. Elder, S. A., *J. Acoust. Soc. Am.*, **31**, 54 (1958).
22. Connolly, C. C., private communication.
23. Nyborg, W. L., *J. Acoust. Soc. Am.*, **30**, 329 (1958).
24. Connolly, C. C., thesis, Univ. of Vermont, 1963.
25. Kinsler, L. E., and R. R. Frey, *Fundamentals of Acoustics*, Wiley, New York, 1950.
26. Jackson, F. I., and W. L. Nyborg, *J. Acoust. Soc. Am.*, **32**, 1243 (1960).
27. Devin, C., *J. Acoust. Soc. Am.*, **31**, 1654 (1959).
28. Minnaert, M., *Phil. Mag.*, **7**, 240 (1933).
29. Kay, E., N. Simmons, and A. Dounce, *J. Am. Chem. Soc.*, **74**, 1724 (1952).
30. Block, R. J., and D. Bolling, *Amino Acid Composition of Proteins and Foods*, Thomas, Springfield, Ill., 1951, p. 47.
31. Ceriotti, G., *J. Biol. Chem.*, **214**, 59 (1955).
32. Zimm, B. H., and D. M. Crothers, *Proc. Natl. Acad. Sci.*, **48**, 905 (1962).
33. Drummond, D. S., N. J. Pritchard, V. F. W. Simpson-Gildemeister, and A. R. Peacocke, in press.
34. Peacocke, A. R., and B. N. Preston, *Proc. Roy. Soc. London*, **153B**, 90 (1960).
35. Zimm, B. H., and R. E. Harrington, *J. Phys. Chem.*, **69**, 161 (1965).
36. Howkins, S., *J. Acoust. Soc. Am.*, **37**, 504 (1965).

Copyright © 1970 by the American Association for the Advancement of Science

Reprinted from *Science*, **169**, 869–871 (Aug. 28, 1970)

Hemolysis Near an Ultrasonically Pulsating Gas Bubble

Abstract. *A small volume of an erythrocyte suspension was subjected to the action of a manipulated gas bubble set into stable oscillation at 20 kilohertz. Release of hemoglobin occurred when the oscillation amplitude exceeded a critical threshold. Hydrodynamic stresses resulting from acoustically induced small-scale eddying motion near the bubble may be the mechanism of hemolysis.*

The sonic interactions with biological suspensions characteristically occur by means of cavitation, a complicated phenomenon involving sonically activated bubbles. To understand the mechanism, the situation may be simplified by maintaining a single stable oscillating gas bubble in a suspension of cells within a Plexiglas vessel attached to a vibrating bar. Observations in this system led to two primary conclusions: (i) stresses associated with acoustic streaming constitute an important mode of action for sonic effects on cell suspensions, and (ii) measurement of the critical threshold for release of cell contents provides a method for determining the mechanical strength of cell membranes.

The gas bubble is formed in a stainless steel tube (260 μm inside diameter by 2.0 cm) which is connected to a gas reservoir through a 30-cm length of stainless steel tubing (50 μm inside diameter). The latter small tubing prevents large volume changes of the bubble by providing resistance to rapid changes in pressure. By observing with a microscope the operator adjusts the pressure in the gas reservoir so that the gas-liquid interface is hemispherical. Upon application of the sound field the bubble starts to oscillate and also flattens as the enclosed air tends to retreat up the tubing. The reservoir pressure is then increased until the bubble is again hemispherical. In this system the increase in reservoir pressure can be used to determine the radial oscillation amplitude ξ_0 of the hemispherical air-liquid interface (*1*).

Upon application of the sound field one observes the onset of small-scale vortex motion near the bubble, called acoustic microstreaming (*2*), as shown in Fig. 1. A thin boundary layer is present near the bubble, and, although acoustic streaming velocities are not large, a high velocity gradient exists in this boundary layer (*2*).

In earlier experiments, with the use of bubbles trapped in holes drilled in the face of an oscillating metal bar, difficulties were encountered because of instabilities. These instabilities seemed to be related to the onset of surface waves on the bubbles and the ejection of microbubbles (*3*). In order to avoid these complications a 13 percent solution of dextran 500 in physiological saline, which stabilizes the bubble by damping surface waves, was used; possibly also the dextran acts by forming a "skin" on the bubble, in line with suggestions by Elder (*2*) and Fox and Herzfeld (*4*). This solution has a viscosity of 0.31 poise at 25°C as measured with a capillary-type viscometer. To 10 ml of the dextran-saline solution 0.05 ml of whole blood was added.

In order to study the effects of a single bubble it was necessary to remove gas from the sample, which might exist in the form of unwanted bubbles or bubble nuclei in the sound field. The degassing was accomplished by placing the irradiation vessel under reduced pressure (a few centimeters of mercury) for 20 minutes. Any fluid that was lost was replaced, and the sample was degassed for an additional 10 minutes. The 0.2-ml Plexiglas vessel (internal dimensions, 1.0 cm by 0.2 cm horizontally and 1.0 cm in height) was set into vertical oscillation for 5 minutes at 20 khz by attachment to a Branson Sonifier transducer excited at low amplitudes. After sonation, the irradiation vessel was centrifuged for 20 minutes at 1150g. The supernatant was then drawn off and placed in a microcell of a spectrophotometer, and the absorbance was measured at 552 nm. This absorbance was proportional to the concentration of released hemoglobin.

Absorbance of the supernatant of the treated cells was studied as a function of the amplitude ξ_0 of bubble oscillation, with results as shown in Fig. 2. There was no significant release of hemoglobin at low amplitudes; above a fairly well-defined threshold the absorbance rises steeply. If the experimental procedure was repeated when the bubble was absent, there was a maximum absorbance of 0.07 for ξ_0 values up to 29 μm. With no sound, the background absorbance was 0.03. Thus the absorbance was much greater when the bubble was present, for the amplitudes used in these experiments. An extrapolation of the rising part of the curve yields a threshold of 18.2 μm for the amplitude ξ_0.

A plateau in the absorbance which corresponds to less than complete hemolysis is reached at higher amplitudes. In explanation of this, it is observed that there is relatively little transfer of material from the top and bottom volumes of the sample into the region of the bubble. Thus, the plateau seems to result from depletion of intact cells in the active region of the bubble. For the situation described by Williams, Hughes, and Nyborg (*5*), a similar plateau was not observed. There a slow streaming motion was observed along the length of the vibrating wire which aids in the transfer of material within the sample.

An experiment was done to test whether the metal tubing itself (without a bubble) could affect the cells. For this purpose a stainless steel rod 400 μm in diameter was placed in the cell suspension. A slow acoustic streaming motion occurred near the tip of the rod but only the background absorbance (about 0.03) was observed.

Tests for the presence of sonically produced free iodide radicals were made with 4N methyl iodide in dextran-saline solution containing 1 percent soluble starch. When cavitation was present the starch was colored by the sonically produced iodine and the effect was readily observed (*6*). This means that the free radical concentration produced sonically was much greater than the minimum concentration detectable (10^{17} radical/ml) by

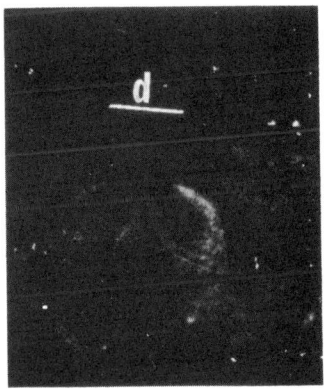

Fig. 1. Bubble-associated streaming pattern obtained by timed-exposure of side-lighted polystyrene spheres (7 to 14 μm in diameter). Amplitude ξ_0 is 26 μm. Dimension d shown on photograph gives the tube diameter (520 μm). See Fig. 2 insert for line drawing of arrangement.

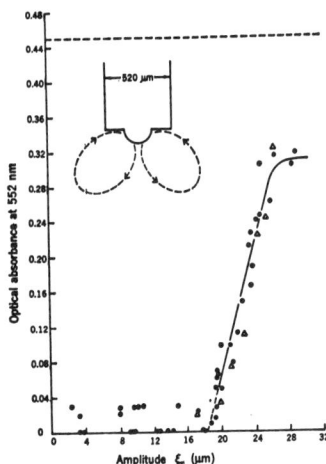

Fig. 2. Hemolysis resulting from a single stable bubble. Dashed line on top of figure indicates complete osmotic hemolysis. In the insert the direction of streaming is shown by the arrows. The circles and triangles are for canine and human erythrocytes, respectively.

this method (7). Under the acoustic conditions of the present experiment no free radicals were detected.

Because the bubble remained stable, no shock waves or pronounced local maxima in temperature ("hot spots") occurred in the sample. At these amplitudes of bubble oscillation, gross heating should not be important. Experiments were done at 25°C, and sample temperatures never rose more than 2°C during sonic irradiation.

It is believed that stresses resulting from acoustic streaming are the important mechanism in this case. Theory does not exist for the specific kind of acoustic streaming described here. The acoustic streaming velocity gradients and stresses may be estimated from the theory derived by Nyborg, who gives an expression (Eq. 147, in 8) for the acoustic streaming velocity $u(z)$ tangential to the end surface of the metal tube holding the bubble as a function of z, the distance above the surface. Differentiating u with respect to z and letting $z = 0$ the velocity gradient G at points near the surface of the bubble is

$$G = 2\pi f \xi_0^2 / a\delta \quad (1)$$

Here δ is the boundary layer thickness defined below, a is the bubble radius (130 μm), and f is the frequency.

The boundary layer thickness can be calculated from the expression

$$\delta = (\eta/\pi\rho f)^{\frac{1}{2}} \quad (2)$$

where η and ρ are, respectively, the shear viscosity and the fluid density. If $\eta = 0.31$ poise, $\rho = 1.0$ g/cm³, and $f = 2 \times 10^4$ hz, then $\delta = 22.0$ μm. The viscous stress is given by ηG. For the critical threshold of hemolysis, G is 1.4×10^4 sec^{-1} and the viscous stress S_c is 4500 dyne/cm². Taking into account experimental errors, one would expect that the standard deviation for S_c would be about 5 percent. However the velocity gradient is nonuniform for the acoustic streaming situation. As a result the maximum uncertainty in S_c is greater, very likely of the order of 1500 dyne/cm².

A comparison of values of critical stress for hemolysis obtained by the ultrasonic technique can be made with those obtained with hydrodynamic methods. Samples of blood treated with heparin have been sheared in a closed concentric cylinder viscometer with the bottom of the bob machined to a conical shape. Using such a device, Nevaril et al. (9) found a threshold stress for hemolysis of 3000 dyne/cm². In other experiments Blackshear et al. (10) have injected jets of saline into suspensions of red cells. The critical velocity gradients observed are of the order of 10^6 sec^{-1}, from which a lethal stress for normal erythrocytes of 40,000 dyne/cm² was calculated. Thus, my results fall within the range of values obtained by others.

Williams et al. (5) describe results for hemolysis caused by acoustic streaming near a vibrating wire. Similarity of results by both ultrasonic techniques demonstrated that details of the ultrasonic interaction with biological materials occurring in a stable bubble field can be elucidated using the vibrating wire apparatus. The comparable results also support the hypothesis that viscous stresses associated with acoustic microstreaming are the important mechanisms involved, since no bubble activity is present near the vibrating wire.

JAMES A. ROONEY

Physics Department,
University of Vermont,
Burlington 05401

References and Notes

1. W. L. Nyborg and J. A. Rooney, J. Acoust. Soc. Amer. 45, 384 (1969).
2. S. A. Elder, ibid. 31, 54 (1959).
3. C. Chuongvan, thesis, University of Vermont (1965); G. W. Willard, J. Acoust. Soc. Amer. 26, 933A (1954); W. L. Nyborg and A. Rodgers, Biotechnol. Bioeng. 11, 235 (1967); W. L. Nyborg and D. E. Hughes, J. Acoust. Soc. Amer. 42, 891 (1967).
4. F. E. Fox and K. F. Herzfeld, J. Acoust. Soc. Amer. 26, 984 (1954).
5. A. R. Williams, D. E. Hughes, W. L. Nyborg, Science, this issue.
6. A. R. Williams and W. L. Nyborg, Ultrasonics 8, 36 (1970).
7. A. Weissler, J. Acoust. Soc. Amer. 25, 651 (1953).
8. W. L. Nyborg, in Physical Acoustics, W. P. Mason, Ed. (Academic Press, New York, 1965), vol. 2, part B, pp. 265–331.
9. C. G. Nevaril, E. C. Lynch, C. P. Alfrey, Jr., J. D. Hellums, J. Lab. Clin. Med. 71, 784 (1968).
10. P. L. Blackshear, Jr., F. D. Dorman, J. H. Steinback, E. J. Mayback, A. Singh, R. E. Collingham, Trans. Amer. Soc. Artif. Intern. Organs 12, 113 (1966).
11. Supported in part by a NASA traineeship [NSG(T)28-S3] and by NIH research grant GM 08209.

22 April 1970; revised 15 June 1970

30

Reprinted from *Science*, **169**, 871–873 (Aug. 28, 1970)

HEMOLYSIS IN A TRANSVERSELY OSCILLATING WIRE

A. R. Williams, D. E. Hughes, and W. L. Nyborg

Abstract. *Erythrocyte suspensions were subjected to hydrodynamic forces generated by a partially submerged tungsten wire set into transverse oscillation at 20 kilohertz. Free hemoglobin appears in solution when the oscillation amplitude exceeds a critical threshold value. The hemolysis probably results from stresses exerted on cell by a microstreaming field established near the wire.*

Rooney has shown that hemoglobin is released from both human and canine erythrocytes when they encounter small scale acoustic streaming in which velocity gradients are sufficiently high (*1*). In his experiments these gradients were produced near an oscillating gas bubble of about 250 μm diameter under conditions where undesirable concomitants of cavitation (such as the production of shock waves and high temperature "pulses") were avoided. Specifically, it was found that erythrocytes in physiological saline containing 13 percent dextran 500 require a minimum velocity gradient (G) of 14,300 sec^{-1} for hemolysis. For this dextran-saline solution, the shear viscosity coefficient (measured by capillary viscometry) is about 0.31 poise, so that the critical shear stress (ηG) becomes about 4500 dyne/cm^2.

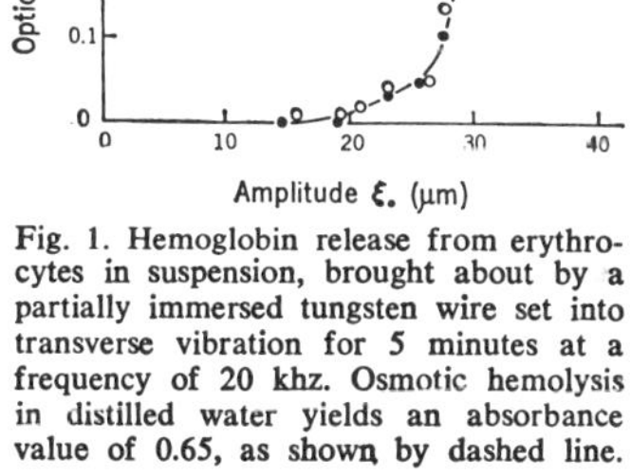

Fig. 1. Hemoglobin release from erythrocytes in suspension, brought about by a partially immersed tungsten wire set into transverse vibration for 5 minutes at a frequency of 20 khz. Osmotic hemolysis in distilled water yields an absorbance value of 0.65, as shown by dashed line.

In view of these results, we were led to consider other arrangements with which one might obtain similar results. An acoustic streaming situation which has probably received more theoretical attention than any other is that which occurs near a transversely vibrating cylinder [see, for example, Schlichting (*2*); Holtzmark, Johnson, Sikkeland, and Skavlem (*3*); Raney, Corelli, and Westervelt (*4*); and a review by Nyborg (*5*)]. Near a transversely oscillating cylinder eddying motions are established (in planes perpendicular to the axis, for an infinite rigid cylinder), with relatively high velocity gradients in a boundary layer very near the cylindrical surface, in the absence of any form of bubble activity. If only the simple approximate expression given by Schlichting is considered, the magnitude G of the maximum velocity gradient at the boundary is

$$G = 2\pi f \xi_o^2 / a\delta \qquad (1)$$

Here f is the frequency in hertz, a is the radius of the cylinder, and ξ_0 is the displacement amplitude of the cylindrical surface; the parameter δ is equal to $(\eta/\pi f \rho)^{\frac{1}{2}}$, where ρ and η are, respectively, the density and the shear viscosity coefficient for the liquid. This velocity gradient applies to fluid motion along the boundary, the gradient being directed perpendicular to it. Thus if the velocity parallel to the surface is $u(z)$ at any distance z from the boundary, then G in Eq. 1 refers to the derivative $\partial u/\partial z$ at $z = 0$. The expression for G in Eq. 1 has precisely the same form as that used by Rooney (*1*) for microstreaming near a vibrating bubble. This is essentially a coincidence; the streaming patterns for the two situations are, in general, quite different.

Proceeding from this expression, we find that a practical arrangement for use with cell suspensions should be possible at ultrasonic frequencies in the vicinity of 20 khz, if cylinders of diameter the same order as that of the bubble discussed earlier, namely, about 250 μm, are used. Thus for cylinders of radius a equal to 0.0125 cm driven at a frequency of 2×10^4 hz in a liquid of viscosity 0.31 poise one obtains

$$G = 4.5 \times 10^9 \xi_0^2$$

and

$$\eta G = 1.4 \times 10^9 \xi_0^2 \quad (2)$$

where ξ_0 is the (oscillatory) displacement amplitude in centimeters. From Eqs. 2 we see that a displacement amplitude ξ_0 of 20 μm would yield a value of 5600 dyne cm^{-2} for ηG, somewhat greater than the threshold value reported by Rooney. A calculation of the acoustic pressure amplitude near the cylinder at this amplitude yields a value of about 0.4 atm, well under the values which one expects to be required for sonically generated cavitation.

We know of no practical means for setting a reasonable length of unsupported 0.25-mm diameter cylinder into vibration transversely at 20 khz, in such a way that the cylinder vibrates as a rigid object. However, it is a relatively easy matter to set up ultrasonic transverse waves in a wire of this diameter, with amplitudes equal to or greater than those required. This has been done by attaching one point of the wire in question to a piezoelectrically or magnetostrictively driven rod. Two kinds of arrangements have been used successfully. In one of these a free end of wire projects into the solution of interest; in the other the wire is maintained under tension while a portion is immersed in the solution. We describe here only results obtained with the "free wire", in which the vibration wavelength λ depends on the Young's modulus Y of the material. For tungsten Y is 36×10^{11} dyne cm^{-2}, and the density ρ is 19 g cm^{-3}; for a wire of 0.0125 cm radius, λ is 0.9 cm. The vibration pattern is generally similar to that of Morse (6) and specifically to that of the transversely driven tube of Williams and Nyborg (7). To achieve maximum displacement amplitude, the length of the wire was adjusted to be an odd multiple of λ/4. Preliminary results presented below were obtained with a resonant length (2.05 cm) of 0.025-cm diameter General Electric type 218 CS tungsten

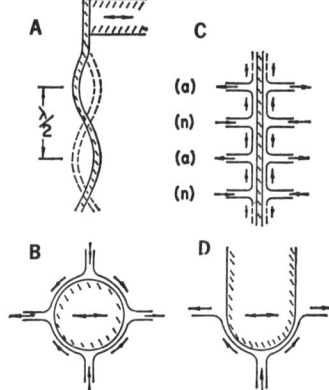

Fig. 2. Schematic of vibrating wire device and circulatory patterns. (A) Attachment of wire to driver. (B) Streaming as it occurs near a transversely oscillating rigid cylinder in plane perpendicular to axis. Arrows near boundary show U_L. (C) Streaming associated with flexural vibrations of wire in the plane of vibration; displacement nodes are at n, and antinodes are at a. (D) Streaming associated with the tip in plane of vibration.

wire (8), clamped at one end to the tip of a stainless steel velocity transformer, and driven at 20 khz by a barium titanate ceramic.

The sonication vessel was that described by Rooney (1). The free end of the tungsten wire was ground to an approximately hemispherical shape with radius of curvature about equal to that of the wire shank. A traveling microscope fitted with a ×20 objective and ×10 eyepiece was positioned with the wire in its focal plane; by its use, displacement amplitudes were directly measured during the actual sonic process. Except at the ends, maxima of displacement amplitude occur at points separated by a distance λ/2, each having the magnitude A. The greatest amplitude ξ_{om} occurs at the free end where ξ_0 has the value $(2)^{\frac{1}{2}} A$. After sonication, the microspectrophotometer cells were centrifuged, and the optical density of the supernatant was measured as described (1).

In Fig. 1, abscissae give the maximum displacement amplitude at the free end. Human or canine erythrocytes were suspended in a solution of physiological saline containing 13 percent dextran 500 (giving a solution viscosity of 0.31 poise). Hemolysis occurred when the displacement amplitude exceeded a threshold value of about 20×10^{-4} cm According to Eqs. 2 this corresponds to an acoustic streaming velocity gradient at the boundary of 18×10^3 sec^{-1} and a shear stress S_c of 5600 dyne cm^{-2}. This value for the critical shear stress required to hemolyze human or canine erythrocytes is in good agreement with the values obtained with the bubble-associated microstreaming (1). It is possible that the agreement is in part fortuitous; the details of the disruption process are not yet fully understood. Uncertainty in S_c also arises from the fact that velocity gradients are not uniform in acoustic streaming situations.

The hemolysis-amplitude curves obtained with the stable vibrating bubble and with the transversely oscillating wire are similar. Both curves show a well-defined threshold amplitude, above which the rate of hemolysis rises very steeply. A difference in the curves appears at high amplitudes. The plateau noted by Rooney does not appear here, evidently because the wire produces more mixing than the bubble does.

Examinations were made with optical and electron microscopes of both control and sonicated erythrocytes to determine whether hemoglobin release was due to complete cell rupture. In the course of our study we observed some novel ultrasonic interactions with erythrocytes, such as hemoglobin-filled microspheres having a diameter of the order of a micrometer.

Equation 1 is obtained from theory for acoustic streaming near an infinite cylinder which oscillates as a rigid body. The nature of the predicted streaming is suggested by Fig. 2B. Arrows along the surface show the direction of a so-called "limiting velocity" U_L, explained in (5, p. 303). Return flow occurs in the outer region.

Observations of the actual streaming reveal features resulting from the fact that the cylindrical wire is neither rigid nor infinite. Thus a relatively slow large-scale circulation occurs, which is associated with the flexural vibrations of the wire (Fig. 2C). This motion transports liquid from displacement node n to antinode a in a region very near the cylindrical surface, with return occurring in the main body of the liquid. The nature of the flow is suggested in Fig. 2C; again arrows near the surface show the direction of the limiting flow U_L. A more rapid eddying occurred near the free end or tip of the wire; its nature is sensitive to the geometry of the tip. The tip was usually rounded to an approximately hemispherical shape. The tip-associated flow (Fig. 2D) is then closely related to that near a cylinder (Fig. 2B).

Full mathematical treatment of these motions has not been given except for the rigid cylinder. However, the main features of the motion can be explained qualitatively in terms of an approximate result given by Schlichting (2) and Nyborg (5, p. 303). This result, based on a "thin boundary layer" approximation, is expressed in terms of a "limiting" velocity U_L already mentioned, which is characteristic of the streaming velocity near the cylinder just outside the boundary layer. This velocity U_L is parallel to the boundary; we take U_L to be along the x direction, with a choice of meanings for x. Thus for rigid-cylinder streaming (Fig. 2B), x measures arc length along a circle around the wire, perpendicular to the axis. For streaming associated with the flexing vibrations (Fig. 2C), x measures arc length along the wire, parallel to the axis. For tip-associated streaming (Fig. 2D), x measures arc length along a great circle formed by intersecting the hemispherical tip with a plane passing through the wire axis.

Let u_o be the amplitude of oscillatory irrotational motion (that is, oscillatory motion as it would be in the absence of viscosity) along the x direction. The Schlichting result is then

$$U_L = -(3/8\,\omega)\partial(u_o^2)/\partial x \qquad (3)$$

This expression may be applied qualitatively to any of the situations in Fig. 2.

It can readily be verified that the direction of U_L given by Eq. 3 is in agreement with the direction indicated by arrows near the wire surface in Fig. 2, B and D.

According to Eq. 3 the magnitude of U_L depends on the magnitude of $\partial(u_o^2)/\partial x$ or $2u_o\partial u_o/\partial x$. For purposes of rough comparisons suppose that A is a typical magnitude for u_o and that in a given situation u_o decreases from A to zero in a distance l. Then U_L is roughly proportional to A^2/l, and for given A the characteristic streaming velocity U_L varies inversely with l.

For both the situations of (Fig. 2, B and D), the length l may be taken as $\pi a/2$, one-fourth the wire circumference; for the situation of Fig. 2C we choose l as $\lambda/4$. Hence we expect U_L to be of the same order of magnitude for Fig. 2, B and D, while for Fig. 2C the magnitude of U_L will be less by about a factor of $2\pi a/\lambda$. For our typical situation ($a = 0.0125$ cm; $\lambda = 0.9$ cm), this factor is about (1/70). Since the maximum viscous stress near the wire is proportional to U_L, we see that effects of such stresses arise primarily from "cylinder streaming" (Fig. 2B) and tip-associated streaming (Fig. 2D), and not appreciably from the relatively large-scale streaming of Fig. 2C. The last-mentioned is nevertheless significant; it plays the role of transporting suspension from the outer fluid to the high-stress region near the wire; it probably explains the absence of a plateau in Fig. 1 analogous to that noted by Rooney (1).

A. R. WILLIAMS
D. E. HUGHES

Department of Microbiology, University College, Cardiff, South Wales

W. L. NYBORG

Physics Department, University of Vermont, Burlington 05401

References and Notes

1. J. A. Rooney, *Science*, this issue.
2. H. Schlichting, *Boundary Layer Theory* (McGraw-Hill, New York, 1955), p. 194.
3. J. Holtzmark, I. Johnson, T. Sikkeland, S. Skavlem, *J. Acoust. Soc. Amer.* **26**, 26 (1954).
4. W. P. Raney, J. C. Corelli, P. J. Westervelt, *ibid.*, p. 1006.
5. W. L. Nyborg, in *Physical Acoustics*, W. P. Mason, Ed. (Academic Press, New York, 1965), vol. 28, chap. 11.
6. P. M. Morse and K. U. Ingard, *Theoretical Acoustics* (McGraw-Hill, New York, 1968), pp. 175 ff.
7. A. R. Williams and W. L. Nyborg, *Ultrasonics* **8**, 36 (1970).
8. The wire was donated by the General Electric Company through R. M. Cogan. Providence, R.I.
9. W.L.N. was Visiting Scientist at the microbiology department, University College, Cardiff, when most of the present work was being done. Supported in part by the Medical Research Council and in part by NIH grant GM-08209.

22 April 1970; revised 26 June 1970

31

Copyright © 1972 by the Acoustical Society of America

Reprinted from *J. Acoust. Soc. Amer.*, 52(6), 1718–1724 (1972)

Shear as a Mechanism for Sonically Induced Biological Effects

JAMES A. ROONEY

Physics Department, University of Vermont, Burlington, Vermont 05401

(Received 12 May 1972)

> By controlled irradiation of erythrocyte suspensions at 20 kHz it is demonstrated that shear associated with acoustic microstreaming can be an important mechanism for biological effects of sound. Two effective sources of acoustic microstreaming are stable oscillating gas bubbles and transversely oscillating wires. The threshold displacement amplitude for achieving critical shear can be reduced by increasing the solvent viscosity and reducing the radius of the source of acoustic streaming. The threshold stress was found to decrease by 55% or more when the sample was heated to 45°C or higher for 10 min. This suggests that synergism exists between mechanical and thermal mechanisms for sonic effects. Mass transfer associated with small-scale acoustic streaming controls the rate of cell disruption.

SUBJECT CLASSIFICATION: 16.4.

INTRODUCTION

Many investigators have demonstrated biological effects of ultrasound, and various mechanisms for the action have been proposed. Much emphasis has been given to sonically induced temperature elevation, and its consequences.[1-4] Investigators[5-9] have also observed biological effects attributed to transient ("collapse"-type) cavitation with its concomitant shock waves, localized high temperatures ("hot spots"), and free radical formation. Cell disruption has also been reported in experiments in which mechanical resonances of the cells were believed to be sonically excited.[10,11] Recent preliminary studies[12,13] have emphasized the importance of ultrasonically induced shearing stresses in causing hemolysis. In general, the different mechanisms become dominant at different threshold acoustic amplitudes for any given biological system. Fry et al.[14] have elicited details of the various thresholds for the case of mammalian brain. It is possible that such thresholds are different for animal tissues, cells in suspension, and plants. Data will be presented in this paper further demonstrating biological damage to cells in suspension resulting from stresses associated with acoustic streaming. The possibility of synergism between thermal and mechanical effects of ultrasound will also be considered.

Acoustic streaming may be classified into two types[15]: (I) "quartz wind" occurring in bulk liquids as the result of attenuation, and (II) boundary-associated streaming resulting from oscillation of inhomogeneities in the sound field relative to the fluid or excitation of a small portion of a membrane. The type II streaming is typically small-scale, with greatest velocities occurring in regions of greatest tangential oscillatory velocity gradients. It is this latter kind which is relevant here.

Some previous experiments have suggested that small-scale acoustic streaming can be an important mechanism for the interaction of sound with suspensions of biological materials. Hughes and Nyborg[16] were able to hemolyze red blood cells (erythrocytes), that is, to cause the release of hemoglobin from them, and also cause protein release from the bacterium *E. coli*, when a small vibrating solid horn was placed in the cell suspension. Acoustic streaming was set up in the vicinity of the tip of the horn in the suspension. Hughes and Nyborg[16] were also the first to use bubbles trapped under a metal bar vibrating at 20 kHz for disruption of cells in suspension. When the bar was lowered into the suspension, bubbles formed by the air trapped in holes in the face of the bar served as effective sources of acoustic streaming. Protein release from suspensions of *E. coli* was observed well below the threshold for transient cavitation (as indicated by chemical tests for free iodine). Pritchard, Hughes, and Peacocke,[17] using a similar arrangement, measured a decrease in viscosity of suspensions of DNA after treatment with sonic amplitudes below the threshold for transient cavitation. They were able to establish that the number of breaks occurring in the DNA molecules could be directly related to velocity gradients calculated from acoustic streaming theory.

I. SONIC VISCOUS STRESS

We now report results which further demonstrate the importance of acoustic streaming as a mechanism

202

of sonic interaction with biological-cell suspensions. Two effective sources of type II acoustic streaming described in earlier studies[12,13] were used in this research. These sources complement each other. The first is a single stable oscillating bubble driven at 20 kHz. This system relates closely to the usual acoustic arrangements for irradiating cell suspensions. Typically sonic irradiation is carried out in media where bubbles are present that generate motions similar to those occurring in our experiments. However, the necessity for stability of the bubble used in our experiments requires the use of dextran or similar material in the cell-suspending medium. The dextran stabilizes the bubble by damping surface waves on the bubble. This probably occurs partly by increasing the viscosity of the suspending medium, and partly by forming a "skin" on the bubble in line with suggestions by Elder[18] and by Fox and Herzfeld.[19] The second source of streaming is a vibrating tungsten wire. This system is easier to use in some research because dextran is not necessary. Using the wire system one can easily demonstrate that acoustically induced biological effects occur when no bubbles or cavitation are present. Temperature increases are small and therefore not an important consideration in these acoustic systems. Theory for the temperature rise in the bubble situation is given in Appendix A.

Assuming that viscous stresses associated with acoustic microstreaming are important, details of their action can be elucidated by varying the viscosity of the suspending medium. A series of experiments was conducted in which the displacement amplitude threshold ξ_c for hemolysis was determined for canine erythrocyte suspensions with various values of the viscosity η. The range of η has a lower limit when a single stable oscillating bubble is used because a certain minimum concentration of dextran (11% w/v) is necessary to avoid surface-wave instabilities, as men-

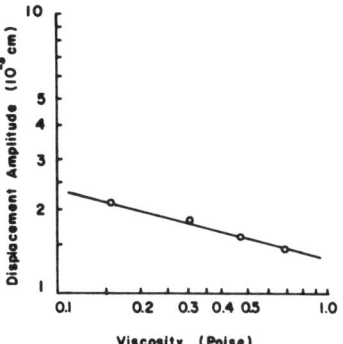

FIG. 2. Log-log plot of threshold displacement amplitude as a function of solvent viscosity. The slope of the line fitted to the data is $-\frac{1}{2}$. Scatter in data is within the area of the circles.

tioned earlier. Threshold amplitudes can be determined when physiological saline alone is used as the suspending medium for the cells using the vibrating wire apparatus. There is also an upper limit to η in that solutions of high viscosity prevent adequate transport of cells into the region of the bubble or wire. The viscosity was changed by varying the concentration of Dextran 500 (Pharmacia Chemicals, New Market, N. J.) in the physiological-saline solutions in which the erythrocytes were suspended; for each concentration, η was determined at 25°C using a Cannon-Manning capillary-type viscometer. It has been demonstrated for dextran that its viscosity is independent of shear rate over a wide range.[20] Results of these experiments are shown in Fig. 1.

We note that ξ_c decreases with increasing η. A theoretical relationship between ξ_c and η can be developed if we assume that viscous stress associated with acoustic streaming is the important mechanism involved in the observed release of hemoglobin. By expressing ξ_c in terms of the critical stress S_c and the viscosity, using theory previously developed,[12] we easily obtain

$$\xi_c^2 = (2^{\frac{1}{2}}bS_c)/(\omega^{\frac{1}{2}}\rho^{\frac{1}{2}}\eta^{\frac{1}{2}}). \tag{1}$$

In this equation b is the bubble or wire radius, ω is the angular frequency, and ρ is the density of the suspending medium. The curve in Fig. 1 shows a straight-line relationship between ξ_c^2 and $\eta^{-\frac{1}{2}}$, as expected from Eq. 1; a linear regression has been performed to fit the data. A value for the critical stress may be determined from the slope of the line; we find 4400 dyn/cm² for S_c, which compares favorably with values previously measured.[12,13] For this experiment the greatest error comes from threshold amplitude determinations. The resulting standard deviation in the stress is about five percent. However, the velocity gradient is nonuniform for the acoustic-streaming situation. As a result the maximum

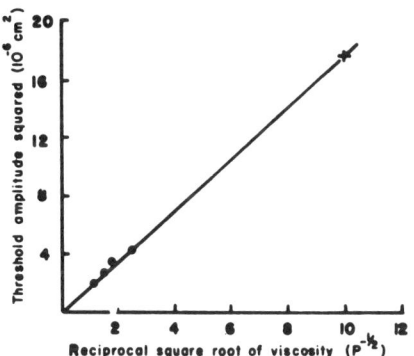

FIG. 1. Square of threshold amplitude as a function of reciprocal square root of viscosity of suspending medium. Data points are indicated by open circles obtained using stable oscillating bubble. Cross indicates data obtained with vibrating wire.

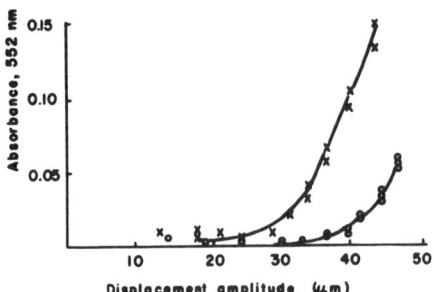

FIG. 3. Hemoglobin release as a function of displacement amplitude for tungsten wires of 125-μm (×) and 250-μm (○) diameter.

uncertainty is that S_c is greater, very likely of the order of 1500 dyn/cm².

II. AC OR DC STRESSES

We have previously calculated the steady or dc shearing stresses[12,13] associated with acoustic streaming to be in the range 3000–4500 dyn/cm² at the threshold amplitude for hemoglobin release. A calculation of the magnitude of the oscillatory or ac stress S_1 associated with the fluid motion near a vibrating object may be made using theory similar to that developed by Schlichting[21] and Lamb.[22] The oscillatory velocity gradient G_1 in the boundary layer near the surface of the bubble or wire may be approximated by the quotient of the velocity amplitude and the boundary-layer thickness δ. The viscous stress S_1 associated with this gradient is just the product of the solvent viscosity η and G_1:

$$S_1 = \eta G_1 = \eta \omega \xi_c / \delta, \qquad (2)$$

$$\delta = (2\eta/\omega\rho)^{\frac{1}{2}}. \qquad (3)$$

In a dextran solution with a viscosity of 0.31 P, density of 1.0 g/cm², and frequency of 2×10^4 Hz we obtain δ as 22.0 μm. At the threshold displacement of 18 μm we find that the oscillatory viscous stress is 32 000 dyn/cm². As can be readily seen, the ac stresses are an order of magnitude higher than those associated with the dc streaming. Therefore, we might expect that the former would be more important in the observed hemolysis. However, the opposite seems to be true according to experimental results.

We refer to two kinds of experiments in determining the relative importance of the two types of stress. From Eq. 1 used for dc stress calculations, we see that for a given critical stress ξ_c^2 is proportional to $\eta^{-\frac{1}{2}}$. In contrast we note from Eqs. 2 and 3 that, for a given critical stress in the ac case, ξ_c is proportional to $\eta^{-\frac{1}{2}}$. A log-log plot of data of the change in ξ_c as a function of η is shown in Fig. 2. The slope is $-\frac{1}{2}$ which is what we would expect if dc stresses are dominant in our experiment.

Again comparing Eqs. 1 and 2 we see that the dc stress depends on the radius of the bubble or wire while the ac stress does not. If dc stresses dominate, it should then be possible to obtain the critical stress for hemolysis at smaller amplitudes if a wire or bubble of smaller radius is used in the experiment. In fact, if the radius of the sound source is halved, the critical displacement amplitude should be reduced to 0.7 times its original value if dc stress is the important mechanism. An experiment to test this prediction was performed using the vibrating wire equipment. As mentioned earlier, this equipment has the advantage over the vibrating bubble arrangement in that surface waves are not a problem and hence the use of dextran is not required. For this study the critical amplitude for hemoglobin release was determined for wires of 125 μm (0.005 in.) and 250 μm (0.010 in.) in diameter.

Procedures are as described previously,[12] but are discussed here briefly, for convenience. In order to eliminate unwanted bubbles or bubble nuclei from the sound field, the sample was degassed by placing the irradiation vessel under reduced pressure (a few centimeters of mercury) for 10 min. Any liquid that was lost was replaced and the sample degassed for an additional 5 min. The wire was placed in the sample and driven at the desired displacement amplitude for 5 min. After sonation the suspension of erythrocytes was withdrawn and centrifuged for 5 min at 1500 g. The supernatant was withdrawn and placed in a microcell of a spectrometer and its absorbance read at 552 nm. This absorbance is proportional to the concentration of released hemoglobin.

Results of the experiment are shown in Fig. 3 and demonstrate that the threshold amplitude for hemoglobin release is indeed lower for the smaller wire; hence dc stresses associated with acoustic microstreaming rather than ac stresses seem to be the dominant mechanism.

The fact that the dc stresses dominate even though they are an order of magnitude less than the ac stresses probably results from the longer duration of stress in the dc case. It has been shown by R. P. Rand[23] and Leverett et al.[24] that for the erythrocyte the critical stress necessary for hemolysis increased with shorter duration of stress. The duration of the applied stress in our acoustic experiments may be estimated from theory. Referring to Fig. 4, the cells are brought into

FIG. 4. Path of typical erythrocyte near vibrating bubble.

the region of the bubble or wire near $\theta=0°$ and leave after passing over approximately one quarter of the circumference near $\theta=90°$. The maximum duration of the stress to which the cell is subjected is this length ($\pi b/2$) divided by the velocity with which the cell travels. As a typical velocity we cite $\tfrac{1}{2}u$, where u is $\omega\xi_o^2/b$, the limiting streaming velocity derived in our earlier paper.[12] Thus the duration D of stress expressed in terms of wire radius b and oscillation amplitude is

$$D=\pi b^2/(\omega\xi_o^2). \qquad (5)$$

For the threshold amplitude of 18.2 μm obtained for the single-stable bubble[12] and with b equal to 125 μm, one finds D to be 1300 μsec. In comparison the ac stress duration is of the order of half the sonic period, or 25 μsec. Thus we would expect that a higher ac stress is needed for hemolysis because of its short duration. The shortest duration of stress used by Rand in his experiment was 2 sec, so that no direct comparison of results can be made. The results quoted by Leverett *et al.* indicate that the critical shearing stress required for a stress duration of 25 μsec may be greater than 10^6 dyn/cm². This value is two orders of magnitude greater than the ac stresses achieved in these experiments.

We note that the extremes of dc stress duration for the experiments with the vibrating bubble (results shown in Fig. 2) are 2 msec for a viscosity of 0.05 P and 1 msec. when the viscosity was 0.7 P. Based on Leverett *et al.*, we should expect a variation in the critical stress of the order of 500 dyn/cm² because of the different stress durations. However, as discussed earlier, the uncertainty in the critical stress may be as high as 1500 dyn/cm². Thus, any variation in the critical stress is within the limits of our experimental arrangement.

III. TIME DEPENDENCE OF HEMOLYSIS RATE

The time dependence of the rate at which hemolysis occurs near a single oscillating bubble provides additional information on the type of mechanism involved in sonic interactions with biological suspensions. Suspensions of canine erythrocytes were treated using a single-stable oscillating bubble and data obtained following the usual procedures,[12] with results shown in Fig. 5. An estimate may be made of the volume rate V at which the sample passes through the boundary layer, extending theory developed by Nyborg.[25] For the purpose of order-of-magnitude estimates let us make three simplifying trial assumptions: first (an extreme condition), that all cells passing through the treatment region are hemolyzed in one passage; second, that the effective treatment region is in the boundary-layer region near the bubble. As indicated in Fig. 4 the cells enter this region near the metal tube holding the bubble. The cross-sectional area of an "entrance" through which the cells enter for treatment is roughly

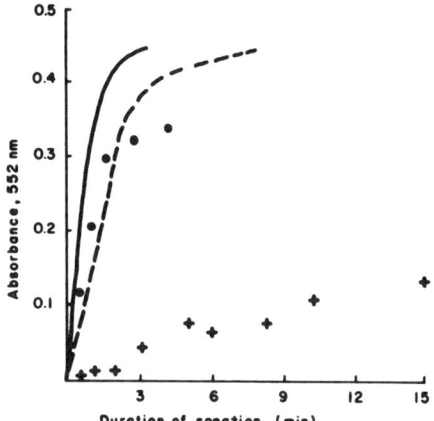

FIG. 5. Time dependence of rate of hemolysis. The solid curve is a theoretical plot corresponding to a displacement amplitude of 30 μm. The circles indicate data points for this amplitude. The dashed curve is a theoretical plot corresponding to an amplitude of 20 μm with the crosses representing data points for this amplitude.

just the product of the circumference of the base of the bubble $2\pi b$ and the boundary-layer thickness δ. The third assumption made is that a typical velocity for a cell in the treatment region is $\tfrac{1}{2}u$ where u is $\omega\xi_o^2/b$. Using these assumptions and Eq. 4 we find the volume rate of treatment $\dot{V}$ to be $\pi b\delta u$, or

$$\dot{V}=\delta\pi\omega\xi_o^2. \qquad (6)$$

The concentration of untreated cells C at any time t is just

$$C(t)=C_0-\left[\int_{-\infty}^{t}\dot{V}C(\tau)d\tau\right]/V_0, \qquad (7)$$

where V_0 and C_0 are sample volume and initial concentration, respectively. A solution of Eq. 7 is

$$C(t)=C_0 e^{-\dot{V}t/V_0}. \qquad (8)$$

Since the absorbance A is proportional to the concentration (C_0-C) of the hemolyzed cells we have

$$A=A_\infty(1-e^{-\dot{V}t/V_0}). \qquad (9)$$

Here A_∞ is the absorbance for complete hemolysis. Theoretical plots using Eq. 9 are shown in Fig. 5 for amplitudes of 20 and 30 μ. There is order-of-magnitude agreement between theory and experiment at the higher amplitude. The large discrepancies at the lower amplitude of 20 μm seems to indicate that our first trial assumption (that damage results from a single pass near the bubble) is wrong under these conditions. Hence, near threshold amplitudes, cells must apparently pass through the effective treatment region several

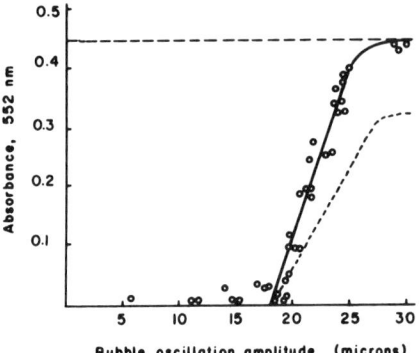

FIG. 6. Absorbance as a function of displacement amplitude for the case of two bubbles is shown with a curve fitted to data indicated by circles. Single bubble results are shown by the dashed curve. The horizontal dashed line indicates absorbance for complete osmotic hemolysis.

times receiving only some slight irreversible damage during each passage.

One also notes that the maximum extent of hemoglobin release varies with amplitude. Thus, for a displacement amplitude of 30 μm the absorbance tends to level off near 0.33; for the 20 μm case the limiting absorbance is close to 0.13. This observation has two explanations. First, the observed extent of the acoustic streaming pattern increases with amplitude; that is, while in theory the fluid velocity extends to infinity, one notes in the experiment that only fluid in the vicinity of the bubble or wire is involved significantly in the eddying motion and that the region in which acoustic streaming velocities are significant increases with displacement amplitude. Thus, at higher amplitude more cells from the sample are in the region of treatment with the resulting greater limiting absorbance. Second, the cell population in the sample is heterogeneous. The blood samples are obtained by venous puncture. The average life span of an erythrocyte is 120 days, and our blood samples contain a random distribution of cells with different ages. If cells of different ages have different susceptibilities to shearing stress, the none would expect the maximum extent of hemolysis to vary with amplitude.

The rate of sample treatment can be increased by increasing the number of sources of acoustic streaming. To demonstrate this possibility, two bubbles were formed in the 0.2-ml sample. The sample holder, i.e., treatment vessel, is 1 cm in height and length and 0.2 cm in width. The bubbles in this experiment were formed 0.1 cm from the front (and also from the back) of the vessel and 0.5 cm from the top; one was 0.25 cm from the left side, the other 0.25 cm from the right side. Results of such an experiment are shown in Fig. 6. First, one notes that the threshold amplitude is the same as for a single source of acoustic streaming. Second, the slope is 1.7 times that for the single bubble indicated by the dashed line. One might have expected a slope of 2.0. In explanation of the smaller slope, observation of the streaming pattern during sonation indicated an overlapping of the streaming patterns from the two bubbles; hence one doesn't obtain the simple doubling of the rate of treatment by forming two bubbles which are too close together.

IV. THRESHOLDS AT ELEVATED TEMPERATURES

In a recent study, C. C. Connolly[26] irradiated suspensions of erythrocytes with ultrasound in the megahertz frequency range. He found that when transient cavitation was not present he could detect no hemolysis at a temperature of 37°C using intensities up to 330 W/cm². On the other hand, he was able to detect the release of hemoglobin in this intensity range when the cell suspensions were sonated at temperatures of 45°C or higher. In fact he found that the threshold intensity for hemolysis decreased with increasing temperatures at all frequencies used in his experiment.

It seemed therefore that a systematic study should be conducted of the critical sonic stress necessary for hemolysis as a function of temperature. For this part of the research the vibrating-wire apparatus was used with a 254 μm (0.01 in.) diameter wire driven at 20 kHz. Physiological saline was used as the suspending medium for the erythrocytes. At each of the temperatures used in the experiment a 0.2-ml sample was placed in a constant temperature bath and allowed to equilibrate for 1 min. The temperature of the sample was maintained constant to within ±0.2°C. The wire was placed in the test vessel and driven at 20 kHz for 10 min. The samples were then centrifuged for 5 min, the supernatant withdrawn, and the absorbance of the supernatant read at 552 nm. Threshold amplitudes were determined for each temperature, and the corresponding critical shearing stress was calculated using

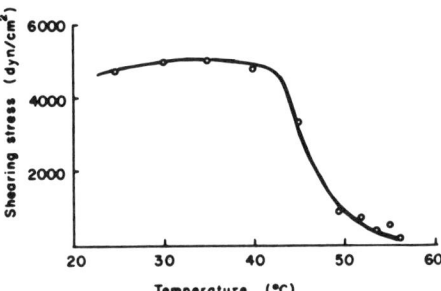

FIG. 7. Critical shearing stress as a function of sample temperature.

Eq. 1. Results are shown in Fig. 7; as can be seen there, susceptibility of erythrocytes to shearing stress decreases rapidly above 45°C with complete hemolysis occurring at 55°C. These results can be compared with those of a study of autohemolysis as a function of temperature. A sample of blood was maintained in the temperature bath for 10 min after a 1-min equilibration period, the sample was centrifuged, and the absorbance was read at 552 mm. No autohemolysis was noted up to 40°C for the 10-min heating period. It was found that 18% of the cells were hemolyzed at 50°C, the percentage increasing to 22% at 53°C, 36% at 55°C, and 67% at 56°C.

The results for the increasing susceptibility of the erythrocytes to shear stress may be compared with the qualitative results on osmotic fragility obtained by Ham[27] et al. They observed no significant change in the fragility of the cells until the temperature had reached 49°C; then the fragility increased rapidly until the erythrocytes hemolyzed completely in physiological saline in the temperature range 55°–56°C. The same investigators made observations on morphological changes occurring in the cells as function of temperature. The cells appeared normal up to temperatures of 46°C. Above this temperature the erythrocytes became spherical, abandoning their normal biconcave disk shape; "buds" and protrusions were also seen on the erythrocytes.

Other studies by Champion[28] et al. and Williams[29] have shown that erythrocytes suspended in a hypotonic medium (0.54% NaCl) are more susceptible to hydrodynamic shear[28] or acoustically induced shearing[29] than those suspended in isotonic saline. The erythrocytes in such a hypotonic medium are typically sphered, with no significant autohemolysis occurring. It seems, therefore, that the increased susceptibility of the cells to acoustically induced shear at elevated temperatures is the result of two effects. The first is enlargement of "pore" size and weakening of the erythrocyte membrane as indicated by the autohemolysis data. The second is morphological change induced by the temperature.

A related question that is of interest is how the threshold shearing stress for a biological effect changes with the length of time a sample is maintained at a given elevated temperature. To answer this, a study was conducted of the threshold stress for hemolysis of erythrocytes maintained at 50.0°C±0.2°C for various durations of heating. The procedure followed was to allow the sample to reach thermal equilibrium, maintain the sample at 50.0°C for a given period, sonate the sample using the vibrating wire for 5 min, and then use the usual analytical techniques to determine the threshold displacement amplitude from which the stress can be calculated. The results of the study are shown in Fig. 8. The threshold stress is shown as a function of length of time the sample was exposed to a given temperature; this includes the 5-min sonation period plus the time for preheating of the sample. We note the rapid decrease in threshold shear stress after an exposure of 10 min. These results parallel the qualitative measurements of Ham[27] et al. His group found that the osmotic fragility of erythrocytes maintained at 50°C increased after 10 min of exposure to the heat.

ACKNOWLEDGMENTS

The author is indebted to Dr. Wesley L. Nyborg for discussions and technical advice. He wishes to thank Dr. Bert Kusserow for his suggestions as well as for providing the canine blood samples. This work is based in part on a PhD thesis completed at the University of Vermont. The research was supported in part by the National Institutes of Health via GM-08209, by the HAS fund at the University of Vermont and by Public Health Service Predoctoral Fellowship GM 43838-01.

APPENDIX A: TEMPERATURE RISE IN A SAMPLE RESULTING FROM IRREVERSIBLE HEAT PRODUCTION NEAR THE BUBBLE

Gould and Nyborg[30] have calculated the rate of viscous heat production in the liquid external to a pulsating bubble. They found the total rate of heat production $\langle \dot{w} \rangle$ in the liquid external to a bubble of equilibrium radius b to be

$$\langle \dot{w} \rangle = \eta \dot{v}_0^2 / 2\pi b^3. \quad (A1)$$

Here $\dot{v}_0$ is the magnitude of the volume velocity. An appropriate approximation for the volume velocity $\dot{v}$ for the bubble is

$$\dot{v} = 4\pi b^2 \dot{R}, \quad (A2)$$

where $\dot{R}$ is the bubble surface velocity. The bubble radius is just

$$R = b + \xi_0 \cos\omega t. \quad (A3)$$

Substituting Eq. A3 into A2 we see that

$$\dot{v}_0^2 = 16\pi^2 b^4 \omega^2 \xi_0^2. \quad (A4)$$

Thus for Eq. A1 one has

$$\langle \dot{w} \rangle = 8\pi \eta b \omega^2 \xi_0^2. \quad (A5)$$

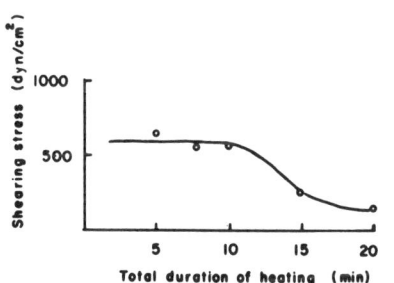

FIG. 8. Threshold shearing stress as a function of duration of heating. Duration includes a 5-min sonation period.

Using this expression for $\langle \dot{w} \rangle$ we can estimate the maximum temperature rise in the liquid. First, consider that all of the heat is produced at the bubble surface. This heat will be lost by way of conduction as well as convection associated with acoustic streaming. In order to calculate the maximum rise in temperature that can occur in the liquid, we will consider the rate of heat transfer when conduction is the only mechanism. The rate at which heat is conducted away from a hemispherical surface is given by

$$\langle \dot{w} \rangle = -2\pi b^2 K (dT/dr)_b. \qquad (A6)$$

Here T is the temperature and K the thermal conductivity of the liquid; r measures distance from the center of the bubble. The temperature gradient is evaluated at the bubble surface $r=b$. Using a solution of the steady-state heat conduction equation for the case of spherical symmetry,[31] we evaluate the temperature gradient in Eq. A6 and find

$$\langle \dot{w} \rangle = 2\pi K b (T_b - T_a). \qquad (A7)$$

Here T_b and T_a are, respectively, the bubble surface and ambient temperatures.

In order to calculate the temperature rise in a typical experiment we consider conditions similar to those for the threshold of hemolysis using the stable oscillating bubble with dextran as the suspending medium. For this case $\eta=0.3$ P, $\omega=125\,000$, $\xi_0=0.002$ cm, $b=0.13$ cm, and $K=0.00143$ cal/(sec) (cm) (°C). The temperature rise calculated from Eq. A7 for these conditions is 1.7°C. We note that even this conservative upper limit for the temperature rise is not of physiological significance in these experiments which are typically conducted at 25°C.

Heat transfer by convection associated with the acoustic streaming would tend to reduce this temperature increase. Theory does not exist for the effect of acoustic streaming on heat transfer for the present arrangement. However, results obtained by Gould[32] for a similar situation indicate that increases in heat transfer up to tenfold are possible.

[1] P. A. Nelson, J. F. Herrick, and F. H. Krusen, Arch. Phys. Med. **31**, 687–695 (1951).
[2] T. P. Anderson, K. G. Wakim, J. F. Herrick, and W. A. Bennett, Arch. Phys. Med. Rehabilitation **32**, 71–83 (1951).
[3] J. F. Lehmann, Arch. Phys. Med. Rehabilitation **34**, 139–152 (1953).
[4] P. P. Lele, Exp. Neurol. **8**, 47–83 (1963).
[5] H. Freundlick and D. W. Gillings, Trans. Faraday Soc. **34**, 649–660 (1938).
[6] D. E. Goldman and W. W. Lepeshkin, J. Cell Comp. Physiol. **40**, 255–268 (1952).
[7] J. P. Horton, J. Acoust. Soc. Amer. **25**, 40–484 (1953).
[8] J. F. Lehmann, Arch. Phys. Med. Rehabilitation **34**, 86–98 (1953).
[9] W. T. Coakley, D. Hampton, and F. Dunn, J. Acoust. Soc. Amer. **50**, 1546–1553 (1971).
[10] E. Ackerman, J. Cell Comp. Physiol. **39**, 167–190 (1952).
[11] E. Ackerman, Bull. Math. Biophys. **19**, 107 (1957).
[12] J. A. Rooney, Science **169**, 869–871 (1970).
[13] A. R. Williams, D. E. Hughes, and W. L. Nyborg, Science **169**, 871–873 (1900).
[14] F. J. Fry, G. Kossoff, R. C. Eggleton, and F. Dunn, J. Acoust. Soc. Amer. **48**, 1413–1417 (1970).
[15] W. L. Nyborg, "Acoustic Streaming," in *Physical Acoustics*, W. P. Mason, Ed. (Academic, New York, 1965), Vol. II, Part B, pp. 265–331.
[16] D. E. Hughes and W. L. Nyborg, Science **138**, 108–114 (1962).
[17] N. J. Pritchard, D. E. Hughes, and A. R. Peacocke, Biopolymers **4**, 259–274 (1966).
[18] S. A. Elder, J. Acoust. Soc. Amer. **31**, 54–64 (1959).
[19] F. E. Fox and K. F. Herzfeld, J. Acoust. Soc. Amer. **26**, 984–989 (1954).
[20] A. R. Williams (private communication).
[21] H. Schlichting, *Boundary Layer Theory* (McGraw-Hill, New York, 1968), 6th ed., pp. 7, 85, 396.
[22] H. Lamb, *Hydrodynamics* (Dover, New York, 1945), 6th ed., Art. 345.
[23] R. P. Rand, Biophys. J. **4**, 303–316 (1964).
[24] L. B. Leverett, J. D. Hellums, C. P. Alfrey, and E. C. Lynch, Biophys. J. **12**, 257–273 (1972).
[25] W. L. Nyborg, Rep. Int. Congr. Acoust., Liège, 5th, D. E. Commins, Ed., Vol. 1b, Paper 1 K43 (1965).
[26] C. C. Connolly, PhD thesis, University of London (1969).
[27] T. Ham, Shan, E. Flemming, and W. W. Castle, Blood **3**, 373–403 (1948).
[28] J. V. Champion, P. F. North, W. T. Coakley, and A. R. Williams, Biorheology **8**, 23–29 (1971).
[29] A. R. Williams, PhD thesis, University College, Cardiff, Wales (1970).
[30] R. K. Gould and W. L. Nyborg, J. Acoust. Soc. Amer. **32**, 775 (1960).
[31] H. S. Carslaw, *Introduction to the Mathematical Theory of The Conduction of Heat in Solids* (Macmillan, New York, 1921).
[32] R. K. Gould, J. Acoust. Soc. Amer. **40**, 219–225 (1966).

Effects of Intense Noncavitating Ultrasound on Selected Enzymes*

R. M. MACLEOD† AND F. DUNN

Biophysical Research Laboratory, University of Illinois, Urbana, Illinois 61801

The denaturation of enzymes in solution by cavitating ultrasound has been reported previously. This report presents the results of an extensive study of the effects of noncavitating ultrasound on solutions of α-chymotrypsin, trypsin, aldolase, lactate dehydrogenase, and ribonuclease. In one set of experiments, the solutions were irradiated and then analyzed to determine the effects on the physical and chemical properties of the enzyme molecules. Irradiations were carried out at different pH values and temperatures using 1-MHz ultrasound at an intensity of 75 W/cm², 10-min continuous exposure, and 11-MHz ultrasound at an intensity of 1000 W/cm², 2000 0.1-sec pulses. Analytical procedures employed included measurements of enzyme activity, specific optical rotation, uv absorption spectrum, and sedimentation coefficient. In a second set of experiments, enzyme-catalyzed reactions were irradiated with ultrasound and simultaneously monitored spectrophotometrically. Ultrasound in the intensity range 0.5–35 W/cm² at the frequencies 1, 9, and 27 MHz were employed with the temperature and pH held constant. Comparison of the results of this study with those from studies employing cavitation shows that cavitation is a necessary condition for ultrasonic denaturation of the five enzymes of this study.

INTRODUCTION

IT is well known that microorganisms and cellular structures of higher organisms can be damaged extensively when subjected to ultrasonic cavitation.[1] It is less widely appreciated that selective alteration can be produced in tissue of the mammalian central nervous system by intense noncavitating ultrasound.[2] Developments in this latter field have provided an advantageous lesion-making method for neuroanatomical research,[3] and ultimately a versatile brain modifying means for therapy.[4] However, the need for a more adequate explanation of the physical mechanisms of interaction of intense noncavitating ultrasound and tissue structures remains. Experimentation with mammals has shown that physiological changes, e.g., limb paralysis,[5] resulting from intense ultrasonic irradiation of the spinal cord can be detected within seconds after exposure to a 1-sec pulse but that histological evidence of tissue alteration does not appear until approximately 10 min after irradiation, with progressive lesion formation following.[6] It has been established, for such interactions, that cavitation,[5] the temperature increase occurring in the irradiated region of the specimen,[5] and unidirectional forces that might produce elastic failure of structural components when displacements from equilibrium positions occur,[7] are not responsible for the observed tissue changes. The temporary absence of detectable structural alterations in a functionally altered tissue suggested that the primary site of action of the ultrasound was a submicroscopic structure, possibly a macromolecule. Reports that organic polymers,[8] the enzymes trypsin[9] and α-amylase,[10] and DNA[11] could be degraded or denatured by exposure of solutions of these polymers to intense noncavitating ultrasound strengthened this suggestion.

In the study reported here, the enzymes α-chymotrypsin, trypsin, aldolase, lactate dehydrogenase, and ribonuclease were selected for a detailed investigation of the interactions of noncavitating ultrasound with proteins in solution. The selection of enzymes was prompted by

---

* Portions of this work were extracted from the thesis submitted by the first-named author in partial fulfillment of the requirements for the PhD degree in biophysics, University of Illinois, Urbana, 1966.
† Present address: Dep. of Biochem., School of Basic Med. Sci., Univ. of Tennessee, Memphis, Tenn. 38104
[1] I. W. El'piner, *Ultrasound: Physical, Chemical, and Biological Effects* (Consultants Bureau, Inc., New York, 1964), pp. 149–231.
[2] W. J. Fry, Advan. Biol. Med. Phys. 6, 281–348 (1958).
[3] W. J. Fry, F. J. Fry, R. Malek, and J. W. Pankau, J. Acoust. Soc. Amer. 36, 1795–1835 (1964).
[4] W. J. Fry and R. Meyers, Confinia Neurol. 22, 315–327 (1962).
[5] F. Dunn, Amer. J. Phys. Med. 37, 148–151 (1958).

[6] J. W. Barnard, W. J. Fry, F. J. Fry, and R. F. Krumins, Comp. Neurol. 103, 459–484 (1955).
[7] F. Dunn, J. Acoust. Soc. Amer. 29, 395–396 (1957).
[8] H. W. Melville and A. J. R. Murray, Trans. Faraday Soc. 46, 996–1009 (1950).
[9] V. Stefanovic, I. Kostic, M. Bresjanac, and D. Zivanovic, Bull. Soc. Chim. Belgrade 24, 175–178 (1959).
[10] V. Stefanovic, A. Djukanovic, K. Velasevic, and D. Zivanovic, Experientia 14, 486–487 (1960).
[11] S. A. Hawley, R. M. Macleod, and F. Dunn, J. Acoust. Soc. Amer. 35, 1285–1287 (1963).

several considerations. *First*, calculations based on acoustic absorption data for protein solutions[12–14] indicate that sufficient energy is available for the inactivation of enzyme molecules.[15] *Second*, previous investigations[9,10] indicated that enzymes were susceptible to inactivation by noncavitating ultrasound. *Third*, enzyme inactivation is a plausible explanation for the effects of intense noncavitating ultrasound on tissues of the mammalian central nervous system[5] since the vital activities of each cell depend upon the normal functioning of its enzymes. *Fourth*, it was felt that more could be learned by studying these relatively well-characterized proteins than any other class of biomacromolecules, e.g., their specific catalytic abilities are very sensitive to any alteration of structure that effects the "active site," and this provides a sensitive indicator for even very subtle changes in the nature of the enzyme's conformation. In addition, denaturation often is reliably reflected by changes in such physicochemical properties as sedimentation behavior, optical rotation, and the uv absorption spectrum. *Finally*, the selected enzymes were available in purified form, thus homogeneous samples of uniform size, shape, and molecular weight were studied rather than distributions of these characteristics.

Two types of studies were conducted. *First*, enzyme solutions were irradiated with intense noncavitating ultrasound and subsequently analyzed by various techniques (uv absorption spectra, optical rotation, sedimentation velocity analysis, chromatography) for changes in the physical and chemical properties of the enzymes. Temperature and pH of the solutions were varied in some cases to determine the effects of these variables on the interaction phenomena. *Second*, irradiations were carried out during enzyme-catalyzed reactions and these reactions were monitored continuously to determine whether reversible changes in enzyme structure occur that are completed too rapidly for detection by the above-mentioned techniques.

I. MATERIALS AND METHODS

A. Biochemical Materials and Methods of Analysis

All inorganic reagents were prepared from reagent-grade chemicals using distilled deionized water.

Trypsin ($2X$ crystallized), α-chymotrypsin ($3X$ crystallized), aldolase ($2X$ crystallized from rabbit muscle), and lactate dehydrogenase ($2X$ crystallized from rabbit muscle) were obtained from the Worthington Biochemical Corporation. Bovine pancreatic ribonuclease ($3X$ crystallized) was obtained from the Sigma Chemical Company. All the enzymes were used as obtained, without further purification. Fructose-1, 6-diphosphate and sodium pyruvate were obtained from Mann Research Laboratories. Reduced nicotinamide–adenine dinucleotide (NADH$_2$) was obtained from P-L Biochemicals, Inc., and cytidine $2':3'$ cyclic phosphate was obtained from Schwarz Bio Research, Inc. N-acetyl-L-tyrosine ethyl ester (ATEE) and N-benzoyl-L-arginine ethyl ester (BAEE) were obtained as "Determatubes" from the Worthington Biochemical Corporation.

Tris[tris(hydroxymethyl)aminomethane] buffer[16] and phosphate buffer[17] were prepared according to established procedures.

Trypsin,[18] α-chymotrypsin,[18] aldolase,[19] lactate dehydrogenase,[20] and ribonuclease[16] activities were assayed spectrophotometrically with a Beckman DU spectrophotometer equipped with a thermostated (25.0° ±0.1°C) cuvette compartment and a Sargent SRL recorder. Enzyme concentrations were determined spectrophotometrically, using published absorptivities for trypsin,[21] α-chymotrypsin,[21] aldolase,[22] lactate dehydrogenase,[23] and ribonuclease.[24]

All pH measurements were made with a Beckman Zeromatic pH meter.

Ultraviolet absorption spectra, in the wavelength range 240–320 mµ were read either with the Beckman DU spectrophotometer used for activity assays (without the recorder) or with a Cary model 14 automatic scanning and recording spectrophotometer.

Optical rotation measurements were made with a Rudolph model 70 precision polarimeter using a 10.00-cm microtube (0.7-ml volume) and a filtered sodium-vapor light source.

Sedimentation velocity analyses were performed with a Spinco model E analytical ultracentrifuge equipped with schlieren optics and a RTIC unit. An An-D rotor was employed in all experiments. Both a standard cell with a Kel-F centerpiece and a valve-type synthetic boundary cell were used. The rotor speed setting for all experiments was 56 100 rpm. The photographic plates were measured with a comparator and sedimentation coefficients were determined in the usual way.[25] The sedimentation coefficients were converted to the standard conditions of water as solvent and 20°C.

---

[12] E. L. Carstensen and H. P. Schwan, J. Acoust. Soc. Amer. 31, 305–311 (1959).
[13] P. D. Edmonds, Biochim. Biophys. Acta 63, 216–219 (1962).
[14] S. A. Hawley, PhD thesis in biophysics, Univ. of Illinois, Urbana, Ill. (1967).
[15] R. M. Macleod, PhD thesis in biophysics, Univ. of Illinois, Urbana, Ill. (1966).
[16] E. M. Crook, A. P. Mathias, and B. R. Robin, Biochem. J. 74, 234–238 (1960).
[17] G. Gomori, *Methods in Enzymology*, S. P. Colowick and N. O. Kaplan, Eds. (Academic Press Inc., New York, 1955), Vol. I. pp. 138–146.
[18] G. W. Schwert and Y. Takenaka, Biochem. Biophys. Acta 16, 570–575 (1955).
[19] V. Jagannathan, K. Singh, and M. Damodaran, Biochem. J. 63, 94–105 (1956).
[20] A. Kornberg, *Methods in Enzymology*, S. P. Colowick and N. O. Kaplan, Eds. (Academic Press Inc., New York, 1955), Vol. 1, pp. 441–443.
[21] B. C. W. Hummel, Can. J. Biochem. Physiol. 37, 1393–1399 (1959).
[22] T. Baranowski and T. Niederland, J. Biol. Chem. 180, 543–551 (1949).
[23] J. B. Neilands, J. Biol. Chem. 199, 373–381 (1952).
[24] J. J. Hermans and H. A. Sheraga, J. Amer. Chem. Soc. 83, 3283–3292 (1961).
[25] H. K. Schachman, *Methods in Enzymology* (Academic Press Inc., New York, 1957), Vol. IV, pp. 32–71.

Thin-layer chromatography experiments were performed using the Eastman Chromagram Developing apparatus and Eastman Chromagram Sheets, type K301R. The solvent system used was n-butanol: acetic acid: water (3:1:1). Spots were located first by examining under uv light and then by spraying with a solution of ninhydrin in absolute ethanol (0.3g/100ml), and heating at 90°C for 15 min.

B. Ultrasonic Instrumentation: Irradiation of Enzyme Solutions

Two basically similar ultrasonic systems were employed, each consisting of an X-cut quartz transducer having a fundamental thickness resonant frequency of 1 MHz intimately coupled by a thin layer of oil to a planoconcave polystyrene lens. The sound-transmitting medium for all experiments was degassed distilled water. Each system contained a castor oil acoustic absorption chamber, separated from the transmitting medium by an acoustically transparent rubber membrane, in order to assure that all specimens were exposed to traveling-wave fields only. Both systems were provided with mechanical three-coordinate positioning devices enabling the specimen container to be located in the focal volume of the sound beam within ±0.01 cm in each orthogonal direction. For System A, the specimen container is moved while the transducer remains fixed in space, and for System B, the specimen container is placed at a fixed position and the transducer is moved. The transducer of System A was $1\frac{1}{2}$-in. in diameter and driven by a power amplifier (capable of delivering approximately 8000 V to the quartz plate) controlled by a signal generator. The transducer of System B was $2\frac{1}{2}$-in. in diameter and excited by an electronic driver possessing a feedback arrangement to ensure constant predetermined voltage applied to the quartz plate and, consequently, a constant acoustic intensity in the specimen chamber. The focal volume (the volume over which the acoustic intensity is not less than one-half of the peak value) is a function of the sound wavelength[26] and at 1 MHz was determined experimentally to be approximately 0.5 cm in diameter and 1 cm long while at 11 MHz, the corresponding values are 0.05 and 0.3 cm, respectively. With the very high intensities employed with System B, it was necessary to use a pulsing regime in order that the acoustic lens not overheat and alter its focal properties. Thus, an automated arrangement was used whereby the transducer delivered a 0.1-sec pulse every 6 sec. At 11 MHz, complete automation was employed because the combination of small focal volume and short irradiation time required that many pulses be given to ensure that an appreciable fraction of the sample be irradiated. The sample size for this system was 2 ml. Experiments with suspensions of small particles showed that a 0.1-sec pulse resulted in energetic particle motion inside the sample container. Thus, since the acoustic pulse provided good stirring and since there was 5.9 sec between pulses, it appears to be a good assumption that complete mixing occurred after each pulse. With this assumption of complete mixing, a conservative estimate of 2 mm³ in the focal volume, and a total volume of 2 ml, calculation shows that after 2000 pulses the probability of every particle being in the focal volume at least once is 0.86.

C. Sample Containers

For System A, where the focal volume at 1 MHz was about 0.8 ml, sample containers were made of Pyrex tubing 32 mm long and 16 mm in i.d. (volume about 6.4 ml). A groove was ground on the outside of each end of the tube to a depth sufficient to hold a $\frac{1}{16}$-in.-thick O ring. Both ends of the tube were closed with pieces of 0.0005-in.-thick Saran, which were clamped in place with the O rings. With practice, it is possible to fill this container with degassed sample solution and to close it so that air is not trapped inside. For the vertical System B, the containers were similar but of lesser volume (2.0 ml) and were made of Pyrex tubing of 9-mm i.d. and 32 mm long. The ends were closed in the manner described above. Each type of container was clasped in a small three-fingered laboratory clamp for mounting. The position of the focal volume in the transmitting medium was located by the thermocouple probe method.[26] The acoustic intensity in the focal volume was determined by the radiation pressure technique.[26]

D. Equipment for Direct Irradiation of Enzyme Activity Assays

An instrument designed especially for the purpose of monitoring the ultrasonically irradiated enzyme-catalyzed reactions was constructed to fit a Beckman model DU spectrophotometer. The cuvette compartment of this spectrophotometer was replaced by an ultrasonic irradiation system consisting of: (a) a 1-MHz X-cut quartz transducer, the piezoelectric element which produces ultrasound when excited electrically; (b) a right circular cylindrical reaction cell of stainless steel, volume approximately 20 ml, with quartz windows to transmit the spectrophotometer light beam normal to the direction of ultrasonic wave propagation, and with water from a constant temperature bath circulating through its walls; and (c) an acoustic absorption chamber (filled with castor oil and separated from the reaction cell by a 0.001-in.-thick polyethylene membrane) whose function is to ensure the absence of standing waves by absorbing all incident acoustic energy. The uv light passing through the reaction cell was detected by a 1P28 photomultiplier tube and the electrical output was fed through a Beckman energy recording adapter to a Sargent model SRL recorder which was equipped with logarithmic gears to give a

---

[26] W. J. Fry and F. Dunn, *Physical Techniques in Biological Research*, W. L. Nastuk, Ed. (Academic Press Inc., New York, 1962), Vol. 4, Chap. 6, pp. 261–394.

plot of absorbance versus time. The slope of the absorbance-versus-time curve gives the rate of the enzyme-catalyzed reaction and was converted, using the appropriate definition for a unit of activity, to the specific activity of the enzyme. More detailed specifications and a description of the construction of the system are given elsewhere.[15]

E. Degassing Procedure

Below 10 MHz, the acoustic intensity cavitation threshold for water containing dissolved gas is lower than for degassed water.[26] Data for the acoustic-intensity cavitation threshold for most aqueous solutions are not available, but the general principle that the threshold increases with viscosity[27] means that the data for water can be considered minimum values for aqueous solutions more viscous than water. Thus, to avoid cavitation for ultrasonic frequencies below 10 MHz, the sample solution should be degassed and the intensity kept below the acoustic cavitation threshold for degassed water.

Degassing was carried out in a vacuum desiccator using a laboratory vacuum pump. The sample solution, 7–20 ml depending on the sample container size and the number of samples, was placed in a 50-ml Erlenmeyer flask and the flask was covered with a piece of perforated Parafilm to prevent loss of the sample due to the splashing and foaming that occur during degassing. The flask was then placed in the vacuum desiccator and degassed for 10 min. All the visibly detectable degassing was usually completed within 5 min. After 10 min under vacuum, the desiccator was slowly brought to atmospheric pressure, the sample solution was removed and the volume lost due to evaporation of water was replaced by carefully adding degassed distilled water.

II. EXPERIMENTAL PROCEDURE

A. Enzymes in Solution: Irradiations and Analyses

The composition of the enzyme solutions studied, the characteristics of the ultrasound to which they were exposed, and the analyses of the irradiated solutions are given in Table I.

The solutions were degassed, the sample containers were filled and positioned in the focal volume of the ultrasonic irradiation systems and for each experiment, a second container was filled and placed in the transmitting medium compartment, but out of the sound beam, to serve as a control. Following irradiation, samples and controls were refrigerated at 5°C until the various analyses could be performed. Appropriate dilutions of aliquots of the irradiated samples and controls were assayed for enzymatic activity and the other properties given in Table I.

The results of the analyses presented in Table I show that there are no significant differences between

[27] T. F. Hueter and R. H. Bolt, *Sonics* (John Wiley & Sons, Inc., New York, 1955), pp. 225–241.

the irradiated samples and their unirradiated controls in any of the experiments and, thus, the enzymes are unaffected in any permanent way by the intense noncavitating ultrasound. A more detailed consideration of each enzyme follows.

Alpha-chymotrypsin is known to be most stable in solution at pH 3–4 and at temperatures in the range 0°–5°C; α-chymotrypsin solutions of higher pH and solutions stored at higher temperatures undergo gradual autolysis.[28] These facts explain why the specific activities observed for both sample and control in Expts. 3 and 4 (Table I) are less than the corresponding activities at the same pH values but lower temperatures in Expts. 1 and 2. Additional control samples for Expt. 3 (pH 3.0) and Expt. 4 (pH 7.1), which were degassed but not subjected to the $3\frac{1}{3}$-h irradiation time at 37°C showed the same specific activities as the controls for Expt. 2 (pH 3.0) and Expt. 1 (pH 7.1), respectively. The interesting conclusion drawn from these results is that α-chymotrypsin is not affected by intense noncavitating ultrasound even under conditions that cause gradual denaturation of the enzyme. The $A_{280\ m\mu}$ and the ratio $A_{280\ m\mu}/A_{250\ m\mu}$ for 1:100 dilutions of irradiated sample and control are given in Table I in order to compare the absorption spectra. The sample spectra were essentially congruent to those of the controls and this congruence is reflected in the near identity of the $A_{280\ m\mu}$ and $A_{280\ m\mu}/A_{250\ m\mu}$ values. In the α-chymotrypsin experiments, the specific optical rotations of the irradiated samples consistently showed slightly more levorotation than the controls. The accepted value of $[\alpha]_D$ for α-chymotrypsin in $0.1M$ NaCl, pH 3 is $[\alpha]_D = -66°$.[29] The discrepancy between this value and the values for the control in Table I is probably mostly due to the difference in the method used for determining concentration, viz., for the value reported in the literature, concentrations were determined by a micro-Kjeldahl method assuming the α-chymotrypsin to be 16.06% nitrogen; the α-chymotrypsin concentrations in Table I were determined spectrophotometrically using an absorptivity at 282 mμ for α-chymotrypsin of 1.85/cm/mg/ml.[21] However, it is not unusual for $[\alpha]_D$ values reported by different investigators for proteins under similar conditions to differ slightly.[30] When proteins are denatured, they usually show a large increase in levorotation. If α-chymotrypsin is denatured with $8M$ urea in $0.1M$ NaCl, pH 3, $[\alpha]_D = -111.6°$.[29] Therefore, though the small differences between the specific rotations of α-chymotrypsin samples and controls in Table I were consistently found, because they are small differences they are not interpreted as evidence of denaturation. In the sedimentation velocity analysis of Expt. 1, the irradiated sample and control, which were

[28] M. Kunitz and J. H. Northrup, J. Gen. Physiol. **18**, 433–458 (1935).
[29] J. A. Schellman, Compt. Rend. Trav. Lab. Carlsberg **30**, 450–461 (1958).
[30] P. Urnes and P. Doty, Advan. Protein Chem. **16**, 401–544 and 486–489 (1961).

TABLE I. Treatment and analyses of ultrasonically irradiated enzyme solutions.

Enzyme	Experiment No.	Conc. (mg/ml)	pH, solvent	Temp. (°C)	Freq. (MHz)	Intensity (W/cm²)	Duration	Specific activity[a] (U/mg)		$A_{280\,m\mu}$ (of 1:100 dil.)		$A_{280\,m\mu}/A_{260\,m\mu}$		$[\alpha]_D$ [at T(°C)]		$s_{20,w} \times 10^{13}$ (sec.)	
								Sample	Control	Sample	Control	Sample	Control	Sample	Control	Sample	Control
α-Chymotrypsin	1	10.20	pH 7.1, tris buffer	26.4	1	75	10 min continuous	15 300	15 300	0.185	0.185	2.83	2.83	−54° (22°C)	−45°	2.63	2.65
	2	10.48	pH 3.0, 0.1M NaCl	23.8	11	1000	1800 0.1-sec pulses	13 500	13 500	0.191	0.191	2.65	2.65	−52° (24°C)	−48°		
	3	9.62	pH 3.0, 0.1M NaCl	37.0	11	1000	2000 0.1-sec pulses	11 100	11 100	0.177	0.176	2.60	2.67	−53° (25°C)	−47°		
	4	10.32	pH 7.1, tris buffer	37.0	11	1000	2000 0.1-sec pulses	9 300	9 300	0.191		2.55				2.37	
Trypsin	5	9.05	pH 7.1, tris buffer	20.2	1	75	10-min continuous	4 600	4 600	0.142	0.141	2.60	2.62	−23° (22°C)	−24°	1.80	1.54
	6	9.24	pH 3.0, 0.1M NaCl	22.7	11	1000	2000 0.1-sec pulses	5 200	5 200	0.145	0.145	2.50	2.59	−48° (24°C)	−55°		
	7	8.40	pH 3.0, 0.1M NaCl	37.0	11	1900	2000 0.1-sec pulses	3 800	3 800	0.132	0.132	2.6	2.64	−56° (23°C)	−62°	1.41	1.46
Lactate dehydrogenase	3	8.19	pH 5.8, 17% sat. (NH₄)₂SO₄	23.0	1	75	10 min continuous	34	34	0.122	0.122	2.60	2.54	−43° (22°C)	−37°	5.40	5.41
	9	11.20	pH 5.8, 17% sat. (NH₄)₂SO₄	23.7	11	1000	2000 0.1-sec pulses	29	29	0.167	0.166	1.25	1.29	−38° (24°C)	−36°		
Aldolase	10	11.00	pH 7.8, 25% sat. (NH₄)₂SO₄	24.4	11	1000	2000 0.1-sec pulses	11	11	0.100	0.100	3.00	2.94	−20° (23°C)	−18°	6.26	6.32
Ribonuclease	11	9.22	pH 6.8, 0.1M KCl	24.2	11	1000	2000 0.1-sec pulses	1 100	1 100	0.062	0.063	2.10	2.14	−53° (24°C)	−68°		

[a] Specific activity defined for five enzymes.

Enzyme	U, unit of activity[a]
α-Chymotrypsin	$\Delta A_{237\,m\mu/min} = 0.001$, 25°C, pH 7.0
Trypsin	$\Delta A_{237\,m\mu/min} = 0.001$, 25°C, pH 7.0
Lactate dehydrogenase	Initial rate of oxidation of 1 μM of NADH₂/min, 25°C, pH 7.4
Aldolase	$\Delta A_{340\,m\mu/min} = 1.000$, 25°C, pH 7.5
Ribonuclease	$\Delta A_{286\,m\mu/min} = 0.001$, 25°C, pH 7.1

dialyzed versus cold (approximately 5°C), $0.1M$ NaCl (pH 6.5) before sedimentation, display identical sedimentation behavior; not only are the sedimentation coefficients nearly equal, but the schlieren patterns are congruent. The value of $s_{20,w}$ is reasonable for the pH and concentration of the solutions.[31] In Expt. 4, the sedimentation coefficient of the irradiated sample was determined in tris buffer, pH 7.1, and, as expected,[31] it is less than the sedimentation coefficients of the sample and control in Expt. 1 at pH 6.5.

The specific activities and uv absorption spectra of the irradiated samples are identical to those of the controls in all experiments with trypsin. The specific optical rotations of samples and controls do not differ significantly and since trypsin is reported to have a $[\alpha]_D = -40°$ at pH 5.2 and a $[\alpha]_D = -69°$ at pH 1.3, the values at pH 7.1 and pH 3 given in Table I seem reasonable. The sedimentation coefficients for Expts. 5 and 7 though approximately equal, are not characteristic of native trypsin. The $s_{20,w}$ for 1% trypsin in tris buffer should be approximately 2.35 S ($S = 10^{-13}$ sec) and for 1% trypsin in $0.1M$ NaCl, pH 3, $s_{20,w}$ should be approximately 2.45 S.[32] The low $s_{20,w}$ value has been reported before[33] and is attributed to changes in molecular size due to autolysis. The changes in size, however, do not make trypsin susceptible to damage by intense noncavitating ultrasound since the controls were essentially the same as the irradiated samples. In Expt. 7, exposure to a temperature of 37°C lowered the specific activity of trypsin but trypsin did not greatly affect the other properties of the enzyme. This was shown by analysis of a second control for Expt. 7, which was degassed but not subjected to the $3\frac{1}{4}$ h at 37°C. The results of the analyses on this "degassed only" control, for comparison with the results in Table I are: specific activity = 5950 $\Delta A_{237\ m\mu/min/mg\ enzyme}$, $A_{280\ m\mu}$ (of 1:100 dil.) = 0.132, $A_{280\ m\mu}/A_{250\ m\mu} = 2.68$, $[\alpha]_D^{23°C} = -58°$, $s_{20,w} = 1.53\ S$. The heated control and heated irradiated sample each have 64% of the specific activity of this control and, thus, as with α-chymotrypsin, intense noncavitating ultrasound did not affect the enzyme even in an environment that gradually denatures it. The sedimentation patterns for experiments revealed that the boundary is asymmetric to the slower sedimenting side of the peak and broadens during sedimentation in each case. This indicates that the enzyme solutions were heterogeneous, as would be expected if autolysis occurred.

The specific activities and uv absorption spectra for irradiated LDH samples and their controls were identical for both experiments in Table I. The irradiated sample and control of Expt. 9 were slightly turbid. This turbidity caused an abnormally high absorption in the 240–260 mμ region of the uv absorption spectra, which in turn yielded low values for the $A_{280\ m\mu}/A_{250\ m\mu}$ ratios. The turbid solutions were centrifuged at 24 000 g for 10 min to obtain clear solutions for optical rotation studies. The specific optical rotation results given are close to a reported value of $-43°$ for LDH in pH 5.6, 19% saturated $(NH_4)_2SO_4$.[34] The irradiated sample and control of Expt. 8 show the same sedimentation behavior. The solutions were sedimented in 17% saturated $(NH_4)_2SO_4$ and at this high salt concentration, LDH is disassociated into two subunits each of molecular weight approximately 72 000.[35] The $s_{20,w}$ values found for LDH subunits (at a concentration of 8.19 mg/ml) are in agreement with a reported value of $s_{20,w} = 5.5\ S$ (at a concentration of 2.6 mg/ml) for, if the $s_{20,w}$ of the subunits decreases with increasing concentration (as does undissociated LDH), then an $s_{20,w}$ at 8.19 mg/ml should be slightly lower than one for a 2.6-mg/ml solution.[35]

The irradiated aldolase sample and its control have identical specific activities and uv absorption spectra. The specific rotations are very nearly equal and are close to a reported value $[\alpha]_D = -23°$.[36] The sedimentation behavior of irradiated sample and control (both dialyzed versus $0.1M$ NaCl, pH 6.5 prior to sedimentation) were also practically identical but the sedimentation coefficients were lower than the value 6.80 S[37] expected for the concentration and pH of the solutions indicating the possibility of a slight expansion of the aldolase molecules. Such an expansion could have occurred since the sample and control were held at 24.4°C for $3\frac{1}{4}$ h during the irradiation. That partial denaturation does occur in the sample and control was shown by analysis of the native solution and a control that was immediately refrigerated after degassing. The results of these analyses are shown in Table II. During degassing, surface denaturation was observed to cause a small amount of flocculent precipitate. The precipitate removed some protein from solution and this is the reason the degassed-only control, heated control, and irradiated sample are slightly less concentrated than the native solution (see $A_{280\ m\mu}$ results). The surface denaturation also accounts for the specific activity of the degassed only control being lower than that of the native solution. The solutions held at 24.4°C for $3\frac{1}{4}$ h have a lower specific activity than the degassed only control as well as lower sedimentation coefficients but, as with previously discussed enzymes, although aldolase is unstable in the irradiation environment it still was not damaged by intense noncavitating ultrasound.

The results for ribonuclease in Table I, with the exception of the specific optical rotation of the irradiated

---

[31] G. W. Schwert, J. Biol. Chem. 179, 655–664 (1949).
[32] L. W. Cunningham, Jr., F. Tietze, N. M. Green, and H. Neurath, Discussions Faraday Soc. 13, 58–67 (1953).
[33] F. F. Nord and M. Bier, Biochim. Biophys. Acta 12, 56–66 (1953).
[34] B. Jirgensons, Arch. Biochem. Biophys. 85, 532–539 (1959).
[35] D. B. S. Millar, J. Biol. Chem. 237, 2135–2139 (1962).
[36] E. Stellwagen and H. K. Schachman, Biochem. 1, 1056–1069 (1962).
[37] J. F. Taylor and C. Lowry, Biochim. Biophys. Acta 20, 109–117 (1956).

TABLE II. Biochemical analyses of additional aldolase controls.

Aldolase sample	Specific activity ($\Delta A_{240\,m\mu}/min/mg$ enzyme)	$A_{280\,m\mu}$ (1:100 dil.)	$\dfrac{A_{280\,m\mu}}{A_{260\,m\mu}}$	$S_{20,w} \times 10^{13}$ (sec)
Degassed only control	12	0.100	3.07	6.48
Native	13	0.104	3.08	6.50

sample, all indicate that the irradiated sample and control solutions were identical. The $[\alpha]_D = -68°$ observed for the ribonuclease control agrees well with the reported value $-73.3°$.[38] Explaining the $[\alpha]_D$ of the irradiated sample is difficult, because all reported denaturations of ribonuclease, e.g., by oxidation, reduction, high pH, or $8M$ urea,[30] give substantial increases in levorotation, and this sample exhibits a *decrease* in levorotation.

Since some of the solutions of Table I were dialyzed prior to sedimentation velocity analysis, it was necessary to determine if amino acids or small peptides were cleaved from the enzyme molecules, either by the ultrasonic treatment or, in the case of trypsin and α-chymotrypsin, by autolysis. Thus, some of the irradiated samples and controls were analyzed by thin-layer chromatography. The limit of detection for most amino acids by thin-layer chromatography is lower than 0.1 μg.[39] Ten micrograms of each enzyme were applied in small spots to the thin layers and several amino acids in 1-μg quantities were similarly applied for comparison purposes. Trypsin and α-chymotrypsin solutions gave some faintly ninhydrin positive areas distributed between the origin and approximately two-thirds the distance to the solvent front. The irradiated samples and controls gave identical patterns and color intensities, indicating that only autolysis was responsible for the material detected. None of the other enzymes displayed any evidence of degradation.

B. Enzyme-Catalyzed Reactions

The following procedure was employed for observing the effects of noncavitating ultrasound on enzyme-catalyzed reactions. The electronic components were tuned up with distilled water in the irradiation cell; the distilled water was then replaced with the degassed substrate for the reaction to be studied. After setting the absorbance at some arbitrarily selected position on the recorder chart, the substrate was irradiated for 1 min, in the absence of enzyme, to determine whether ultrasonic irradiation produced any change in absorbance, i.e., affected the substrate. The temperature was monitored during the irradiation with a thermistor probe inserted in the filling hole of the sample irradiation cell. When assured that irradiation had no effect upon the substrate (other than the effects due to light diffraction or temperature change, which will be discussed later), the enzyme was added to start the reaction. After the reaction had proceeded 1 min, the ultrasound

[38] F. H. White, Jr., J. Biol. Chem. 236, 1353–1360 (1961).
[39] K. Randerath, *Thin Layer Chromatography* (Academic Press Inc., New York, 1965), pp. 96–97.

was turned on for 1 min while the spectrophotometrically monitored rate of reaction, $\Delta A/t$, was recorded continuously on the chart paper. After 1 min of ultrasonic irradiation, the ultrasound was turned off and the reaction was allowed to continue as long as the reaction rate remained linear. The temperature was monitored continuously and any changes were noted on the recorder chart. Delivery of the enzyme, a careful stirring of the reaction mixture, and positioning the thermistor probe in the filling hole of the reaction irradiation cell could all be accomplished in 20 sec or less. Degassing the substrate for the 1-MHz irradiations was a precaution taken to ensure the absence of cavitation, although the intensity employed at 1 MHz was below the cavitation threshold for aerated water. If cavitation had occurred in the reaction mixture, it would have been detected because the cavitation bubbles scatter light causing an erratic increase in absorbance.

Several interrelated factors had to be considered in the selection of the procedure outlined above, and the following remarks describe the instrumental limitations from which the procedure evolved. The rate of an enzyme-catalyzed reaction depends on the concentrations of enzyme and substrate and on the temperature and pH of the reaction mixture. Temperature and pH are held constant according to the conditions specified by the definitions for a unit of catalytic activity. The sensitivity of the spectrophotometer limits the range of substrate concentrations that may be used, and within the useful range, preliminary experiments were carried out to determine the best combination of enzyme and substrate concentrations that yields a suitable first-order reaction for study. The criteria for suitability are a reaction whose rate is sufficient to enable detection of a change in rate of 5% or less, and a reaction that remains first order for at least 3 min. Three minutes is considered a minimum time for the reaction to remain first order because it is desirable to be able to observe the reaction for one minute both before and after a 1-min ultrasonic irradiation. Irradiations longer than 1 min usually were undesirable at the intensities employed, since they resulted in temperature increases that could not be brought back to the initial value before the reaction ceased to be first order. Since enzyme-catalyzed reactions are temperature-dependent, approximately doubling in rate for a 10°C temperature increase, it was desirable not only to maintain constant temperature during a given enzyme assay but also to assay at the same temperature when comparing the activities of different samples. If, in addition, the enzyme-catalyzed reaction is ultrasonically irradiated,

TABLE III. Conditions of the irradiated reactions.

Enzyme	μg Enzyme assayed	Substrate conc.	pH
α-Chymotrypsin	10	$2.89 \times 10^{-4} M$ ATEE	7.0
Trypsin	10	$7.22 \times 10^{-5} M$ BAEE	7.0
Lactate dehydrogenase	0.62	$3.45 \times 10^{-4} M$ Pyruvate $6.90 \times 10^{-5} M$ NADH$_2$	7.4
Aldolase	10	$4 \times 10^{-3} M$ Fructose-1, 6-diphosphate $2.3 \times 10^{-3} M$ Hydrazine sulfate	7.3
Ribonuclease	100	0.1 mg/ml cytidine 2':3' phosphate $(4.68 \times 10^{-4} M)$	7.1

it is necessary to monitor the temperature throughout the reaction because absorption of ultrasound by the reaction mixture produces a temperature rise. By limiting the irradiation times to 1 min, intensities higher than 35 W/cm² could be employed and the reaction mixture in the thermostated irradiation cell could still be quickly restored to its initial temperature after irradiation.

Two phenomena that complicated the interpretation of the results were a substrate absorbance change with increased temperature and the Debye–Sears effect.

Temperature increases, due to acoustic absorption, changed the light absorbance ($\Delta A/\Delta T$) of solutions of ATEE, BAEE, cytidine 2':3' cyclic phosphate and solutions of their reaction products. For the maximum temperature increases produced during irradiation, the observed ΔA was approximately 5% of the total change observed in the assay. However, since the original temperature was quickly re-established after the acoustic exposure and since irradiation of substrate alone had previously provided knowledge of the time course of $\Delta A/\Delta T$, the effects of $\Delta A/\Delta T$ were readily separated from other ultrasonic effects on the reaction rate. It was particularly easy to detect the effect of $\Delta A/\Delta T$ in the case of the α-chymotrypsin assay since the assay produces a decrease of $A_{237 m\mu}$ while $\Delta A/\Delta T$ produces an increase.

The diffraction of light passing through a transparent medium that is also transmitting a beam of parallel ultrasonic waves is known as the Debye–Sears effect.[27] Diffraction occurs as a result of the slight differences in refractive index between the alternate regions of compression and rarefaction in the medium and, hence, the ultrasonic waves may be considered a three dimensional diffraction grating traveling with the speed of sound. However, since the speed of light is much greater than the speed of sound, the diffraction grating is effectively stationary. In the resulting diffraction pattern, the light intensity of the zeroth-order line is inversely proportional to the acoustic intensity, i.e., the greater the acoustic intensity, the greater is the proportion of light diverted from the zeroth-order to higher orders.[40] Debye–Sears diffraction could affect the form of the experimental results, since the acoustic intensity is sufficient to diffract light into higher orders that do not reach the detector and the result on the recorder

[40] J. Blitz, *Fundamentals of Ultrasound* [Butterworth and Co. (Publishers) Ltd., London, 1963], pp. 137–141.

chart is an apparent increase in absorption. However, these diffraction effects presented no difficulty in the interpretation of data in this study, since they merely produce an abrupt displacement of the absorbance to a higher level marking the commencement of ultrasonic irradiation. The reaction rate, $\Delta A/t$, continues to be recorded, but displaced by a fixed amount. When the ultrasound was turned off, the absorbance abruptly decreases by this same amount while the reaction continues.

Table III gives the amount of enzyme, the substrate concentration, and the pH in the ultrasonically irradiated reaction mixture.

The results of the ultrasonic irradiations of the enzyme-catalyzed reactions are listed, according to the frequency and intensity employed, in Table IV. All the irradiations were continuous for 1 min and the sample irradiation cell was maintained, except for the brief temperature rise indicated, at 25.0°±0.1°C. The 1-MHz ultrasound and the lowest intensity (0.5 W/cm²) at 9 MHz (for the first four enzymes of Table V) were generated by a crystal with a 1-MHz fundamental resonant frequency. The higher intensities at 9 and 27 MHz were generated by a crystal with a 9-MHz fundamental resonant frequency.

It is seen that none of these irradiations (Table IV) had any effect on the enzyme-catalyzed reactions.

III. CONCLUSIONS

The results of this study lead to the following conclusions. Irradiation of approximately 1% solutions of α-chymotrypsin, trypsin, lactate dehydrogenase, aldolase, and ribonuclease with noncavitating ultrasound at dose levels sufficient to cause extensive structural and functional damage in tissues, has no effect on either the structure or function of the enzymes. That the catalytic function of the five enzymes was unaffected by intense noncavitating ultrasound was demonstrated in two ways: The approximately 1% solutions showed full catalytic ability when assayed after irradiation and irradiations failed to inhibit the enzymes while they were in the process of catalyzing a reaction. Structural integrity of the irradiated enzyme samples was demonstrated by the lack of any significant differences between the uv absorption spectra, specific rotations, sedimentation coefficients, and thin-layer chromatographic analyses of the irradiated samples and unir-

TABLE IV. Effects of intense noncavitating ultrasound on enzyme-catalyzed reactions (1-min continuous irradiations at 25°C).

Enzyme of the irradiated reaction	Frequency (MHz)	Intensity (W/cm^2)	Max. temp. rise during irradiation (°C)	% Native activity
α-Chymotrypsin	1	5	0.5	100
	9	0.5, 3, 23	<0.1, 0.5, 2.5	100 100 100
	27	1	1.0	100
Trypsin	1	5	0.5	100
	9	0.5, 4, 23	<0.1, 0.7, 3.3	100 100 100
	27	1	1.0	100
Lactate dehydrogenase	1	5	0.5	100
	9	0.5, 14, 39	<0.1, 0.7, 1.5	100 100 100
	27	1	1.2	100
Aldolase	1	5	0.5	100
	9	0.5, 12, 37	<0.1, 0.8, 1.5	100 100 100
	27	1	1.2	100
Ribonuclease	9	14, 39	0.8, 1.7	100 100
	27	1	1.0	100

radiated controls. Since the denaturation of the enzymes studied in this paper by cavitating ultrasound has been shown in other reports,[41] it is concluded that cavitation is a necessary condition for damage to be produced by ultrasound in the five enzyme solutions *in vitro* of this study.

These conclusions raise several points worthy of further discussion. *First*, the inability of intense noncavitating ultrasound to damage the enzymes, particularly trypsin solutions, contradicts the findings of other investigators.[9] A recent study[42] carried out in conjunction with the work reported here has shown that the reported inactivation of trypsin[9] and α-amylase[10] were not due directly to an interaction between the ultrasonic waves and the protein molecules in solutions, but rather to an unspecified reaction between the solution and rubber materials employed in the specimen container.

Another interesting aspect of the results obtained with intense noncavitating ultrasound is the fact that, despite the expected absorption of acoustic energy being many times greater than the energy required to denature the enzymes, the mechanisms of absorption and dissipation of this energy apparently do not involve permanent changes in the enzyme molecules nor detectable transient changes that in any way affect the normal functioning of the enzyme *in vitro*. The results suggest that the mechanism of acoustic energy absorption does not involve great changes in the secondary and tertiary structure of the enzyme molecules, and this information may be useful to investigators studying the mechanisms of acoustic absorption by polymers in solution.

The molecular biological approach in studying the interactions of intense noncavitating ultrasound with biological structures has revealed that DNA[11] can be degraded *in vitro*, but enzymes cannot. Degradation of DNA *in vivo* cannot account for the effects observed in irradiated tissues since it is unlikely that the loss of cellular control processes involving DNA would result in the rapid functional changes observed. The inactivation of enzymes would provide a more reasonable explanation for the rapidly appearing effects in tissues since they are more directly concerned in the chemical reactions essential to the cell's survival. The fact that intense noncavitating ultrasound has no effect on enzymes *in vitro* does not necessarily imply that there are no interactions in the cellular environment. Some enzymes, such as the enzymes of the Krebs cycle (which are located in mitochondria), apparently are structurally organized in groups. In the case of the Krebs cycle, the enzymes are closely associated with, and possibly bound to, the cristae of the mitochondria.[43] Thus, it is possible that in this structured environment, an enzyme may be more susceptible to denaturation by intense noncavitating ultrasound either directly or by some indirect mechanism involving the adjacent structures. Use of histochemical techniques[44] may prove useful for future studies of the effects of intense noncavitating ultrasound on enzymes *in vivo*.

It is felt that the most promising future research into the nature of the interactions between intense noncavitating ultrasound and tissue will deal with those levels of biological structure that lie between the molecular level and structures observable with the light microscope. A study of the effects of intense noncavitating ultrasound on the properties of membranes would be useful not only because membranes are a ubiquitous structural feature of cells, being found around the nucleus and as a part of such cellular organelles as mitochondria, endoplasmic reticulum, etc., but also because of the importance of excitable membranes in nervous tissue.

ACKNOWLEDGMENTS

This work was supported by the National Science Foundation and the Office of Naval Research, Acoustics Programs.

---

[41] R. M. Macleod and F. Dunn, J. Acoust. Soc. Amer. **42**, 527–529 (1967).
[42] R. M. Macleod and F. Dunn, J. Acoust. Soc. Amer. **40**, 1202–1203 (1966).
[43] A. L. Lehninger, Fed. Proc. **19**, 952–962 (1960).
[44] W. Bostelmann, Acta Neuropathol. **2**, 24–39 (1962).

33

Copyright © 1971 by the Acoustical Society of America

Reprinted from J. Acoust. Soc. Amer., **50**(6), Pt. 2, 1539–1545 (1971)

Degradation of DNA in High-Intensity Focused Ultrasonic Fields at 1 MHz

W. T. COAKLEY

Department of Microbiology, University College, Cardiff, Wales, United Kingdom

F. DUNN

Bioacoustics Research Laboratory, University of Illinois, Urbana, Illinois 61801

Gassy and degassed aqueous solutions of DNA were treated ultrasonically while monitoring the field for discrete cavitation events. At irradiation intensities greater than about 500 W/cm², transient cavitation is shown to be responsible for the observed reduction in molecular weight. At intensities of 200 to 288 W/cm², degradation of DNA was observed which did not depend upon transient cavitation.

INTRODUCTION

Attempts to understand in detail the alterations induced in animal tissues by intense ultrasound have led investigators to examine such interactions at various levels of biological structure. As a result, biopolymers in solution, in particular the unique macromolecular structure deoxyribonucleic acid (DNA), have received considerable attention. While it is recognized that conformation of DNA *in vitro* bears little relation to that *in vivo*, such studies aid in assessing the mechanical stresses to which polymers *in vivo* are subjected during ultrasonic irradiation.

Much interest has arisen around the point of whether or not polymer degradation can occur in the absence of cavitation. The general approach has been to effect damage under certain conditions and then to take steps to suppress cavitation. Persisting degradation at a rate less than that existing before these steps have been taken has been used as an argument for the presence of a mechanism other than cavitation. There is general agreement that the main action of ultrasound on polymers is mechanical, since experiments show that degradation proceeds to a limiting molecular weight if irradiation is continued for long periods.[1] However, when working with cavitation in aqueous solutions where free radicals are produced, Alexander and Fox[2] assessed that 30% of their breakage could be due to chemical effects.

On ultrasonic irradiation of synthetic polymers at intensities from 5 to 700 W/cm² over the frequency range from 210 kHz to 2 MHz and simultaneously suppressing cavitation, several investigators reported no degradation.[3–7] Some investigators have reported observing continued degradation, though at a reduced rate, with suppression of cavitation by irradiating samples under reduced ambient pressure.[8,9] This latter method of suppressing cavitation has not been widely used and no quantitative measurements are available on the efficacy of this technique.

The DNA molecule is of a polymer of two strands coiled in a double helix about a common axis. The backbone of the individual strands consists of pentose residues connected together by phosphate bridges. The pentose residues, in turn, are linked to purine or pyrimidine bases. In the double helical form, the sugars and phosphates are on the outside and the bases on the inside of the helix. The main forces holding the two strands together are the base stacking forces which are the hydrophobic interactions between the heterocyclic bases as they stack in parallel arrays at right angles to the main helix axis.[10]

Electron micrographs of DNA fragments produced by cavitation at 7 kHz have shown that breakage of the molecule occurs by double backbone scission, i.e., the ends of the fragments were double rather than single stranded.[11,12] An investigation of the viscosity of sonic fragments after irradiation with cavitating ultrasound at 800 kHz and 15 W/cm² supports the view that double backbone scission, rather than an accumulation of single strand breaks, is the cause of DNA degradation.[13] Chemical studies suggest that the main bond involved in the backbone cleavage is the C–O bond (90%) with 10% P–O rupture and no appreciable C–C damage.[14] In an attempt to detect a mechanism for DNA breakdown which does not involve the effects of cavitation,[15] degassed samples of DNA were irradiated at 981 kHz and 25–31 W/cm². The sample holder and field geometry were such that the distribution of the acoustic energy over the sample chamber did not vary by more than 10%. A sedimentation coefficient change from 32 to 16, corresponding[16] to a molecular weight change from 2.2×10^7 to 4×10^6, occurred in 15 sec. Irradiation for two minutes produced little additional degradation, showing that a limiting molecular weight had been reached. Several methods were employed simultaneously to detect the presence of cavitation and no evidence was observed suggesting its presence. It was considered then that cavitation was not present and that the degradation resulted from viscous stresses set up due to relative motion resulting from the density difference between the DNA molecule and the solvent molecules.

The degradation of DNA of molecular weight 10^7 exposed to 1-MHz ultrasound of 10 W/cm² and 30-msec repeated pulses has been observed to occur when the spherical sample holder was rotated during the irradiation period.[17] No effect was observed in degassed solutions on irradiation without rotation of the sample holder. The degradation correlated with the sonochemical release of iodine from potassium iodide and the detection of the first subharmonic of the driving frequency. It was concluded that microstreaming around stable oscillating bubbles was responsible for the effect.

In order to study further the mechanism responsible for degradation, calf thymus DNA in solution was irradiated in a focused sound field at 1 MHz at intensities greater than those of previous studies.

I. METHODS AND MATERIALS

The 1-MHz (48-mm radius of curvature) PZT4 ceramic transducer, the irradiation tank, the preparation of degassed water, sound intensity measurement procedure, transient cavitation, and first-order subharmonic detecting systems are described elsewhere.[18] All are established procedures except the cavitation detection scheme. Briefly, when the rf voltage applied to the transmitting transducer is rectified and filtered, transient signals may be detected when isolated transient cavitation occurs at the focal region of the field.

Commercially available highly polymerized calf thymus DNA (British Drug Houses Ltd.) was dissolved over chloroform in sterile B.P.E.S. buffer ($0.006M$ Na_2HPO_4, $0.002M$ NaH_2PO_4, $0.001M$ EDTA, $0.179N$ NaCl, pH 6.8). After one week, the solution was filtered through Whatman G.P. filter paper to remove a gelatinous mass of denatured protein. The protein concentration of the filtrate was determined with the Folin–Ciocalteu reagent[19] and was typically 6% of the DNA concentration. DNA concentrations were estimated at 260 nm ($D = 1.0 = 48$ μg/ml). The stock solution of DNA was stored over chloroform, which was removed by degassing prior to ultrasonic irradiation to avoid any effect of the chloroform on the onset of cavitation, and 0.01% weight/volume of the antibiotic tris(hydroxymethyl)nitromethane was added. The flask was sealed with a cotton wool plug and allowed to stand for days to saturate the solution with air. The solution was not refrigerated, in order to avoid air bubbles on reheating to room temperature before treatment. When it was desired to suppress cavitation, DNA solution (10 ml) in a 20-ml test tube was degassed at approximately 3 cm Hg, during which time the glass was tapped to encourage bubble emission. Degassing was stopped when no further bubble release was produced in a 1-min interval. Dissolved oxygen was estimated with an oxygen electrode[20,21] and the output of the polarizing and measuring circuit was applied to a strip chart recorder.

The DNA was irradiated in two types of containers, viz., a thin-walled aluminum cylinder of 13-mm i.d., 5-mm length, having a volume of 0.65 ml, and a lucite cylinder of 18-mm i.d., 12-mm length, with volume of 3.3 ml. Rubber O rings were used to attach 0.0005-in.-thick "Saran" (Dow Corning) foil windows to the cylinder. The DNA was loaded into the container with a wide-mouth pipet, the container being sealed without trapping air bubbles. The sample container was placed in the focal region of the acoustic field for the irradiation procedure.

The DNA molecular weight was estimated from empirical relations linking molecular weight and intrinsic viscosity.[16] The intrinsic viscosity was determined from relative viscosity measurements in an electromagnetic modification[22] of the Crowthers and Zimm low-shear-rate viscometer at a DNA concentration of 50 μg/ml. Since the main interest of this work was in relative degradations, measurements were made at only one concentration. It was considered that the concentration dependence of the intrinsic viscosity below 50 μg/ml would not alter degradation estimates significantly.

II. RESULTS

Gassy and degassed samples of calf thymus DNA molecular weight 2.2×10^7 (100 μg/ml) irradiated in the aluminum container for 10 sec were diluted with equal volumes of B.P.E.S. buffer and their viscosities measured. The oxygen level in the degassed stock solutions varied from 5% to 12% before loading the DNA into the container. No degradation of degassed solutions was observed below 400 W/cm². In the intensity region from 1500 to 2000 W/cm², degradation in the gassy samples varied from no change in molecular weight to a decrease in molecular weight to 4×10^6. The results for the degassed samples ranged from no observable degradation to a lowest value of molecular weight of 13×10^6. From the most degraded of the three gassy samples, a sharp cracking audible noise was emitted. Bubbles were visible in fewer than four of the eight degassed samples exhibiting degradation. Therefore degradation could occur in degassed samples without obvious signs of cavitation.

The degradation of undegassed calf thymus DNA in the lucite cylinder was later studied with an electroacoustic system[18] capable of detecting single transient cavitation events (Table I). The sound was pulsed with 2 min between pulses (column 2, Table I). All transients greater than 90 mV were counted electronically. The maximum amplitudes of the cavitation events at the different intensities were comparable. The table suggests a correlation between the number of cavitation events and the reduction in molecular weight.

The transducer could be excited continuously below 550 W/cm² without overheating. The molecular weight of undegassed DNA (56 μg/ml) irradiated at 515 W/cm² in the lucite cylinder and the total number of transient cavitation events plotted against time both show a high rate of change in the first few minutes levelling off as the time of irradiation increased (Fig. 1). It will be seen that there is a strong correlation between the decrease in molecular weight and the number of cavitation events.

For the above and for 12 further samples of undegassed DNA from two stock solutions ($M_c = 15.53 \times 10^6$ and 20.0×10^6) irradiated at 515 W/cm², the decrease in molecular weight, ΔM, is expressed as a fraction of the mean control molecular weight, $\Delta M/M_c$, to facilitate the comparison of the degradation from both stocks (Table II). The best-fit straight line when $\Delta M/M_c$ for the 16 samples is plotted against the number of cavitation events N (Fig. 2), independent of irradiation time, is $\Delta M/M_c = 4.67 \times 10^{-4} N + 0.038$. The 95% confidence limits for the intercept of the line on the ordinate are 0.038 ± 0.056 which, since zero is in-

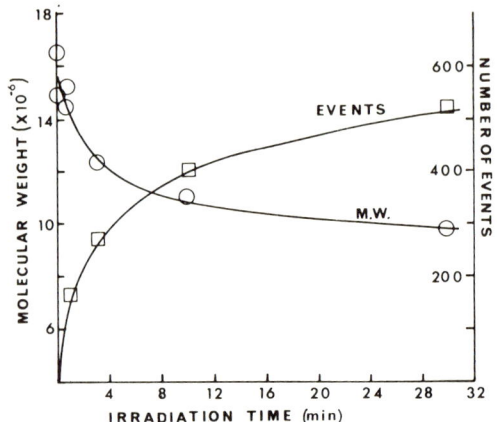

FIG. 1. Time dependence of molecular weight reduction of DNA under ultrasonic irradiation at 515 W/cm² and number of transient cavitation events greater than 85 mV.

cluded in the range, shows that transient cavitation alone could explain the degradation observed. The correlation for the 16 points is 0.68. The best-fit straight line for the 10 points for 30-min irradiation is $\Delta M/M_c = 4.59 \times 10^{-4} N + 0.023$ and r^2 is 0.69. A complete relationship ($r^2 = 1.0$) would not be expected because of the contribution to the variance by visco-

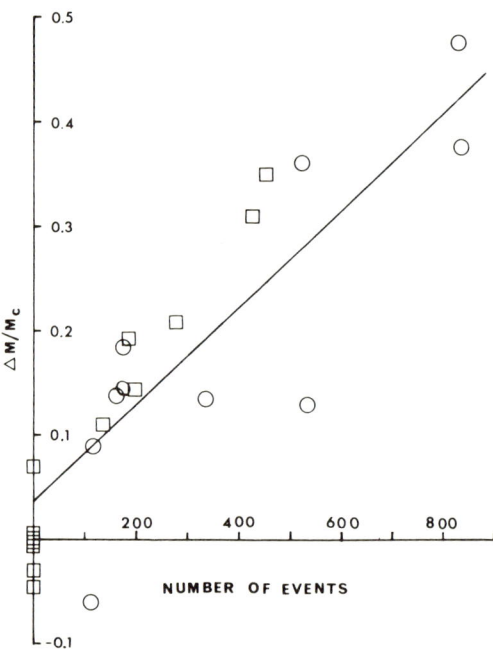

FIG. 2. Relative change in molecular weight $\Delta M/M_c$ as a function of the number of cavitation events for samples irradiated at 515 W/cm² from 0.75 to 30 min. The line is $\Delta M/M_c = 4.67 \times 10^{-4} N + 0.038$. (○: 30-min irradiation points only).

TABLE I. Ultrasonic degradation of undegassed calf thymus DNA (50 μg/ml) and number of transient cavitation events greater than 90 mV.

I (W/cm²)	Time regime (sec)	Total irradiation time (sec)	Number of transients	Maximum amplitude (V)	Molecular wt $\times 10^{-6}$
0	0	0	0	0	18.4 and 17.4
1470	10+8+6+5+6	35	163	6	13.2
1900	5+5+7+6	28	142	12	13.7
2000	5+5+5+6+6	27	293		7.7
2000	6+5+5+5+5	26	283		10.1
2050	8	8	360	7	10.1

TABLE II. Degradation of undegassed calf thymus DNA treated at 515 W/cm² for various times. The control molecular weight estimate on the day of sonication, M_{cl}, and the mean molecular weight estimate, M_c from many measurements of the stock solution are shown. The decrease in M, ΔM, was calculated as $M_c - M$ and expressed as $\Delta M/M_c$. The concentration of the stock of $M_c = 15.53 \times 10^6$ was 56 μg/ml and that of the other stock was 46 μg/ml.

Irradiation time (min)	Control molecular wt ($M_{cl} \times 10^{-6}$)	Mean control molecular wt ($M_c \times 10^{-6}$)	$\Delta M \times 10^{-6}$	$\Delta M/M_c$	Number of transient events	Maximum amplitude of transients greater than 45 mV (V)
0.75	15.7	15.53	1.43	0.092	184	
3.0	15.7	15.53	3.23	0.208	276	
10.0	15.7	15.53	4.83	0.310	424	
30.0	15.7	15.53	5.63	0.362	523	4 V, max
1.0	14.6	15.53	1.73	0.111	134	
3.0	14.6	15.53	2.23	0.144	197	
10.0	14.6	15.53	5.03	0.324	450	4 V, max
30.0	14.6	15.53	5.83	0.375	835	2 V, max
30.0	20.0	20.0	2.6	0.130	535	3×3 V, max
30.0	20.1	20.0	9.5	0.475	839	21×5 V, 13×3 V
30.0	19.9	20.0	2.7	0.135	336	3×5 V, 8×3 V
30.0	20.1	20.0	2.9	0.145	173	1×6 V, 3×2 V
30.0	20.0	20.0	3.7	0.185	175	3×5 V, 2×2 V
30.0	19.9	20.0	1.8	0.09	116	1×5 V
30.0	17.4	17.4	2.4	0.138	164	3×6 V
30.0	17.4	17.4	−1.0	−0.058	106	1×0.8 V

metric estimates of M as evidenced by the cluster of the control points about the origin, and because the calculated correlation is with N alone and does not take account of the variations in transient amplitude with each total number of events. The amplitude distribution was not monitored during this work, but recently irradiation of tap water has shown it to be very positively skew.

Figure 3 shows that the average number of cavitation events in each minute of irradiation, of all 16 samples of Table II, and the average maximum amplitude of the cavitation events, calculated from maximum amplitudes recorded during each two-minute interval both fell by factors greater than 10. Since $\Delta M/M_c = 0.3$ for $N = 570$ (Fig. 2), it is assumed that approximately 0.3 of the volume of the container has been affected at least once by a cavitation event. This leads to an "average effective radius" of 0.77 mm, at 515 W/cm², of a transient event, considering the region of influence to be spherical. It has been shown that treating the transient event as a disturbance moving through a stationary cylinder of liquid does not change the estimate of this radius appreciably.[23]

Two samples of undegassed DNA were irradiated at 1100 W/cm² (rows 3 and 4, Table III). When compared with data from Table I (rows 1 and 2, Table III) and Table II (row 5, Table III), it can be seen that the degree of damage observed over a wide range of intensity and time of irradiation correlates well with a number of cavitation events, allowing for the fact that the transient amplitudes are larger at the higher acoustic intensities. This also supports the view that

TABLE III. Comparison of DNA degradation in samples irradiated over a wide intensity and time range.

I (W/cm²)	Irradiation time (sec)	Number of transient events	Maximum amplitude of transients (V)	Relative change in molecular wt $\Delta M/M_c$
2050	8	360	7	0.436
1470	35	163	6	0.262
1100	9×10	139	2.5	0.133
1100	9×10	157	2.5	0.177
515	1800	835	2	0.375

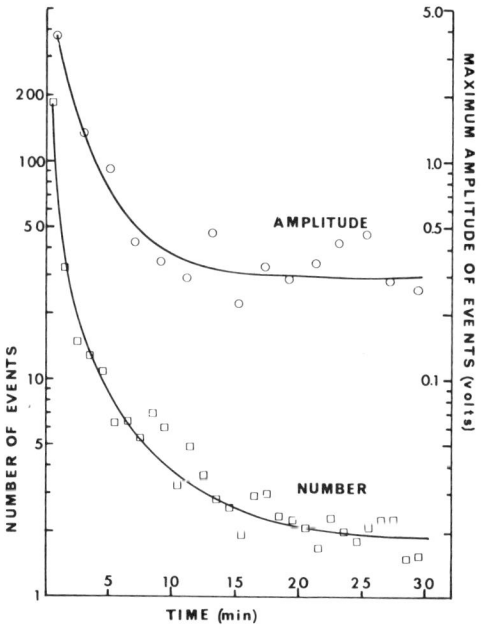

FIG. 3. The number of cavitation events per minute of irradiation (lower curve) and the maximum transient cavitiation event amplitude in each 2-min interval, both averaged over the 16 samples of Table II.

TABLE IV. Intensity dependence of degradation of undegassed DNA (50 μg/ml) at 1 MHz.

I (W/cm²)	Irradiation time (min)	$\Delta M/M_c$	Standard error of mean degradation
72	30	0.030	±0.031
200	35–39	0.352	±0.030
288	30	0.207	±0.031
515	30	0.192	±0.049

transient cavitation events are responsible for the damage at intensities greater than or equal to 515 W/cm².

The degradation of DNA irradiated at 288 W/cm² for 30 min, plotted against number transient cavitation events (Fig. 4), shows one point at $N=266$ isolated from the other nine. If there is no source of degradation other than transient cavitation events, then (0,0) is included as a data point. The curve of Fig. 4 is the best-fit line to those nine points and passes through the origin. The value of the correlation index[24] R for the nine points about the line is 0.10, showing little correlation between damage and transient cavitation. This result was expected since a comparison of Fig. 2 and Fig. 4 shows that the damage at 288 W/cm² was greater by a factor of approximately 2 for the same number of transient events. The amplitudes of the transients were also less by a factor of 2 at 288 W/cm² than at 515 W/cm². Thus a mechanism other than transient cavitation is required to explain the degradation at 288 W/cm².

A series of results at 200 W/cm², after many preliminary experiments to isolate and reduce the sources of experimental variation in the viscosity measure-

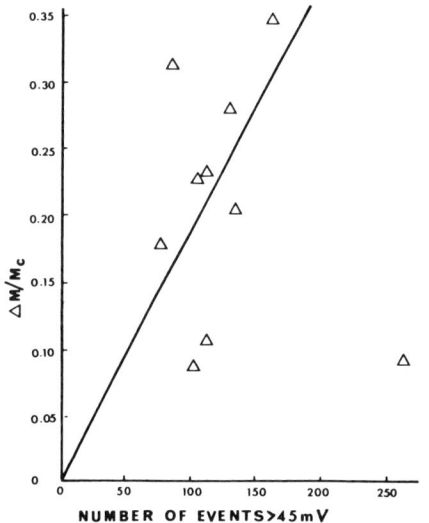

FIG. 4. Molecular degradation of DNA irradiated at 288 W/cm² for 30 min as a function of the number of transient events greater than 45 mV.

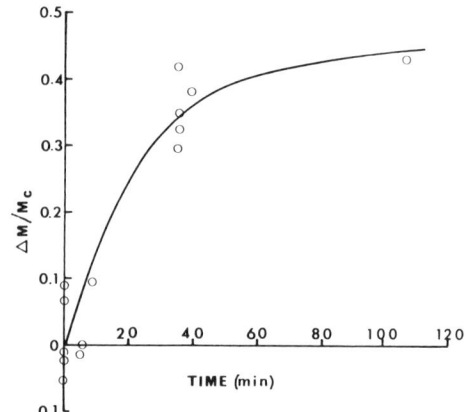

FIG. 5. $\Delta M/M_c$ for calf thymus DNA irradiated at 200 W/cm² vs exposure time.

ments, supports the suggestion of a mechanism which does not depend on transient cavitation events, as there was negligible correlation of degradation with transient cavitation numbers at 200 W/cm². However, a plot of degradation against time (Fig. 5) supports the concept of a time-dependent mechanism, as the three points irradiated for up to 8 min show little degradation while there is marked degradation after 35 min.

A series of seven irradiations at 72 W/cm² for 30 min showed no detectable damage.

To summarize these results, Table IV shows that the degradation which is not strongly associated with transient cavitation events rises to a maximum and then decreases until at 515 W/cm² the breakage may be accounted for by the transient events alone.

A previous study of DNA degradation, in a system designed to stabilize subharmonic-emitting cavitation bubbles at 1 MHz, had an intensity threshold in the range 1–1.5 W/cm², a maximum at a peak intensity in the range 7–10 W/cm², and then decreased on further increase of intensity.[17] The maximum degradation of DNA coincided with a maximum in the detected 500-kHz subharmonic signal. Even though the intensities of the above work were much lower than those presented in this paper, the occurrence of a maximum in the DNA degradation offered a possible explanation of the results of Table IV. This was particularly so since the subharmonic maximum[18] in tap water treated in the open irradiation tank was at 200 W/cm² and the level of DNA degradations was also high at that point (Table IV). However, the 500-kHz subharmonic detected from an undegassed DNA sample when a bowl hydrophone was positioned vertically over the sample container was not very different from the signal from the degassed tank water in the absence of the container (Table V). This was surprising since the subharmonic signal detected from tap water was, at most intensities,

TABLE V. The mean of the zero-to-peak 500-kHz signal, measured every 15 sec in a 2-min period, from an undegassed DNA sample and from the tank in the absence of the sample container, as a function of intensity.

I (W/cm^2)	Mean DNA subharmonic (mV)	Mean tank subharmonic (mV)
8	0.3	0.3
32	0.6	0.59
72	2.0	0.95
128	2.75	2.15
200	3.25	3.6
328	3.95	3.8
515	5.45	5.50

an order of magnitude greater than that from degassed water.[18]

The 500-kHz signal from the DNA samples irradiated at 72, 288, and 515 W/cm^2 was measured at the end of each minute of the 30-min irradiation periods and the results averaged for each sample. The bowl hydrophone monitoring the subharmonic was in a fixed position for the early samples but later a technique was adopted for adjusting the bowl position to that of a maximum activity. The level of subharmonic activity did not show a marked decrease irradiation time—the greatest was one of 16% at 515 W/cm^2 for 30 min. It was not possible to detect a dependence of degradation on subharmonic activity at 72, 288, or 515 W/cm^2 when considering only those results for which the bowl hydrophone was adjusted as above (Table VI). The number of results at each intensity is small but the absence of any trend in these, and in the other experiments in which the bowl position was not adjusted, indicated that monitoring subharmonic emission in focused fields is not as useful in determining mechanisms of breakage as it has been shown to be in plane fields.[17] The difficulties of interpreting the subharmonic signal were compounded because it was not steady during the irradiations, possibly because the amplitude at any instant was a function both of the number of emitting bubbles present and of their amplitude of vibration.

Since the undegassed DNA samples were separated from the degassed tank water by a Saran membrane, there was a gas concentration gradient across the membrane, and gas diffusion was accentuated during irradiation by bulk streaming inside and outside the container. The oxygen content in a stock DNA solution (50 μg/ml), measured with an oxygen electrode, was 95%. On treatment of a sample for 5 min (515 W/cm^2), the oxygen level had fallen to 86%, and it fell to 42% in a sample treated for 30 min. Seventy-six percent of the transient events at 515 W/cm^2 occur in the first 5 min of irradiation (Fig. 3); therefore, the decrease in the rate of transient cavitation is much more rapid than the fall in gas content in the solutions, so it is unlikely that the two phenomena are related. The experiment did show, however, that significant gas depletion occurs in samples treated for periods of the order of 30 min.

III. DISCUSSION

Transient cavitation has been shown to be the cause of DNA breakage at intensities of 515 W/cm^2 and higher. The number and amplitude of transient cavitation events recorded at 288 and 200 W/cm^2 were too small to account for the breakage observed. Since it was not possible to link 500-kHz subharmonic activity with degradation, as previously reported,[17] the possibility of a noncavitation-linked mechanism of degradation cannot be completely discounted.

If a noncavitating mechanism depending upon a time-averaged force existed, the strain S experienced by a molecule in the sound field would be proportional to $I^{\frac{1}{2}}$ and also would depend upon the relaxation time T required for the molecule to respond to the stress. The streaming velocity of the molecule along the sound-beam axis is proportional to I. The value of the relaxation time required of the DNA molecule so that the strain at 200 W/cm^2 (S_{200}) may be greater than that at 400 W/cm^2 (S_{400}) may be calculated since

$$S \sim I^{\frac{1}{2}}(1 - e^{-0.693t/T}),$$

where t is the time the molecule spends streaming along the beam axis. This condition is fulfilled if

$$T > 0.82t.$$

The estimated speed of a cavitation event through the focal region at 515 W/cm^2 is 1 mm/msec.[23] This decreases to about 0.8 mm/msec at 400 W/cm^2. If it is assumed that the DNA molecule moves at the same speed as the event, then the molecule traverses the 13-mm length of the container in 16 msec. Thus T must be greater than (0.82×16) or 13 msec. The DNA molecule extended and oriented in hydrodynamic flow relaxes on stopping the flow suddenly with a spectrum

TABLE VI. First-order subharmonic emission and molecular degradation in undegassed DNA irradiated for 30 min at different intensities.

I (W/cm^2)	Mean sample subharmonic (mV)	Mean tank subharmonic (mV)	$\Delta M/M_c$
72	3.0	4.9	0.13
72	4.6	3.2	0.080
288	5.1	5.2	0.205
288	8.1		0.230
288	7.8	3.2	0.090
288	13.0	8.0	0.280
515	14.1	18.5	0.475
515	9.3	5.2	0.145
515	8.5	7.9	0.185
515	12.0	8.0	0.090

of relaxation times,[25] the longest of which is given as

$$T = 5.0 \times 10^{-14} M^{1.6}.$$

The value of T relevant to the present study is 20 msec, which fulfills the condition that T be greater than 13 msec. A large fraction of the DNA relaxes with the longest relaxation time.[25,26] The fact the DNA relaxation time meets the requirements of a mechanism which would be noncavitational and yet explain the strange intensity dependence of degradation (Table IV) may be fortuitous but further experimentation involving irradiation of degassed DNA will be required before deciding between it and cavitational microstreaming.

ACKNOWLEDGMENTS

The authors are indebted to Professor D. E. Hughes, Dr. A. R. Williams, and Mr. C. James, Department of Microbiology, University College, Cardiff, where this research was performed, for technical advice, for use of the viscometer, and for technical assistance, respectively. W. T. C. thanks the Medical Research Council for support during part of this work. F. D. acknowledges gratefully the National Institutes of Health Special Research Fellowship enabling him to pursue these studies while on sabbatical leave from the University of Illinois, Urbana.

[1] G. Gooberman, J. Polymer Sci. **42**, 25-33 (1960).
[2] P. Alexander and M. Fox, J. Polymer Sci. **12**, 533–541 (1954).
[3] W. Roberts, E. Yeager, and F. Hovorka, Tech. Rep. No. 18, Dep. Chem., Western Reserve Univ., Cleveland, Ohio (1957).
[4] G. Schmid, G. Paret, and H. Pfleiderer, Colloid Z. **124**, 150–160 (1951).
[5] P. Grabar and R. O. Prudhomme, J. Chim. Phys. **46**, 667–670 (1949).
[6] H. W. W. Brett and H. H. G. Jellinek, J. Polymer Sci. **21**, 535–545 (1956).
[7] A. Weissler, J. Appl. Phys. **21**, 171–173 (1950).
[8] H. W. Melville and A. J. R. Murray, Trans. Faraday Soc. **46**, 996–1009 (1950).
[9] M. A. K. Mostafa, J. Polymer Sci. **33**, 311–322 (1958).
[10] J. N. Davidson, *Biochemistry of the Nucleic Acids* (Methuen, London, 1968), p. 77.
[11] C. E. Hall and P. Doty, J. Amer. Chem. Soc. **80**, 1269–1274 (1958).
[12] C. E. Hall and M. Litt, J. Biophys. Biochem. Cytol. **4**, 1–4 (1958).
[13] N. I. Ryabchenko, F. I. Braginskaya, I. E. El'piner, and P. I. Tseitlin, Biofizika **9**, 31–40 (1964).
[14] O. C. Richards and P. D. Boyer, J. Mol. Biol. **11**, 327–340 (1965).
[15] S. A. Hawley, R. M. Macleod, and F. Dunn, J. Acoust. Soc. Amer. **35**, 1285–1287 (1963).
[16] J. Eigner and P. Doty, J. Mol. Biol. **12**, 549–580 (1965).
[17] C. R. Hill, P. R. Clarke, M. R. Crowe, and J. W. Hammick, *Ultrasonics for Industry Papers* (Illiffe, London, 1969), pp. 26–30.
[18] W. T. Coakley, J. Acoust. Soc. Amer. **49**, 792–801 (1971).
[19] E. Layne, in *Methods in Enzymology*, S. P. Colowick and N. O. Kaplan, Eds. (Academic, New York, 1957), Vol. III, pp. 447–450.
[20] L. C. Clark, Trans. Amer. Soc. Artificial Internal Organs **2**, 41 (1956).
[21] W. W. Kielley, in *Methods of Enzymology*, S. P. Colowick and N. O. Kaplan, Eds. (Academic, New York, 1963), Vol. VI, pp. 272–277.
[22] A. R. Williams, J. Phys. E **2**, 279–281 (1969).
[23] W. T. Coakley, PhD thesis, Univ. Wales (1971).
[24] E. B. Mode, *Elements of Statistics* (Prentice Hall, Englewood Cliffs, N. J., 1961), p. 260.
[25] P. R. Callis and N. Davidson, Biopolymers **8**, 379–390 (1969).
[26] D. S. Thompson and S. J. Gill, J. Chem. Phys. **47**, 5008–5017 (1967).

ERRATA

Page 1540, column 1, last line should read: "Na_2HPO_4, $0.002M$ NaH_2PO_4, ..."

Page 1541, column 2, line 3 should read: "correlation index R ($R = r^2$, where r is the correlation coefficient) for the 16 points. ..."

Page 1542, column 1, line 5 should read: "within each total number of events."

Page 1543, column 1, line 4 should read: "for 30 min, plotted against number of transient. ..."

Page 1544, column 1, line 11 should read: "activity did not show a marked decrease with irradiation. ..."

Direct and Indirect Effect of Ultrasound on Bone Marrow Cell Suspensions

I. HRAZDIRA

Department of Medical Physics, Faculty of Medicine, Purkyně University, Brno

Received December 2, 1964

Abstract. The direct and indirect effect of ultrasound (frequency 800 kc, intensity 0.8 and 1.5 W/cm.^{2}) on rat bone marrow cells suspended in saline and in saline containing 5% (10%) Subtosan (3.5% polyvinylpyrrolidone) was studied. The number of surviving cells, the pH change and the amount of free HNO_2 formed as a result of exposure to ultrasound were determined in every experimental sample. The experiments conclusively demonstrated the indirect effect of ultrasound on suspended bone marrow cells and showed that it increased with the time of exposure of the suspension solutions. A connection between this effect and physicochemical changes in the suspension medium was also demonstrated. The presence of a small amount of a colloidal substance limited the formation of chemically active substances and the indirect effect of ultrasound on the suspended bone marrow cells.

The effect of ultrasound on cell suspensions was previously studied chiefly with reference to destructive changes in the cells, resulting from the action of mechanical factors of the ultrasonic field (Hughes and Nyborg 1962). It is an indisputable fact, however, that the direct effect of ultrasound also causes changes in the suspension medium. Chemical changes in aqueous solutions, characterized by the formation of chemically active substances, have been known for a good many years (Beuthe 1933, Schultes and Gohr 1936) but it was not until quite recently that their relationship to biological material was investigated, in the form of an indirect effect. It can be assumed that contact of a physically and chemically changed suspension medium with suspended cells is also bound to affect the cells, despite the fact that some authors deny the existence of a direct connection between the formation of free radicals or other chemically active substances and injury to the cells, leading to their lysis (Hughes and Rodgers 1960, Prudhome and Constantin 1963). The water permeability of connective tissue membranes was found to be a sensitive indicator of the indirect effect of ultrasound (Pospíšilová 1964).

Quantitative determination of the products of the ultrasonic oxidation of nitrogen in water exposed to ultrasound was carried out by Polotsky (1947). The present author also demonstrated a drop in the pH and the formation of free HNO_2 in saline treated with ultrasonic waves of low intensity (Hrazdira and Zelníček 1964). It was also found that in exposure to ultrasound, while freely admitting air, soluble nitrogen, as well as atmospheric nitrogen, participated in the formation of nitrous acid. On the basis of these findings, the author decided to follow on from previous experiments (Hrazdira and Bílková 1963) and study the indirect effect of ultrasound on suspended bone marrow cells and its relationship to certain physicochemical changes in the suspension medium.

Bone marrow was obtained from the femurs of adult Wistar rats with an average weight of 195 g., by the technique described by Drásil and Soška (1957). Marrow from each femur was suspended in 2 ml. saline, giving a basic suspension containing about 15,000 cells/mm.^{3}. Samples for studying the direct and indirect effect of ultrasound were prepared by diluting this suspension in tenfold dilutions.

After staining with 0.1% eosin, the number of dead and living cells was determined in every sample. Since bone marrow cells are very sensitive to a wide range of different factors, the time intervals of the various operations were adhered to strictly: exposure to ultrasound (direct and indirect) five minutes, staining three minutes, counting in a counting cell five minutes. The whole process did not thus take longer than 15 minutes. Control tests showed that the number of suspended cells did not alter either absolutely or relatively during this period. In addition to determining the percentage of surviving cells, the pH was also determined in every sample with a Multoscop pH-meter to within 0.05 pH, before and after exposure to ultrasound. The amount of free HNO_2 was determined by Griess's colour reaction, with sulphanilic acid and α-naphthylamine, on a spectrophotometer at 510 mμ, to within 0.05 μg. HNO_2/ml.

Plain saline and saline containing 5% (10%) Subtosan (3.5% polyvinylpyrrolidone solution) were used as suspension solutions in the actual experiments. Isotonic NaCl was used rather than the recommended Hanks solution (Michalowski and Jasińska 1963), as the simplest suspension medium, with reference to the possibility of chemical reactions after exposure to ultrasound. For the same reason the initial pH of this solution was not adjusted to the optimal pH for bone marrow cells. The source of ultrasound was an adapted clinical TuR US-2 apparatus, with a frequency of 800 kc. The intensities used (0.8 and 1.5 W/cm.^{2}) were controlled radiometrically.

The indirect effect of ultrasound was determined by first exposing 2 ml. plain suspension solution for 1—5 minutes and then immediately adding 0.2 ml. of the basic bone marrow cell suspension. After five minutes' exposure at room temperature, this test suspension was treated in the manner described above.

The temperature of the samples was controlled by a thermocouple. The results of control heat tests were negative.

In the first part of the experiments, survival of the cells, pH changes and the amount of HNO_2 formed by the direct effect of ultrasound were studied. The results, which are the average of ten series of experiments, are given in tab. 1.

These experiments demonstrated that the presence of a small amount of a colloidal substance not only has a protective effect on surviving cells, but also limits chemical ultrasonic reactions. Differences in the initial pH in the suspension solutions and in the character of pH changes after exposure to ultrasound were due to the addition of Subtosan at pH 8. The synthetic colloidal solution acted in this case as a buffer. No reliable explanation has yet been found, however, for the pH increase which occurred after exposure to ultrasound in 10% Subtosan.

In the second part of the experiments, the indirect effect of ultrasound (intensity 1.5 W/cm.^{2}) in saline and 5% Subtosan was studied. The suspension solutions were exposed to ultrasound for 1, 3 and 5 minutes and the connection between the

Tab. 1. Direct effect of ultrasound on bone marrow cell suspension. Time of exposure 1 min. C — unexposed control, 0.8 and 1.5 — intensity in W/cm^2.

Value studied	Saline			5% Subtosan			10% Subtosan		
	C	0.8	1.5	C	0.8	1.5	C	0.8	1.5
Percentual survival	94	0	0	94	92	8	94	94	92
pH	6.3	6	5.8	6.8	6.95	6.75	6.95	7.15	7.2
μg. HNO_2/ml.	0	0.4	0.94	0	0.16	0.75	0	0	trace

survival percentage and physicochemical changes was determined. The results of these experiments are given in fig. 1. They conclusively demonstrated the indirect effect of ultrasound on suspended bone marrow cells. This effect, which was manifested in a decrease in the number

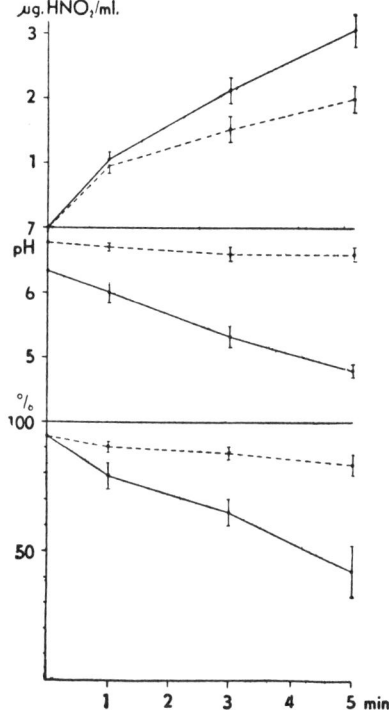

Fig. 1. Correlation of percentual survival, pH changes and HNO_2 formation (y) to the time of exposure of suspension solutions to ultrasound (x). Indirect exposure 5 minutes, unbroken lines — saline, broken lines — 5% Subtosan. The vertical lines denote the standard errors.

of surviving cells, increased in correlation to the time of exposure to ultrasound and was connected with physicochemical changes determined after five minutes' exposure of the suspension solutions to ultrasound. As in the case of the direct effect, the presence of a colloidal substance again modified the changes observed.

When evaluating the results, two series of control experiments were carried out.

In the first, the percentage of surviving bone marrow cells was determined after five minutes' exposure in saline at different pH, adjusted with a small amount of freshly prepared equimolar HNO_2 solution. The results are illustrated in fig. 2. In the second series, the survival percentage was determined after exposure for the same length of time in saline containing different nitrite concentrations. The results showed that nitrite ions alone, up to a concentration of 5 µg./ml., at a constant pH, did not significantly influence the survival of suspended bone marrow cells.

The direct effect of ultrasound on a cell suspension is the function of mechanical factors of the ultrasonic field and the properties of the suspension medium. Comparison with the indirect effect shows that the average participation of physicochemical changes in the final result is only 2—3%.

In the case of the indirect effect, the course of survival after exposure of the medium to ultrasound is far more dependent on the hydrogen ion concentration than on the nitrite ion concentration, but is not completely explained by either. HNO_2, of course, is only an indicator of chemically active substances formed by the action of ultrasound in water and aqueous solutions (Levinson and Kovrov 1959). These substances of a highly oxida-

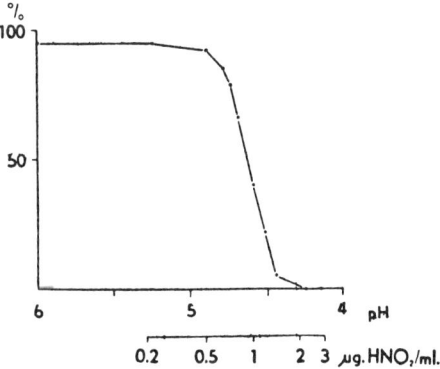

Fig. 2. Survival of bone marrow cells in saline at different pH, corrected by HNO_2. Exposure 5 minutes.

tive character, which are probably formed by ultrasonic ionization of water, disturb the isotonicity and isoionic character of biological systems and can have a direct toxic effect on surviving suspended cells.

In conclusion, it can be claimed that the above study demonstrated the indirect effect of ultrasound on surviving isolated bone marrow cells and its close connection with the formation of chemically active substances. The detailed mechanism of this effect requires further study, however.

References

Beuthe, H.: *Über den Einfluss von Ultraschallwellen auf chemische Prozesse.* Zschr. phys. Chem., A, *163*:161, 1933.
Drášil, V., Soška, J.: *Vliv injekce kostní dřeně na metabolismus DNK v krvetvorných orgánech po ozáření (Effect of injection of bone marrow on DNA metabolism in haematopoietic tissue after irradiation).* Čs. biologie, *6* : 343, 1957.
Hrazdira, I., Bílková, B.: *Effect of ultrasound on bone marrow cell suspensions in vitro.* Fol. biol. (Praha), *9* : 397, 1963.
Hrazdira, I., Zelníček, E.: *Fysikálně-chemické účinky ultrazvuku. (Physico-chemical effects of ultrasound.)* Scripta Medica, *37* : 1, 1964.
Hughes, D. E., Nyborg, W. L.: *Cell disruption by ultrasound.* Science, *138* : 108, 1962.
Hughes, D. E., Rodgers, A.: *Some experiments on the disintegration of bacteria by ultrasound.* In: *Medical Electronics.* Ed. Illiffe, London 1960 (p. 397).
(Levinson, M. S., Kovrov, B. G.) Левинсон, М. С., Ковров. Б. Г.: *Действие ультразвука на дистиллированную воду.* Биохимия, *24* : 535, 1959.
Michalowski, A., Jasińska, J.: *In vitro survival of rat bone marrow and mouse Ehrlich ascites tumour cells in fluid media.* Rad. Biol. Ther., *4* : 25, 1963.
(Polotsky, I. G.) Полоцкий, И. Г.: *Определение NO_2^-, NO_3^- и H_2O_2 в воде, экспонированной в ультразвуковом поле.* Журнал общей химии, *17* : 649, 1947.
Prudhomme, R. O., Constantin, T.: *Action des ultrasons sur Saccharomyces cerevisiae. I. Courbes de survie, en présence de différents gaz.* C. R. Soc. Biol., *157* : 489, 1963.
Pospíšilová, J.: *Indirect effect of ultrasound on water permeability of connective tissue.* Experientia, *20* : 120, 1964.
Schultes, H., Gohr, H.: *Über chemische Wirkungen der Ultraschallwellen.* Angew. Chem., *49* : 420, 1936.

Прямое и косвенное действие ультразвука на суспензии клеток костного мозга

И. ГРАЗДИРА

Резюме

Исследовалось прямое и косвенное действие ультразвука при частоте 800 кгц и мощности 0,8 и 1,5 вт/см² на клетки костного мозга крыс в форме суспензии в физиологическом растворе или отдельно, или с примесью 5% (10%) субтосана (3,5% раствора поливинилпирролидона). В каждом взятом в опыт образце определяли количество выживающих клеток, изменения pH и содержание свободной HNO_2, возникающей в результате обработки ультразвуком.

В опытах было доказано существование косвенного действия ультразвука на клетки костного мозга в суспензии. Это действие усиливалось с продлением обработки ультразвуком. При этом была обнаружена связь между количеством выживающих клеток и физико-химическими изменениями среды суспензии. Присутствие небольших количеств коллоидного вещества ограничивает как прямое, так и косвенное действие ультразвука на суспензию клеток костного мозга.

Physical and Chemical Aspects of Ultrasonic Disruption of Cells

P. R. Clarke and C. R. Hill

Physics Department, Institute of Cancer Research, Belmont, Sutton, Surrey, United Kingdom

An assessment is made of the relative contributions of mechanical and sonochemical effects to damage suffered by mammalian cells when sonicated in aqueous-suspension culture. In a 1-MHz progressive wave beam, under conditions providing stabilization of the cavitation field, the intensity threshold and intensity optimum for cell disintegration occur at about 1 and 5 W cm^{-2}, respectively, and the same pattern is found for the liberation of free iodine in solutions of KI. The cell culture medium is, however, found to have a chemiprotective effect on the cells, probably as a result of successful competition for OH radicals on the part of glucose and amino acid constituents of the medium. Thus, although prolonged sonication of the medium can eventually reduce its ability to support cells, such indirect effect is insignificant in the relatively short irradiation times required to kill 99% of cells and the primary mechanism of damage here appears to be mechanical. Pulsing conditions have an important influence on both cell disintegration and iodine-release rates, which both show maxima for 30-msec pulse duration and drop to zero for pulse durations below 1 msec. The relevance of this finding to the kinetics of cavitation mechanisms is discussed.

INTRODUCTION

When a liquid system is exposed to ultrasonic irradiation, a number of widely different phenomena are known to occur. These include heating of the liquid, the acceleration or induction of chemical reactions, the production of luminescence and a characteristic noise spectrum, and a variety of biological effects including the disruption of microorganisms and cells. It would be interesting to know to what extent these effects arise independently and to what extent they constitute a chain of linked reactions.

In the program of research of which the present study forms a part, the generation of a characteristic subharmonic signal, the release of free iodine from potassium iodide solutions, the degradation of DNA molecules, and the disruption of mouse lymphoma cells have all been observed and are believed to arise as a consequence of processes covered by the general term *cavitation*.[1] The irradiation system, together with some of these results, have been described by Hill *et al.*,[2] and the present study was designed to investigate the relationship between the chemical effects of ultrasound and the death of cells in sonicated cell suspensions.

The term *cavitation* is used here in the general sense that includes both the transient and stable form. The latter, in particular, refers to the process involving steady growth, from preexisting nuclei and under the action of acoustically stimulated rectified diffusion, of a population of gaseous microbubbles. At some point in the growth of each microbubble, it passes through a size at which it is resonant in the applied field and, in this condition, with relatively large vibration amplitude, it constitutes an effective means of converting acoustic energy into energy of both unidirectional microstreaming fields and altered chemical bonds. The microstreaming fields constitute localized regions of high shear stress within a liquid.

Despite the fact that cell death due to ultrasonic irradiation has formed the basis of several studies and that the method is widely used as a tool for cell disruption,[3] the mechanisms involved are not fully understood. Hughes and Rogers found that iodine release from potassium iodide was not paralleled by cell breakage either at 20 kHz or at another, unspecified, frequency[4] and considered it unlikely, therefore, that free radicals were in any simple way connected with cell

---

[1] H. G. Flynn "Physics of Acoustic Cavitation in Liquids," in *Physical Acoustics*, W. P. Mason, Ed. (Academic Press Inc., New York, 1964), Vol. 1B, Chap. 9, pp. 58–172.
[2] C. R. Hill, P. R. Clarke, M. R. Crowe, and J. W. Hammick, "Biophysical Effects of Cavitation in a 1-MHz Ultrasonic Beam," in *Ultrasonics for Industry 1969* (Iliffe House, Guildford, England, 1969), pp. 26–30.
[3] D. E. Hughes and W. L. Nyborg, Science 138, 108–114 (1962).
[4] D. E. Hughes and A. Rodgers, "Medical Electronics," Proc. Intern. Conf. Med. Electron. 2nd, Paris (Iliffe, London, 1960), pp. 397–400.

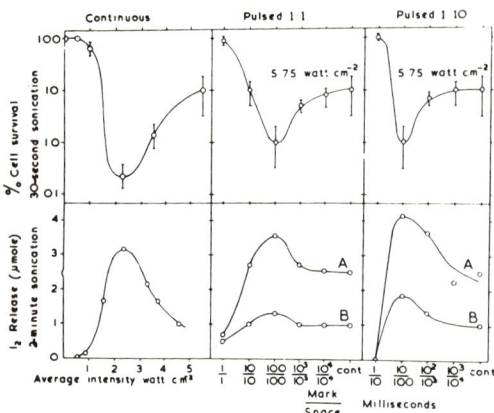

FIG. 1. Variation of cell survival and iodine release with ultrasonic intensity and pulsing conditions at 1 MHz. (A: 3.1 W cm^{-2}; B: 4.6 W cm^{-2}).

disintegration. Hrazdira considered that contact with a physically and chemically changed suspension medium was bound to affect cell metabolism in a manner that could lead to loss of reproductive integrity, although such change would not necessarily be accompanied by morphologically visible changes corresponding to "cell disruption."[5] He estimated that this indirect effect participated in cell killing only to the extent of 2%–3%, but he did not consider the possible action of very short-lived species that may be created in a cavitation field.

The present study was designed to assess the relative contributions of chemical and physical factors to cell death in sonicated liquid suspensions, and also to investigate further the nature of the physical and chemical processes involved.

I. MATERIALS AND METHODS

The cells used were a radiosensitive strain of the mouse *lymphoma* L5178Y, maintained *in vivo* on the full medium devised by Fischer.[6] These cells, and the measurement of cell survival using a criterion of reproductive integrity (which may be impaired equally by mechanical or chemical forms of stress), have been described by Courtenay.[7] This method involves drawing growth curves for treated cells and comparing these with control curves and will not be described in detail here. It is interesting to note, however, that delayed effects of the type found following ionizing irradiation of cells are absent following sonication. The process of ultrasonic cell killing appears to be an "all or nothing" effect, survivors appearing unaffected with respect to growth rate, microscopic appearance, and even progression through the cell cycle.[8] Incubation and ultrasonic irradiation were both carried out at 37°C in sterile disposable polystyrene tubes. During sonication, such tubes containing the culture medium were introduced into a 1-MHz ultrasonic beam propagated through distilled water, as described by Hill *et al.*[2] The methods of ultrasonic dosimetry are described elsewhere.[9] Intensities quoted are the average intensity over the sample: investigation of beam crosssections within the specimen container show that peak intensities at the beam center are approximately three times these values. For pulsed beams, the intensity quoted is similarly the average intensity over the duration of the pulse, and the sonication time is the total irradiation time multiplied by the duty factor. The sample volumes are 3 ml throughout.

Sample tubes were rotated during irradiation. This feature was originally incorporated into the system in order to provide agitation of cell suspensions during long runs, but it was found subsequently that cell disruption was reduced below the level of detectability, i.e., by a factor of at least 100, when rotation was not employed. At the same time, it was established that the ultrasonic field conditions within the container were not affected by the rotation. This "rotation effect" is discussed further in Sec. III.

Temperature changes during sonication were minimized by the large volume of temperature-controlled water surrounding the sample tubes. At the intensities used in these experiments, irradiated suspensions were found to reach equilibrium not more than 2°C above tank temperature. The effect of a temperature change of this magnitude was found to be negligible in itself.

Chemical effects of ultrasound were investigated in an identical irradiation system, using the liberation of free iodine from 0.005N potassium iodide, in the presence of carbon tetrachloride, as the observable end point. Estimation was by titration against sodium thiosulphate, using starch as indicator.

II. RESULTS

Figure 1 shows the variation of cell killing and iodine release with ultrasonic intensity and pulsing conditions. Both these effects exhibit an intensity threshold in the 0.5- to 1-W cm^{-2} region and a maximum in the 2.5-W cm^{-2} region, followed by a falloff at higher intensities. They also show a maximum response when the ultrasound is pulsed with a pulse length of 10 or 100 msec, longer pulses or continuous irradiation being less effective and very short pulses, of the order of 1 msec, being virtually ineffective. Detailed discussion of the shapes of these curves is reserved for Sec. III, but a point of immediate interest is the close correlation between iodine release and cell death. This correlation may be due to one of two causes:

---

[5] I. Hrazdira, Folia Biol. (Prague) **11**, 330–333 (1965).
[6] G. A. Fischer, Ann. N. Y. Acad. Sci. **76**, 673 (1958).
[7] V. D. Courtenay, Int. J. Radiat. Biol. **9**, 581–592 (1965).
[8] P. R. Clarke and C. R. Hill, "Biological Action of Ultrasound in Relation to the Cell Cycle," Exptl. Cell Res. **58**, 443–444 (1969).

[9] C. R. Hill, "Calibration of Ultrasonic Beams for Bio-Medical Applications," Phys. Med. Biol. (1970) (to be published).

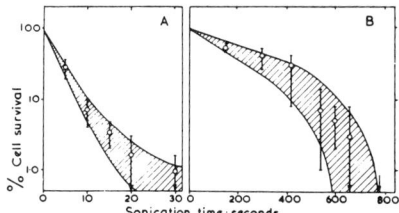

FIG. 2. Variation of cell survival with irradiation time: 2.5 W cm^{-2}, 1 MHz. (A: cells sonicated in suspension; B: cells added immediately after sonication of medium.)

(1) Cell death is a consequence of a chemical effect of ultrasound.
(2) Both cell death and the chemical effects of ultrasound depend on a common cause.

The first possibility was examined as follows.

Any chemical change in the cell culture medium brought about by sonication could lead to cell death and the possibility of an indirect effect of this kind was investigated by exposing samples of medium to continuous sonication in the normal way for varying periods of time and immediately adding untreated cells. The results of such an experiment are shown in Fig. 2, with the survival curve for cells during sonication shown for comparison. It is clear that, during the short time required to kill 99% of cells *in situ*, the indirect effect contributes less than 10%. However, longer sonication seriously reduces the ability of the medium to support the cells and this factor might become an important consideration in other systems involving extended irradiations.

The action of short-lived species remains unknown in an experiment of this type, but an approach to this problem was made by observing pH changes, in addition to iodine release, in various sonicated solutions. Unfortunately, the "cavitation chemistry," even of pure water, is by no means fully understood. It is believed[10] that sonicated water undergoes the reaction

$$H_2O \rightleftharpoons H + OH,$$

and that, in the absence of CCl_4, the production of I_2 from KI is largely due to the action of the OH radical,[11] possibly via H_2O_2. When CCl_4 is present, the yield of I_2 is increased by the action of oxidizing chlorine.[12] In the present study, this increased yield in the presence of CCl_4 was found to be of the order of 20 times and the pH change shown in Table I indicates that an acidic species is also formed.

The most interesting information in this Table is that, in the presence of the culture medium, no release of free iodine is observed. The culture medium itself is found to be without effect on the stability of the starch–iodine complex and nor is the free-iodine release simply masked by the color of the medium: Sonication of equal quantities of KI and phenol red solution gives clearly visible I_2 release. The most likely explanation is that components of the medium compete successfully for OH radicals. Consideration of the concentrations of various substances in the medium and their known rate constants with OH indicate that glucose, and the amino acids present, might fulfill this function.

Thus it is suggested that the medium acts to a considerable extent as a protection for cells against direct free-radical attack but that, in the process, vital elements of the medium undergo chemical reaction that eventually renders the medium incapable of supporting cell growth. However, during the short time required to kill 99% of cells in this system, chemical effects, either direct or indirect, do not play a major role in cell disruption.

Thus it appears that, in the system described, bulk heating and chemical effects are not the primary causes of cell death and that, of the two possible explanations for the correlation between cell death and iodine release given above, the second is the correct one; that is, both effects depend on a common cause. The group of phenomena covered by the general term "cavitation" is generally agreed to be important in this respect. Flynn[1] has made the useful theoretical distinction between "stable" and "collapse" cavitation, the former involving the vibration of bubbles over many sound cycles, the latter involving their complete collapse during a single compression phase of the vibration. Although this distinction cannot be rigidly applied in practice, it would appear that the present system is more closely allied to the stable-cavitation situation, as the results using pulsed ultrasound indicate that the processes under observation are initiated only after the completion of some 1000 or more sound cycles. Further, sonoluminescence, a phenomenon often associated with the high compression ratios occurring in cavity collapse, could not be detected using either the dark-adapted eye or a sensitive photomultiplier system.[2]

TABLE I. Iodine release and pH changes during 2-min sonication at 2.5 W cm^{-2} (1 MHz).

Solution	I_2 release?	pH Before sonication	pH After sonication
Cell culture medium	—	7.4	7.4
Medium + KI	no	7.4	7.4
Medium + CCl_4	—	7.4	6.8
Medium + CCl_4 + KI	no	7.4	6.8
KI + CCl_4	yes	6.7	3.4

---

[10] E. Webster, Ultrasonics **1**, 39–48 (1963).
[11] O. Lindström, J. Acoust. Soc. Amer. **27**, 654–671 (1955).
[12] A. Weissler, N. W. Cooper, and S. Snyder, J. Amer. Chem. Soc. **72**, 1769–1775 (1950).

III. DISCUSSION

If the biological and chemical effects observed here do indeed arise independently as a result of a vibrating bubble phenomenon, consideration must be given to the mechanisms involved. Griffing presents convincing evidence that all primary sonochemical reactions take place within cavitation bubbles and have a thermal origin, owing to the heat developed during adiabatic compression.[13] Fitzgerald, working at 2 MHz, has shown that collapse of the bubble is not necessarily involved and that sufficient heat can be generated inside a vibrating bubble.[14] Despite the presence of such "hot spots" within vibrating bubbles, the bulk temperature of the sonicated liquid can be controlled and a different mechanism is suggested for cell disruption. This is the phenomenon of bubble-associated streaming.

Hughes and Nyborg have shown that the streaming produced near trapped gas bubbles vibrating at 20 kHz is of sufficient magnitude to cause cell disruption.[3] This is perhaps not surprising in view of the calculations of Elder,[15] which show that velocity gradients of up to about 1.5×10^5 sec^{-1} may be attainable at the boundary of an air bubble undergoing forced vibration in water at 10 kHz and that this limit will tend to increase with frequency.

The "rotation effect" and the shape of the curves in Fig. 1 can be conveniently explained in terms of these mechanisms. It should be remembered that a bubble vibrating under the influence of a sound field is a continuously developing system. It begins as a nucleus,[10] probably a stable microbubble, and grows by rectified diffusion, with gas moving preferentially into the bubble from liquid solution.[16,17] At one particular size in this process, depending only on the frequency of the applied sound field, such a bubble exhibits resonance, with vibration amplitude several orders of magnitude greater than that of the particles of the liquid in the absence of the bubble.[18] It is at this stage that the associated streaming caused in the vicinity of the bubble and mentioned above is believed to lead to sufficiently high shear and tensile stresses to disrupt cells. Similarly, in this situation the compression inside the bubble will be particularly great, giving rise to the temperatures necessary for thermal dissociation. However, at earlier and later stages in the development of the bubble (i.e., away from resonance), the vibration will be less violent, and the associated biological and chemical effects less marked.

Bearing this in mind, the "rotation effect" is believed to be due to a stabilization of the bubble field caused by the rotary motion. In our irradiation system, the contents of the sample tube are subject to a traveling wave system and the induced unidirectional streaming would be expected to carry bubbles and nuclei to the side of the container distant from the transducer, where further growth and interaction with the main body of the solution would be inhibited. Rotation of the container gives rise to centripetal forces on the bubbles and will also lead to continuous change in effective beam direction, both of which effects will cause the bubbles to move in circular or spiral paths, rather than linearly, and will thus tend to maintain the bubble population within the body of the liquid.

Turning to the shape of the curves in Fig. 1, the theory of rectified diffusion implies that, for a bubble of a certain size, there will be a lower limit of acoustic pressure amplitude below which bubble growth does not occur but above which it continues until the bubble is in some way removed from the field. The sharp intensity thresholds observed are in accord with this idea. The falloff of cell killing and iodine release at higher intensities is open to a number of interpretations. Bubbles may be swept out of the central region by the higher streaming forces, despite rotation; surface waves may be created, leading to premature disruption or reduced microstreaming; or the entire bubble development may be more rapid, giving a shorter period of resonance vibration. Other workers have observed anomalous intensity dependence of this kind, although none has suggested any reason for it.[12, 19–21]

The shape of the curves obtained when pulsed ultrasound is employed enables some time values to be assigned to the development of bubbles in this particular system. The ineffectiveness of very short pulses suggests that times of the order 10^3 cycles (1 msec at 1 MHz) elapse before a bubble constitutes an efficient center for free-radical formation and cell disruption. In the latter case, a fraction of this time may be taken up by the time required for the bubble activity to affect the surroundings, for example, by the setting up of microstreaming. The ineffectiveness of short pulses has been further demonstrated by longer cell irradiations up to 5 h (27 min "ON time") at 5 W cm^{-2}, pulsed 1 msec ON, 10 msec OFF, which led to 100% survival of cells, as did a number of 2-h runs using a commercial diagnostic machine delivering 455 pulses of 0.5-μsec duration every second. For this latter irradiation, the nominal frequency was 2 MHz and the measured peak

---

[13] V. Griffing, J. Chem. Phys. **18**, 997–998 (1950); **20**, 939–942 (1952).
[14] M. E. Fitzgerald, V. Griffing, and J. Sullivan, J. Chem. Phys. **25**, 926–933 (1956).
[15] S. A. Elder, J. Acoust. Soc. Amer. **31**, 54–64 (1959).
[16] E. N. Harvey, D. K. Barnes, W. D. McElroy, A. H. Whiteley, D. C. Pease, and K. W. Cooper, J. Cellular Comp. Physiol. **24**, 1–21 (1944).
[17] D. Y. Hsieh and M. S. Plesset, J. Acoust. Soc. Amer. **33**, 206 (1961).
[18] W. L. Nyborg, "Acoustic Streaming" in *Physical Acoustics*, W. P. Mason, Ed. (Academic Press, New York, 1965), Vol. 2B, Chap. 11, pp. 265–332.

[19] O. Nomoto and S. Okni, J. Jap. Phys. Soc. **304**, 47–52 (1948).
[20] E. A. Neppiras and D. E. Hughes, Biotech. Bioeng. **6**, 247–270 (1964).
[21] I. Hrazdira, Proc. Intern. Symp. "SIDUO II," Brno, 1967, pp. 147–151.

intensity in the pulse was 98 W cm^{-2}.[22] Other workers have also found that pulse lengths less than 1 msec inhibit ultrasonic cavitational activity.[21,23].

The curves indicate that the bubbles are maximally efficient after 10^4–10^5 cycles, longer pulse lengths or continuous beams being less efficient, presumably as the result of continued growth of the bubbles beyond resonant size. Clearly, the relative effectiveness of various pulsing conditions depends not only on the pulse length but also on the space length. The behavior of the bubbles during the OFF time is uncertain, but they will surely lose gas to the surroundings by diffusion and so be reduced in size. Comparison of the mark:space ratios 1:1 and 1:10 indicate that 100 msec is a particularly effective OFF period, and one may imagine that, under optimum pulsing conditions, a bubble could be taken repeatedly through resonant size without developing to the less active larger size.

Experiments of this kind clearly show the value of pulsing as a means of understanding processes involving vibrating bubbles. They may be particularly relevant to the field of large-scale ultrasonic cell processing since they show that, if a stable bubble field can be created, the efficiency of the operation can be increased by using relatively low intensity and pulsed beams.

---

[22] C. R. Hill, "Acoustic Intensity Measurements on Ultrasonic Diagnostic Devices," Proc. Intern. Congress Ultrasonic Diagnostics, 1st, Vienna Academy of Medicine (1970) (to be published).

[23] K. H. Rust, Angew. Chem. 6, 162 (1952).

Quantitative Relationships between Ultrasonic Cavitation and Effects upon Amoebae at 1 MHz

W. T. COAKLEY AND D. HAMPTON

Department of Microbiology, University College, Cardiff, Wales, United Kingdom

F. DUNN

Bioacoustics Research Laboratory, University of Illinois, Urbana, Illinois 61801

An amoeba, *Hartmannella castellanii*, which possesses many features typical of higher-order animal cells, was irradiated with 1-MHz ultrasound while suspended in ordinary growth medium and in one with increased viscosity. The ultrasonically produced cavitation was monitored and a strong correlation is found between the number of discrete cavitation events occurring and the decrease in cell numbers, on irradiating at 515 W/cm² for 10 min. The growth of treated cells was also examined.

INTRODUCTION

Various interactions between megahertz ultrasound and mammalian tissues have been reported which have been claimed to have arisen from causes other than heating due to acoustic absorption or to cavitation.[1-4] This paper reports on the investigation of growth and structural alterations, sought in ultrasonically irradiated amoebae, as an indication of fundamental interactions between the acoustic field parameters and cells in suspension. Microorganisms possess the following advantages as experimental material for such studies: (1) The short generation, or doubling time, ranging from 30 min for some bacteria to about 20 h for cultured mammalian cells, enables post-irradiation growth to be examined in a much shorter time than for multicellular animals. (2) A continuous supply of material may be produced under rigorously controlled conditions at low cost. (3) Since all cells treated are of the same type, it is possible to apply specific tests of cellular activity to the entire population.

Numerous reports have appeared regarding the effect of ultrasound on microorganisms and viruses, mostly concerned with release of intracellular components by acoustically induced cavitation.[5]

Electron micrographs of *Acanthamoeba* (*Hartmannella castellanii*) exhibit intracellular structures such as the nucleus and nucleolus, Golgi apparatus, "smooth" and "rough" endoplasmic reticula, endocytotic vacuoles, numerous mitochondria, and water expulsion vesicles.[6] Its outer boundary is a double layer of lipoprotein with some suggestion of areas of mucopolysaccharide on the outer surface of the membrane. Thus, in its ultrastructure and in its diameter of 20–30 μ, this amoeba is typical of many animal cells and was considered an appropriate choice for this study. The most obvious features of the cell distinguishable with the light microscope are the nucleus, vacuoles, and pseudopodia.

I. MATERIALS AND METHODS

The cells were grown in 50-ml volumes of 4% mycological peptone (Oxoid) in 250-ml flasks in a reciprocal shaking water bath at 30°C and irradiated ultrasonically in midlog phase after 30-h growth. For growth studies on treated samples, benzyl penicillin (Crystapen Glaxo Ltd.) was added to the cooled autoclaved medium to a final concentration of 400 units/ml. The ratio of flask volume to culture volume (5:1) and the ratio of innoculum to culture volume (1:10) were kept constant in the growth experiments.

The cells were irradiated in two types of containers, viz., a thin-walled aluminum cylinder of 13 mm i.d., 5 mm long and having a volume of 0.65 ml, and a lucite cylinder of 18 mm i.d., 12 mm long, with a volume of 3.3 ml. Rubber O rings were used to attach 0.0005-in.-

thick "Saran" (Dow Corning) foil windows to the cylinders. The cells were loaded into the container with a Pasteur pipet. A drop was removed and the cells were counted in a Fuchs Rosenthal haemacytometer, to serve as a control. The container was then sealed without trapping air bubbles, and placed carefully in the focal region of the acoustic field. The cells were counted again after the irradiation procedure.

Cells examined in the haemacytometer after irradiation showed three stages of damage: (1) At high intensities all cells were completely destroyed, leaving only tiny visible fragments. (2) At intermediate intensities, intact whole cells, cells with apparently intact cytoplasmic membranes with much of their interior organization destroyed, "ghost" cells apparently emptied of their organelles, large fragments of outer membrane, and many free organelles were visible. (3) At lower intensities most of the cells were apparently intact, though a few "ghosts" and cells with their internal organization affected were visible. Only those cells judged to be intact without visible damage were counted when estimating cell survival.

When growth of irradiated cells was planned, the container and its windows were sterilized in ethanol before loading and control cells were placed in a second container for the duration of the irradiation procedure. After irradiation, the number of cells in the treated and control samples were counted, and a 2-ml sample from each container was loaded into 18 ml of sterile medium in 100-ml flasks with added penicillin. Care was taken, by thorough mixing, to minimize counting errors due to cell sedimentation at all stages. Nevertheless, when the cell counting technique was checked by comparing pairs of counts from four containers loaded as above, the variance was 1036 when the average count was 408. This indicated that sources of variance greater than those expected from the Poisson distribution of cell counts were present and also suggested that

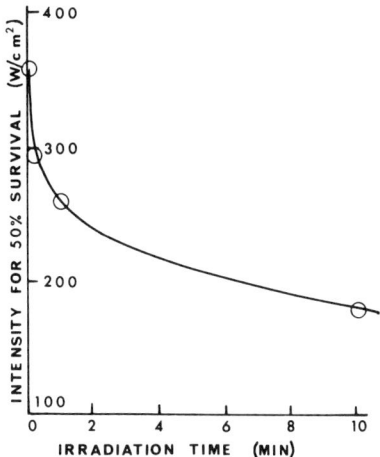

FIG. 2. Acoustic intensity for which 50% of the cells remained intact after treatment for various times (derived from Fig. 1).

little was to be gained by counting more cells to reduce the "Poisson distribution" contribution to the variance. Routinely, 500 to 1000 cells were counted.

Cells were also treated utilizing a viscosity increasing agent, viz., Methocel HG 400 (Dow Corning). Amoebae from a culture in midlog phase were sedimented in a sterilized centrifuge tube and resuspended in a Methocel-mycological peptone–penicillin mixture. Irradiation and subsequent culturing were as described. Because heat sterilization of the Methocel was not possible, only 50% of samples taken through the above procedure grew without subsequent contamination.

The 1-MHz 48-mm radius of curvature 48-mm chord diameter PZT4 ceramic transducer, the irradiation tank, the preparation of degassed water, sound-intensity measurement procedure, transient cavitation, and first-order subharmonic detecting systems are described elsewhere.[7] All are established procedures except for the cavitation detection scheme. Briefly, when the rf voltage applied to the transmitting transducer is rectified and filtered, transient signals may be detected when isolated transient cavitation events occur at the focal region of the field.

II. RESULTS

A. Intensity and Time Dependence of Cell Breakage

Intact cells from the one stock culture were counted, after irradiation at different intensities and exposure times in the aluminum container, and expressed as a percentage of the control cells in Fig. 1, from which Fig. 2 was derived. The latter figure shows that as the intensity decreased, the effect of the destructive mechanisms became much less severe. It was observed that more noticeable cell clumping and sedimentation occurred in samples irradiated for some minutes than

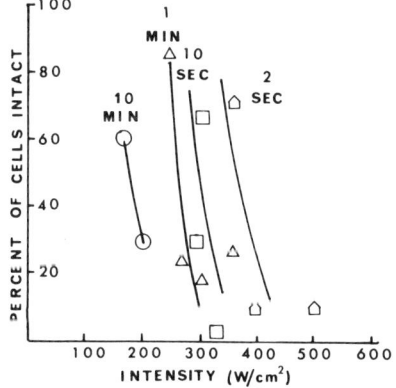

FIG. 1. Percentage of cells intact after irradiation in the aluminum cylinder at different intensities and exposure times.

TABLE I. Correlation of cell damage with transient cavitation events at 515 W/cm² for 10 min.

Batch	Medium[a]	Percent of cells surviving (microscope)	Percent of cells surviving (growth)	Number of transient events greater than 0.5 V	Number of transient events between 0.1 and 0.5 V
1	m. pep.	69		90	
1	m. pep. & Meth.	86		8	21
1	m. pep. & Meth.	100		3	22
2	m. pep.	26		162	
2	m. pep. & Meth.	87		0	14
2	m. pep. & Meth.	112		0	29
3	m. pep.	91	60	20	
3	m. pep. & Meth.	103	91	0	31
3	m. pep. & Meth.	115	90	0	15
4	m. pep.	9		212	
5	m. pep.	83	50	21	
5	m. pep. (degassed)	62	21	99	
5	m. pep. (degassed)	67	30	63	
6	m. pep.	99	88	4	
6	m. pep.	75		27	
6	m. pep. (degassed)	91		33	
7	m. pep.	1		greater than 360	
8	m. pep.	34	28	152	
8	m. pep.	83	62	25	
8	m. pep.	88	64	6	
8	m. pep.	58	41	86	

[a] m. pep.—4% mycological peptone, $\eta = 0.9$ cp; m. pep. & Meth.—0.4% Methocel in 4% mycological peptone, $\eta = 6.0$ cp.

in controls. On occasions, the clumps formed a ring on the front container window approximately in the position of the first off-axis zero of the sound beam. While not many of the cells were involved in this clumping, for long time treatments, the irradiation was interrupted after 2 min and the container was shaken to resuspend the cells.

The intensity required to destroy more than 97% of the cells in a 2-sec pulse in the aluminum container varied from 390 to 740 W/cm² in samples treated on ten different days, but was repeatable in cells from the same stock irradiated during any one day. Exhaustive checks on the sound-intensity calibrations, and monitoring of the voltage produced at the terminals of a small hydrophone placed at the end of the irradiation tank farthest from the transducer, showed that the wide range of intensities required for cell destruction was not due to any misbehavior of the sound-generating systems.

A number of bubbles of approximately 0.5-mm diameter were observed in some of the irradiated samples but, while they indicated that the medium had cavitated, they were not a reliable guide to the degree of cell damage inflicted.

When cells in growth medium were centrifuged and resuspended in degassed growth medium, the intensity required to damage 50% of the cells in 2 sec increased from 430 to 660 W/cm². The cells in the degassed medium became spherical, probably changing the mechanical properties of the cell boundary, so it could not be concluded that more cells survived because of any reduction in cavitation resulting from degassing.

Pulsing the sound increased the intensity required to break 50% of the cells from 390 W/cm² in a single 2-sec pulse to 790 W/cm² when the sound was delivered in 90-μsec pulses, mark:space 1:9 (duty cycle) for 20 sec. While pulsing the sound increases the cavitation threshold,[8] the above effect might also have been due to a time-dependent response of the cell to an ultrasonic stress.

Attempts to use the chemical starch–iodine test to detect cavitation in the irradiated medium[9] were unsuccessful because the chemical complex formed between starch and released iodine was interfered with by reaction with some growth medium components, as demonstrated by adding growth medium to an iodine–starch solution, which immediately loses its blue color.

B. Correlation of Cell Damage with Cavitation

The results which follow were obtained by irradiating samples in the lucite cylinder of volume 3.3 ml. An extensive study was carried out at an irradiation intensity of 515 W/cm², since a reasonable number of cells remained intact for subsequent growth studies, and yet this intensity was as high as practicable to identify any direct action between the sound field and the cells.

The cells were irradiated for a total exposure time of 10 min, in 2-min pulses, separated by 1 min to resuspend the cells by shaking the container. The electroacoustic system for detecting and counting single cavitation events was employed for all of the results reported from this point on.[7] However, the electronic counter for the single events had not been incorporated when the following initial experiments were performed. Those transient cavitation events which produced an amplitude greater than 0.5 V were counted by eye from the screen of the oscilloscope which monitored the filter section of the cavitation detecting circuit. The threshold of 0.5 V was chosen, as it was the only way of counting accurately by eye from the oscilloscope screen when profuse cavitation was present. The number of events

TABLE II. Total number of transient events greater than 0.5 V counted in the second half of each of the five 2-min pulses, in 14 samples.

Pulse number	Number of transient events greater than 0.5 V
1	399
2	156
3	125
4	92
5	100

occurring in the second half-minute of each of the five 2-min pulses was noted and totalled (Table I). Degassing of the solutions was affected in this instance by transferring 25 ml of cell suspension containing penicillin into a sterile 100-ml round-bottomed flask and degassing at a pressure of approximately 3 cm Hg. Controls were drawn from the same degassed stock solution.

The comparison of the results for cells irradiated in Methocel and in mycological peptone on the same day (Table I) showed that the protective effect of the former was one of reduction in the generation of transient events. However, Methocel is used in laboratory tissue culture apparatus to protect mammalian cells from shear stresses produced by mixing propellors.[10] Nevertheless, the stresses on erythrocytes suspended in a Methocel solution and sheared in a cone and plate viscometer have been shown to be greater than the stresses produced in isotonic saline.[11] It was possible to count the cavitation transients with amplitudes between 0.1 and 0.5 V in the Methocel samples. It can be seen also from Table I, that degassing, as described above, was not a very efficient way of reducing cavitation since the number of cavitation events increased on the two occasions on which degassing was attempted. The variation of number of transient events from day to day probably had its origins in the amount of dissolved gas and the distribution of nuclei in the suspensions.

The number of transient cavitation events produced decreased as the irradiation continued (Table II). This effect was not observed when irradiating tap water in the open tank,[7] where fresh gas and nuclei were constantly streaming through the focal region, probably because of the limited number of nuclei available in the 3.3-ml volume of the sample container.[12]

Figure 3 shows the result of plotting column 3 of Table I against column 5, i.e., a very strong relationship between the amount of damage and the number of cavitation events. Since the curve may reasonably be drawn through the 100% survival point when no cavitation was detected, it leads to the conclusion that there is no detectable damage to cells, observed under 100× magnification, in the absence of transient cavitation events greater than 0.5 V on irradiation at 515 W/cm² for 10 min at 1 MHz.

TABLE III. Number of transient events and cell survival at different acoustic intensities.

I (W/cm²)	No. of transient events in 10 min	Maximum amplitude	Mean subharmonic amplitude	Percent of cells surviving
580	21 720	4.0	6.8	3.5
450	11 500		7.4	33
340	8196	2.0	6.0	60
268	5620	0.4	4.3	89

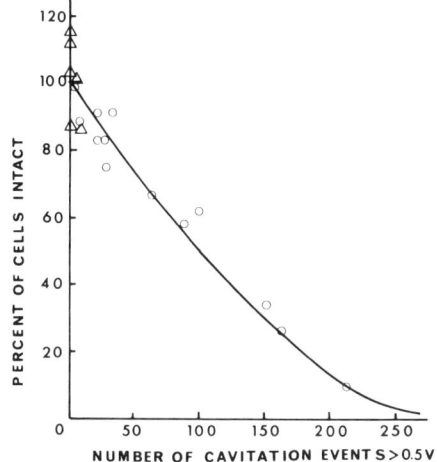

FIG. 3. Percentage of cells intact after irradiation at 515 W/cm² for 10 min versus number of events greater than 0.5 V, counted by eye in the second half-minute of each 2-min acoustic pulse. ○—Mycological peptone; △—mycological peptone and Methocel.

A series of experiments was performed at different sound intensities with five 2-min sound pulses for which cavitation activity was monitored with the complete detecting system. The subharmonic activity from the sample was monitored on the bowl hydrophone. Cavitation transients were counted over the full 10-min period, with the counter threshold set at 45 mV, and the total count was corrected for the counter dead time. The amplitude of the largest events recorded was noted from the oscilloscope monitoring the filter system. The relationship between cavitation event amplitude and duration was the same as that for irradiation in tap water.[7]

The mean subharmonic amplitude was averaged from the readings of the subharmonic at the end of each minute of the 10-min exposure period. Table III shows that, as the sound intensity decreased by a factor of 2, the number of cells surviving irradiation increased by a factor of 25. The number of transient cavitation events decreased by a factor of 4, but there was a much more marked decrease in the maximum amplitude of the transients recorded, by a factor of more than 10. The distribution of amplitude of the transient events at any one intensity was continuous, but was not measured. Subsequent measurement of the amplitude distribution of transients in tap water showed it to be very positively skew.

At 515 W/cm², 33% of the cells survive 150 cavitation events greater than 0.5 V (Fig. 3). Since these events were counted over a quarter of the 10-min irradiation period, about 600 events greater than 0.5 V would occur in 10 min. A total of 11 500 events greater than 45 mV were required to produce the same survival figure at 450 W/cm² (Table III), showing that the amplitude distribution of the transients is very positively skew. Since 89% of the cells survive 5620 events less than 0.4 V (Table III), it seems likely that the high-amplitude events are those mainly contributing to the damage. Correlation between cell damage and subharmonic activity previously reported[13] for low intensities (ca. 5 W/cm^2) is not evident in the results presented in Table III obtained at intensities of the order of 500 W/cm^2.

Since 33% of the cells survive 11 500 transients (Table III), then 67% of the cells must have been affected at least once by a cavitation event. The fact that some cells may have been affected more than once is ignored in the following calculation of an "effective volume" for a cavitation event. Since two-thirds of the cells were altered, about 2.2 ml of the suspension was affected by 11 500 events, giving a value of 2×10^{-4} ml for the "average effective volume" of a transient event.

An approximate value of the streaming velocities v_f in the cell suspension along the axis of the sound beam is given by[14]

$$v_f = \alpha \rho r^2 u^2 / 2\eta, \quad (1)$$

when

$$2\alpha L \leq 3, \quad (2)$$

where α is the linear acoustic pressure absorption coefficient, ρ is the density, η is the shear viscosity of the suspension, u is the fluid particle velocity, r is the radius of the sound beam, and L is its length. When applied to the focused field with r taken as the half-intensity radius in the center of the focal region, i.e., 0.75 mm in this case, then $v_f = 4$ cm/sec. This can only be an order-of-magnitude estimate, owing to the uncertainty in choosing an appropriate value for r and to the nonuniform intensity distribution of the sound field. It is more informative to use the results of Willard[15] in deriving the velocity of the cavitation event in the field. He irradiated at a maximum intensity of 1800 W/cm^2 in a focused field at 2.5 MHz where the first zero was 1.3 mm from the beam axis. His cavitation bursts moved at a speed of 10 m/sec at the focus, presumably at 1800 W/cm^2. Utilizing the dependence of streaming on intensity, absorption, and beam radius (Eq. 1) converts Willard's value of 10 m/sec to one of 1 m/sec for our conditions of intensity of 515 W/cm^2, beam axis to first zero distance of 2.0 cm, and ratio of absorption coefficients[14] of $1/(2.5)^2$.

Two cases are now considered, one in which the cells move in the suspension at the same speed as the cavitation event, and the second in which the cavitation event moves much faster than the liquid under second-order forces. For the first case, the average effective volume v_{ef} is $4\pi R_{ef}^3/3$, and since it has been shown above that $v_{ef} = 2 \times 10^{-4}$ ml, this leads to a value of R_{ef}, the average effective radius of 0.36 mm. For the second case, if the event travels along the focus at a speed of 1 m/sec, it would take 13 msec to travel the 13-mm length of the container. The time duration of transients varied from 0.6 to 3.0 msec, mostly about 1 msec. The effective volume would then be a cylinder of approximate volume $\pi R_{ef}^2 \times 1$ mm^3, leading to a value of 0.25 mm for R_{ef}. If the 600 events greater than 0.5 V of average duration 2 msec (Fig. 3) are producing most of the damage, the two estimates for R_{ef} become 0.9 and 0.7 mm. Thus, all of the above estimates of R_{ef} are approximately two orders of magnitude greater than the 3.3-μ radius of a resonant bubble at 1 MHz.[16]

Measurements from Willard's[15] photographs of cavitation events show that the radius of the microbubble region is about 0.9 mm. Examination of his movie films indicates that these bubbles were moving away from the beam axis and did not persist for longer than 120 μsec. It is not possible to say whether the bubbles were produced in a very narrow cylinder along the beam axis and were then shot rapidly away from the axial zone, or whether they were produced by shock waves in a wider cylinder and then moved slowly away from this region. Whether the microbubbles and the cavitation damage are both products of the shock waves from the unique cavitating center or whether these microbubbles are acoustically active during their short lifetime and are contributing to the biological damage, must remain a matter for conjecture at the moment.

Since 11 500 events of average duration 1 msec were required at 450 W/cm^2 to damage 67% of the cells, this corresponds to one event every 52 msec. Thus, for 98% of the time of irradiation not even one event was occurring in the container. This situation is in marked contrast to the commonly observed drastic breakdown of a liquid, with the production of visible bubbles and audible noise at 20 kHz, when using total acoustic powers similar to those employed in this work.

The fact that irradiations were performed in a container of volume 3.3 ml with a limited supply of cavitation nuclei has already been advanced to explain the decrease in cavitation rate as irradiation continued. This effect may also explain the small number, and in some cases the complete absence of visible bubbles in the container even though cavitation had occurred. Elsewhere[7] it has been reported that only one in 15 bubbles emitting a strong subharmonic signal in tap water subsequently grew to a visible size. The limited supply of gassy water in the container would further decrease the possibility of this occurring.

The cyclic form of cavitation claimed to indicate the breakdown of subharmonic emitting bubbles[7] was absent in the samples treated in this work. It was found that it required 1350 W/cm^2 to produce the cyclic behavior in a sample of tap water in the container compared with about 180 W/cm^2 in tap water in the open tank.

On decreasing the sound intensity from 270 W/cm^2 and increasing the magnification of the microscope to 400×, damage was observed down to 17 W/cm^2.

Because of the variance in the counts of cells surviving treatment, it was more useful, in this region of little damage, to count only damaged cells, i.e., cells whose outer membranes were largely intact, but which had lost a large amount of intracellular material. At least 50 of these cell "ghosts" were counted. The average number of ghosts per hemacytometer square was then expressed as a percentage of the number of intact and ghost cells per square. The variation in cell damage correlates better with transient cavitation activity than with sound intensity in the range 270–127 W/cm² (Table IV). The three differing results at 127 W/cm² are believed to be due to the fact that another mechanism, possibly related to the activity of subharmonic-emitting bubbles, may be making a slight contribution.

The results of irradiating cells in 1% mycological peptone compared with irradiations in a cell suspension which also contained 0.5% Methocel (Table V) again demonstrate the ability of Methocel to suppress transient cavitation and cell damage. Cells which were not as badly damaged as the "ghosts" (Table IV), but which had an over-all granular appearance, no noticeable loss of intracellular organelles, and vacuoles that were not as refractile as in normal cells, were included in the count of damaged cells here.

C. The Growth of Cells Irradiated at 515 W/cm² for 10 min at 1 MHz

Figure 4 shows that the optical estimate of cell survival is greater than the growth estimate (columns 3 and 4, Table I). If the optical method had been capable of detecting all cavitation damage, and if a direct effect of ultrasound on cell metabolism were present so that, say, 50% of the cells surviving obvious damage were incapable of further growth, then it would be expected that the growth curve would intersect

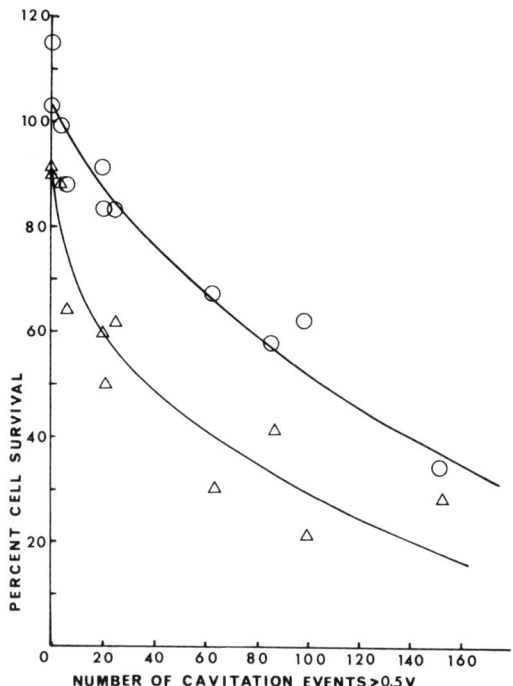

FIG. 4. Percentage of cells estimated to be intact by light microscopy after irradiation at 515 W/cm² for 10 min and the cell survival estimate from growth studies of these cells versus number of events greater than 0.5 V. ○—Microscope; △—growth.

the ordinate at 50% and would have a slope of one-half that of the optical-estimate curve at all other points. If, on the other hand, the optical examination was not capable of detecting all of the damage inflicted by cavitation, both estimates should have been, for no cavitation events, 100% ($N=0$), and the growth estimate of survival should have decreased more rapidly than the optical estimate as N increased. Figure 4 supports the latter rather than the former view. The two curves do not intersect at (0,100), but since the slope of growth estimate curve is large as N approaches zero, there is uncertainty as to the point of

TABLE IV. Percentage of "cell ghosts" produced in the intensity range 32–270 W/cm² as a function of mean subharmonic activity, number of transient cavitation events counted electronically, and the amplitude of the largest event recorded during the 10-min irradiation.

I (W/cm²)	Mean subharmonic amplitude (mV)	Number of transients	Maximum amplitude (V)	Percent "ghost cells"
270	4.7	7532	1.0	9.4
200	4.9	4160	0.8	4.5
185	4.9	10 000	0.4	7.8
127	3.8	4700	0.25	3.5
127	4.2	1687	0.25	2.8
127	3.4	868	0.25	1.6
72	2.5	1650	0.15	1.9
72	2.1	570	0.10	1.6
32	1.7	776ᵃ	0.04	0.7

ᵃ Counted by eye.

TABLE V. Percentage of cells damaged in mycological peptone and in mycological peptone +0.5% Methocel.

Mediumᵃ	I (W/cm²)	Subharmonic amplitude (mV)	Number of transients	Maximum amplitude (mV)	Percent of cells damaged
m. pep.	200	3.5	6920	2500	20
m. pep. & Meth.	200	4.6	873	200	3
m. pep.	128	2.7	3542	320	7.4
m. pep. & Meth.	128	1.6	330	150	3
m. pep.	72	0.72	1906	200	10.8
m. pep. & Meth.	72	1.4	80	100	0

ᵃ m. pep.—Mycological peptone; m. pep. & Meth.—mycological peptone and Methocel.

intersection with the ordinate. Further data in the range of N less than 20 would improve the estimate of the position of the intercept. This information would place an upper limit on the amount of cell damage not explained in terms of cavitation events with an amplitude greater than 0.5 V. However, the effects of transients of lesser amplitude would also need to be considered since Table V shows significant damage to cells when the amplitude of transients was low. Column 7 of Table I shows that events with amplitudes between 0.1 and 0.5 V were observed when no event greater then 0.5 V was observed. To summarize, the failure of both curves to intercept at (0,100) in this work could not be taken as conclusive evidence for a noncavitational interaction between ultrasound and the cells in suspension.

III. DISCUSSION

It seems appropriate, in view of the increased use of ultrasound as a clinical diagnostic tool, to compare the results of this study with selected reports of the effects of ultrasound on tissues, though it must be recognized that significant differences exist between the situations where cells are irradiated in suspension and in the more highly organized structural arrangements. Cells in suspension are heated to a much lesser degree than are cells in tissue, owing to the higher absorption coefficient of tissue, and to the fact that water streaming past a container of cells in the irradiation tank acts as a heat transfer system. Cells in suspension are free to rotate and change their shapes to accommodate any ultrasonic stress to which they are subjected, while cells in tissue are more constrained by the established architecture. Finally, cells in suspension are free to stream out of the high-intensity region of the field, depending upon the sample container size and shape.

The following calculation is made in order to compare the results of this study with those of whole animal irradiations studies. The half-intensity beamwidth of the focused field employed herein was 1.5 mm. The first zeros were 2 mm normal to the beam axis, so the intensity in a cylinder of radius about 1 mm was equal to, or greater than, 100 W/cm^2 when the peak intensity was 515 W/cm^2. The radius of the cylindrical container was 9 mm, so that 1/81 of the volume of cells was being treated at an intensity greater than 100 W/cm^2 at any one time, if it is assumed that acoustic streaming in the container produced complete mixing. The fraction of 1/81 can be considered equivalent to pulsed radiation with a mark space of 1:80; as the total radiation time was 10 min, the cells were exposed on the average to sound for 600/81 or 7.4 sec. It has been estimated that the speed at which a transient event moves along the axis was 1 mm/msec, and if it is assumed that the cells move at approximately this same speed, a cell would traverse the 13-mm container length in 13 msec, if it were on the beam axis, and at slower speeds for distances away from the axis. It may then be considered that these cells have been irradiated with 13-msec sound pulses at an intensity greater than 100 W/cm^2 with a mark space ratio of 1:80 for 7.4 sec.

An early study, concerned with the roles of heating and cavitation in the production of irreversible effects on living systems, showed that repeated pulses of subthreshold intensities could be integrated to produce suprathreshold effects (with doses similar to that of the above calculation), but in so doing the total irradiation time increased, depending upon pulse width, mark:space, etc.[1] At the risk of appearing specious, it can be argued that the increased irradiation time increased the probability for the requisite number of cavitation events to occur, although these investigators also showed that increasing the hydrostatic pressure of the irradiated system, to values greater than the acoustic pressure amplitudes employed, merely made manifest the pressure coefficient for the observed process, viz., production of paralysis of the hind legs of frogs.[17] A more recent investigation, in which threshold doses to produce structural changes were studied by detailed histological examinations of the treated tissue, appeared to show that cavitation occurs beyond a critical high acoustic intensity.[18] These investigators, however, determined the presence of cavitation by the unique appearance of the lesions and did not monitor for discrete events.

Suspending cells in a gel, so that cavitation would be suppressed and the cells less free to move out of the field, produces conditions more comparable with irradiation in tissue. Mouse lymphoma cells have been irradiated in a gel for 5 min at 15 W/cm^2 without significant ill effects.[19] The temperature rise in this work was small, indicating that the technique may be used at greater sound intensities. Further such studies at higher intensities and frequencies may well be necessary to elucidate the physical mechanisms involved in the acoustic alteration in tissues.

ACKNOWLEDGMENTS

The authors are indebted to Dr. A. J. Griffiths and Professor D. E. Hughes, Department of Microbiology, University College, Cardiff, where this research was performed, for technical advice. W. T. C. thanks the Medical Research Council for support during part of this work. F. D. acknowledges gratefully the National Institutes of Health Special Research Fellowship enabling him to pursue these investigations while on sabbatical leave from the University of Illinois, Urbana.

[1] W. J. Fry, V. J. Wulff, D. Tucker, and F. J. Fry, J. Acoust. Soc. Amer. **22**, 867–876 (1950).
[2] F. Dunn, Amer. J. Phys. Med. **37**, 148–151 (1958).
[3] M. Dyson and J. B. Pond, Physiotherapy, 136–142 (April 1970).
[4] K. J. W. Taylor, J. Pathol. **102**, 41–47 (1970).
[5] I. E. El'Piner, *Ultrasound: Physical, Chemical and Biological Effects* (Consultants Bureau, New York, 1964).
[6] B. Bowers and E. D. Korn, J. Cell Biol. **39**, 95–111 (1969).
[7] W. T. Coakley, J. Acoust. Soc. Amer. **49**, 792–801 (1971).

[8] T. F. Hueter and R. H. Bolt, *Sonics* (Wiley, New York, 1955), p. 237.
[9] C. R. Hill, P. R. Clarke, M. R. Crowe, and J. W. Hammick, in *Ultrasonics for Industry Conference Papers* (Iliffe, London, 1969), pp. 26–30.
[10] J. C. Bryant, Biotechnol. Bioeng. **11**, 155–179 (1969).
[11] J. B. Champion, P. F. North, W. T. Coakley, and A. R. Williams, Biorheology **8**, 23–29 (1971).
[12] G. Iernetti, Acustica **24**, 191–196 (1971).
[13] P. R. Clarke and C. R. Hill, J. Acoust. Soc. Amer. **47**, 649–653 (1970).
[14] F. Dunn, P. D. Edmonds, and W. J. Fry, in *Biological Engineering*, H. P. Schwan, Ed. (McGraw-Hill, New York, 1969), p. 242.
[15] G. W. Willard, J. Acoust. Soc. Amer. **25**, 669–686 (1953).
[16] C. Devin, Jr., J. Acoust. Soc. Amer. **31**, 1654–1667 (1959).
[17] W. J. Fry, D. Tucker, F. J. Fry, and V. J. Wulff, J. Acoust. Soc. Amer. **23**, 364–368 (1951).
[18] F. J. Fry, G. Kossoff, R. C. Eggleton, and F. Dunn, J. Acoust. Soc. Amer. **48**, 1413–1417 (1970).
[19] P. R. Clarke and C. R. Hill, Exp. Cell Res. **58**, 443–444 (1969).

ERRATUM

Page 1548, second line of Table II legend should read: "... the second half of each minute of the five 2-min pulses...."

Part III
INTERACTION OF ULTRASOUND WITH BIOLOGICAL TISSUES AND ORGANS

As Part III demonstrates, interest in the interaction of ultrasound with biological tissues and organs has been dominated by determination of the role of thermal events in the production of irreversible structural changes. Although the controversy continues, the topical areas for investigation become more sharply defined with increased inquiry. The motivation for these pursuits has, of course, been the probable application to medical problems, and the choice of central nervous tissue as an often-employed tissue specimen has also been promoted by its relatively static acoustic and biological properties.

For an interesting example of the speed with which investigators embraced the new form of energy as an investigative tool, see Harvey (1929).

REFERENCE

Harvey, E. N. 1929. "The Effect of High Frequency Sound Waves on Heart Muscle and Other Irritable Tissues." *Amer. J. Physiol.*, **91**, 284–290.

Editors' Comments
on Papers 37 Through 53

37 FRY et al.
Physical Factors Involved in Ultrasonically Induced Changes in Living Systems: I. Identification of Non-Temperature Effects

38 FRY et al.
Physical Factors Involved in Ultrasonically Induced Changes in Living Systems: II. Amplitude Duration Relations and the Effect of Hydrostatic Pressure for Nerve Tissue

39 WULFF et al.
Effects of Ultrasonic Vibrations on Nerve Tissues

40 MAZOUÉ et al.
Nerve Excitation Due to High-Frequency Ultrasound

41 LEHMANN and BIEGLER
Changes of Potentials and Temperature Gradients in Membranes Caused by Ultrasound

42 FRY et al.
Ultrasonic Lesions in the Mammalian Central Nervous System

43 WELKOWITZ and FRY
Effects of High Intensity Sound on Electrical Conduction in Muscle

44 FRY and DUNN
Ultrasonic Irradiation of the Central Nervous System at High Sound Levels

45 DUNN
Physical Mechanisms of the Action of Intense Ultrasound on Tissue

46 HUETER et al.
Production of Lesions in the Central Nervous System with Focused Ultrasound: A Study of Dosage Factors

47 FRY et al.
Production of Reversible Changes in the Central Nervous System by Ultrasound

48 ROBINSON and LELE
An Analysis of Lesion Development in the Brain and in Plastics by High-Intensity Focused Ultrasound at Low-Megahertz Frequencies

49 HAWLEY and DUNN
UHF Acoustic Interaction with Biological Media

50 POND
The Role of Heat in the Production of Ultrasonic Focal Lesions

51 FRY et al.
Threshold Ultrasonic Dosages for Structural Changes in the Mammalian Brain

52 DYSON et al.
The Production of Blood Cell Stasis and Endothelial Damage in the Blood Vessels of Chick Embryos Treated with Ultrasound in a Stationary Wave Field

53 TAYLOR and POND
A Study of the Production of Haemorrhagic Injury and Paraplegia in Rat Spinal Cord by Pulsed Ultrasound of Low Megahertz Frequencies in the Context of the Safety for Clinical Usage

The research group established and guided by the late W. J. Fry was most influential in calling attention to and contributing meaningfully to the basic problems in the elucidation of the mechanisms responsible for producing changes in tissues and organs, and several of their writings are included here. Papers 37 and 38 are early attempts at quantitative biophysical studies. In Paper 37, the role of ultrasonically produced temperature changes is explored and compared with other means of introducing heat

Editors' Comments on Papers 37 Through 53

into systems. Temperature rise is not considered to be responsible for the observed effects.

In Paper 38, the Fry group shows that the observed functional effect in frogs is not due to cavitation, as the end point was observed under a hydrostatic pressure sufficient to prevent its occurrence. The existence of threshold phenomena associated with ultrasonic effects in biological systems is introduced here and arguments against time rate of change of temperature and against local "hot" regions are presented.

Paper 39 reports acoustically induced effects on nerve activity and shows that temperature increase is not responsible.

The role of heat in the production of ultrasonic effects involved many early investigators, and the basic biophysical study reported in Paper 40 further argues against its being primarily responsible for the observed effects on nerve preparation.

In Paper 41, Lehmann and Biegler recognize that important biophysical events may occur at, or otherwise involve, the ubiquitous biological membrane and strongly urge that these events be studied. In this early work both mechanical and thermal events were identified as possible explanations for the observed potential changes across the membrane. Mechanical events are attributed to a stirring effect, which depletes concentration gradients at the membrane surface. Thermal events affected electric charges on membrane proteins.

Papers 42 through 45 are four additional contributions by the Fry group: the first shows the kinds of localized effects that can be produced selectively in gray and white matter of cat brain when more sophisticated radiation techniques (focusing) are employed.

Paper 43 shows that under appropriate ultrasonic dosage conditions the propagated action potential of excised striated muscle can be permanently reduced or completely blocked in the absence of damaging temperature levels. Histological examination showed no evidence of cavitation having been present.

Papers 44 and 45 show that precision data can be obtained with a suitable biological preparation. The effects are obtained in the absence of damaging temperature levels, and temperature-increase data comparing continuous-wave and pulse regimes of exposure are presented. Arguments against cavitation processes being involved in producing functional changes in the embryonic preparation are also given.

The MIT group, under the leadership of T. F. Hueter, concludes in Paper 46, after a detailed study of dosage factors, that the mechanism(s) of cell destruction is a temperature-dependent mechanical effect originating at weak points in the tissue. In arriving at this conclusion, the acoustic intensity, exposure, frequency, pulse width, and duty cycle were treated in an attempt to delineate the influence of thermal and mechanical factors.

Paper 47 reports the first demonstration of reversible functional changes in an *in vivo* preparation. Here, the lateral geniculate nucleus of the cat brain was irradiated ultrasonically while cortical potentials, evoked by a light flash, were monitored.

Paper 48, a comprehensive treatment from the MIT group, describes detailed studies with the mammalian central nervous system, which lead them to conclude that thermal processes are a most important mechanism in developing lesions in the dosage regions considered. As part of these investigations, the authors found it necessary to determine the ultrasonic absorption properties, as a function of temperature, of their specimens and these are included in Paper 12.

In Paper 49, the search for the identification of the physical mechanisms responsible for effects on living systems is conducted at such high frequencies that a clear demonstration is provided that cavitation is not essential. The unimportance of thermal processes is also argued.

In Paper 50, Pond describes detailed studies in which temperature cycles are produced in mammalian brain by ultrasound and by electric currents; this allows determination of the role to be ascribed to thermal processes in the observed tissue alteration.

It is shown in Paper 51 that threshold dosages for irreversible structural changes produced by focused ultrasound have only a very small dependence upon frequency in the low-megahertz frequency range (see also Dunn et al., 1975). The idea that three dosage regions exist, dominated by three mechanisms and exhibiting three kinds of lesion characteristics, is also present. Agreement on the threshold dosage region for these lesions is found among three independent research groups and is shown in Paper 51.

Paper 52 presents a clear demonstration of what might be a pure mechanical effect, without contribution from thermal or cavitation processes. The authors make a serious attempt at elucidation of the various mechanisms that may contribute to the observed stasis phenomena and methods for its avoidance.

Effects produced by ultrasound, which are contrary to involvement of thermal processes, are reported in Paper 53. The authors define a quantity that describes the ability of the sound exposure to produce tissue damage, and they exhibit its frequency dependence. A synergistic effect involving ultrasound and hypoxia is also demonstrated.

REFERENCE

Dunn, F., J. E. Lohnes, and F. J. Fry. 1975. "Frequency Dependence of Threshold Ultrasonic Dosages for Irreversible Structural Changes in Mammalian Brain." *J. Acoust. Soc. Amer.*, **58**, 512–514.

37

Copyright © 1950 by the Acoustical Society of America

Reprinted from *J. Acoust. Soc. Amer.*, **22**(6), 867–876 (1950)

Physical Factors Involved in Ultrasonically Induced Changes in Living Systems: I. Identification of Non-Temperature Effects*

W. J. FRY, V. J. WULFF, D. TUCKER, AND F. J. FRY
University of Illinois, Urbana, Illinois
(Received May 3, 1950)

The results of the first step in a systematic investigation of the mechanism of the action of ultrasound on tissue are reported. The temperature changes resulting from absorption of acoustic energy were determined while irradiation was in progress.

Experimental evidence is presented which demonstrates the existence of non-temperature effects in various nerve tissue preparations.

INTRODUCTION

THIS is the first of a series of reports concerned with the physical factors involved in changes produced in the physiological characteristics of tissues by ultrasound. It appears desirable to consider briefly those changes which take place in a medium in which any ultrasound field exists and which appear to require consideration for understanding the biological effects produced by ultrasound. Accordingly, we will first discuss changes in physical variables which may be of general importance. Then, as the first step in a systematic program of investigation, we will discuss the results of a series of experiments which demonstrate that temperature change, caused by absorption of acoustic energy, cannot account for the observed changes which occur in certain nerve tissue preparations exposed to ultrasound.

Numerous reports have appeared in the literature concerned with the effects of ultrasound on biological material.[1,2] With respect to its effect on nerve tissue, Harvey[3] has reported some work on excised frog peripheral nerve; Pohlman, Richter, and Parow[4] and Parow[5] have presented results obtained on human subjects with peripheral nerve disorders; and Lynn and Putnam[6] have observed effects on the brains of mammals.

GENERAL DISCUSSION

The changes in physical variables which occur in a medium in which a sound field exists and which appear to require consideration to understand the possible effects manifested in tissues are: changes of temperature and pressure; forces resulting from radiation pressure and viscosity; and cavitation and its concomitants.

---

* This research was supported by Contract N6ori-71, Task XXI with the Physiology Branch of the ONR.
[1] E. E. Gregg, Jr., *Medical Physics* (Year Book Publishers, Inc., Chicago, 1944), pp. 91–96.
[2] E. N. Harvey, Biol. Bull. **59**, 306–325 (1930).
[3] E. N. Harvey, Am. J. Physiol. **91**, 284–290 (1929).
[4] R. R. Pohlman, R. Richter, and E. Parow, Deutsche Med. Wochenschr. **65**, 251–254 (1939).
[5] E. Parow, Zeits. f. ärztliche fortbildung. **39**, 362–366 (1942).
[6] J. S. Lynn and T. J. Putnam, Am. J. Pathol. **20**, 637–649 (1944).

A. Temperature Changes in the Absence of Cavitation

The changes in temperature which occur in a liquid medium propagating a periodic acoustic disturbance of constant amplitude in the absence of cavitation are of two types; a periodic temperature change and a monotonic temperature change caused by absorption of acoustic energy. However, it is quite readily shown that the amplitude of the periodic changes is small even for high intensity sound and would, therefore, be of minor significance in producing biological effects. For example, for a sound wave in water with a pressure amplitude of 10 atmos., the amplitude of the periodic temperature change is only of the order of 0.01°C.

However, the magnitude of the temperature changes which result from the absorption of acoustic energy are much greater than the periodic changes for the media with which we are concerned. They are great enough to produce changes in the functional characteristics of living systems and, therefore, must be considered in relation to the effects produced by ultrasound.

Consider a plane wave traveling in the positive direction of the x-axis. We can write the following expression for the intensity, I,

$$I = I_0 e^{-\alpha x}, \quad (1)$$

where α is the intensity absorption coefficient. The energy absorbed per unit volume per second at the position x_0 can be expressed as $\alpha I(x_0)$. This energy manifests itself as a change in temperature. Acoustic absorption coefficients of some of the biological materials of interest in this investigation are of such magnitude that time rates of temperature rise up to 50°C/sec. have been observed under radiation intensities of the order of 30 watts/cm². (The ultrasonic frequency was 0.98 Mc.) Under continuous irradiation, the temperature increases and approaches an equilibrium value which is determined by the processes of conduction and radiation if there is no mass movement. Both of these processes account for transportation of energy because of temperature gradients. At the time when irradiation is initiated, there is no temperature gradient as the result of sound absorption. It is, therefore, possible to evaluate the absorption coefficient of the material by experimental determination of the shape of the heating curve at the time of initiation of irradiation. Specifically, we can relate the slope of this curve at this instant of time to the acoustic absorption coefficient as follows:

$$(dT/dt)_0 = (\alpha/\rho C_p)I, \qquad (2)$$

where I is the sound intensity at the location under consideration, C_p is the heat capacity of the material at constant pressure, and ρ is the density. This method of determining the coefficient α is extremely useful when the dimensions of the material are small. A fine wire thermocouple can be used to indicate the temperature changes. The results of experiments carried out during this investigation are indicated in Fig. 1. The sound intensity at the thermocouple will depend on the reflection coefficient at the water interface and the thickness of the material through which the sound has traveled. Assuming a reflection coefficient of zero, a density of 1 g/cc, and a heat capacity of 1 cal./g/deg. C, a value for α of 0.2 is obtained for nerve and muscle. This value may be compared with absorption coefficients obtained by Hüter:[7] 0.35 for heart muscle; 0.57 for tongue muscle, sound propagation normal to fiber direction; 0.25 for tongue muscle, sound propagated in the direction of muscle fibers.

Various mechanisms are involved in the determination of acoustic absorption and are certainly of interest with respect to biological media. Their relative importance is dependent on the sound frequency and the structure of the medium.[8] For pure non-metallic liquids, it appears at present that acoustic absorption is caused principally by viscous damping. For many liquids, the dilatational viscosity is of much greater importance in this respect than the shear viscosity.[9,10] However, in solutions, absorption of sound energy may result from the fact that equilibrium is not maintained between the different chemical species as the pressure changes because of the presence of an acoustic disturbance. Water solutions of $MgSO_4$ show such anomalous behavior.[11]

In addition to these two mechanisms which are certainly of importance in understanding sound attenuation in biological material, there are other possibilities which arise from the fact that such material is not a homogeneous liquid. Hüter[7] has shown for various bovine tissues that the absorption coefficients as a function of frequency are approximately linear relations over the range 1.5 to 4.5 Mc. The mechanisms discussed above do not account for such a dependence. It has been suggested that plastic flow and viscous slip at grain boundaries (elastic hysteresis) might account for such a relation in metals.[8,12] Consideration of similar

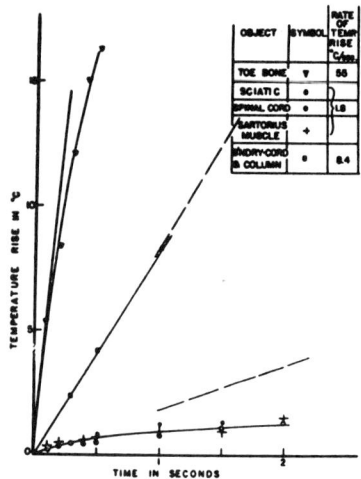

FIG. 1. Temperature rise as a function of time in various tissues under ultrasonic irradiation at a frequency of 980 kc and an intensity ~ 35 w/cm².

mechanisms in biological systems would appear to be of interest.

B. Pressure Changes in the Absence of Cavitation

A second variable of interest in any consideration of the mechanisms of the effect of ultrasound on biological material is the pressure. For a traveling wave, the sound pressure amplitude, P, is related to the intensity, I, by

$$P^2 = 2\rho VI, \qquad (3)$$

where V is the velocity of sound in the medium and ρ is its density.[13] At the maximum intensity used in the experiments reported upon later in this paper (~ 35 w/cm²), the corresponding sound pressure amplitude from (3) is about 10.0 atmos. A static hydrostatic pressure of 10 atmos. has a negligible effect on the electrical activity of nerve axons.[14] A similar situation exists with respect to the effect on gel sol transformations.[15,16] Stresses of this magnitude varying at the rate of $(10)^6$ cycles per second may, of course, act somewhat differently. Definite evidence has been obtained by Freundlich and Gillings and others which shows that in the absence of cavitation and after elimination of temperature change as a possibility, one still obtains major changes in the structural viscosity of some colloidal solutions.[17]

---

[7] Th. Hüter, Naturwiss. 35, 285–287 (1948).
[8] C. Kittel, Reports on progress in physics 9, 205–247 (1948).
[9] G. Eckart, Phys. Rev. 73, 68–76 (1948).
[10] L. N. Liebermann, Phys. Rev. 75, 1415–1422 (1949).
[11] L. N. Liebermann, Phys. Rev. 76, 1520–1524 (1949).
[12] W. P. Mason and H. J. McSkimin, J. Acous. Soc. Am. 19, 464–473 (1947).
[13] P. M. Morse, *Vibration and Sound* (McGraw-Hill Book Company, Inc., New York, 1948).
[14] H. Grundfest, Cold Spring Harbor Symposia on Quantitative Biology 4, 179–187 (1936).
[15] D. A. Marsland and D. E. S. Brown, J. Cell. Comp. Physiol. 8, 167–178 (1936).
[16] D. A. Marsland and D. E. S. Brown, Anat. Rec. 75, 141 (1939).
[17] H. Freundlich and D. W. Gillings, Trans. Faraday Soc. 34, 649–660 (1938).

In addition to the forces resulting from periodic pressure changes, there are two other forces which require consideration from the viewpoint of possible biological effects. The first of these is the so-called radiation pressure which manifests itself as a unidirectional force at an acoustically reflecting interface. This force causes migration or movement of reflecting particles in a sound field. In a traveling wave field, the force on a rigid sphere, large compared to the wave-length of the sound, is given by the expression[18]

$$F = (I/V)\pi r^2 Y. \quad (4)$$

The quantity Y is close to unity if $(2\pi r)/\lambda > 10$. For a sound intensity ~ 35 w/cm², we obtain a force of $\sim 2\pi r^2$ g (r is the radius expressed in cm). For rigid spheres, small compared to the wave-length, King[19] has shown that in a traveling wave field, the forces are of a much smaller order of magnitude than in a standing wave field.

In order to determine the order of magnitude of the rate of migration to be expected in a standing wave field, we consider rigid spherical particles of small enough size that the inertia forces can be neglected compared to the viscous forces. The condition which must be satisfied in order to insure this is $vr \ll \eta/\rho$, where v is the velocity and r is the radius of the particle, η is the coefficient of viscosity and ρ is the density of the fluid medium.[20] The viscous force is given by the expression $f = 6\pi \eta r v$. The formulas of King show then that the velocity of such particles in a standing wave field is independent of their size. For such a field of stored energy per unit volume equal to that for a traveling wave of intensity ~ 35 w/cm² and for a frequency of 1 mc, one obtains for the maximum velocity of migration

$$v = [3(10)^3/\eta] \text{ cm/sec,}$$

where the velocity v is subject to the above restriction. For a viscosity coefficient of the order of 1000 poises, one obtains a velocity of the order of 3 cm/sec.

In the experiments reported upon later in this paper, the standing wave ratio was of the order of five percent. We can thus divide the total sound field into two parts, a pure traveling wave and a standing wave of amplitude about 1/20 the traveling wave amplitude. The traveling wave component does not contribute appreciably to migration of small particles through the mechanism of radiation pressure. The standing wave component would, however, yield a velocity of migration at the maximum sound intensity (~ 35 w/cm²) of the order of 1/400 the value given by the above formula.

The second system of forces is a result of the fact that the medium has viscosity. Eckart[9] has shown that the viscous forces account for the flow of a homogeneous liquid in a sound field. He has also shown that the velocity of streaming involves only the ratio of the bulk and shear viscosity coefficients.

These two systems of forces may act to produce movement and reorientation of cell contents which could cause changes in biologic functions.

D. Physical Factors Associated with Cavitation

Cavitation is present in many experiments involving high intensity sound in liquid media. It is, therefore, necessary to consider the possible role of cavitation and its concomitants in discussions of mechanisms of the effects of ultrasonic irradiation on biological materials. In the experiments to be described later, no cavitation was present in the medium surrounding the test object.

A detailed understanding of the effect produced by cavitation involves a study of conditions promoting formation, growth, and collapse of cavities in liquid media. In this discussion we will review only briefly some of the features of this phenomenon which have been studied by other investigators and which appear to be of importance in considerations of the biological effects produced by intense ultrasound. We are interested both in cavities which contain only relatively small amounts of gas (the dynamics of which are not dependent appreciably on the diffusion of gas through the liquid) and in those where the diffusion of gas into the cavity from the surrounding liquid is an important factor in the growth process.[21]

Harvey and co-workers[22,23] have demonstrated that cavities result from growth of gas nuclei, and in the absence of such gas nuclei, cavity or bubble formation cannot occur even when the water is under considerable tension. The formation of such bubbles is accompanied by pressure and temperature changes. Knapp and Hollander[24] studied the history of individual bubbles and from their data on the velocity of bubble collapse, maximum pressures of several thousand atmos. may be calculated. The pressures produced by cavitation in an ultrasonic field at a frequency of 1.0 Mc are probably much lower than this because of the relatively short time between pressure reversals. The bubbles studied by Knapp and Hollander spent 0.001 sec. in the low pressure field. In a sound field of 1 Mc, the bubble would grow for only 1/1000 of this period before compression took place. It has been suggested that bubbles grow in a step-like manner and Harvey has suggested a possible mechanism for such growth. The pressure variations accompanying this type of bubble

---

[18] F. E. Fox, J. Acous. Soc. Am. 12, 147–149 (1940).
[19] L. V. King, Proc. Roy. Soc. London 147A, 212–240 (1934).
[20] H. Lamb, *Hydrodynamics* (Cambridge University Press, London, 1932).
[21] For a comprehensive discussion of various aspects of cavitation, see F. G. Blake, Tech. Memo. No. 12 (1949), Acous. Res. Lab., Harvard University.
[22] Harvey, Barnes, McElroy, Whiteley, Pease, and Cooper, J. Cell. Comp. Physiol. 24, 1–22 (1944).
[23] Harvey, Whiteley, McElroy, Pease, and Barnes, J. Cell. Comp. Physiol. 24, 23–34 (1944).
[24] R. T. Knapp and A. Hollander, Trans. A.S.M.E. 70, 419–431 (1948).

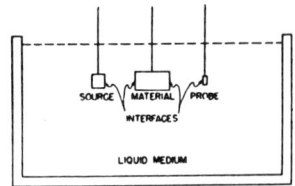

FIG. 2. Experimental arrangement for detecting the presence of cavitation.

formation are certainly smaller than the above except for the possibility discussed by Smith and others.[25,26]

Temperatures at or near the surface of collapsing bubbles were obtained indirectly by Marinesco[27] who immersed powdered explosives in liquids which do not wet them and subjected the suspension to ultrasound. The known detonation temperatures of the explosives enabled him to obtain an approximate idea of the temperature. In a sound field of intensity 20 w/cm^2 and at a frequency of 1 Mc, he obtained values up to 230°C.

Much research with ultrasound has been concerned with differentiating between effects which appear to require cavitation and those that do not. For example, Freundlich and co-workers[17,28] and others[29] have demonstrated that (1) emulsification of non-metallic systems by ultrasound (200 kc, intensity unknown) requires cavitation; (2) liquefaction of thixotropic gels by ultrasound requires cavitation (not caused by temperature rise since the gels studied do not soften on heating); (3) reduction of the structural viscosity by ultrasonic radiation requires cavitation for some materials. For other materials, the changes may be greater when cavitation is present but it is not required.

It has been shown that the disruptive effects of ultrasound on single cellular organisms are in many cases associated with cavitation[30] and it has also been reported that disruptive effects occur in the absence of cavitation.[2] The acceleration of chemical reactions in an ultrasound field has also been related to cavitation.[26] The splitting of the macromolecule haemocyanin has been reported.[31] The irreversible reduction of the molecular weight of macromolecules by ultrasound can take place in the absence of cavitation.[32]

The arrangement indicated in Fig. 2 is a convenient method for detection of cavitation either in the interior of media or at interfaces between them. It consists of a sound source and a pick-up probe and means for suspending the test object between them. In the absence of cavitation, the voltage measured across the pick-up probe is a linear function of the driving voltage across the source. However, if cavitation occurs, the probe voltage readings will fall below the values indicated by this linear relation and fluctuating values will result. The minimum intensity for cavitation is the point at

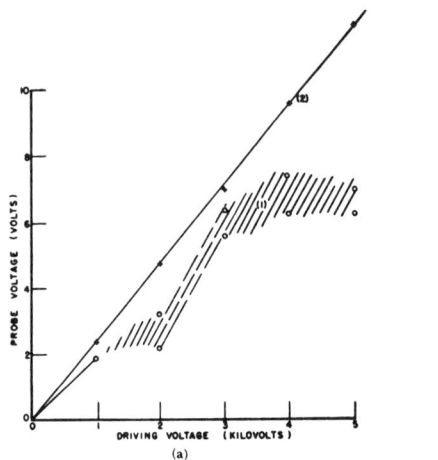
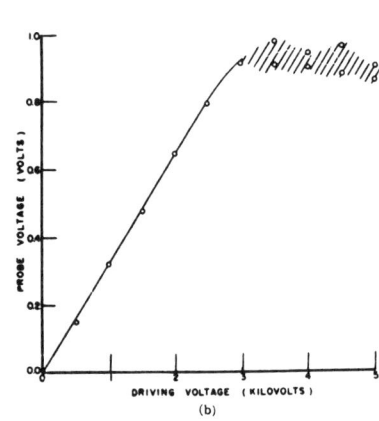

FIG. 3. (a) Sound probe voltage as a function of driving crystal voltage for degassed distilled water and for tap water.
(b) Sound transmission through a frog immersed in degassed water.

[25] F. D. Smith, Phil. Mag. **19**, 1147–1151 (1935).
[26] W. T. Richards, Rev. Mod. Phys. **11**, 36 (1939).
[27] N. Marinesco, Comptes Rendus **201**, 1187–1189 (1935).
[28] H. Freundlich and K. Sollner, Trans. Faraday Soc. **32**, 966–970 (1936).
[29] C. Bondy and D. Sollner, Trans. Faraday Soc. **31**, 835–843 (1935).
[30] F. O. Schmitt and B. Uhlemeyer, Proc. Soc. Exp. Biol. Med. **27**, 626 (1930).
[31] S. Brohult, Nature **140**, 805 (1937).
[32] See for example, H. Mark, J. Acous. Soc. Am. **16**, 183–187 (1945).

ULTRASONICS AND LIVING SYSTEMS

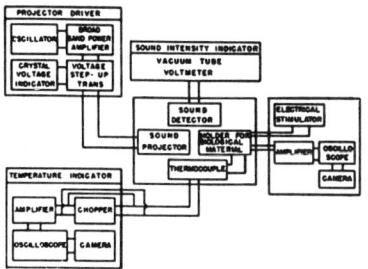

FIG. 4. Block diagram of the experimental setup.

which deviation from a straight line occurs. Curve 1 of Fig. 3a indicates that cavitation occurred in tap water at a crystal driving voltage of the order of 1.0 kv. (X-cut quartz crystal operating in thickness mode at 1.0 Mc.) Curve 2, of Fig. 3a is obtained when degassed water (boiled) is the coupling medium between the source and the probe. There is no indication of cavitation below 5.0 kv. The insertion of a frog between generating and pick-up crystal (all in previously boiled distilled water) yielded the curve of Fig. 3b, indicating that cavitation has occurred above 3 kv.

II. EFFECTS OF ULTRASOUND ON NERVE TISSUE (NON-TEMPERATURE EFFECTS)

As the first step in a systematic program to determine the mechanism of the effects of ultrasound on nerve tissue, it was felt desirable to evaluate the temperature factor associated with the ultrasonic propagation. The following experimental results are presented as evidence to show that there exist reversible and irreversible effects produced in nerve tissue by ultrasound which are not the result of temperature changes.

A. Materials and Methods

1. Ultrasonic Generation and Detection

A schematic diagram of the instrumentation is given in Fig. 4. A detailed description of all of the apparatus indicated in this diagram will not be given in this paper.[33]

The ultrasonic frequency used in the following experiments was between 975 and 980 kc. The source was an X-cut quartz crystal, one-inch diameter, vibrating in thickness mode. Because quantitative results with respect to sound intensity were desired, a voltage measuring arrangement was incorporated into the electrical generator which drives the crystal, so that the voltage across the crystal was known and could be set at any desired level. The intensity of the acoustic radiation is proportional to the square of this measured voltage.

The measurement of bio-electric potentials and other

[33] Tech. Report I—ONR Contract N6ori-71, Task XXI, University of Illinois (1949).

relatively small voltages in the immediate neighborhood of the sound generator made complete electrical shielding necessary. The face of the generator crystal was at ground potential. To achieve consistent and quantitative measurements, it was required that there be no continuously changing geometry in the sound field, such as a variable liquid surface. This was accomplished by arranging the crystal holder and sound tank so that propagation took place in a horizontal direction. For the work presented here, it was desirable to utilize a traveling wave field. This insures uniformity of irradiation of small samples by eliminating critical positioning in a standing wave field.

An important feature of the sound generator used in these studies is the possibility of either continuous or pulsed operation. In addition to continuous variation in amplitude, one has available a range of pulse durations from 0.5 msec. to continuous operation, and a repetition rate variable from 500 cycles per sec. to as low a frequency as desired. In addition, it is possible to obtain delay or precession times from 100 μsec. to 1.0 sec. relative to a pulse generator which is used to apply electrical stimuli to nerve tissues.

The crystal holder is illustrated in Fig. 5, upper. It will be described in detail elsewhere.[34]

The crystal probe is illustrated in Fig. 5, lower. It consists of a small insulated piezoelectric crystal mounted on the end of a hypodermic needle. In practice, it is supported and moved about by a coordinate system which permits accurate measurement of field distribution. Horizontal and vertical field distributions are given in Fig. 6 for a plane 2⅝ in. from the crystal face. This corresponds to the approximate position of materials under observation.

In these studies, the relative sound intensity was determined by means of the voltage generated by the crystal probe inserted in the sound field and by the

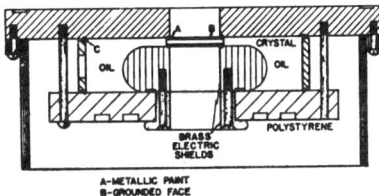

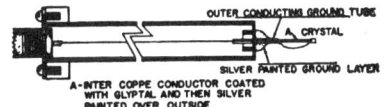

FIG. 5. (Upper) Detailed diagram of the sound projector. (Lower) The acoustic probe.

[34] F. J. Fry, Rev. Sci. Inst. 21, 940 (1950).

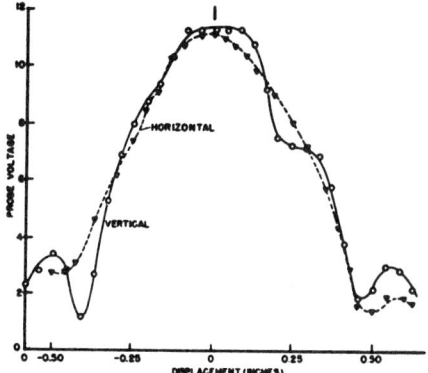

FIG. 6. Horizontal and vertical beam pattern distributions.

driving crystal voltage. The absolute sound intensity was determined by the method described by Fox and Griffing.[25] The measured free field sound intensity at the positions used in the experiments at maximum crystal voltage (5500) is between 30 and 40 w/cm^2. The uncertainty is due partially to the variation in the position of the test object in the sound tank. A calculated value of the intensity based on the assumption of a pure thickness mode yields a value of 10 w/cm^2. In this calculation, account was taken of the energy loss in the crystal holder. The high values of the measured sound intensity may be attributed to focusing at a distance of about two inches in front of the crystal. Measured intensity values as a function of distance from the crystal face show a maximum in this region.

2. Types of Preparations

(a). Fresh preparations of sciatic nerves from Rana pipiens were employed in these studies. The excised nerves were bathed continuously in cold-blooded vertebrate Ringer's solution at room temperature, 21° to 25°C.

(b). Walking leg nerves of the crayfish were prepared after the method of Welsh and Gordon.[26] This nerve contains two large motorneurons, one to the flexor of the dactyl and the second to the extensor. After excision, these nerves were continuously bathed in van Harreveld's solution at room temperature, 21° to 25°C.

(c). The ventral abdominal nerve cord of the crayfish was dissected out in its entirety and immersed in van Harreveld's solution. The sixth abdominal ganglion was always removed and either part or all of the remaining cord was used.

(d). The electrode chamber consisted of a stainless steel shell containing four glass tubes filled with

---

[25] F. E. Fox and V. Griffing, J. Acous. Soc. Am. 21, 352–359 (1949).
[26] J. H. Welsh and H. T. Gordon, J. Cell. Comp. Physiol. 30, 147–171 (1947).

Ringer's solution. These salt bridges made contact with small calomel half-cells or with coils of silver-silver chloride wire. The salt bridges terminated in small jets against which the nerve preparation was fixed. This chamber was always used in the vertical position. The nerve was immersed in physiological salt solution. It was found undesirable to use plastic nerve holders because of the large temperature change induced in the plastic by ultrasound.

(e). Whole frogs were placed in the path of the sound beam by mounting the animal (dorsal surface down) as firmly as possible on a piece of plywood cut to fit the sound tank. A 2.0-cm hole, centered in front of the crystal, served to admit the sound. The frogs were oriented on the board so that the region of the lumbar enlargement was approximately centered in the aperture. Intact frogs were employed for the most part. The sound tank was filled with distilled water which had been boiled for 10 minutes to drive off most of the gas.

(f). Temperature changes in the spinal cords of intact animals, in excised spinal columns containing the cord, in excised sciatic nerves of the frog, and in the ganglia of excised crayfish ventral abdominal nerve cord were measured with constantan-copper thermocouples. For the sciatic nerve and crayfish ventral nerve cords, a soldered junction made of 0.013-mm copper and 0.038-mm constantan was used. For the spinal cord of the frog, a soldered junction of 0.25-mm copper and constantan was employed. The sciatic nerve and the ventral nerve cord of the crayfish were threaded over the junction and the ultrasound was incident on the preparation in the region of the junction. The thermocouple in the spinal cord was introduced laterally through the foramina of the column through which the peripheral nerves pass. The foramina selected were in the region exposed to the ultrasound. The frogs used in making the temperature measurements were not utilized for any other purpose.

The thermocouple output was interrupted by a mechanical chopper, amplified, and recorded by photographing the trace of a cathode ray beam. The chopping frequency was 10 cycles per sec. A sudden temperature change, produced by thrusting the junction into a cold water bath, produced a deflection reaching 0.7 of maximum in 0.030 sec. This measurement was made without the chopper.

RESULTS

A. Effect of Ultrasound on the Ventral Nerve Cord

Spontaneous activity recorded from the commissure between two adjacent ganglia (1 and 2) of the crayfish ventral abdominal nerve cord, immersed in physiological salt solution is illustrated in Record 1, Fig. 7. Superimposed on a low level background activity is a series of periodically occurring spike potentials at an average frequency of 7.8 per sec. Exposure of the two ganglia and the commissure to ultrasound ($\sim$35 w/cm^2) produces a characteristic sequence of changes in the activity, a typical example of which is illustrated in Records 2, 3, and 4 of Fig. 7. Note that the frequency of the spike potentials at first increases (Record 2, center), then decreases (Record 2, left), and is followed by total disappearance of the large spike potentials after 43.5-sec. exposure to the sound. Twenty-five seconds after the ultrasound was turned off, the large spike potentials reappeared, first slowly, then more rapidly (Record 5, Fig. 7), finally reaching a stable frequency of 6.7 per sec. (Record 6, Fig. 7). Subsequent treatment of the same preparation with ultrasound produced a similar sequence of events. Similar observations were made on six other preparations.

Measurements of temperature changes in the ganglion of a preparation similar to the above, under identical conditions, indicated a maximal rise of 1°C.

B. Effect of Ultrasound on Peripheral Nerves

Prolonged application of ultrasound to excised frog sciatic nerves suspended in Ringer's solution and in contact with salt bridges, produced no detectable changes in wave form, magnitude of the spike potential, and excitability. The nerves were exposed to ultrasound in the region of the stimulating electrodes, in the region of the recording electrodes, and in the intermediate region. Similar experiments on the excised leg nerves of crayfish gave negative results. Temperature changes in excised frog sciatic nerves recorded from the region of the incident ultrasound indicated a rise of 2°C.

C. Physiological and Morphological Changes Produced by the Action of Ultrasound on the Frog Spinal Cord

Twelve intact normal frogs, suspended under water at room temperature, 21° to 25°C, and placed so that

FIG. 7. The effect of ultrasound on the frequency of discharge of spontaneously occurring spikes in the excised crayfish ventral nerve cord.

the sound beam was incident on the center of the back over the lumbar enlargement, showed complete paralysis of the hind legs with exposures of 4.3-sec. duration. Shorter exposures to sound either produced no paralysis or a temporary partial paralysis which disappeared after a variable time interval.

Experiments similar to the above were performed with frogs cooled and maintained at 1° to 2°C. Ultrasound incident on the back over the lumbar enlargement of these cooled frogs produced paralysis of the hind limbs after exposures of 7.3 sec. The paralysis was permanent in all of 50 frogs so treated. Exposures of shorter duration produced no paralysis or a temporary paralysis which disappeared.

Examination of histological preparations of sciatic nerves obtained one week after complete paralysis, fixed and stained with osmic acid vapor, revealed considerable degeneration of axons. A typical example is illustrated in Fig. 8, No. 2 (compare with control nerve, Fig. 8, No. 1). In all preparations examined, considerable degeneration of axons was evident.

Gross examination of the spinal cord of frogs immediately after exposure to a paralyzing dose of ultrasound revealed a loss of surface configuration and a definite change in appearance of the white matter in six frogs examined. Examination of the cord one and two weeks after ultrasound treatment indicated marked degeneration of the cord tissues in the treated region. In some cases, this degeneration was indicated by a constriction in the cord, the superficial white matter appearing intact, and in others, the cord was completely divided in two.

Histological examination of the spinal cords of ultrasound treated frogs revealed marked abnormality of the large motor neurons of the ventral horn of the gray matter. These abnormalities are evident in spinal cords dissected out and fixed 20 minutes after treatment and stained with thionine ($C_{12}H_9N_3S$) (Fig. 8, No. 4). Note the ragged cell outlines, the very intense stain, and enlarged nuclei (compare with control, Fig. 8, No. 3).

D. Temperature Changes Produced by Ultrasound

1. Spinal Cord of the Frog

Changes in temperature of the spinal cord of intact frogs were measured during and following the period of ultrasound treatment. In experiments at room temperature, a rapid rise occurs with onset of ultrasound, usually reaching a maximum between 40° and 50°C at the end of the exposure, and, thereafter, exhibiting a typical decline in temperature (Fig. 9, Graph 1). Temperature measurements on frogs at 1° to 2°C indicate again a sharp increase with onset of ultrasound (Fig. 9, Graph 2), which usually reached a level between 25° to 30°C at the end of the 7.3-sec. exposure. After the

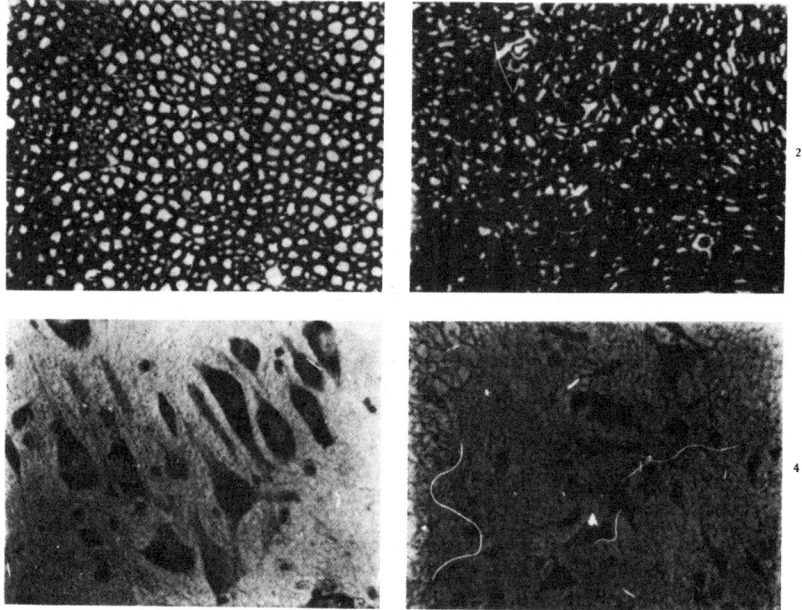

FIG. 8. Photomicrographs of sections through sciatic nerves (1 and 2) and spinal cords of frogs (3 and 4). Magnification 350×. (1) and (3) are controls. (2) and (4) illustrate the results of a damaging exposure to ultrasound.

exposure, the spinal cord exhibits a decrease in temperature. Experiments on isolated spinal column and cord preparations gave similar results (Fig. 9, Graph 3).

To determine the influence of temperature on the production of paralysis of the hind legs of frogs, twelve experiments were performed using brief repetitive pulses of ultrasound. Frogs, cooled to 1° to 2°C, were exposed to ultrasound for 4.3 sec. This is a sub-paralytic dose. The temperature change (max. temp. 15°C) produced by this exposure is indicated in Fig. 9, Graph 4. This exposure was followed by a four-minute interval to permit the cord temperature to return to the previous level. Then the frog was subjected to a second 4.3-sec. dose of ultrasound, which produced a temperature change similar to the first (Fig. 9, Graph 4). Frogs subjected to a similar procedure (without insertion of the thermocouple) exhibited permanent paralysis of the hind legs after the second exposure.

In six experiments in which a 3.3-sec. exposure followed by a four-minute cooling-off period was used, paralysis was produced after the sixth exposure. A third experiment, in which an exposure of 2.8 sec. and a four-minute interval was used, produced paralysis after 50 exposures. Another set of experiments utilizing sound pulses of shorter duration were performed with frogs at room temperature, 22° to 24°C. Sound pulses of 0.08-sec. duration delivered at a rate of two per sec. produced a rapid rise in temperature, reaching equilibrium at about 36°C after four minutes of treatment. Five frogs so treated were permanently paralyzed after three minutes. Sound pulses 0.01-sec. duration delivered at a rate of 20 per sec. produced an equilibrium temperature of 40°C after four minutes. Five frogs treated in this manner were not paralyzed even after ten minutes.

The effect of temperature on the spinal cord was assessed in the absence of ultrasound. The posterior half of frogs immersed in water baths at 35°C for twenty minutes and then raised to 38°C for twenty minutes showed no obvious abnormality in behavior. Frogs immersed in 40°C water for various periods up to six minutes showed a paralysis which gradually disappeared. Temperature measurements in the lumbar region of the spinal cord of frogs immersed in 40°C water indicated a level of 40°C after six minutes.

2. Bone and Muscle

The initial rate of rise of temperature of the frog spinal cord appeared to be constant for different preparations exposed to equal sound intensities, regardless of the starting temperature level. This suggested a similar series of measurements for frog bone and muscle. The rate of change of temperature as a function of time after beginning of exposure to ultrasound of $\sim$35 w/cm^2 is shown in Fig. 1. Note that the curves fall off with time, indicating flow of heat from the tissue into the environment. An approximate value for the initial

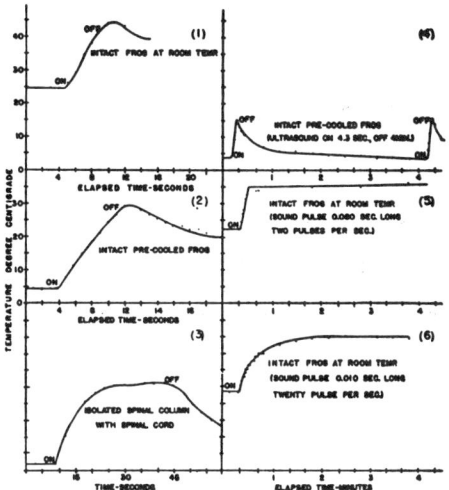

FIG. 9. Temperature changes in frog spinal cord during and after ultrasonic irradiation as a function of time for various experimental conditions.

rate of change of temperature of nerve and muscle is 1.8°C/sec., while for bone it is 55°C/sec.

DISCUSSION

A. Biological Effects Produced by Ultrasound

The temperature changes produced by absorption of acoustic energy may be of a magnitude sufficient to produce changes in the functional characteristics of the living systems studied. Ultrasound incident on ganglia of the crayfish ventral nerve cord containing spontaneously active neurones caused a reversible depression of this spontaneous activity. A maximal and rapid temperature increase of 1°C was measured. Prosser[27] has shown that increasing the temperature 1°C between 26° to 30°C may produce an increase in the frequency of discharge of single units of about 4 to 5 per sec. The effect of ultrasound in depressing the frequency of discharge is in a direction opposite to the effect of the temperature change. It is concluded, therefore, that the effect of ultrasound on these spontaneously active neurones is mediated by physical factors other than the simultaneously occurring but slight temperature change.

The marked increase in temperature of the spinal cords of frogs exposed to ultrasound suggests, at first glance, that the paralysis may be caused by heat. Further examination of the data, however, indicates that paralysis can occur in the absence of high (35° to 40°C) temperature levels. Observations on frogs cooled to 1°C and subjected to pulses of sound separated by four-minute intervals demonstrate summation when the maximum temperature did not rise above 15°C.

[27] C. L. Prosser, J. Gen. Physiol. **19**, 65–73 (1935).

SUMMARY

A general discussion of the changes in physical variables which accompany a high intensity ultrasonic disturbance in liquid media and which appear important in understanding the effects of such disturbances on tissue is presented. As the first step in a systematic investigation, the role of the temperature changes produced in an ultrasound field is investigated experimentally.

The results indicated that ultrasound ($\sim$35 w/cm^2, frequency 1 mc) was without effect on excitability, wave form of the spike potential, or propagation velocity of excised peripheral nerve, even after prolonged exposures. The excised crayfish ventral nerve cord exposed to ultrasound exhibited a reduction of spontaneous activity after several seconds exposure and recovered its original activity about one minute after the ultrasound was turned off. Frogs positioned so that ultrasound was incident, on the dorsal surface over the lumbar enlargement of the spinal cord exhibited paralysis of the hind legs after 4.3-sec. exposure (at room temperature) and exhibited paralysis after 7.3-sec. exposure (at 1° to 2°C). Histological examination of the sciatic nerves showed extensive degeneration of nerves and examination of the spinal cord showed marked pathology of the lower motor neurones.

Temperature measurements indicated that peripheral nerve and crayfish ventral nerve cord exhibited a maximal rise of 1° to 2°C. The spinal cord of intact frogs exhibited temperature increases of the order of 25°C. By using frogs cooled to 1°C and reducing the ultrasound exposure to two 4.3-sec. pulses interrupted by four-minute cooling-off periods, it was demonstrated that temperature rises did not exceed 15°C and that paralysis of the hind legs occurred during the second 4.3-sec. exposure. Similar experiments on frogs (room temperature) indicated paralysis upon exposure to ultrasound pulses of 0.080 sec. delivered at a rate of 2.0 per sec. and no paralysis upon exposure to sound pulses of 0.010 sec. delivered at a rate of 20 per sec., yet, the latter procedure produced a higher cord temperature than the former.

It was concluded that the effect of ultrasound on the system studied is produced by physical factors other than temperature. Of these factors, cavitation is the one most easily controlled and will be investigated in the future.

A method is presented for obtaining acoustic absorption coefficients by measuring the initial rate of change of temperature in various test objects.

ACKNOWLEDGMENT

We would like to thank Dr. Warren McCulloch and Dr. C. L. Prosser for their interest in this research.

Physical Factors Involved in Ultrasonically Induced Changes in Living Systems: II. Amplitude Duration Relations and the Effect of Hydrostatic Pressure for Nerve Tissue*

W. J. FRY, D. TUCKER, F. J. FRY, AND V. J. WULFF
University of Illinois, Urbana, Illinois
(Received January 20, 1950)

The results of experiments with frogs under a hydrostatic pressure demonstrate that cavitation is not an important factor in the mechanism of production of paralysis of the hind legs of frog by ultrasonic (frequency one megacycle) irradiation over the lumbar enlargement region of the spinal cord. Experimental results indicate that a linear relation exists between the reciprocal of the minimum exposure time for paralysis and the acoustic amplitude. This result is readily described in terms of a one factor rate process. On the basis of this experimentally determined relation, it is shown that time rate of change of temperature cannot be correlated with the observations. It is concluded on the basis of a theoretical calculation that absorption of ultrasound at interfaces in the spinal cord does not result in minute hot regions.

Further work on summation of subparalytic doses, spaced apart at various time intervals, indicates that the recovery process following exposure to a subparalytic dose of ultrasonic radiation may not be a monotonic function of time.

INTRODUCTION

IN the first paper of this series concerned with the physical factors involved in ultrasonically induced changes in nerve tissue, we presented experimental data which demonstrated the existence of non-temperature effects.[1] Observations have been made, during ultrasonic irradiation, on the random activity of the ventral nerve cord of crayfish and on the reflex discharge of the frog spinal cord.[1,2] Under appropriate experimental conditions, permanent paralysis of the hind limbs of frog is observed after irradiation.[1]

In this paper, we are concerned with the possible role of cavitation as a factor in the mechanism producing paralysis of the hind legs of frogs. For this purpose, observations were made on paralysis in frogs under a hydrostatic pressure sufficiently high to suppress all cavitation.

In addition, the time required for paralysis (for a single irradiation) has been determined for various acoustic amplitudes. These data were obtained both at atmospheric pressure and under a hydrostatic pressure sufficient to suppress all cavitation up to an acoustic pressure amplitude of 13 atmospheres. The results indicate that rapid time rate of change of temperature of the tissues of the spinal cord under ultrasonic irradiation is not an important factor in the production of the paralysis. Arguments are also presented which indicate that minute local hot regions cannot exist in the tissue which is under irradiation.

A further study of the summation process for subparalytic doses has been accomplished.

DESCRIPTION OF APPARATUS AND EXPERIMENTAL PROCEDURE

The experiments described in this paper were all performed at an ultrasonic frequency of 0.98 mc. The acoustic pressure amplitude was continuously variable from zero to about 15 atmospheres. The absolute sound level and field distribution were determined by the method described in a previous paper.[1] The electronic part of the ultrasonic generator is identical with that used in earlier experiments.

However, in order to irradiate the frogs in an environment under a hydrostatic pressure, it was necessary to design a different sound chamber from that used for previous experiments. The essential design features of the new system are illustrated in Fig. 1. There are three ports of entry to the stainless steel chamber, each bolting into position. The port A gives access to the left-hand section B which is isolated from the rest of the chamber so that the gas composition in this part is independent of the composition in the other section. Since B contains the sound head C which requires high voltage for its operation, it is necessary, for stability and safety, to introduce a dry inert atmosphere. Port D is used to introduce the sound absorber E which is filled with castor oil and is capped on its left-hand side by a $\frac{1}{32}$-in. sheet of "ρc" rubber. Under normal oper-

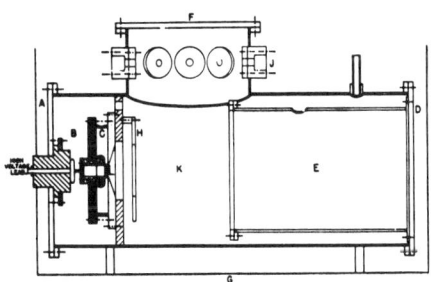

FIG. 1. Sound chamber for pressure studies.

---
* This research was partially supported under a contract with the Physiology Branch of the ONR.
[1] Fry, Wulff, Tucker, and Fry, J. Acoust. Soc. Am. **22**, 867–876 (1950).
[2] Tucker, Wulff, and Fry, "Reversible suppression of the reflex discharge of frog spinal cord by ultrasound" (in preparation).

259

ating conditions, port F is used for introducing the object to be irradiated into the chamber.

The entire pressure chamber is enclosed in a container G supporting a bath which furnishes temperature control for the system.

Electrical lead-ins are provided through a radial array J of pressure tight seals.

In practice, the sequence of events for a run is as follows. The bath in container G is brought to the desired temperature, and degassed water is introduced into the region K through F. When the degassed water has attained the desired temperature, the specimen is introduced on support H which accurately positions it in the sound field. After a sufficient time has elapsed for temperature equilibrium of the specimen and bath to be attained (which may be quite short since the specimen can be brought to the test temperature in an external bath), the desired gas compositions are simultaneously introduced into both sections of the pressure chamber under controlled conditions. There never exists more than a few pounds per square inch pressure difference. This control is necessary for mechanical protection of the crystal-crystal holder assembly L. The specimen is irradiated, and the pressures are then simultaneously released.

The sound projector used in this arrangement is similar to those described previously.[3]

The rate of compression is adjusted so that a pressure change from 1 atmosphere to 13 atmospheres was accomplished in about one to two minutes. Decompression is accomplished in an equal period of time. The frogs were irradiated on the dorsal surface over the region of the lumbar enlargement. The coupling medium between the frog and the projector is water. The ultrasonic beam width at half-amplitude is approximately 5 mm. The frogs were rigidly supported and accurately positioned in the sound field. The average weight of the frogs used in these experiments was about 23 grams. The range of weights was approximately 15 to 32 grams.

The irradiation procedure usually consisted in setting the sound level to some given value, and then subjecting frogs to this constant level for increasing lengths of time. In this way, the maximum time for which "no" frogs are paralyzed and the minimum time for which "all" frogs are paralyzed can be determined. The criterion of "all" or "none" is arbitrarily specified by observations on from three to five frogs. Such a criterion leads to results which can be duplicated.

EXPERIMENTAL RESULTS

The results of the experiments on the relation between the sound amplitude and the time required for paralysis are shown graphically in Fig. 2 and Fig. 3. The results obtained with the frogs at atmospheric pressure, at a temperature between 0° and 1°C and coupled to the

[3] F. J. Fry, Rev. Sci. Instr. 21, 940–941 (1950.)

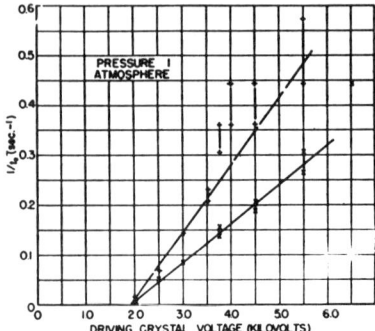

FIG. 2. Reciprocal of minimum irradiation time for paralysis and maximum irradiation time for no paralysis as a function of relative sound amplitude for frogs at atmospheric pressure and 0°–1°C.

sound source by degassed water, are presented in Fig. 2. The rms driving crystal voltage, which is proportional to the acoustic amplitude, is plotted along the horizontal. A voltage of 5.5 kilovolts corresponds to a free field pressure amplitude of about 12 atmospheres. The reciprocal of the time is plotted along the vertical.

The ratio of the acoustic pressure amplitude within the cord to the free field pressure amplitude was determined by direct probe measurements in several cords. A probe whose sensitive element consists of a small insulated crystal was used for these measurements. The average value of this ratio for a set of random positions of the probe in the lumbar enlargement region of the cord was computed for each cord. These averages for different cords agree closely. The measurements indicate that the average acoustic pressure amplitude in the cord is 0.8 of the free field pressure amplitude.

The solid line of the graph designates the relation between the sound amplitude and the minimum time for which "all" frogs are paralyzed. The dashed line is the relation between the sound amplitude and the

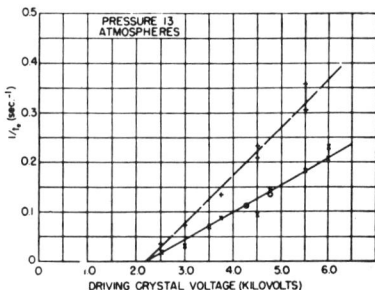

FIG. 3. Reciprocal of minimum irradiation time for paralysis and maximum irradiation time for no paralysis as a function of relative sound amplitude for frogs at 13 atmospheres pressure and 0°–1°C.

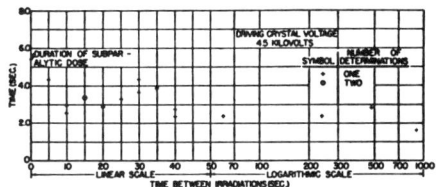

FIG. 4. Difference between the duration of a single irradiation for paralysis and the duration required of a second irradiation to produce paralysis when the first irradiation is subparalytic as a function of the time between the first and second irradiations. The sound amplitude is equal for all irradiations.

maximum time for which "no" frogs are paralyzed. The solid vertical lines connecting two points are not an indication of variability in the experimental results, but rather the result of the fact that a step timer was used to control the duration of the sound and that timer settings, corresponding to durations intermediate between the points indicated on the graph, were not possible. In general, the minimum step in irradiation time was 10 percent if this was not less than 0.5 second.

With the frog under a hydrostatic pressure of 13 atmospheres and a temperature of 0° to 1°C, the results illustrated in Fig. 3 were obtained. As in the previous case, two sets of data are shown, one corresponding to the maximum time for no paralysis and the other corresponding to the minimum time for paralysis. Most of the determinations were made with the frog's head in pure oxygen. However, two determinations were made under pure nitrogen. These are indicated by the circles on the lower curves.

Observations on about 500 frogs were required to obtain the data for Figs. 2 and 3.

Results of previous experimental work on summation of successive subparalytic doses to produce paralysis were presented in reference 1. These previous studies were confined to a fixed period of four minutes between irradiations. A more complete picture of the decay process which follows an exposure of the frog to a subparalytic dose of ultrasound at one acoustic amplitude and dose duration is indicated by the results illustrated in Fig. 4. The sound level is held constant at a pressure amplitude of 10 atmospheres, which corresponds to a driving crystal voltage of 4.5 kilovolts. This level corresponds to a paralysis time of 7.8 seconds for a single irradiation. The duration of the initial dose is 5.4 seconds. The frogs were under a hydrostatic pressure of 13 atmospheres and at a temperature of 0° to 1°C. On this graph, the difference between the required duration of a single irradiation for paralysis and the duration required of a second irradiation to produce paralysis, when the first irradiation is subparalytic, is plotted as a function of the time between irradiations. The plotted points correspond to irradiation times for which paralysis is obtained. An irradiation time shorter by 0.4 second did not produce paralysis.

Because of the complex appearance of the relationship, as indicated graphically, more than one determination was made at some of the time intervals between irradiations. The symbols used in plotting indicate the number of times a particular point was determined by independent experiments. These results indicate that the over-all decay process, determined as described above and following exposure of the frog to a subparalytic dose of ultrasound, is not monotonic for at least some conditions of irradiation. It is possible that the decay process is monotonic for shorter subparalytic doses or for different acoustic amplitudes.

A curve is not drawn through the points of Fig. 4, since it is felt that many more data would be necessary to specify a reasonably quantitative relationship. The graph is included in the paper simply to illustrate the complexity of the aftereffects of a subparalytic exposure. Irradiation of from 12 to 15 frogs is required for the determination of one point.

DISCUSSION

(1) Cavitation

One of the purposes of this study was the investigation of the possible role of cavitation as the agent responsible for the observed paralysis of the frogs when ultrasound is incident on the dorsal surface over the region of the lumbar enlargement. That such is not the case is shown by the results of the experiments in which a hydrostatic pressure sufficient to completely prevent cavitation at acoustic pressure amplitudes equal to or less than 13 atmospheres was used (see Fig. 3). This pressure amplitude corresponds to a driving crystal voltage of 5.7 kilovolts. Experimentally, paralysis was obtained at voltages as low as 2.5 kilovolts.

It is of interest to note that other experimental observations indicate that frog tissues will withstand considerable tension before cavitation occurs.[1] It is inferred that cavitation probably does not occur in the frog spinal cord at the highest driving crystal voltage, 6.0 kilovolts, indicated on the graph of Fig. 3. At atmospheric pressure, cavitation may cause a deviation from linearity at the higher sound levels, as suggested by the results of Fig. 2.

An interesting result of this investigation is the linear relations, presented graphically in Figs. 2 and 3, between the reciprocal of the minimum irradiation time for paralysis ("all" frogs paralyzed under a single irradiation) and the relative acoustic amplitude, and also between the reciprocal of the maximum irradiation time for no paralysis ("no" frogs paralyzed under a single irradiation) and the relative acoustic amplitude. The form of the relationship can be expressed mathematically as

$$1/t_e = m(A - A_0), \quad (1)$$

where t_e is the exposure time, and A is the relative acoustic amplitude (the voltage across the projector).

A_0 is the value of A at the intersection of the curve and the horizontal axis. This relationship is valid under 1 atmosphere pressure of air and under 13 atmospheres pressure of O_2. The existence of a definite threshold (acoustic amplitude below which no paralysis can occur, irrespective of the duration of irradiation, and above which paralysis always occurs if the period of irradiation is sufficiently long) is established. This threshold increases slightly with pressure (14 percent for a pressure change from 1 to 13 atmospheres). The slope of the line, which designates the relation between the reciprocal of the time for paralysis and the relative acoustic amplitude, is smaller at the higher pressure, i.e., the time for paralysis under pressure is always greater than the time for paralysis under 1 atmosphere for equal acoustic amplitudes. That the environment of pure oxygen (13 atmospheres) does not account for these differences is demonstrated by the fact that when the frog is immersed in an environment of pure nitrogen (13 atmospheres), the time for paralysis is unchanged from that obtained for an environment of pure oxygen. This was checked at two different sound levels, as indicated by the two points designated by circles in Fig. 3.

(2) Temperature

a. *Time Rate of Change of Temperature*

Results reported in reference 1 showed that the paralysis was not produced by excessive temperatures. This was proved by direct measurement of the temperatures of the cord under irradiation and by experiments demonstrating summation. It has been suggested that the time rate of change of the temperature may be a factor. That this is not the case can be seen from the following. Consider two relative sound levels near threshold A_1 and A_2 such that

$$(A_2 - A_0) = n(A_1 - A_0), \qquad (2)$$

where n is any arbitrary positive number. Then, from Eq. (1), letting t_e designate the paralysis time, we have $t_{e2} = t_{e1}/n$. Now the intensity of the sound is proportional to A^2, and the time rate of change of temperature is proportional to the intensity. It is clearly possible to choose A_1 and A_2 in such a fashion that the difference between their squares is as small as we wish and yet preserve the relation (2). We can thus make the difference in the rates of change of temperature in the two cases as small as we wish and yet the times for paralysis differ by the factor n. This indicates that the time rate of change of temperature is not a contributing factor in the mechanism determining paralysis.

b. *Minute High Temperature Regions*

It has also been suggested that another way in which the temperature factor might enter into the mechanism of the effects produced by ultrasound on tissue is through localized heating at interfaces. It is felt, on the basis of the following discussion, that local heating is not important in producing the results which we have observed on nerve tissue.

As reported previously,[1] we observed no appreciable change in the electrical threshold in excised frog sciatic nerves subjected to ultrasound at 1.0 megacycle up to intensities of the order of 35 watts/cm^2. The change in average temperature of the sciatic nerve in the sound field is about 2°C. This indicates that the temperature at the interface across which the nerve demarcation potential exists is not appreciably different from the average temperature of the preparation.

When an excised ventral nerve cord of the crayfish is irradiated with ultrasound, it is first observed[1] that the electrical activity of a single neuron increases and is then depressed. The increase in activity can be accounted for by the rise in average temperature of the preparation, as measured by a small thermocouple.[4] This suggests that the temperature in the interfaces in close proximity to the active neuron does not differ much from the average temperature of the portion of the nerve cord under ultrasonic irradiation.

It is possible to calculate an upper limit for the difference between the temperature of the interfaces and the average temperature of the tissue under ultrasonic irradiation. We assume that all sound absorbed is absorbed at interfaces and consider a unit volume of material filled with cells of a cylindrical shape of average diameter 10μ. Calculations based on spherical cells would differ by only a small factor from those based on cylindrical cells, and the calculated temperature difference would be smaller. If we choose a smaller size cell, the calculated temperature differences would be smaller. We assume that there is one absorbing interface per cell. If the difference in temperature between the interfaces and the average temperature of the tissue is ΔT, then the heat H conducted from the interfaces per second per unit volume is given by

$$H = KA(\Delta T/L), \qquad (3)$$

where A is the total area of the interfaces per unit volume, $\Delta T/L$ is the temperature gradient, and K is the coefficient of heat conductivity. The time rate at which heat is absorbed from the sound wave in a slab of unit area and thickness Δx (Δx along the direction of propagation) is given by

$$\Delta H_a = \alpha I_0 \Delta x, \qquad (4)$$

where I_0 is the incident sound intensity, and α is the intensity absorption coefficient per unit path length. At equilibrium, $H\Delta x = \Delta H_a$ if we neglect convection which would act only to further decrease the temperature difference. Expressions (3) and (4) yield the

---

[4] C. L. Prosser, J. Gen. Physiol. **19**, 65–73 (1935).

following equation for the temperature difference:

$$\Delta T = \alpha I_0 L / KA. \qquad (5)$$

To obtain a numerical estimate for ΔT, we assume that the heat conductivity of the tissue is equal to that of water, and the absorption coefficient α is equal to 0.4 per cm. This value of α is greater than that calculated from data given in reference 1. We insert the radius of the cells for L, i.e., 5μ. At a sound intensity of 50 watts/cm^2, the calculated value of the difference between the temperature at the interfaces and the average temperature of the tissue is, roughly, $6(10)^{-4}$ degree C. Even if only 1 percent of the cells have membranes which absorb the sound, the temperature difference would still be less than 0.1°C. It thus appears that temperature differences between the interfaces in the nerve tissues and the average temperature of the tissue are not important in producing the observed results.

c. Summation

The experimental results on summation plotted in Fig. 4 indicate that more than one process is important in determining the rate of change with time of the aftereffects of a subparalytic dose of radiation. Many more experimental data are necessary before a complete picture of the summation process can be presented.

d. One Factor Theory

The linear relations obtained experimentally between the reciprocal of the time for paralysis and the acoustic amplitude can be described in terms of a rate process as follows. We assume that a factor x in the spinal cord tissue satisfies the relation

$$dx/dt = K(A - A_0), \qquad (6)$$

which has the solution

$$x = K(A - A_0)t + b. \qquad (7)$$

The quantity A is proportional to the acoustic amplitude; for example, it may be the driving crystal voltage. A_0 is the value of this quantity at threshold. The quantity K is a proportionality constant which is dependent on the hydrostatic pressure and the base temperature of the cord† and may be dependent on the acoustic frequency.

We let x_0 designate the value of the factor x before ultrasonic irradiation and let x_p be the value necessary for paralysis. We assume that there is negligible decay in the factor during irradiation.

We obtain from Eq. (7) the result

$$1/t_p = [K/(x_p - x_0)](A - A_0), \qquad (8)$$

where t_p is the time for paralysis under a single irradiation. This is the experimentally observed relation (1), both at atmospheric pressure and at 13 atmospheres pressure. We let $m = K/(x_p - x_0)$ and observe that m is a function of hydrostatic pressure, base temperature, and ultrasonic frequency.

Further study of the mechanism of the effects of ultrasound on nerve will involve a determination of the dependence on acoustic frequency. Experiments will be designed to differentiate between the roles of the periodic variables; pressure, particle velocity, and particle acceleration.

SUMMARY

By applying a hydrostatic pressure sufficient to prevent cavitation, it is shown that the primary physical factor involved in the mechanism producing permanent changes in nerve tissue, as manifested by paralysis of the hind limbs of frogs, is not cavitation. (It has been previously shown that the average temperatures produced by absorption of the sound in the tissue cannot account for the observed results.)

It is determined experimentally that a linear relation exists between the reciprocal of the minimum time for paralysis ("all" frogs paralyzed) and the acoustic amplitude. The existence of a definite threshold is established. A linear relation also exists between the reciprocal of the maximum time for no paralysis and the acoustic amplitude. An argument is formulated which shows that the "time rate of change of temperature" is not important in the mechanism of paralysis.

Calculation shows that interface absorption of the ultrasound in the spinal cord cannot result in local hot regions. The single-dose paralysis results are described in terms of a one factor rate process.

---

† In reference 1 it is shown that the time required for paralysis is shorter when the base temperature of the frog is higher.

Effects of Ultrasonic Vibrations on Nerve Tissues.*

V. J. Wulff, W. J. Fry, Don Tucker, Frank J. Fry, and Carlton Melton.

From the Departments of Physiology and Electrical Engineering, University of Illinois

Numerous studies have already been published concerning the effects of ultrasound on living systems [Gregg(1), Harvey(2,3)]. Among these there are several kinds of experiments on the effects of ultrasound on nerve tissues, both excised(2) and *in situ* (4,5). Further, Lynn and Putnam(6) determined the effect of ultrasound on the brains of mammals, and demonstrated functional disturbances, neurone damage and even death of organisms.

The propagation of ultrasonic vibrations through living tissues is accompanied by a variety of physical factors such as: (1) heating caused by absorption of acoustic energy; (2) periodic pressure changes; (3) radiation pressure; (4) streaming or flow in viscous media; and, (5) high temperatures and pressures associated with cavitation, defined as the formation of holes in liquid media. Any one or all of these factors may produce significant and measurable changes in the state of a living system. The experiments described above do not afford the opportunity to determine how ultrasound produces its effect. The experiments to be described were designed to determine whether tissue heating is a major factor contributing to the effect of ultrasound on nerve tissues.

Materials and methods. The following types of preparations were employed:

(1) The ventral abdominal nerve cord of the crayfish was dissected out in its entirety and immersed in van Harreveld's solution. The sixth abdominal ganglion was always removed and either part or all of the remaining cord was used. The excised nerve cord was mounted in an electrode chamber in contact with 2 glass tubes filled with a salt solution. These salt bridges made contact with small calomel half cells which were, in turn, connected to the amplifier. The spike potentials were amplified by a condenser coupled amplifier and recorded by photographing the trace of a cathode ray tube.

(2) Intact frogs were placed vertically in the path of the sound beam by mounting the animal (dorsal surface down) as firmly as possible on a piece of plywood cut to fit the sound tank. A 2.0 cm hole, centered so

---

* This research was supported by the Biophysics Branch of the Office of Naval Research under Contract N6ori-71-Navy Task XXI. It is now under the cognizance of the Physiology Branch of the Office of Naval Research.

1. Gregg, E. E., Jr., Medical Physics, 1944, 1591-96, Chicago, Year Book Publishers.
2. Harvey, E. N., *Am. J. Physiol.*, 1929, v91, 284.
3. Harvey, E. N., *Biol. Bull.*, 1930, v59, 306.
4. Pohlman, R., Richter, R., and Parow, E., *Deutsche Med. Wochenschr.*, 1939, v65, 251.
5. Parow-Souchon, E., *Z. f. ärtzliche Fortbildung*, 1942, v39, 362.
6. Lynn, J. S., and Putnam, T. J., *Am. J. Path.*, 1944, v20, 637.

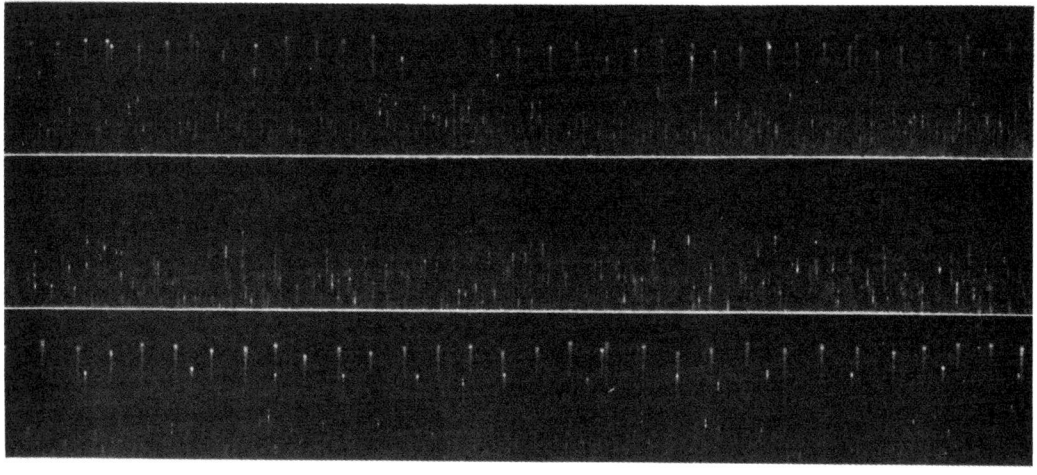

FIG. 1.
Spontaneously occurring spike potentials recorded from the commissure between the first and second abdominal ganglia of an excised crayfish ventral nerve cord. Table I indicates the frequency of discharge (average values) as a function of the elapsed time during and after exposure to ultrasound. Records should be read from right to left.

as to fit directly over the crystal, served to admit the sound. The frogs were oriented on the board so that the region of the lumbar enlargement was approximately centered in the aperture. The approximate center of the lumbar enlargement was estimated by adjusting the frog so that a line 12-14 mm behind the posterior edge of the tympanum bisected the hole in the board. The sound tank was filled with distilled water which had been boiled for 10 min. to drive off most of the gas.

(3) Temperature changes in the spinal cords of intact animals, in excised spinal columns containing the cord, and in the ganglia of excised crayfish ventral abdominal nerve cord were measured with constantan-copper thermocouples. For the crayfish ventral nerve cord, a soldered junction made of 0.013 mm copper and 0.038 mm constantan was used. For the spinal cord of the frog, a soldered junction of 0.25 mm copper and constantan was employed. The ventral nerve cord of the crayfish was threaded over the junction and the ultrasound was incident on the preparation in the region of the junction. The thermocouple in the spinal cord was introduced laterally through the foramina of the column through which the peripheral nerves pass. The foramina selected were approximately 12 mm behind the posterior edge of the tympanum. The frogs used in making the temperature measurements were not utilized for any other purpose. The thermocouple output was interrupted by a mechanical chopper, amplified by a condenser coupled amplifier and recorded by photographing the trace of a cathode ray beam. The chopping frequency was 10 per sec. A sudden temperature change, produced by thrusting the junction into a cold water bath, produced a deflection reaching 0.7 of maximum in 0.030 sec. This measurement was made without the chopper. The ultrasound used was at a frequency of about 1 mc and the maximum intensity was 35 watts/cm^2. The vibrations were propagated through water from the generating crystal surface to the preparation. For further details concerning the sound generator, see Fry, et al.(7).

Results. 1. *Effect of ultrasound on the ventral nerve cord.* Spontaneous activity recorded from the commissure between 2 adjacent ganglia (1 and 2) of the crayfish ventral abdominal nerve cord, immersed in a balanced

7. Fry, W. J., Wulff, V. J., Tucker, D., and Fry, F. J., *J. Acous. Soc. Am.*, 1950, v22, 867.

TABLE I.

Time sequence	Frequency
Control	7.8
Sound on 15"	12
Sound on 17"	7
Sound on 43"	0
Sound off 50"	5.3
Sound off 65"	6.7

salt solution, is illustrated in Record 1, Fig. 1. Superimposed on a low level background activity is a series of periodically occurring spike potentials at an average frequency of 7.8 per sec. Exposure of the 2 ganglia and the commissure to ultrasound (~ 35 watts/cm^2) produces a characteristic sequence of changes in the activity, a typical example of which is illustrated in records 1, 2, and 3 of Fig. 1 and tabulated in Table I. Note that after the sound is turned on, the frequency of the spike potentials at first increases, then decreases and is followed by total disappearance of the large spike potentials after 43 sec. exposure (Record 2). Twenty-five seconds after the ultrasound was turned off, the large spike potentials reappeared, first slowly, then more rapidly, finally reaching a stable frequency of 6.7 per sec (Record 3, Fig. 1). Subsequent treatment of the same preparation with ultrasound produced a similar sequence of events. Similar observations were made on 6 other preparations. Measurements of temperature changes in the ganglion of a preparation similar to the above, under identical conditions, indicated a maximal rise of 1°C.

2. The effect of ultrasound on the frog spinal cord. A. Physiological and morphological effects. Twelve intact normal frogs suspended under water at room temperature, 21°-25°C, and placed so that the sound beam was incident on the center of the back over the lumbar enlargement, showed complete paralysis of the hind legs with exposures of 4.3 sec. duration. Shorter exposures to sound either produced no paralysis or a temporary partial paralysis which disappeared after a variable time interval. Experiments similar to the above were performed with frogs cooled and maintained at 1°-2°C. Ultrasound incident on the back over the lumbar enlargement of these cooled frogs produced paralysis of the hind limbs after exposures of 7.3 sec. The paralysis was permanent in all of 50 frogs so treated. Exposures of shorter duration produced no paralysis or a temporary paralysis which disappeared. Stimulation of the sciatic nerves of a frog just after the production of paralysis with ultrasound resulted in muscular response. Similar stimulation of sciatic nerves one week after the paralyzing sound treatment produced no muscular response. Examination of histological preparations of sciatic nerves fixed and stained with osmic acid vapor 2 weeks after irradiation, revealed considerable degeneration of axones. A typical example is illustrated in Fig. 2, No. 2 (compare with control nerve, Fig. 2, No. 1). In all preparations examined, considerable degeneration of axones was evident, as well as degeneration of the region of the spinal cord exposed to the ultrasound.

Histological examination of the spinal cords of ultrasound treated frogs revealed marked abnormality of the large motor neurons of the ventral horn of the gray matter. These abnormalities are evident in spinal cords dissected out and fixed 20 minutes after treatment and stained with thionine ($C_{12}H_9N_3S$), (Fig. 2, No. 4). Note the ragged cell outlines and the very intense stain (compare with control, Fig. 2, No. 3). Normal motor neurons do not stain intensely and have smaller nuclei, usually centrally located. Examination of preparations made four and eight days after ultrasound treatment show neurons which stain intensely and exhibit peripherally located nuclei (Fig. 2, No. 5 and No. 6). These changes are evident in sections both above and below the degenerated regions. A reduction in population of neurones is evident. The eight-day-old lesion, Fig. 2, No. 6, exhibits a marked abnormal appearance. Lower motor neurons from the spinal cords of frogs exhibiting temporary paralysis of the hind legs indicate some abnormality (Fig. 2, No. 7) and those from a frog treated at low temperature for 5 sec. and exhibiting no paralysis of the hind legs (Fig. 2, No. 8) do not show any obvious abnormality.

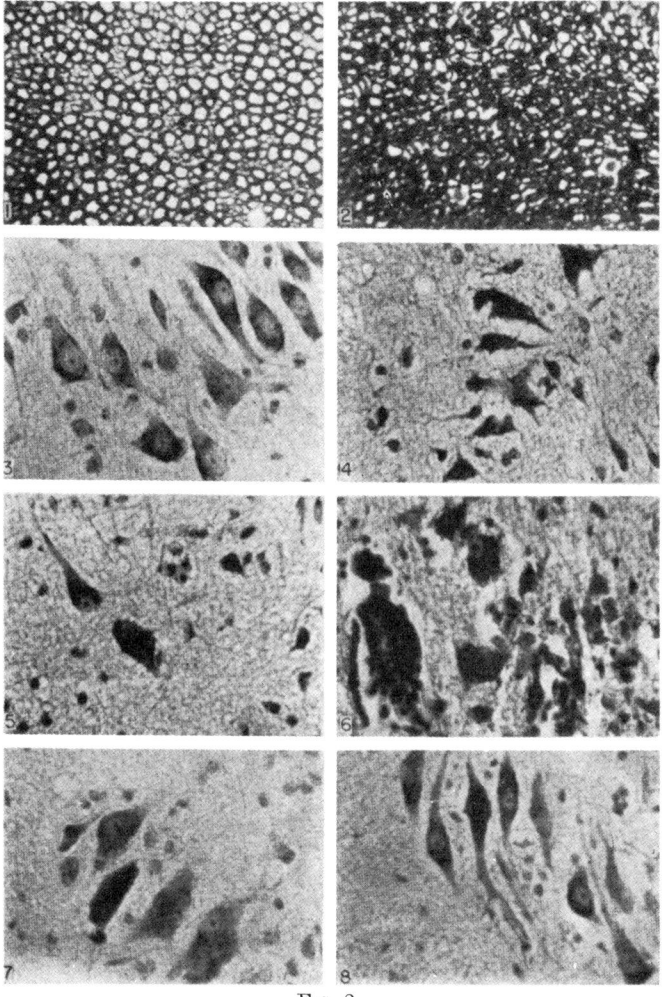

Fig. 2.
Photomicrographs of sections through sciatic nerves (1 and 2) and spinal cords of frogs (3-8) showing neurones in the ventral horns of the gray matter. ×175.

B. Temperature changes produced by ultrasound. Changes in temperature of the spinal cord of intact frogs were measured during and following the period of ultrasound treatment. Ultrasound incident on frogs at 1°-2°C produced a sharp rise in temperature (Fig. 3, Graph 1), which reached a level between 25°-30°C at the end of the 7.3 sec. exposure. After the exposure, the spinal cord exhibits a decrease in temperature. Experiments on isolated spinal column preparations gave similar results. To determine the influence of temperature on the production of paralysis of the hind legs of frogs, 12 experiments were performed using brief repetitive exposure to ultrasound. Frogs, cooled to 1°-2°C, were exposed to ultrasound for 4.3 sec. This is a sub-paralytic dose. The temperature change (max. temp. 15°C) produced by this exposure is indicated in Fig. 3, Graph 2. This exposure was followed by a 4 minute interval to permit the cord temperature to return to the previous level. Then the frog was subjected to a second 4.3 sec. dose of ultrasound, which produced a temperature change similar to the first (Fig. 3, Graph 2). Frogs subjected to a similar procedure (without insertion of the thermocouple) exhibited

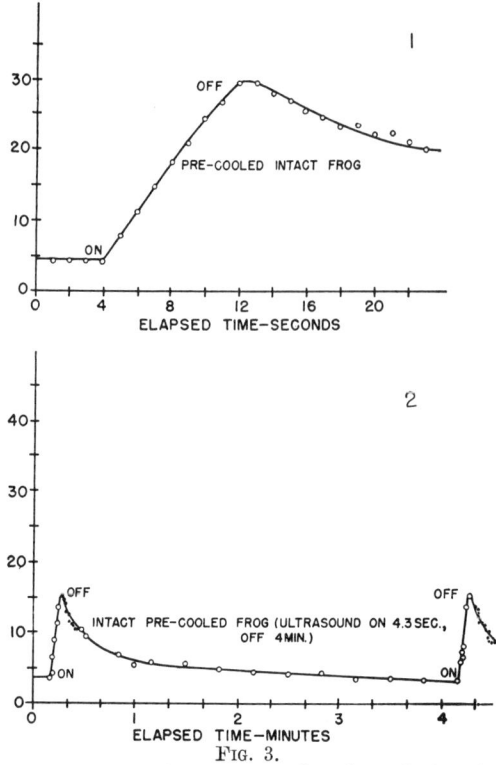

Fig. 3.

Temperature changes as a function of time in the frog spinal cord during and after sound treatment.

perature increase of 1°C was measured. Prosser(8) has shown that increasing the temperature 1°C between 26°-30°C may produce an increase in the frequency of discharge of single units of about 4-5 per sec. The effect of ultrasound in depressing the frequency of discharge is in a direction opposite to the effect of the temperature change. It is concluded, therefore, that the effect of ultrasound on these spontaneously active neurones is mediated by physical factors other than the simultaneously occurring but slight temperature change.

Examination of the data pertaining to neurone damage in the frog indicates that paralysis can occur in the absence of high (35°-40°C) temperature levels. For instance, frogs pre-cooled to 1°-2°C and subjected to a 7.3 sec. dose of ultrasound never exhibit temperature levels in excess of 30°C (Fig. 3). Further, exposure of pre-cooled frogs to successive sub-paralytic doses (4.3 sec. exposure, followed by a four minute interval) show temperature maxima of 15°C (Fig. 3), and yet, paralysis occurred after two exposures. The initial rates of change of temperature measured in the intact frog spinal cord, are rather constant, 1.8°C/sec. It is possible that the rapid increase in temperature occurring in the spinal cord upon incidence of ultrasound may produce physiological changes even though the final level of temperature remains below 30°C. Since rates of temperature change of the order of 1.8°C/sec. could not be achieved except by ultrasound, it was not possible to assess this factor. Unless a rate of change of temperature of the order of 1.8°C/sec. exerts a specific effect leading to paralysis it is apparent that the temperature changes measured cannot account for the experimental results in all cases. It is concluded that ultrasound produces effects on the neurones of the spinal cord which are primarily dependent upon physical factors other than temperature, such as periodic pressure changes, particle acceleration, radiation pressure and temperature and pressure changes associated with cavitation.† The evaluation of some of these factors must be

permanent paralysis of the hind legs after the second exposure. The effect of temperature on the spinal cord was assessed in the absence of ultrasound. The posterior half of frogs immersed in water baths at 35°C for 20 minutes and then raised to 38°C for 20 minutes showed no obvious abnormality in behavior. Frogs immersed in 40°C water for various periods up to 6 minutes showed a paralysis which gradually disappeared. Temperature measurements in the lumbar region of the spinal cord of frogs in 40°C water indicated a level of 40°C after 6 minutes.

Discussion and conclusions. Whenever ultrasound is transmitted through living tissues the existence of damaging temperature levels, produced by absorption of acoustic energy, must be ascertained. Ultrasound incident on ganglia of the crayfish ventral nerve cord containing spontaneously active neurones caused a reversible depression of the spontaneous activity. A maximal and rapid tem-

---

8. Prosser, C. L., *J. Gen. Physiol.*, 1935, v19, 65.

left to future research.

Summary. The excised crayfish ventral nerve cord exposed to ultrasound (~35 watts/cm^2, frequency 1 mc) exhibited a reduction of spontaneous activity after several seconds exposure and recovered its original activity about one minute after the ultrasound was turned off. Frogs positioned so that ultrasound was incident on the dorsal surface over the lumbar enlargement of the spinal cord exhibited paralysis of the hind legs after 4.3 sec. exposure (at room temperature) and exhibited paralysis after 7.3 sec. exposure (at 1°-2°C). Histological examination of the sciatic nerves showed extensive degeneration of nerves and examination of the spinal cord indicated destruction of the lower motor neurones. It was concluded that the effect of ultrasound on the systems studied is produced by physical factors other than temperature.

We would like to extend our appreciation to Professor C. L. Prosser for his continual interest and helpful suggestions during the course of this research. We also wish to thank Professor Warren McCulloch for his interest in this work.

---

† Cavitation is a name applied to the phenomenon of formation of holes in liquid media.

Received December 19, 1950. P.S.E.B.M., 1951, v76.

40
NERVE EXCITATION DUE TO HIGH-FREQUENCY ULTRASOUND

H. Mazoué, P. Chauchard, and R.-G. Busnel

This article was translated expressly for this Benchmark volume by Floyd Dunn and W. D. O'Brien, Jr., University of Illinois at Urbana–Champaign, from "L'Excitation nerveuse par les ultrasons de haute fréquence," J. Physiol. (Paris), 45, 179–182 (1953)

We have demonstrated, by chronaximetry, the appreciable sensitivity of the nervous system, which reacts to the application of ultrasound with a state of latent excitation (Busnel et al., 1952). In our preliminary study, we were able to establish that the degree of nervous excitation (amplitude of metachronosis and its duration) increased with the intensity and duration of application of the ultrasound. A threshold of intensity exists below which ultrasound is ineffective whatever the time period of application. A threshold of duration of ultrasonic application also exists, which is longer the weaker the intensity of the dissipated sound. At the same time, we provided proof of a specific ultrasonic effect on nervous tissue, probably of mechanical origin, which has no relation to the temperature elevation that can be produced by high-intensity ultrasound.

Here we shall present the preliminary results of a detailed study of the relationships between ultrasonic nervous excitation and the two factors intensity (i) and time of application of ultrasound (t). As a criterion for the ultrasonic effect, we have adopted the duration of the state of excitation, that is, the time required for chronaxy to return to normal.

We employed an ultrasonic generator that emitted a frequency of 1 MHz dispersed by a truncated cone with an emission surface of 0.13 cm^2 (0.78-cm diameter), or by a tube with an area of 0.11 cm^2 (0.38-cm diameter). The apparatus allows variable intensities to be obtained, which are checked by calorimetry. These values are expressed in acoustical watts (W).[1]

The ultrasonic applications were carried out on normal, awake rats immersed in water at 35°C. The hind paw was always treated; the cone of application was in contact with the depilated skin at the level of the point of measurement for the chronaxies of extension and flexion of the toes, opposite the two branches of the sciatic nerve above the joint.

Accordingly, there was a direct ultrasonic effect on the nerves.

---

[1] The intensities reported here are expressed in watts for the surface actually irradiated.

RESULTS

1. *Threshold of action:* If one seeks, for various intensities, the minimum duration of the ultrasonic application which permits the threshold to be obtained (i.e., a state of excitation lasting only 2 to 3 min, which is the time necessary for the determinations and associated with a small metachronotic amplitude), a law is found that is very similar to the classical law of excitability obtained as a function of time for electrical stimulation or specific sensory stimuli (Weiss's law).

The curve of Figure 1, plotted as a function of the two parameters i and t, characterizes this phenomenon. A minimum intensity (0.18 W) exists below which no effect is obtained.

At 0.18 W, the threshold is observed for a relatively long time of treatment (10 s), and it remains unchanged for increased treatment periods. Above the minimum intensity, it is apparent that the greater the intensity, the more the treatment time must be decreased to obtain the threshold. For example, for 3.2 W, the threshold is 0.75 s. The minimum effective intensity can be considered as a *piezobase*.[1] The law obtained may be compared with that for mechanical stimulation of nerves defined by Blair (1936), who studied the effect of pressure from a jet of air.

2. *Maximum effectiveness:* It is apparent that, proceeding from the threshold and increasing the time of treatment, the duration of the excitation state is prolonged. This increase, which is not produced at the piezobase, is all the more rapid the greater the intensity.

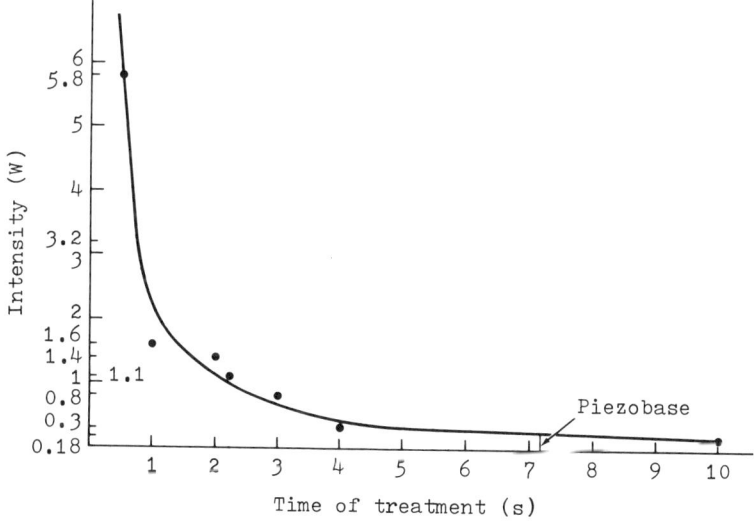

Figure 1 Law of ultrasonic nervous excitation. Reproduced from *J. Physiol. (Paris)*, **45,** 180 (1953), published by Masson & Cie, Paris.

[1] The term piezobase was first defined by Fontaine (1930) in his study of the effect of pressure on imbibition.

However, it is found that this increase reaches a limiting value rapidly. Thus, for 0.18 W, the treatment must be for 20 s for the state of excitation to last 4 min. The latter time cannot be exceeded even if the time of treatment is increased. The phenomenon is also found for other values at other intensities (e.g., for 0.3 W, 30 s are required to obtain a maximum duration of 25 min; for 3.2 W, 100 s are required to obtain a duration of 38 min; and for 5.8 W, 100 s produce a maximum duration of 40 min; (see Figure 2).

Within the limits of our experiments, that is, without appreciable temperature elevation in the tissue and without paralytic or necrotic phenomena, the following facts are evident:

a. A characteristic limit exists for each intensity studied, which is characterized by greater values for the duration of excitation and longer treatment times the greater the intensity.

b. There is no proportionality between this limit duration and the value corresponding to the treatment times with the values of the intensities used.

c. The maximum duration of excitation tends to be 40 min obtained for 100-s treatment time. Longer times of treatment or greater intensities do not increase the effectiveness.

CONCLUSIONS

The study of the nerve response to ultrasonic stimulation has led to the definitions of two series of phenomena, one of which corresponds to the classic laws of nervous excitability.

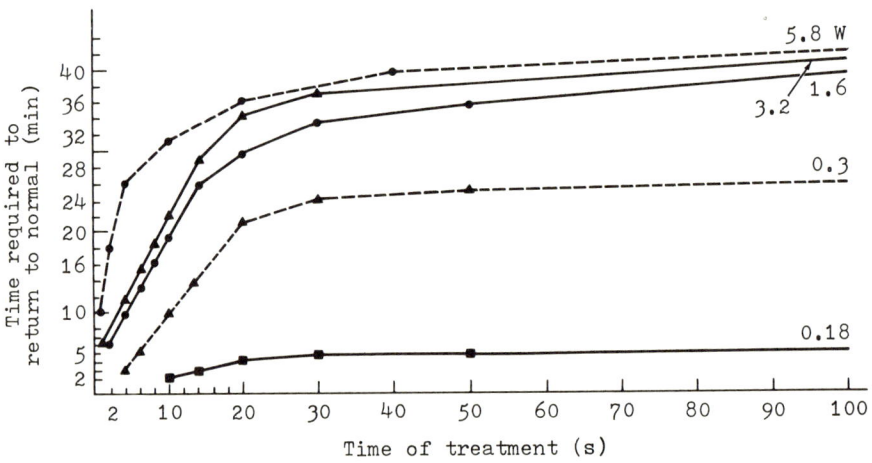

Figure 2 Return of chronaxies to normal as a function of time of treatment and of ultrasonic intensity. Reproduced from *J. Physiol. (Paris)*, **45**, 180 (1953), published by Masson & Cie, Paris.

With regard to the threshold, an excitation curve of the type obtained by Weiss and Lapique can be established as a function of the factors i and t, and a piezobase determined.

The duration of the nervous excitation increases above the threshold but tends to be limited for each intensity, being greater the greater the intensity. It reaches a maximum that is not exceeded regardless of the intensity and treatment period.

REFERENCES

H. A. Blair, *Amer. J. Physiol.*, **114**, 586 (1936).
R. G. Busnel, Gligorijévic, P. Chauchard, and H. Mazoué, *Compt. Rend. Acad. Sci. (Paris)*, **235**, 1535, 1684 (1952).
M. Fontaine, *Ann. Inst. Océanog. (Monaco)*, **7**, fasc. 1 (1930).

41

Copyright © 1954 by the American Congress of Rehabilitation Medicine

Reprinted from *Arch. Phys. Med. Rehabil.*, 35, 287–295 (1954)

Changes of Potentials and Temperature Gradients in Membranes Caused by Ultrasound

Justus F. Lehmann, M.D. and Rolf Biegler, M.D.
ROCHESTER, MINN.   FRANKFURT, GERMANY

Introduction

According to many investigators[1-24], various biologic and therapeutic ultrasonic reactions are of rather complex origin. The thermal component of the ultrasonic effect is quantitatively dominant. However, the heating effect is highly specific, because ultrasonic energy increases the temperature at the interfaces of the tissues selectively. This fact is of special interest in the therapeutic application of ultrasound. It has been demonstrated that there are in addition nonthermal, mechanical effects. For example, the permeation of ions is not only enhanced by the selective rise of temperature at these interfaces; it is also augmented by a stirring effect of ultrasonic energy. This stirring effect diminishes the diffusion layer and increases the gradient of concentration of the ions at the interface.

The effect of ultrasonic energy on the potentials of various biologic membranes is of great interest, because these potentials are characteristic of the properties of the living tissue at the interfaces and surfaces. Therefore the changes of these potentials caused by ultrasound may give valuable information concerning the effects on the living organism under therapeutic conditions. The isolated frog skin was especially suitable for our investigation, because it was known from previous experiments[12] that the properties of the skin, such as its permeability to the passage of ions, are altered by ultrasound. Furthermore, the potentials measured across the skin and their location in the tissues have been studied in detail by many physiologists. According to Amberson and Klein[25], and to Sumwalt, Amberson and Michaelis[26], the potential measured across the frog's skin is the sum of at least three different potentials. The main potential or so-called biopotential is located in the exterior layers of the skin, probably in the epithelium[26,27]. Ussing[27] has shown that the biopotential is caused by an active transport of ions through the membrane. It can be measured if Ringer's solution is applied to both sides of the skin. Huf[28-30] demonstrated that this potential is based on the intact metabolism of the skin. It disappears when there is hypoxia of the tissue and after poisoning with potassium cyanide (KCN), sodium sulfide (Na_2S) or carbon monoxide (CO)[31-35].

Another potential is produced by the electrical charge of the proteins, which are essential for the physico-chemical structure of the skin. Therefore the potential is dependent on the pH value of the surrounding solutions and it has an isoelectric point characteristic of the properties of the proteins involved. This potential is also located in the epithelium[26]. Finally Sumwalt, Amberson and Michaelis[26] have shown that there exists a third diffusion potential. It is encountered only when saline solutions of different concentration are applied to the interior and exterior surfaces of the membrane. According to Höber[36], this potential is located in the interior layers of the skin.

Because of the different ways in which the potentials described are produced, it was expected that information could be obtained concerning the primary ultrasonic effect on the various structures of such a biologic membrane. This assumption was made especially because the

---

Fellow in Biophysics, Mayo Foundation, Rochester, Minn.

The experimental work on which this paper is based was performed at the 1 Medizinische Universitäts Klinik, Frankfurt/Main, Germany.

potentials are based on certain vital functions of the cells, such as the maintenance of the physicochemical structure of the proteins and the selective permeability for ions with an electrical charge. Furthermore, it was of interest to determine whether there was specificity of the heating effect of ultrasonic energy, by studying whether gradients of temperature were created in a membrane, which might be biologically effective along distances of the order of the microscopic structures of the tissues.

Methods

The ultrasonic generator used in this study applied a constant voltage to the crystal of the applicator. This voltage was modulated only by the ultrasonic frequency. This frequency was crystal controlled. Tuning was not necessary. The radiating surface of the applicator was 10 sq. cm., the frequency 1 megacycle. The ultrasonic output was measured with a calorimeter and with an ultrasound pressure balance. In addition, the intensity was checked with a special sound pressure balance and a calibrated probe[19]. The standard deviation of the measurement was approximately ± 5 per cent. The skin was exposed to ultrasound in the far field[8]. The output was 30 watts (3 watts/sq. cm.). The ventral skin of female *rana esculenta* weighing approximately 25 gm. was used for experimentation. The frogs were decapitated and pithed. Then the skin was stripped off and inserted into the apparatus. The latter was made of two glass tubes connected by a ground joint. The circular opening (2 cm. diameter) between the tubes was covered tightly with the skin. The skin was always inserted in such a manner that the ultrasonic waves entered the interior part first. The glass tube was closed at the end nearest the ultrasonic applicator by a thin rubber diaphragm. The other end of the tube consisted of a glass wall, which was covered by a layer of absorbing glass wool. The whole apparatus was submerged in a water bath of constant temperature. Agar bridges were inserted into each of the two sections of the glass tube. Two openings were made, which served as inlet and outlet for changing the solutions. The agar bridges connected the liquids in the glass tube with the saturated calomel half-cells, the temperature of which was kept constant during the experiments. The potentials across the skin were measured with a potentiometer with an accuracy of 0.07 millivolts. The observed potentials were drawn as positive curves in the diagrams when the charge of the inside of the skin was positive and that of the outer surface negative.

In order to measure the biopotential a buffered Ringer solution was applied to both sides of the skin according to the technic of Huf[28-30]. The metabolism of the skin was inhibited with potassium cyanide (KCN) according to the method of Huf. The isoelectric point of the proteins was measured with the technic of Sumwalt, Amberson and Michaelis[26]. Sodium acetate was used as a buffer. The free diffusion potentials encountered do not interfere with the result, since they are small and constant. The pH of the solutions was checked before use. Maximal deviations of ± 0.01 pH from the desired value were allowed.

The so-called diffusion potentials were measured in the following way. The biopotential was determined as already described. Then, the Ringer's solution applied to the exterior surface of the skin was replaced through rinsing by a 1:100 diluted Ringer's solution. The potential was measured one minute later. Afterwards the biopotential was measured again after the diluted solution had been removed. The biopotential, as compared with the main potential of the controls, was not changed essentially by these manipulations. However, the observed potential decreased when the diluted solution was applied. The difference between the values of the biopotential, measured before and after the change of the solution, and the potential obtained during application of the diluted solution was called the diffusion potential because it is caused in some way or other by a diffusion process from the concentrated to the diluted solution. The active transport of sodium ions from the outside to the inside of the membrane (Ussing[27]) is hampered presumably by the gradient of concentration which is established. This potential also was plotted in the diagrams as positive. The average temperature of the skin was measured during exposure to ultrasound. However, no significant deviation of the readings was observed as compared with the measurements taken in the experiments on the biopotential without changing the solutions. All curves shown in this paper are based on an average of fifteen experiments each. The standard error of the mean is indicated in one of the diagrams (fig. 4). The quotient

$$A = \frac{D}{\sqrt{\sigma_{M_1}^2 + \sigma_{M_2}^2}}$$

was determined for all corresponding values of the curves to be compared. The standard errors of the mean of the values to be compared are σM_1 and σM_2. D is the difference between two compared values of the curves. This difference was defined for the given number of experiments as statistically significant if the quotient A were greater than 4 (likelihood 99.7 per cent) (Koller[37]). It was assumed that two curves were statistically different if the statistical evaluation, just described, had a significant result in at least five pairs of corresponding values. Values were regarded as identical within the limits of the statistical error if the quotient A were smaller than 2.

The thermoelectric measurements of temperature were performed as described previously[12,20].

Experiments

It was necessary first to study the changes of the biopotential during and after exposure to ultrasound. There were three distinct periods of observation in these experiments. The first period consisted of control observations. During the second period (period of treatment) ultrasound, heat or stirring was applied. The third period was for the purpose of observing those effects which remained. The temperature of the surrounding solutions was 30 C. As shown in figure 1, the potential dropped during application of ultrasound. The change was irreversible during the period of observation. The average temperature of the skin was measured with a thermocouple in another series of experiments during exposure to ultrasound. The temperature was raised to 36 C. in the tissue.

The increase of average temperature of the skin obtained during application of ultrasound was not sufficient to explain the effect of ultrasonic energy. If the skin were heated to 36 C. by increasing the temperature of the surrounding water bath, a curve of the potential was obtained which was identical with that of the controls (fig. 1). This result could be interpreted in the following way. The

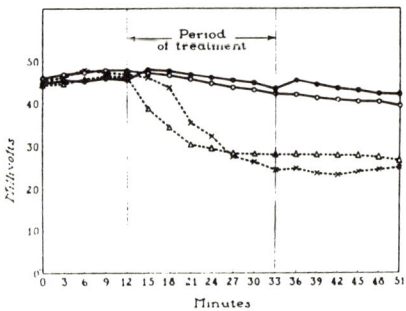

Fig. 1 — The effect of ultrasound and heat on the biopotential: (1) Controls at 30 C. o———o; (2) Effect of ultrasound (3 watts/sq. cm., 1 megacycle) when the temperature of the surrounding liquids was 30 C. △------△; (3) Effect of heating to 36 C. ●———●; (4) Effect of heating to 41 C. x------x. In (3) and (4), the heat was applied during the period of treatment only; otherwise the temperature was 30 C.

soldered junction of the thermocouple can measure only the average temperature because of its relative large size. Therefore the question was raised as to whether the temperature of the epithelial layer of the frog skin was selectively increased. The epithelium is considered to be responsible for the production of the main potential. In consequence of this assumption an effort was made to determine first which temperature could reproduce the ultrasonic effect. Heating to 41 C.: $\sigma M = \pm 0.1$ C. created the same decrease of the biopotential as ultrasonic energy (fig. 1). The temperature of 39 C.; $^{\sigma}M = \pm 0.1$ C. had less effect. But heating to 44 C.; $^{\sigma}M = \pm 0.1$ C. produced a more marked decrease of the curve than was observed after ultrasonic treatment (fig. 2).

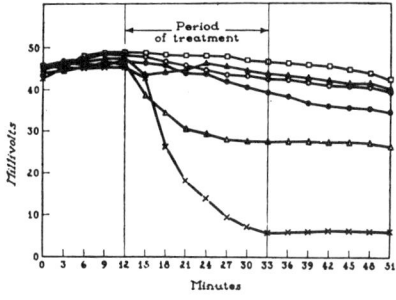

Fig. 2 — The effect of ultrasound, heat and stirring on the biopotential: (1) Controls at 30 C. o———o; (2) Effect of ultrasound (3 watts/sq. cm., 1 megacycle) when the temperature of the surrounding liquids was 30 C. △———△; (3) Effect of ultrasound (3 watts/sq. cm., 1 megacycle) when the temperature of the surrounding liquids was 25 C. ●———●; (4) Effect of ultrasound (3 watts/sq. cm., 1 megacycle) when the temperature of the surrounding liquids was 20 C. ▲———▲; (5) Effect of heating to 44 C. x———x [Here, in (5), the heat was applied during the period of treatment only; otherwise the temperature was 30 C.]; (6) Effect of stirring at 30 C. □———□.

Therefore, question arose as to whether this temperature of 41 C. was the actual temperature of the epithelial layer of the skin. This assumed temperature was decreased in the epithelium to a biologically ineffective degree of 36 C. by lowering the temperature of the surrounding liquids during exposure to ultrasound from 30 C. to 25 C. No ultrasonic effect was observed under these conditions (fig. 1). This result suggested that, in the foregoing experiment, the ultrasonic energy actually produced a temperature of 41 C. in the epithelium.

If ultrasound were applied when the temperature of the surrounding liquids was 20 C., the observed curve was also identical with that of the untreated controls. We concluded from this experiment that the observed decrease of the biopotential was due to the thermal component of the ultrasonic effect and not to another nonthermal effect dependent on temperature. It is very unlikely that such a nonthermal effect had the same dependence on temperature and the same temperature threshold as the effect of heat itself.

According to previous investigations[12], ultrasonic energy increases the diffusion of ions through the isolated frog skin by a stirring effect. Since Höber[26] and Ussing[27] had assumed that the biopotential is based on a selective passing of ions through the membrane, it was conceivable that the stirring effect of ultrasonic energy could influence the biopotential. However, experimentally, no significant effect of stirring alone was found (fig. 2) even if 450 cc. solution per minute passed the glass tube on each side of the skin. This result is in good agreement with the fact that the temperature rise was sufficient to explain the ultrasonic effect.

Finally it seemed necessary to prove that the observed ultrasonic effects were actually connected with the biopotential which is located in the epithelium of the skin and based on the intact metabolism of the cells. No ultrasonic effect on the potential curve was observed if the aerobe glycolysis was inhibited by potassium cyanide (KCN) (fig. 3). Therefore, it was concluded that the observed drop of the potential curve during exposure to ultrasound was indeed an effect on the biopotential.

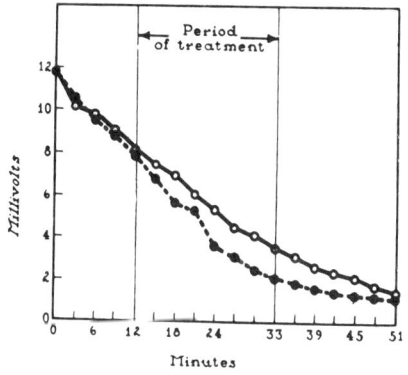

Fig. 3 — The effect of ultrasound after poisoning of the biopotential with potassium cyanide: (1) Controls at 30 C. o———o; (2) Effect of ultrasound (3 watts/sq. cm., 1 megacycle) when the temperature of the surrounding liquids was 30 C. ●----●.

Furthermore, it was of interest to

study in greater detail the manner in which the ultrasonic effect on this epithelial layer of the skin was produced. It had already been demonstrated that the ultrasonic effect on the biopotential could not be observed if the aerobe glycolysis were inhibited. Next an effort was made to determine whether the structure of the proteins of the membrane was altered by exposure to ultrasound. To accomplish this an investigation was made to determine whether the electrical charge of the proteins was influenced. The isoelectric point of the skin was measured. The skin was inserted into the apparatus and remained there for five minutes before it was exposed to ultrasound for ten minutes. After another period of four minutes, the measurements were performed according to the method of Sumwalt, Amberson and Michaelis[26]. The controls remained in the glass tube for the same time as that required for the measurement to be performed. However, no ultrasound was applied. The temperature of the solutions was raised for the same period of time as that employed when ultrasound was applied in studying the effect of heat. These measurements were made at pH values of 4.8 and 5.4 for the surrounding solutions. These pH values were very close to the isoelectric point. Therefore it was possible to interpolate the curve by a straight line and to obtain the isoelectric point. As a result of these experiments it was found that, after ultrasonic irradiation, the isoelectric point of the membrane was shifted from pH 5.0 to pH 5.23 (fig. 4). It is very likely that this result is based on denaturation of the proteins involved.

Since the epithelium is responsible for the biopotential as well as for the isoelectric point of the membrane[26] it was to be expected that heating to 41 C. would produce the same shift of the isoelectric point as ultrasonic energy provided the actual temperature during the exposure to ultrasound was 41 C. (see above). Indeed the ultrasonic effect could be reproduced by heating to 41 C. (fig. 4).

The study of the effect of ultrasound on the diffusion potential showed that the potential dropped only slightly during application of ultrasound (fig. 5). However the potential was markedly less afterwards. The potentials of the controls did not show any significant change (fig. 5).

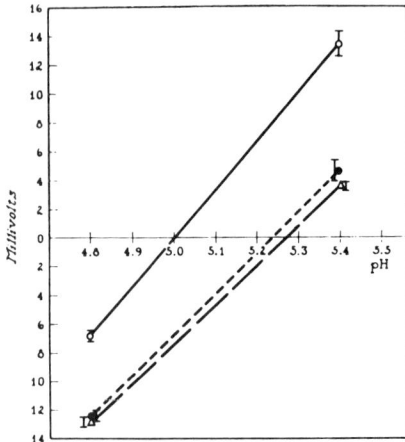

Fig. 4 — The effect of ultrasound and heat on the isoelectric point of the skin: (1) Controls at 30 C. o———o; (2) Effect of ultrasound (3 watts/sq. cm., 1 megacycle) when the temperature of the surrounding liquids was 30 C. ●------●; (3) Effect of heating to 41 C. △------△ [Here, in (3), the heat was applied during the period of treatment only; otherwise the temperature was 30 C.].

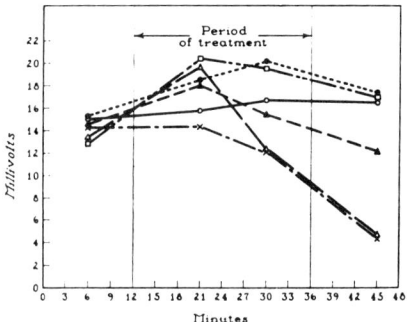

Fig. 5 — The effect of ultrasound and heat on the diffusion potential: (1) Controls at 30 C. o———o; (2) Effect of ultrasound (3 watts/sq. cm., 1 megacycle) when the temperature of the surrounding liquids was 30 C. x—·—x; (3) Effect of ultrasound (3 watts/sq. cm., 1 megacycle) when the temperature of the surrounding liquids was 22 C. □—·—□; (4) Effect of heating to 36 C. ●------●; (5) Effect of heating to 41 C. ▲----▲; (6) Effect of heating to 44 C. △---△. In (4) (5) and (6), the heat was applied during the period of treatment only; otherwise the temperature was 30 C.

If the skin were heated to 36 C. no

irreversible effect was observed after application of ultrasound. The temperature of 41 C. caused a smaller drop of the potential than did the ultrasonic treatment. However, this temperature could reproduce quantitatively the ultrasonic effect on the potentials located in the epithelial layer of the skin. Heating to 44 C. reproduced exactly the ultrasonic effect. Therefore the question arose again as to whether 44 C. was the actual temperature of the interior layers of the membrane during exposure to ultrasound. The temperature of the surrounding liquids was lowered from 30 C. to 22 C. so that the assumed temperature of 44 C. in the interior layer of the skin was depressed to a biologically ineffective degree of 36 C. As expected, no irreversible effect on the potential was observed (fig. 5). Therefore it was concluded that 44 C. was the actual temperature of the interior layer of the skin during application of ultrasound.

The diffusion potentials showed a slight increase, at least during the first part of the period of treatment, when the tissues were heated to 36 C. and to 44 C. and when ultrasound was applied while the temperature of the surrounding liquids was 22 C. (fig. 5). This increase of the potentials is probably the result of several processes occurring simultaneously. It can be assumed that the diffusion potential is increased when the temperature is raised, because the diffusion coefficient is dependent on temperature. On the other hand, as has been previously shown[12], a longer persistent and sufficiently high rise of temperature markedly increases the permeability. Therefore the selective passing of ions is abolished and in consequence the diffusion potential is diminished. Furthermore, the potential might be increased by application of ultrasonic energy because the stirring effect of ultrasound augments the selective passing of ions with an electrical charge if the effect of heat is prevented by cooling. The last mentioned assumption can be checked experimentally[12]. Observations were made relative to the effect of stirring of the diluted liquid on the exterior

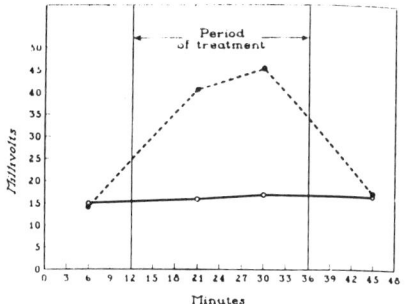

Fig. 6 — The effect of stirring on the diffusion potential: (1) Controls at 30 C. o———o; (2) Effect of stirring at 30 C. ●------●.

surface of the skin. It was evident that stirring markedly increases the diffusion potential (fig. 6).

Discussion

As a result of the experiments described it can be assumed that the effect of ultrasound on the potentials measured across the frog skin is created in the following way: The thermal effect of ultrasonic energy was quantitatively dominant. In agreement with the investigations of several observers[15,19,38-42], this heating effect was found to be highly specific because the temperature is selectively raised at the interfaces between tissues of different acoustic impedance. The selective rise of temperature had been measured for instance in the bone[38] or in the nerve (Rosenberger[43]). However, the question as to which gradients of temperature were created by ultrasonic energy in the tissues could not yet be answered. It was not known whether gradients of temperature which were biologically effective could be expected along a distance which was of the order of the microscopic structures of the tissues. It was found that the ultrasonic effect on the biopotential and on the isoelectric point of the membrane could be reproduced quantitatively by heating to 41 C. Both potentials are located in the epithelium of the skin. It could be shown that 41 C. was the actual temperature of the layer of the tissue during exposure to ultrasound. The ultrasonic effect on the diffusion potential could be reproduced by heating to 44 C., which was likewise the actual temperature at the interior layer of the skin during application of ultrasound. From these experiments it was evident that the distribution of temperature in this thin membrane is not at all uniform. One might conclude that a gradient of temperature of approximately 3 C. along a distance of the order of $5 \cdot 10^{-2}$ cm. is created by the absorption of the ultrasonic energy. The assumption that the observed effects were owing to the thermal component can be supported by the fact that the threshold and the dependence on temperature of the ultrasonic reaction were the same as those of a heating effect. If the assumed temperatures of 41 C. and 44 C. in the different layers of the skin during exposure to ultrasound were lowered to a biologically ineffective degree by decreasing the temperature of the surrounding liquids, no effect on the curve of the potentials was observed. When the temperature was decreased further during application of ultrasound, the dependence on temperature of the ultrasonic reaction could be studied. The latter was compared with the dependence on temperature of a heating effect.

Furthermore, during the exposure to ultrasound, a nonthermal effect was noted. The ultrasonic energy decreased the diffusion layer at the membrane by stirring and therefore augmented the selective passing of ions with an electrical charge and, in this way, increased the diffusion potential.

Having discussed the biophysical mode of action of ultrasonic energy on the potentials, it was of interest for a better understanding of the ultrasonic effect on the tissues to know which structures and physiologic reactions were altered by the ultrasound. It could be shown that the biopotential, which is dependent on the intact metabolism of the epithelium, was lowered by application of ultrasound. This effect could no longer be observed after inhibiting the aerobe glycolysis by potassium cyanide (KCN). In addition, the isoelectric point of the proteins of this epithelial layer of the membrane was shifted from pH 5.0 to pH 5.23. These findings can be interpreted by the assumption that the proteins are probably denatured by ultrasound. Therefore the physico-chemical structure of the cells is altered. Probably as a result the metabolism is decreased. This might lead to a better understanding of certain previous findings[17,20] which indicated that there was a decrease of the aerobe glycolysis produced by the thermal effect of ultrasonic enegry.

The ultrasonic effect on the diffusion potential probably can be explained as resulting from an increase of the permeability of the membrane by the heating effect of ultrasound as previously described[12]. Therefore one can surmise that the selective permeability for certain ions with an electrical charge (Ussing[27]) is diminished. As a result the diffusion potential drops. But, when the selective rise of temperature in the membrane during exposure to ultrasound increases the diffusion coefficient and a stirring effect of ultrasound augments the passing of ions, and when the heating effect is not sufficient to cause an increased permeability of the membranes for all sorts of ions, then the potential is raised.

Summary

The result of the investigation suggests that the ultrasonic effect on the potentials measured across the frog skin is created by both a thermal and a mechanical component. The heating effect of ultrasound is quantitatively dominant. The reaction is irreversible over a rather long period of time. The temperature is selectively raised at the interfaces. It might be concluded from the effect on the different potentials located in different layers of the membrane that a gradient of temperature of approximately 3 C. is created by the ultrasonic energy along a distance of the order of $5 \cdot 10^{-2}$ cm.

The mechanical effect of ultrasound could be explained by a stirring effect which diminished the diffusion layer at the membrane.

The thermal component of the ultrasonic energy altered the charge of the proteins of the membrane. The isoelectric point of the membrane was changed. The biopotential, which is based on the intact metabolism, decreased during exposure to ultrasound, but no effect was observed if the aerobe glycolysis were poisoned by potassium cyanide (KCN). Therefore it seemed likely that the potential was affected through the metabolism of the cells. Cellular metabolism is altered probably as a consequence of the change of the physicochemical structure of the cell proteins. The diffusion potential measured across the membrane dropped after application of ultrasound. This phenomenon may be produced by an increase of permeability, which results in a diminished selective permeability for ions with an electrical charge. The thermal component of the ultrasonic effect is decisive for this reaction. A stirring effect of ultrasonic energy was observed. It was encountered only during exposure to ultrasound. As a result the selective permeation of ions is enhanced not only by this phenomenon but also by the increase of the diffusion coefficient following the selective rise of temperature at the membrane.

References

1. Barth, G.: Über das Auftreten von Ultraschallschäden und das Dosierungsproblem. In Matthes, K., and Rech, W.: Der Ultraschall in der Medizin, Zürich, S. Hirzel Verlag, 1949, p. 207.
2. Barth, G., and Wachsmann, F.: Über den Wirkungsmechanismus biologischer Ultraschallreaktionen. III. Untersuchungen bei Verschiedener Umgebungstemperatur. Strahlentherapie 81:649 (May 5) 1950.
3. Born, H.: Über den Einfluss des Interferensfeldes auf die Dosierung bei der Ultraschalltherapie. In Matthes, K., and Rech, W.: Der Ultraschall in der Medizin, Zürich, S. Hirzel Verlag, 1949, pp. 40-44.
4. Dittmar, Carl: Über die Wirkung von Ultraschallwellen auf tierische Tumoren. Strahlentherapie 78:217 (Nov. 12) 1948.
5. Lehmann, J. F.: Über die Temperaturabhängigkeit therapeutischer Ultraschallreaktionen unter besonderer Berücksichtigung der Wirkung auf Nerven. Strahlentherapie 79:543 (July 13) 1949.
6. Lehmann, Justus: Beitrag zur therapeutischen Ultraschall-Wirkung. Strahlentherapie 82:281 (July 5) 1950.
7. Lehmann, J. F.: Die Therapie mit Ultraschall und ihre Grundlagen. Ergebn. d. phys.-diätat. Therapy 4:196, 1951.
8. Lehmann, J. F.: The Biophysical Basis of Biologic Ultrasonic Reactions With Special Reference to Ultrasonic Therapy. Arch. Phys. Med. & Rehab. 34:139 (Mar.) 1953.
9. Lehmann, J. F.: The Biophysical Mode of Action of Biologic and Therapeutic Ultrasonic Reactions. J. Acoustical Soc. America. 25:17 (Jan.) 1953.
10. Lehmann, Justus, and Becker, Georg: Über die permeabilitätssteigernde Wirkung von Ultraschallwellen auf biologische Membranen. Strahlentherapie 79:553 (July 13) 1949.
11. Lehmann, Justus, and Becker, Georg: Über histologische Veränderungen nach Ultraschall und anderen physikalischen Einwirkungen bei Permeabilitätsuntersuchungen an der Froschhaut. Strahlentherapie 84:306 (Mar. 22) 1951.
12. Lehmann, Justus; Becker, Georg, and Jaenicke, Walther: Über die Wirkung von Ultraschall-Wellen auf den Ionendurchtritt durch biologische Membranen als Beitrag zur Theorie des therapeutischen Wirkungsmechanismus. Strahlentherapie 83:311, 1950.
13. Lehmann, Justus; Becker, Georg, and Otto, Joachim: Thermische und mechanische Wirkungen des Ultraschalles auf einzelne Zellen, untersucht am Beispiel der eosinophilen Leukozyten. Strahlentherapie 87:550, 1952.
14. Lehmann, Justus, and Feissel, H. J.: Über die Abhängigkeit biologischer Ultraschallreaktionen von der Teilchenamplitude als Beitrag zum therapeutischen Wirkungsmechanismus. Strahlentherapie 82:293 (July 5) 1950.
15. Lehmann, Justus F., and Feissel, H. J.: Inwieweit lässt sich die Ultraschall-Leistung durch Zufuhr von Wärmeenergie in ihrer biologischen Wirkung ersetzen? Strahlentherapie 85:615 (Oct. 30) 1951.
16. Lehmann, J. F., and Herrick, J. F.: Biologic Reactions to Cavitation, a Consideration for Ultrasonic Therapy. Arch. Phys. Med. & Rehab. 34:86 (Feb.) 1953.
17. Lehmann, Justus, and Hohlfeld, Rolf: Der Gewebestoffwechsel nach Ultraschall und Wärmeeinwirkung. Strahlentherapie 87:544, 1952.
18. Lehmann, J. F., and Koester, E.: Die Abhängigkeit der hyperaeminierenden Ultraschallwirkung von der Beschallungszeit. Strahlentherapie (In press).
19. Lehmann, Justus, and Nitsch, Willi: Über die Frequenzabhängigkeit biolo-

20. Lehmann, Justus, and Vorschütz, Rosemarie: Die Wirkung von Ultraschallwellen auf die Gewebeatmung als Beitrag zum therapeutischen Wirkungsmechanismus. Strahlentherapie 82:287 (July 5) 1950.
21. Lehmann, Justus, and Vorschütz, Rosemarie: Über die biologische Wirksamkeit des Impulsschalles als Beitrag zum therapeutischen Wirkungsmechanismus der Ultraschallwellen. Strahlentherapie 81:639 (May 5) 1950.
22. Pfander, Friedrich: Durch Ultraschall ausgelöster thermischer Nystagmus. HNO, Beihefte Z. Ztschr. f. Hals-Nasen- und Ohrenheilkunde 1:512 (Nov.) 1949.
23. Theismann, Hans: Beitrag zur Frage der biologischen Wirkung des Ultraschalls. Strahlentherapie 79:559 (July 13) 1949.
24. Treanor, W. J.; Lambert, E. H.; Herrick, J. F., and Krusen, F. H.: Comparative Study of the Effects of Heat and Ultrasound on Nerve Conduction. Unpublished data.
25. Amberson, W. R., and Klein, Henry: The Influence of pH Upon the Concentration Potentials Across the Skin of the Frog. J. Gen. Physiol. 11:823 (July 20) 1928.
26. Sumwalt, Margaret; Amberson, W. R., and Michaelis, Eva: Factors Concerned in the Origin of Concentration Potentials Across the Skin of the Frog. J. Cell. & Comp. Physiol. 4:49 (Dec.) 1933.
27. Ussing, H. H.: The Active Ion Transport Through the Isolated Frog Skin in the Light of Tracer Studies. Acta physiol. scandinav. 17:1, 1949.
28. Huf, Ernst: Versuche über den Zusammenhang zwischen Stoffwechsel, Potentialbildung und Funktion der Froschhaut. Arch. ges. Physiol. 235:655, 1935.
29. Huf, Ernst: Über den Anteil vitaler Kräfte bei der Resorption von Flüssigkeit durch die Froschhaut. Arch. ges. Physiol. 236:1, 1935.
30. Huf, Ernst: Über aktiven Wasser-und Salztransport durch die Froschhaut. Arch. ges. Physiol. 237:143, 1936.
31. Francis, W. L.: The Electrical Properties of Isolated Frog Skin. Part II. The Relation of the Skin Potential to Oxygen Consumption, and to the Oxygen Concentration of the Medium. J. Exper. Biol. 11:35 (Jan.) 1934.
32. Francis, W. L., and Pumphrey, R. J.: The Electrical Properties of Frog Skin. Part I. Introductory. J. Exper. Biol. 10:379 (Oct.) 1933.
33. Lund, E. J.: Relation Between Continuous Bio-electric Currents and Cell Respiration. II. J. Exper. Zool. 51:265 (Aug.) 1928.
34. Lund, E. J.: Relation Between Continuous Bio-electric Currents and Cell Respiration. III. Effects of Concentration of Oxygen on Cell Polarity in the Frog Skin. J. Exper. Zool. 51:291, 1928.
35. Taylor, A. B.: Studies of the Electromotive Force in Biological Systems. IV. The Effect of Various Nitrogen-oxygen and Carbon Monoxide-oxygen Mixtures on the E. M. F. and Oxygen Consumption of Frog Skin. J. Cell. & Comp. Physiol. 7:1 (Oct.) 1935.
36. Höber, Rudolf: Physical Chemistry of Cells and Tissues, Philadelphia, Blakiston Company, 1945.
37. Koller, S.: Graphische Tafeln zur Beurteilung statistischer Zahlen, Dresden. Theodor Steinkopf, 1940.
38. Herrick, J. F.; De Forest, R. E.; Janes, J. M., and Krusen, F. H.: Effects of Ultrasonic Energy on Growing Bone. (Abstr.) Fed. Proc. 10:62 (Mar.) 1951.
39. Horvath, J.: Experimentelle Untersuchungen über die Verteilung der Ultraschallenergie im menschlichen Gewebe. Arzt. Forsch. 1:357, 1947.
40. Hüter, T: Messung der Ultraschallabsorption in tierischen Gewebe und ihre Abhängigkeit von der Frequenz. Naturwissenschaften 35:285, 1948.
41. Hueter, T. F.: On the Mechanism of Biological Effects Produced by Ultrasound. Chemical Engineering Progress Symposium Series 47:57, 1951.
42. Lehmann, Justus: Die Spezifität der biologischen und therapeutischen Ultraschallwirkung. Arch. phys. Therap. 3:57 (Mar.-Apr.) 1951.
43. Rosenberger, H.: Uber den Wirkungsmechanismus der Ultraschallbehandlung, insbesondere bei Ischias und Neuralgien. Chirurg 21:404 (July) 1950.

Reprinted from *Science*, **122**, 517–518 (Sept. 1955)

Ultrasonic Lesions in the Mammalian Central Nervous System

Early histological studies of nerve tissue of animals irradiated with intense focused ultrasound at this laboratory indicated that nerve cell bodies were more susceptible than nerve fibers to changes by the ultrasound (*1*). These preliminary histological results have not been substantiated in subsequent studies. Rather it has been found, as was previously reported (*2*), that white matter is more readily affected by the sound and that higher ultrasonic dosages are required for producing changes in gray matter. It can be readily seen that this selectivity provides a unique tool for basic neurological studies. Recent publications of this laboratory present results on the production and time sequence of changes in relatively large white-matter lesions of controlled shape (*2, 3*). This paper, however, is concerned primarily with small ultrasonic lesions in both gray and white matter (*4*).

Selective, accurately positioned lesions as small as 2 to 3 mm in maximum diameter can be produced. The lesions, which can be localized at any desired depth in the brain without affecting intervening tissue, are quantitatively reproducible from one animal to another, so that dosage studies made on a series of animals can be used as a guide in choosing the conditions of irradiation for neuroanatomical or functional studies. The blood vessels are most resistant to the action of the sound. It is, therefore, possible to interrupt fiber tracts without destroying neighboring gray matter and without breaking blood vessels even within the site of the lesion. It is also possible, by appropriate choice of the ultrasonic dosage, to affect irreversibly the nerve tissue (fibers and cell bodies) in gray matter without causing hemorrhage.

The results reported here were obtained from histological studies of ultrasonically irradiated cats and monkeys. Extensive dosage studies have been completed, and the time course of development of the lesions has been followed in animals sacrificed from immediately after irradiation (5 min) up to 30 days. The preparation of the animal and the technique of irradiation are described in previous papers (*2, 3*). Results of investigations concerned with the physical mechanism of the action of the sound on the nerve tissue have been published (*5*).

When a region of the white matter of the central nervous system is irradiated at one spot with a single exposure of ultrasound at a dosage just above the minimum required to produce an effect, a small lesion about 2 to 3 mm in maximum diameter is produced. Figure 1 illustrates such a lesion in the subcortical white matter 12 days after irradiation (dose 51 atm acoustic pressure and $4.8(10)^3$ cm/sec acoustic particle velocity for 1.00 sec). It shows a sharp boundary between the affected white matter (lower end) and the neighboring unaffected gray matter.

A lesion such as that shown in Fig. 1 is first seen 10 to 15 min following irradiation in tissue sections prepared with Weil's myelin stain. The lesion area is first recognized as a light-staining matrix as compared with normal tissue. One hour after irradiation the myelin sheaths appear beaded. The perivascu-

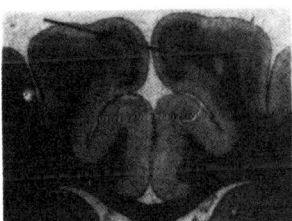

Fig. 1. Small ultrasonic lesion in the subcortical white matter of the brain of a cat. Dosage used selectively affects the fiber tracts, and no damage is produced in the neighboring gray matter. (PTAH stain)

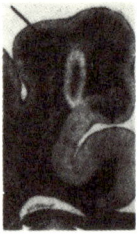 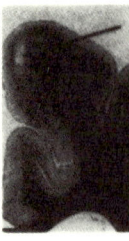

Fig. 2. Ultrasonic lesion in the subcortical white matter of a cat brain exhibiting a central region of dark-staining fibers and some invasion of neighboring gray matter. (PTAH stain) Fig. 3. Small ultrasonic lesion in the cortical gray matter of a cat brain. (PTAH stain)

lar spaces are dilated, and some separation appears between the fibers. Within 6 to 12 hr (depending on the dosage) the myelin sheaths break down into separated spheres. During this same period axis cylinder fragments increase in the lesion area. These changes are followed by the hematogenous and microglial responses until all of the debris is cleared away. Other neuroglia then form a glial scar.

A greater dosage (53 atm acoustic pressure and $4.9(10)^3$ cm/sec acoustic particle velocity for 1.50 sec) produces a slightly larger lesion containing a central normal staining area or island of myelinated fibers surrounded by a zone or moat containing completely disrupted nerve tissue and large clear fluid-filled spaces (Fig. 2). No hemorrhage is present. These more severe lesions may involve neighboring gray matter, causing changes that are described in the following paragraphs. Lesions of the same order of size can be produced in fiber tracts at any depth in the brain without affecting the intervening nervous tissue.

Figure 3 illustrates a small lesion produced by ultrasound in the cerebral cortex of a cat. To produce such a lesion in gray matter, greater dosages of ultrasound are required than for white matter. When a region of gray matter is irradiated with a single exposure at a dosage (53 atm acoustic pressure and $4.9(10)^3$ cm/sec acoustic particle velocity for 2.50 sec) above the minimum required to produce a lesion, the effects that appear first (10 min after exposure) are a lightening in the staining ability of the background matrix and a slight dilation of the perivascular spaces. Nerve cells stain more faintly than normal within 1 hr. Many contain large clear vacuoles in their cytoplasm; others have ruptured cell membranes, and only ragged strands of cytoplasm remain around the still intact nucleus. The nerve cells have disappeared by the end of 1 day. The background contains many clear spaces, and in the more severe lesions large fluid-filled clefts may appear in the tissue. The myelin sheaths and axis cylinders of nerve fibers begin to break down within 1 hr and undergo the afore-described changes for white matter. Some blood-filled capillaries are present at 1 hr. The hematogenous response is manifest within 6 hr by the presence of leucocytes. Microglial multiplication is evident at 4 days, and 12 days after irradiation the glial response is well developed.

The ultrasonic method of producing localized selective lesions in the central nervous system constitutes a unique and potent tool for experimental neurological and neurosurgical applications (6). The technique is currently being used in this laboratory in a variety of neurological studies.

W. J. FRY
J. W. BARNARD
F. J. FRY
R. F. KRUMINS
J. F. BRENNAN

Bioacoustics Laboratory,
University of Illinois, Urbana

References and Notes

1. P. D. Wall et al., *Science* 114, 686 (1951).
2. W. J. Fry et al., *Neurosurg.* 11, 471 (1954).
3. J. W. Barnard et al., *J. Comp. Neurol.*, in press.
4. This study was partially supported by the Biophysics Section of the Physiology Branch of the Office of Naval Research under contract Nonr 336(00)-NR 119-075.
5. W. J. Fry, *J. Acous. Soc. Amer.* 25, 1 (1953).
6. No commercial equipment is yet available.

11 April 1955

ERRATA

Page 283, column 2, line 14 from the bottom: "51" should read "40"; line 13 from the bottom: "4.8(10)³" should read "3.9(10)²."

Page 284, column 1, line 12, and column 2, line 1: "53" should read "41"; column 1, line 13, and column 2, line 2: "4.9(10)³" should read 4.0(10)²."

43

Copyright © 1956 by The Wistar Press

Reprinted from *J. Cell. Comp. Physiol.*, **48**(3), 435–457 (1956)

EFFECTS OF HIGH INTENSITY SOUND ON ELECTRICAL CONDUCTION IN MUSCLE [1]

WALTER WELKOWITZ [2] AND WILLIAM J. FRY

Bioacoustics Laboratory, University of Illinois, Urbana, Illinois

TWELVE FIGURES

INTRODUCTION

In recent years there has been considerable experimentation on the biological and medical applications of ultrasound. Pertinent literature can be located by reference to bibliographies (Naimark et al., '51; Curry et al., '51) and symposium publications (Matthes and Rech, '49; Giacommi, '50; Fry et al., '53). Some of this research has dealt with effects on muscle tissue (Gersten, '53 and '54; Gary and Gerendas, '49; Harvey, '29). In much of the reported work there is no adequate description of the characteristics of the sound field so that quantitative physical interpretations are difficult.

In this paper, the results of a study of the effect of intense acoustic radiation on the propagation of the action potential in muscle tissue are presented. Temperature measurements in the muscle were accomplished during irradiation, and the effect of pure heating on the electrical conduction process was investigated. The results indicate that irreversible suppression of the propagated action potential by the sound can be accomplished in the absence of a damaging temperature level. Although no experimental work was performed under a hydrostatic pressure sufficiently high to insure that no tension forces existed in the muscle, it is possible to conclude

---

[1] Supported by Contract AF33(038)-20922 with the Aero-Medical Laboratory of the Wright Air Development Center, Ohio.

[2] Submitted in partial fulfillment of the requirements for the degree of Doctor of Philosophy at the University of Illinois, 1954. Present address: Gulton Mfg. Corp., Metuchen, New Jersey.

on the basis of indirect observations that no extensive cavitation occurred in the muscle tissue.

EXPERIMENTAL METHODS

1. *Ultrasonic generation*

A travelling wave sound field, frequency 991 kc/sec., was used for most of the experimental work. This field was pro-

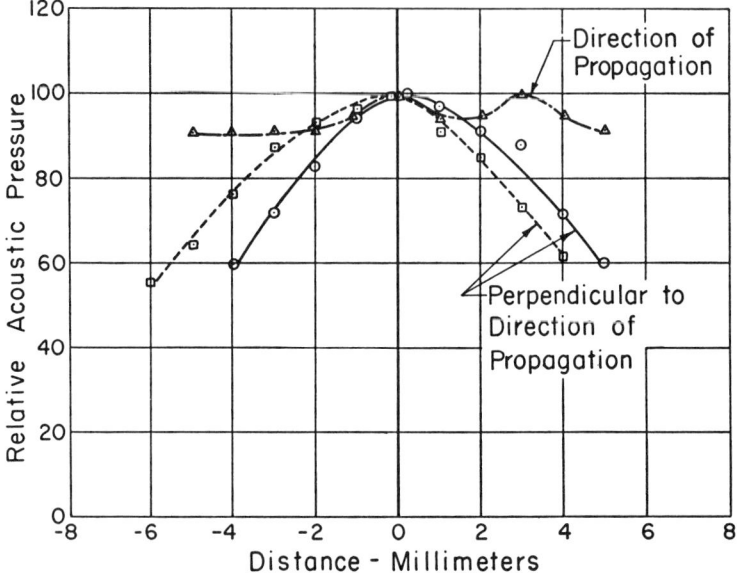

Fig. 1 Acoustic pressure distribution in the sound field. The zero of the coordinate system is the point at which the muscles were centered in the field. The value of the relative pressure amplitude at this point is arbitrarily taken equal to 100.

duced in degassed Ringer's solution by a flat, one inch diameter x-cut quartz crystal mounted as previously described (Fry, '50). The sound field distribution produced by the crystal system and measured at the muscle location is shown graphically in figure 1. The system was calibrated with a thermocouple probe described by Fry and Fry ('54). The intensity and acoustic pressure amplitude, as determined from measurements made with this device at the spatial peak of the field, are given in table 1 as a function of the voltage applied to the crystal. The biological preparations were subjected to single pulses (square wave envelope) of radiation. The temperature of the solution in which the muscles were immersed could be adjusted to any desired value.

TABLE 1

Calibration data for the crystal systems

SYSTEM	CRYSTAL VOLTAGE	PRESSURE AMPLITUDE	INTENSITY
	volts	*atmospheres*	*watts/cm^2*
1	8000	19.0	123
	7000	16.6	94
	6000	14.2	69
2	8000	22.2	168
	7000	19.4	128
	6000	16.6	94

2. Biological preparation

Excised hind leg biceps muscles from Rana pipiens were utilized in these studies. They were mounted in the holder as shown in figure 2. Irradiation was carried out with the muscle immersed in Ringer's solution. To eliminate cavitation during irradiation, the solution was degassed by boiling for 10 minutes and then cooled rapidly. The preparation was positioned, transverse to the direction of propagation, at the peak of the sound field by means of a probe and pointer system. The positioning procedure was found to give the same location repeatedly within one millimeter. Any nerve or nerve-muscle endplate effects were eliminated by curarizing the preparation to block nerve-muscle stimulation. The muscle fibers were excited directly by a square wave electrical pulse. The curarization was carried out by introducing d-tubocurarine chloride into the Ringer's solution at a concentration of 6 mg per liter. This concentration is about 5 times the dosage found by Kuffler ('42) to be necessary to completely block the nerve-muscle excitation path for single muscle fibers. For complete curarization, checked by electrical test, the muscle was permitted to remain in the solution for 20 minutes before irradiation.

The experimental procedure followed was to excite the muscle with a pair of electrodes at one end, record the propa-

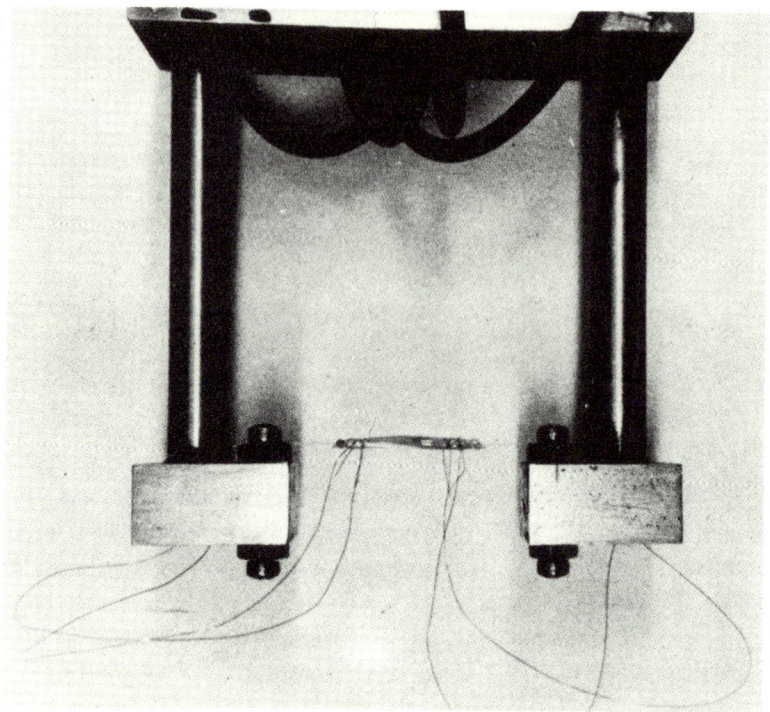

Fig. 2 Muscle mounted in holder. The electrical leads to both ends are shown.

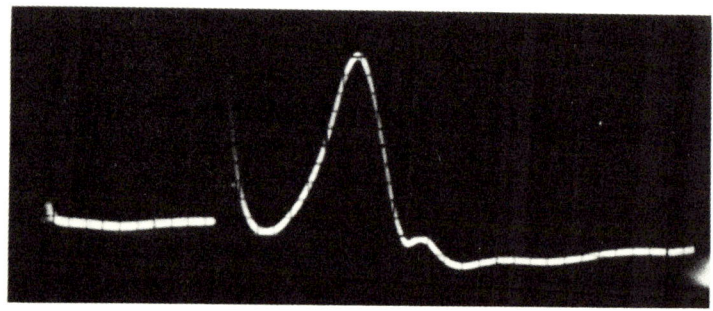

Fig. 3 Propagated muscle action potential obtained with the arrangement illustrated in figure 2.

gated action potential with a pair of electrodes at the opposite end and irradiate with sound between the pairs of electrodes. A propagated muscle action potential obtained under the experimental conditions outlined is shown in figure 3. No special attempt was made to obtain monophasic responses since this would have complicated the experimental procedure. The amplitude of such an action potential was the quantitative measure used throughout the experiments.

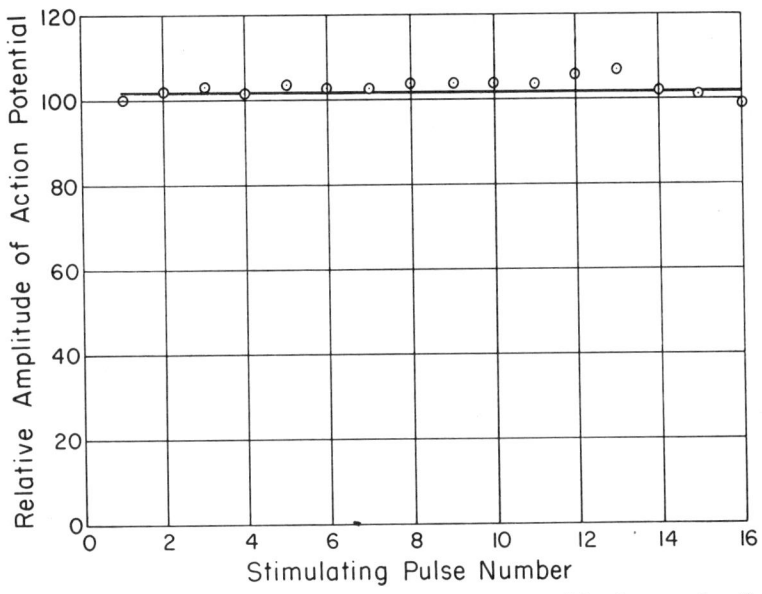

Fig. 4 Variation in the amplitude of the action potential of a muscle stimulated once every five seconds.

EXPERIMENTAL RESULTS

1. *Effects of acoustic irradiation*

Before proceeding to the irradiation studies, electrical control measurements were carried out on a number of muscle preparations. A typical graph of the relative amplitude of the action potential as observed on a muscle stimulated approximately once every 5 seconds for a minute and a half is shown in figure 4. This set of measurements and others like it demonstrate that deviations from an average value of less than about 5% can be obtained by this technique. The stability of the preparation using the experimental procedure described is therefore quite good.

The most pronounced effect obtained from irradiation was a large permanent suppression, even complete block, of the action potential after exposure to a single pulse of sound.

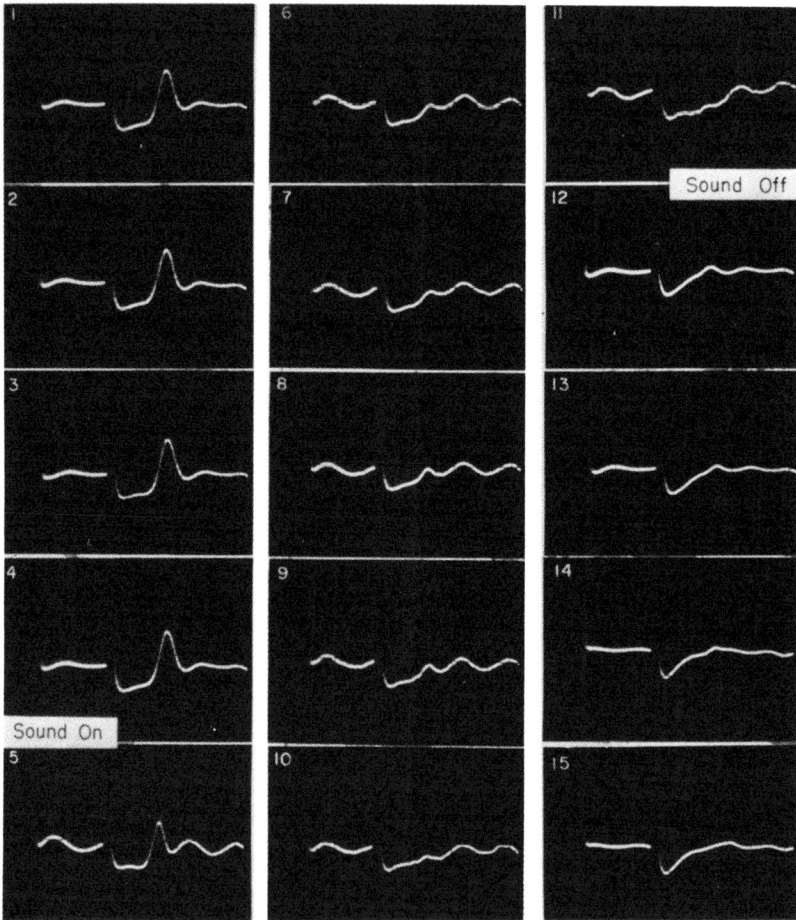

Fig. 5 Effect of acoustic radiation on the action potential of muscle. The propagated electrical responses to successive stimuli are ordered in columns. See figure 6 for a plot of relative amplitude.

This effect was established by measurements made on about 30 muscles. A typical run of data taken at 20°C. with a peak acoustic pressure amplitude of 19 atmospheres and a 40 second pulse of sound is shown in figure 5. The relative amplitude as a function of time is presented graphically in figure 6. Monotonically decreasing curves have been obtained

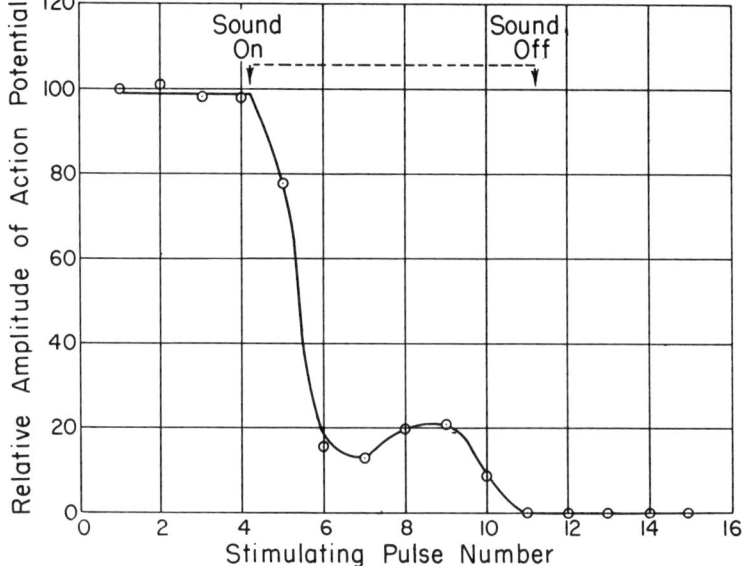

Fig. 6 Relative amplitude of the action potential of muscle as affected by acoustic irradiation. Values obtained from the records of figure 5. A length of the muscle between stimulating and receiving electrodes was subjected to the sound. Acoustic pressure amplitude, 19 atm.; period of irradiation, 40 sec.; bath temperatures, 20°C.

in addition to the type illustrated which shows a small rise during the period of suppression by the sound. No fundamental significance is attached to this small rise. The muscle was stimulated to twitch about once every 5 seconds. Four control measurements were taken before the sound was turned on and 4 more after the sound was turned off. From the figure and the graph it is apparent that propagation of the action potential was completely blocked by the sound. This effect is obtainable under various dosage conditions. Figure 7 shows a graph of results using a peak pressure amplitude of 16.6 atmospheres with a 100 second pulse. Similar effects can be obtained at lower temperatures. Figure 8 shows a run taken at 15°C. The amplitude of the action potential is presented graphically in figure 9. In this case about 75% suppression was obtained. A 60 second pulse of sound at 19

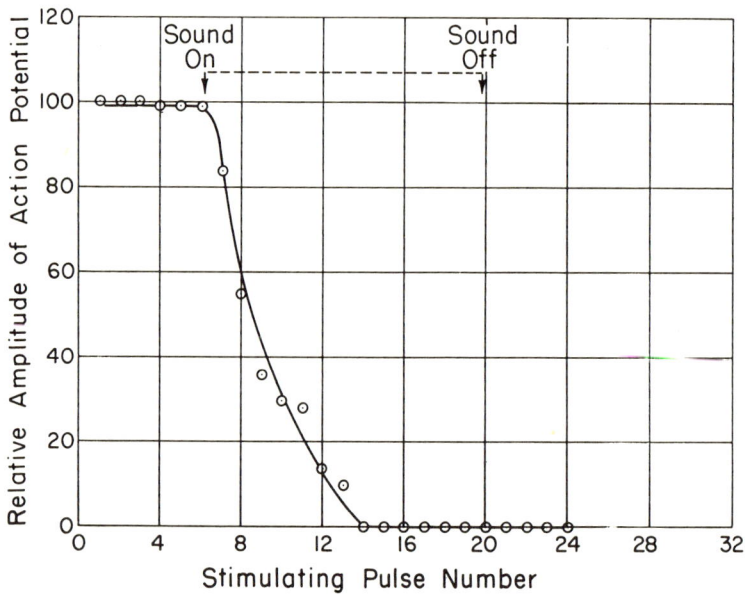

Fig. 7 Relative amplitude of the action potential of muscle as affected by acoustic irradiation. A length of the muscle between the stimulating and receiving electrodes was irradiated. Pressure amplitude, 16.6 atm.; period of irradiation, 100 sec.; bath temperature, 20°C.

atmospheres was used in obtaining this result. This run, in conjunction with others made at lower temperatures, indicates that the dosage conditions required to produce the observed effect on muscle are temperature dependent.

In order to derive more quantitative information from the experiments than that provided by a single dosage condition, data were obtained for a curve of minimum dosage required to produce a prescribed change in the electrical conduction

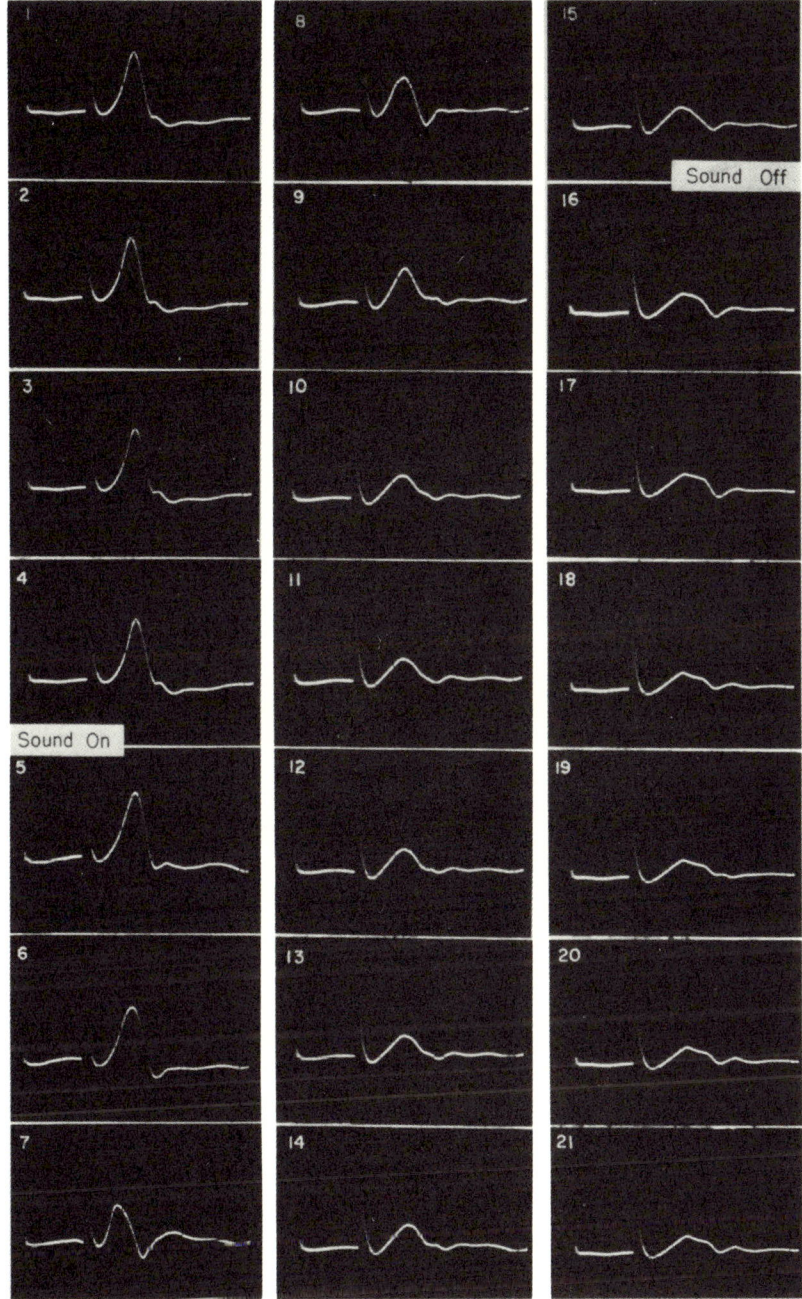

Fig. 8 Effect of acoustic radiation on the action potential of muscle. The propagated electrical responses to successive stimuli are ordered in columns.

characteristic of the biceps muscle. Minimum dosage was defined in the following manner. For each irradiation time, three muscles were subjected to the sound at each of several driving voltages across the crystal. These were spaced 250 volts apart, this increment representing less than 5% of the total driving voltage. The pressure amplitude of minimum

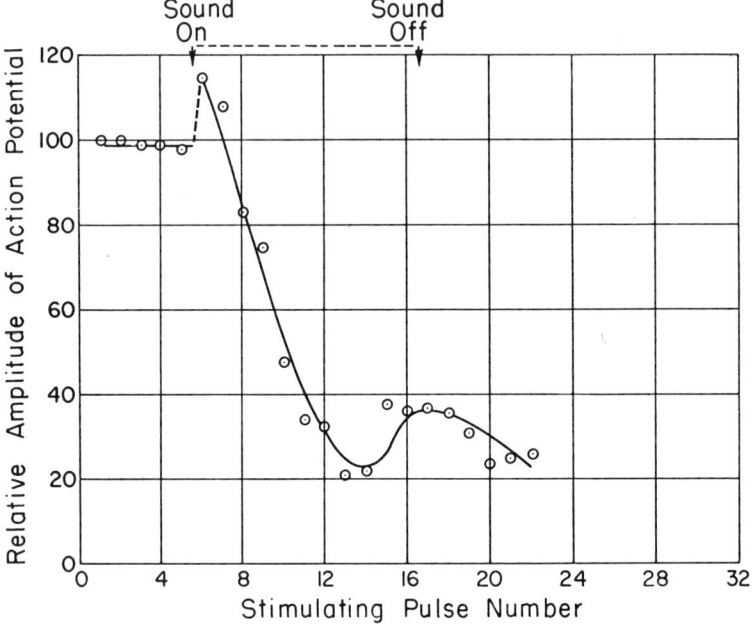

Fig. 9 Relative amplitude of the action potential of muscle as affected by acoustic irradiation. Values obtained from the records of figure 8. Acoustic pressure amplitude, 19 atm.; period of irradiation, 60 sec.; bath temperature, 15°C.

dosage for a specific duration of irradiation was defined to lie between the pressure amplitude corresponding to the voltage at which at least two of the three muscles experienced a reduction of more than 10% in action potential amplitude and that pressure amplitude at which no more than one muscle was so affected. A further proviso was required that at higher pressure amplitudes than the first, a greater reduction in action potential must be observed, and at pressure amplitudes below the second, the observed effects must be smaller. The reciprocal of the irradiation time was plotted against the sound pressure amplitude of minimum dosage to yield the curve of figure 10. The temperature range over which all these data were obtained was between 18 and 20°C.

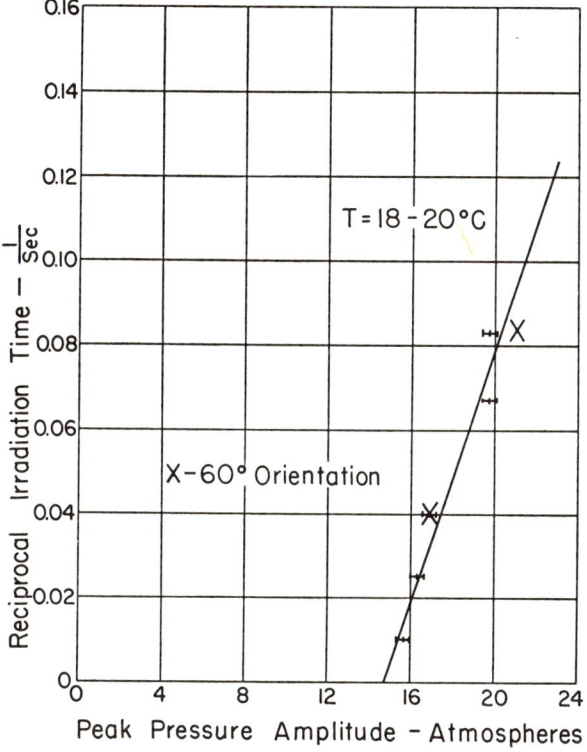

Fig. 10 Reciprocal of the minimum duration of irradiation to obtain a permanent reduction of the muscle action potential as a function of the acoustic pressure amplitude. See text for a precise definition of minimum dosage.

Over 120 frogs were used in obtaining the measurements on minimum dosage. The straight line relationship of the data is similar to the curve obtained by Fry ('51) from data on the paralysis of frogs irradiated in the lumbar enlargement region of the spinal cord. It is of interest to note here that the induction by ultrasound (lower in pressure amplitude than the levels used in the experiments reported in this paper) of reversible changes in muscle which can be ascribed to a non-temperature mechanism have been previously reported (Busnel et al., '53).

It was observed in the course of the experimentation that no difference in effects occurred by switching the muscle end for end. The possible dependence of the observed effects on the geometrical orientation of the muscle with respect to the direction of propagation of the sound was considered. Consequently, experiments were performed at two of the pulse durations used in obtaining the data for figure 10 with the muscle oriented at 60 degrees with respect to the direction of propagation rather than perpendicular to it. These results, plotted in figure 10, indicate that there is no major difference for the orientations tested.

Histological studies were carried out on irradiated muscles stained with hematoxylin-eosin. The tissue was fixed in formalin immediately after irradiation. No differences were observed microscopically between the irradiated and normal tissue. Examination of sections of irradiated tissue indicate no tissue tearing or vacuolization.

Preliminary work on the acoustic irradiation of striated muscle with the vascular system intact (gastrocnemius muscle of Hyla) indicates that results similar to those described in this paper are obtained.

Irradiation of excised sciatic nerve of Rana pipiens at the highest sound level used in the studies reported herein, exposed for much longer periods of time than that required to permanently block muscle action potentials, produces no observable suppression in the nerve action potential.

2. Temperature changes in muscle produced by irradiation

Temperature changes were measured in muscles during irradiation by means of imbedded thermocouples. These thermocouples were made of three mil copper and constantan wires joined with a soldered lap junction. The thermocouple was threaded through the excised mounted muscle. Equilibrium temperature measurements were made. Practical equilibrium was established in less than 10 seconds after the sound was turned on. The thermocouple was placed in the muscle as close to the center line as possible. As indicated previously the muscle was located at the peak of the sound field by means of a probe and pointer. The thermocouple emf was observed for various sound levels so that a curve

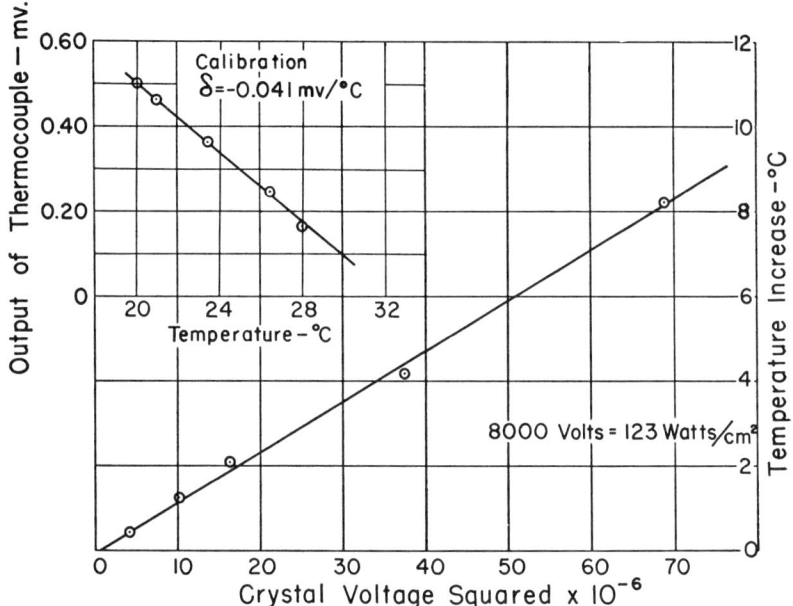

Fig. 11 Equilibrium temperature rise in a muscle at various sound levels determined by measuring the emf generated by an imbedded thermocouple.

could be obtained of temperature change versus acoustic intensity. The thermocouple was calibrated for each muscle while imbedded in the muscle. The holder with muscle and imbedded thermocouple was placed in baths of Ringer's solution at various temperatures and emf readings were taken. The temperatures of the baths were measured with a mercury thermometer and a calibration curve of thermocouple emf as a function of temperature was plotted.

Data of the type described were taken on 10 muscles. A typical thermocouple calibration curve and temperature measurement curve for a muscle are shown in figure 11. In all but one of these instances, a plot of temperature rise as a function of the square of the crystal voltage demonstrated a linear relationship. The one set of non-linear data was therefore disregarded as representing an atypical situation, possibly caused by cavitation at the thermocouple-muscle interface. Table 2 shows values of the equilibrium temperature rises in muscles for peak sound intensities of 123 and 94 watts/cm². At the higher acoustic level it is possible to block completely

TABLE 2

Measured temperature changes in excised biceps muscle under acoustic irradiation

MUSCLE NUMBER	TEMPERATURE CHANGE AT 94 WATTS/CM²	TEMPERATURE CHANGE AT 123 WATTS/CM²
	°C.	°C.
1	5.5	7.1
2	5.9	7.8
3	4.5	5.8
4	4.3	5.6
5	3.2	4.2
6	3.4	4.4
7	4.6	6.0
8	5.3	6.9
9	9.6	12.5

the muscle response in about 40 seconds. At the lower level about 100 seconds are required. It is apparent from table 2 that quite a wide range of values for the equilibrium temperature rise is obtained in various muscles. The reasons for this are readily understood on the basis of the following discussion.

In the problem being considered, the muscle is roughly the shape of a long cylinder and is imbedded in a large bath of Ringer's solution which is maintained at essentially constant temperature. Transverse to the axis of the muscle, the sound intensity varies less than 10% over the muscle. In the axial direction the beam is about 7 mm wide at the half power points. The muscle is generally quite a bit longer than this.

Two simple geometries for the mathematical solution of the problem suggest themselves for the physical conditions. One is an infinitely long, solid, circular cylinder with a constant amount of heat per unit volume generated throughout and with the surface kept at a constant temperature. Since the sound beam is finite and small, the other suggested geometry is a solid sphere with a constant amount of heat per unit volume generated throughout and with the surface kept at a constant temperature. The cylindrical model leads to predicted temperature values in the muscle which are high, since heat is not actually supplied all along the cylinder but only in a small region. The second geometry probably leads to low values of predicted temperature rise, since part of the boundary is then not Ringer's solution but is muscle and does not remain at the fixed temperature of the Ringer's solution.

For the cylindrical situation, under equilibrium conditions, the solution to the general heat flow equation is (see for example Carslaw and Jaeger, '47)

$$T - T_1 = A(a^2 - r^2)/4k \qquad (1)$$

while for the spherical case

$$T - T_1 = A(a^2 - r^2)/6k \qquad (2)$$

In equations (1) and (2), T_1 is the constant temperature of the Ringer's solution and T designates the temperature in the muscle. The symbol r represents the radial coordinate and a is the radius of the muscle. The symbol k designates the heat conductivity coefficient of the muscle and A is the heat generated per second per unit volume resulting from absorption of acoustic energy. These equations demonstrate that the solutions for the temperature distribution as a function of radial coordinate are identical in form and differ for any value of r by only 33% in magnitude. Since these are extreme situations with the physical case falling somewhere between, it seems reasonable to use these solutions to describe approximately the actual condition in the muscle. Comparison of the temperature changes predicted by the equations with the measured values can be made by inserting appropriate values of the constants into equations (1) and (2).

The value of A, the heat generated per unit volume per second, is given by

$$A = I\mu \qquad (3)$$

where I designates the sound intensity and μ the intensity absorption coefficient per unit path length. In order to calculate values corresponding to the measured results given in table 2, intensities of 123 and 94 watts/cm² were used. The value of μ for striated frog muscle was found by Fry et al. ('50) to be about 0.2. For lack of more complete information, the value of the thermal conductivity coefficient of water (k = 0.0060 watts/cm²/°C. at 20°C.) was used in

TABLE 3

Computed temperature changes along the axis of a model of a muscle under sound irradiation

SOUND INTENSITY	TEMPERATURE RISE CYLINDRICAL GEOMETRY	TEMPERATURE RISE SPHERICAL GEOMETRY
watts/cm²	°C.	°C.
123	10.3	6.9
94	7.9	5.3

the calculations. Measurements of the diameters of 50 muscles indicated an average value of a = 0.1 cm. Table 3 shows the computed values of peak temperature rise (r = 0) for the two geometries at the two sound intensities chosen.

A comparison of tables 2 and 3 demonstrates that the agreement between calculation and measurement is good. An examination of equations (1) and (2) helps explain the variation in the measured values. Part of the variation may be accounted for by inaccurate positioning of the thermocouple. For example, if the thermocouple is positioned at half the radial distance away from the axis of the muscle, then the temperature rise will be only three quarters of the peak value (r = 0). Much of the variation can be accounted for by differences in muscle size. Since the equilibrium temperature rise is proportional to the square of the muscle radius, an increase in this value of only 25% leads to a change in the value of the temperature rise of more than 50%.

A pertinent point to note from the equations is that the temperature rise at any radial distance is proportional to the quantity (a^2-r^2), therefore half the muscle fibers undergo a temperature rise of less than half the peak value.

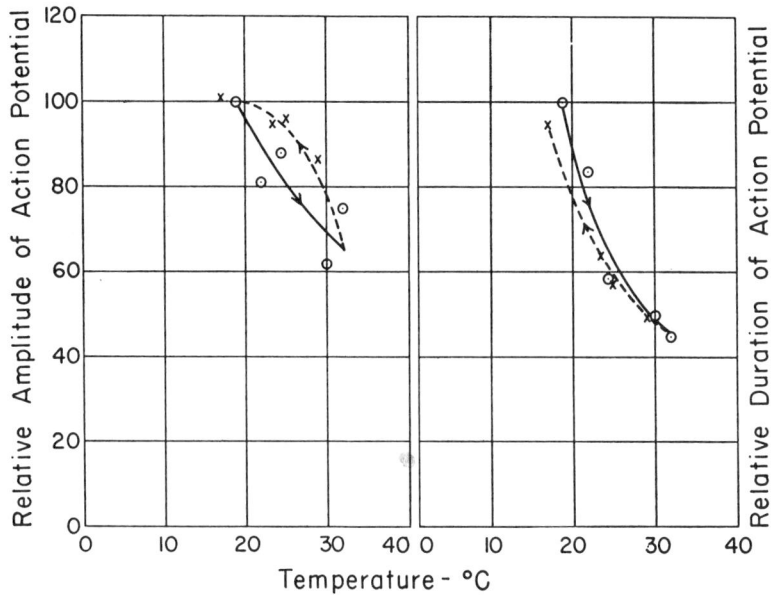

Fig. 12 Relative amplitude and duration of the action potential of frog biceps muscle as a function of the temperature.

3. Effects of temperature changes on muscle action potential

Study of some of the literature in the field (Doi, '20; Walker, '49) yielded little conclusive information on the effects of temperature changes on the action potential of muscle. Experiments were therefore undertaken to determine the effects of temperature changes on the frog muscle action potential under conditions similar to those established for the acoustic irradiation experiments. Measurements were carried out on 7 muscles under the following conditions. A number of salinetubocurarie baths were kept at various fixed temperatures and each muscle was run through the different baths. The muscle was kept in each bath for two minutes to bring about equilibrium conditions before electrical measurements were made. This procedure yielded consistent results. In 6 of the 7 muscles, almost complete reversibility in height and duration of the action potential was measured as the muscle was carried through a temperature cycle from about 20°C. to about 35°C. then back to 20°C. In no case was the amplitude of the action potential reduced to zero. A graph of the relative changes in amplitude and duration of the action potential as a function of temperature is shown in figure 12 for a typical muscle. These measurements of the effects of temperature change on the action potential demonstrate that if acoustic absorption in the muscle does not result in a temperature greater than 35°C. throughout the muscle, permanent reduction or irreversible block of the action potential cannot be attributed to the temperature change.

DISCUSSION OF RESULTS

Before considering the physical mechanism of the action of sound on muscle tissue, it is necessary to eliminate possible artifacts. Hydrodynamical flow, which is present in intense sound fields in fluids, might cause mechanical difficulties at the connection of the electrodes to the muscle. In addition, direct mechanical forces on the electrodes, such as radiation pressure, might contribute to the observed effects.

In investigating the hydrodynamic flow artifact possibility, experiments were carried out with flows produced by ejecting fluid from a jet immersed in the coupling liquid and directing it at the muscle. The effects on the observed action potential, caused by flow velocities 6 to 10 times those present during acoustic irradiation, are slight and indicate that the results obtained during ultrasonic irradiation (large reduction or complete block of the action potential) are not caused by a flow artifact.

Two types of experiments were carried out to eliminate the possibility of direct electrode artifacts. When a balsa wood fixture was mounted on the muscle holder for the purpose of shielding the electrodes from the sound field, large reductions of the height of the action potential were still attainable under ultrasonic irradiation. When a pair of electrodes was placed at the peak of the acoustic field and irradiated with a sound level equivalent to that normally incident on the electrodes for the customary geometry only slight changes in the observed response were evident. Therefore electrode artifacts do not account for the observed ultrasonic effects.

Besides these artifacts, it is always possible that the observed effects can be caused by heating or cavitation in the muscle tissue. Finally, the mechanical forces which are inherent in a sound field travelling through a medium can be involved in the mechanism. These forces can be unidirectional, such as radiation pressure or Oseen type, or sinusoidally varying with time such as the oscillatory viscous force associated with the relative motion between a fluid medium and imbedded particles.

The possibility that the observed irreversible suppression of the action potential is caused by excessive muscle heating has been eliminated. An examination of the irradiation data indicates that it is possible to block permanently or greatly reduce the amplitude of the action potential with the muscles at initial temperatures of 20°C. and 15°C. by irradiating at a peak sound pressure amplitude of 19 atmospheres (123 watts/cm^2). The measured values of the temperature rise in the muscle under the same irradiation conditions show that the temperature in the tissue does not exceed approximately 30°C. for an initial temperature of 20°C., and for an initial temperature of 15°C. the maximum temperature does not exceed 25°C. An examination of the effects on the muscle action potential of a temperature cycle reaching a maximum value of 35°C. proves that blocking does not occur and that the amplitude changes are essentially reversible. The experimentally measured values of the temperature rise in the muscle during irradiation are in agreement with calculated values, and the analysis further indicates that half the fibers in the muscle undergo temperature changes of less than half the maximum value.

In discussing the possible role of cavitation in the mechanism of the observed effects produced in muscle, it is noted that as yet no direct attempt has been made to completely eliminate the cavitation possibility by carrying out the experiments under a hydrostatic pressure sufficiently high to insure that no tension forces exist in the tissue. Such experiments are currently in the planning stage. However, there are a number of indications that cavitation was not present, at least to any appreciable extent, under the experimental conditions which prevailed. The first evidence comes from an examination of the thermocouple measurements of temperature rise in the muscle under sonic irradiation. These measurements indicate that for peak sound pressure amplitudes up to about 20 atmospheres the temperature change is proportional to the square of the crystal voltage. Marked deviation from this linear relationship would be expected in the presence of cavitation. Many investigators have demonstrated that sound level measurements, in a medium in which cavitation is present, become erratic. A second piece of evidence against excessive cavitation in the tissue under irradiation results from an examination of the stained sections of irradiated muscle. Microscopic study shows no vacuole formation or fiber tearing. Another argument against the possibility that cavitation is involved in the mechanism of the effect produced in these experiments is brought out by the work of Esche (52) in which he found no evidence of cavitation at a hydrostatic pressure of one atmosphere and a frequency of 500 kc/sec, in beef muscle up to sound intensities in the tissue of about 400 watts/cm^2.

It appears that neither tissue heating nor cavitation is the primary physical factor involved in producing irreversible block of the action potential in striated muscle under the experimental conditions described in this paper. The effect of the high intensity acoustic irradiation of the muscle may therefore be a result of mechanical forces acting on submicroscopic elements of the tissue. From the experimentally observed form of the minimum dosage curve, it is reasonable to expect the mechanism to take a mathematical form approximating that proposed by Fry et al. ('51). If the threshold value is chosen as unity for the scale along the horizontal axis for the muscle dosage curve and for the frog paralysis dosage relation (Fry et al., '51) then the two relations appear almost identical. This suggests that the mechanism may well be the same in the two cases. Knowledge of the apparent lack of a dependence of the effect on the angle between the direction of the muscle fibers and the direction of propagation of the sound, should be useful in any attempt to establish a mechanical force mechanism.

SUMMARY

The studies presented in this paper demonstrate that under appropriate ultrasonic dosage conditions, the propagated action potential of an excised striated muscle can be permanently reduced or completely blocked. This can be accomplished in the absence of a temperature level sufficient in itself to cause permanent suppression. A quantitative determination of the minimum dosage relation (duration of exposure as a function of sound level) for a 10% permanent reduction of the muscle action potential has been accomplished. The form of this relation is the same as that given by Fry et al. ('51) for the relation between the minimum exposure time for paralysis of the hind legs of frogs, irradiated in the lumbar enlargement region of the spinal cord, and the sound level. Histological examination of stained tissue sections shows no gross tearing or vacuolization which might be expected if cavitation were present. Measurements made during irradiation with an acoustic probe imbedded within the muscle provide supporting evidence that cavitation is absent.

Results similar to those obtained on excised muscles were obtained on muscles with intact vascular systems.

No permanent suppression of the nerve action potential was produced on excised sciatic nerves under irradition conditions similar to those used on excised muscles.

ACKNOWLEDGMENT

J. W. Barnard, F. J. Fry, and L. Dreyer also contributed to this study.

LITERATURE CITED

BUSNEL, R. G., J. GLIGORIJEVIC, P. CHAUCHARD AND H. MAZOUÉ 1953 Contribution à l'etude des effets et des mécanismes d'action des ultrasons sur le système neuro-musculaire. Der Ultraschall in der Medizin, 6: 1–25.

CARSLAW, H. S., AND J. C. JAEGER 1947 Conduction of heat in solids, Oxford University Press, Oxford, England.

CURRY, B., E. HSI, J. S. AMBROSE AND F. W. WILCOX 1951 Bibliography — Supersonics or Ultrasonics, Research Foundation, Oklahoma A. and M. College.

DOI, Y. 1920 Studies on muscular contraction. I. The influence of temperature on mechanical performance of skeletal and heart muscle. J. Physiol., 54: 218–226.

ESCHE, R. 1952 Untersuchung der Schwingungskavitation in Flüssigkeiten. Akustische Beihefte, 4: 208.

FRY, F. J. 1950 An ultrasonic projector design for a wide range of research applications. Rev. Sci. Inst., 21: 940–941.

FRY, W. J., V. J. WULFF, D. TUCKER AND F. J. FRY 1950 Physical factors involved in ultrasonically induced changes in living systems. I. Identification of non-temperature effects. J. Acoust. Soc. Am., 22: 867–876.

FRY, W. J., D. TUCKER, F. J. FRY AND V. J. WULFF 1951 Physical factors involved in ultrasonically induced changes in living systems. II. Amplitude duration relations and the effect of hydrostatic pressure for nerve tissue. J. Acoust. Soc. Am., 23: 364–368.

FRY, W. J., J. F. HERRICK, J. F. LEHMANN, J. J. WILD, ET AL. 1952 Symposium on ultrasound in biology and medicine. University of Illinois. J. Acoust. Soc. Am., 25: 1–25 and 270–285.

FRY, W. J., AND R. B. FRY 1954 Determination of absolute sound levels and acoustic absorption coefficients by thermocouple probes — Experiment. J. Acoust. Soc. Am., 26: 311–317.

GARAY, K., AND M. GERENDAS 1949 Effect of ultrasonic vibration on muscle fibres in vitro. Experientia, 5: 410–411.

GIACOMINI, A. Editor of Atti del Convegno Internazionale di Ultracoustica 1950 Supplement Vol. VII, Ser. IX Nuovo Cimento, Rome, Italy.

GERSTEN, J. W. 1953 Thermal and non-thermal changes in isometric tension, contractile protein, and injury potential, produced in frog muscle by ultrasonic energy. Archives Phys. Med., 34: 675–685.

GERSTEN, J. W. 1954 Ultrasonics and muscle disease. Am. J. Phys. Med., *33*: 68–74.

HARVEY, E. N. 1929 The effect of high frequency sound waves on heart muscle and other irritable tissues. Am. J. Physiol., *91*: 284–290.

KUFFLER, S. W. 1942 Electric potential changes at an isolated nerve-muscle junction. J. Neurophysiol., *5*: 18–26.

MATTHES, K., AND W. RECH Editors of Der Ultraschall in der Medizin 1949 S. Hirzel, Zürich, Switzerland.

NAIMARK, G. M., J. KLAIR AND W. A. MOSHER 1951 A bibliography on sonic and ultrasonic vibration; biological, biochemical, and biophysical applications. J. Franklin Institute, *251*: 279–299 and 402–408.

WALKER, S. M. 1949 Potentiation of twitch tension and prolongation of action potential induced by reduction of temperature in rat and frog muscle. Am. J. Physiol., *157*: 429–435.

Ultrasonic Irradiation of the Central Nervous System at High Sound Levels*

WILLIAM J. FRY AND FLOYD DUNN
Bioacoustics Laboratory, University of Illinois, Urbana, Illinois
(Received October 25, 1955)

High level ultrasound produces, under properly controlled dosage conditions, selective changes in the central nervous system. The physical mechanism of the action of the sound requires elucidation. Some of the problems associated with determining the physical mechanism are discussed and a preparation and procedure are described which are appropriate for accurately determining dosage relations for such a study. The quantitative results obtained with this preparation are presented.

THE high level ultrasonic method for producing selective accurately localized lesions in brain tissue by focusing an ultrasonic beam in the region to be affected now constitutes a unique tool for neurological research and human neurosurgery.[1,2] Extensive histological studies,[3,4] on monkeys and cats, demonstrate that accurately localized selective changes can be made at a chosen region in the brain without destroying tissue between the chosen region and the area of entry of the sound into the brain. Both fibers and cell bodies can be destroyed without interrupting blood vessels traversing the lesion area, and nerve fiber tracts can be interrupted without destroying surrounding or neighboring gray matter.[1–4]

If the full potentialities of this new method of producing selective changes in brain are to be realized the physical mechanism of the action of the sound on the tissue must be understood. Previous publications[5,6] from this laboratory on the physical mechanism reported evidence demonstrating that damaging temperature levels and cavitation are not responsible for the action of the sound. This earlier work was accomplished using frogs as the biological test specimen. To further elucidate the physical mechanism, a comprehensive investigation involving the determination of the ultrasonic dosage relations, i.e., duration of exposure as a function of the acoustic variables (intensity, particle velocity, pressure, etc.), required to realize a given functional end point (paralysis of the hind legs following irradiation of the lumbar enlargement of the spinal cord) has been undertaken. The dosage relations will be determined at a variety of frequencies with the preparation under various hydrostatic pressures and at a number of base temperatures. This is essential in order to separate concomitant secondary effects from the primary action of the sound, for example, the temperature coefficient of the primary action as contrasted with the primary mechanism itself. Since mammals are the most important animal class to which this new method will be applied, it is desirable that a representative of this group be used for this dosage study. The young mouse (strain LaF1), 24

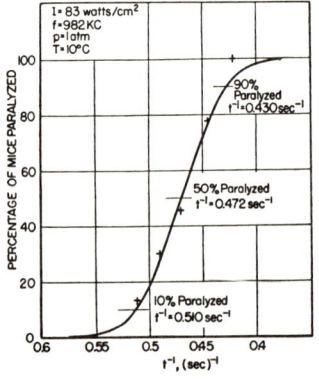

FIG. 1. Sigmoidal distribution of percentage of mice paralyzed as a function of the reciprocal of the duration of exposure at a constant sound intensity.

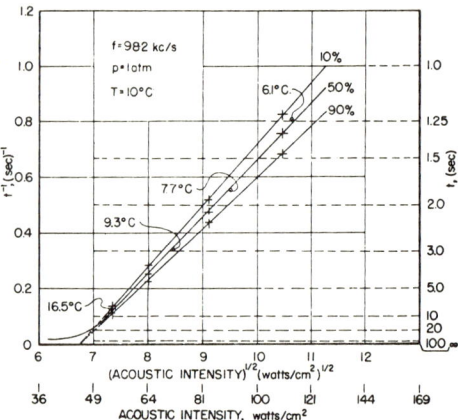

FIG. 2. Threshold region for paralysis of the hind legs of mice under ultrasonic irradiation. The reciprocal of the duration of exposure is shown as a function of the square root of the acoustic intensity for 3 percentages of animals paralyzed. The temperatures shown in the graph are the temperature increases measured by imbedded thermocouples in the spinal cords of the irradiated mice corresponding to the values of acoustic intensity and duration of exposure at the coordinate positions indicated by the arrows.

hours after birth (weight 1.2 to 1.4 g), is a convenient preparation for the following reasons:

(1) It is essentially poikilothermic,[7] that is, it possesses virtually no temperature control mechanism, so that it can be carried through reversible temperature cycles, with its temperature being reduced to nearly 0°C, without producing either physiological or morphological changes.

(2) Ossification is not complete. As determined by standard staining techniques, the tissue overlying the dorsal side of the cord is soft tissue, while that over the lateral and ventral sides shows a slight degree of ossification. Thus, acoustic absorption in the region surrounding the spinal cord is low, and no surgery need be performed in preparing the animal for irradiation.

(3) This animal is small in size so that it is possible to irradiate the desired region with a nearly uniform acoustic field with a single controlled ultrasonic pulse.

In order to realize a dosage study of the type envisioned, i.e., such that the experimental results can be interpreted with some facility to yield information concerning the fundamental physical mechanism, it is essential that the tissue of interest be exposed only once to the radiation. If multiple exposures (both temporal and overlapping spatial-temporal) are used, uncertainty arises regarding the magnitude of the residual effect[6] of the previous exposure. Data obtained with the use of multiple exposures are thus of limited value. However, the technique is useful in limiting the maximum temperature rise by the choice of an appropriate duty cycle and, therefore, may be used during the initial stage in a mechanism study in order to rule out the possibility of a damaging temperature level as the basic factor. For single exposures, the region to be affected by the sound must be irradiated with a beam which is uniform since experimental results obtained by the use of nonuniform beams are more difficult to interpret to elucidate the fundamental mechanism. Therefore, an unfocused quartz crystal is used to develop a traveling wave field. At 5% below the peak intensity, the beam width (along the length of the spinal cord) is 2.6 mm.

In the study undertaken at this laboratory, the mice are irradiated at the third lumbar vertebra. Since motor paralysis of the hind legs is the functional end point observed, the region of the spinal cord which must be altered is that containing the neurons and fibers associated with the femoral, sciatic, and obturator nerves. This region, at which the axis of the acoustic beam is centered, was determined by acoustic means. In the lumbar region, the vertebral segments are 0.67 mm long, measured from corresponding edges. Thus, nearly four vertebral segments are irradiated with an acoustic intensity variation of no more than 5%. The over-all uncertainty of the center of the third lumbar vertebra with respect to the axis of the sound field is 0.25 mm. Since the beam width at 95% of the peak intensity is 2.6 mm, it appears that the over-all accuracy of positioning the animal in the sound field is adequate.

The acoustic field is calibrated each day with a thermocouple probe which has previously been calibrated by a radiation pressure detector[8,9] utilizing a small steel sphere.

In preparing the animal for the irradiation procedure, the mouse is first cooled to render it dormant so that it can be properly positioned in the mouse holder and remain in that position until

TABLE I. Temperature rises measured by imbedded thermocouples in spinal cords of irradiated mice at various values of the sound intensity and duration of exposure.

t (sec)	I (w/cm²)	$I^{\frac{1}{2}}$	ΔT (°C)
7.70	54	7.4	16.5
2.88	71	8.5	9.3
1.80	90	9.5	7.7
1.25	112	10.6	6.1
0.965	140	11.8	6.0
0.865	154	12.4	5.7

it is placed in the sound tank. The mouse holder is designed to prevent movement once the animal is properly positioned in the acoustic field. The positioning of the animal in the mouse holder is accomplished by an optical arrangement.

The holder containing the animal is then placed in the sound tank, which is filled with degassed 0.9% saline. Several minutes are permitted to elapse before irradiation in order that the animal can reach temperature equilibrium which is realized, as checked by measurement with imbedded thermocouples, in this period of time. When the mouse reaches equilibrium, a single acoustic pulse of rectangular envelope (having rise and decay times of several microseconds), predetermined acoustic intensity (plane wave case), and time duration is then initiated. After the cessation of the sound, the animal holder is removed from the tank and the mouse is rapidly warmed to room temperature. The animals are examined for paralysis or overt movements approximately 15 min after exposure and again after 6 hours.

Thus far, results have been obtained at a frequency of 982 kc, a hydrostatic pressure of one atmosphere, and base temperatures of 10°C and 20°C. These results show that a well-defined threshold region exists. This threshold region is defined as follows: If a large number of animals are irradiated with identical values for the acoustic field variables for various periods of time, and the percentage of animals paralyzed at each duration of exposure is plotted as a function of the reciprocal of the time duration of exposure, a sigmoid curve is obtained. A typical sigmoid curve obtained by standard statistical treatment of the data is shown in Fig. 1. Each point represents approximately 20 animals. The threshold range at the chosen values for the acoustic field variables is arbitrarily defined as the range of time durations of exposure from 10% of the animals paralyzed to 90% of the animals paralyzed. The collection of these threshold ranges for various values of a specific acoustic variable, for example, particle velocity amplitude, defines the threshold region for that variable.

Figure 2 exhibits the threshold region (with the square root of intensity as the acoustic field variable) at a frequency of 982 kc, a hydrostatic pressure of one atmosphere, and a base temperature of 10°C. The ordinate is the reciprocal of the exposure time, and the abscissa is the square root of the acoustic intensity. Time and intensities are also indicated on the coordinate axes for convenience. The plotted points are obtained from sigmoid curves similar to that shown in Fig. 1. From Fig. 2, it is seen that in the range of dosages between approximately 50 w/cm² (10-sec duration) and 120 w/cm² (1-sec duration), the relation between the values of the reciprocal of the exposure time and the square root of the acoustic intensity (at which 50% of the animals are paralyzed) appears to be linear. In this linear range, the boundary values of the threshold region, at constant intensity, differ by only 17%. At the extremities of this region, the threshold curves deviate from linearity.

In the course of this study, measurements were made of the temperature rise in the spinal cord of the mice as a function of ultrasonic dosage. This was accomplished by imbedding small thermocouples in the cord. Table I shows the results of these measurements for dosages within the threshold region. These temperature rise values are also indicated at the corresponding dosage coordinates in Fig. 2. Considering these temperature increases in the cord in conjunction with the value for the base temperature of the animal, 10°C, it can be concluded that temperature rise is not the primary factor for the observed alterations in the central nervous system. Pulsed sound with an appropriate duty cycle can be used to suppress the temperature rise resulting from acoustic absorption. The use of a pulse width of 0.4 sec and a duty cycle of 40% as reported by other investigators[10] does not, however, yield a practical advantage in this respect. Experiments were performed at this laboratory to compare single and multiple pulsing methods of irradiation. The results are shown in Table II. Two duty cycles were chosen for the multiple pulsing scheme (40% and 10%) to compare with a single pulse of equal total

TABLE II. Comparison of the temperature rises in spinal cords of irradiated mice under various pulsing procedures. Values are given for two sound intensities and three duty cycles.

I w/cm²	Sd. on (sec)	Sd. off (sec)	Duty cycle (%)	No. of pulses	ΔT (°C)	Diff. (%)
100	0.400	0.60	40	10	15.7	3.2
100	4.000	...	100	1	16.2	
100	0.100	0.90	10	10	4.3	19
100	1.000	...	100	1	5.2	
70	0.400	0.60	40	7	8.9	5.3
70	2.800	...	100	1	9.4	
70	0.100	0.90	10	10	2.5	49
70	1.000	...	100	1	3.8	

irradiation time. The experiments were carried out at two intensities. In Table II it is seen that the 40% duty factor provided a rather negligible decrease in the temperature rise compared with a single pulse of the same total exposure time, the difference being approximately 3%. At the same intensity, the 10% duty factor provides a greater advantage over the single exposure procedure, but the decrease is still only 20%. In any event, since it is possible by suitable choice of preparation to obtain the ultrasonically produced changes in the complete absence of a questionably damaging temperature increase and, since results obtained from multiple pulsing procedures are more complex to interpret, it is felt that the single pulse method constitutes the procedure of choice at our present stage of understanding of the physical mechanism.

* This research was partially supported by Contract AF33(038)-20922 with the Aero Medical Laboratory of the Wright Air Development Center Ohio.
[1] Fry, Mosberg, Barnard, and Fry, J. Neurosurg. 11, 471 (1954).
[2] Fry, Barnard, Fry, and Brennan, Am. J. Phys. Med. 34, 413 (1955).
[3] Barnard, Fry, and Krumins, J. Comp. Neurol. 102 (1955).
[4] Barnard, Fry, Fry, and Brennan, Arch. Neurol. Psychiat. 74 (1955).
[5] Fry, Wulff, Tucker, and Fry, J. Acoust. Soc. Am. 22, 867 (1950).
[6] Fry, Tucker, Fry, and Wulff, J. Acoust. Soc. Am. 23, 364 (1951).
[7] See for example, T. J. B. Stier and G. Pincus, J. Gen. Physiol. 11, 349 (1928).
[8] W. J. Fry and R. B. Fry, J. Acoust. Soc. Am. 26, 294 (1954).
[9] W. J. Fry and R. B. Fry, J. Acoust. Soc. Am. 26, 311 (1954).
[10] Hueter, Ballantine, and Cohen, Quarterly Progress Report, Massachusetts Institute of Technology (July–September) (1954), p. 21 (October–December) (1954), p. 14.

PHYSICAL MECHANISMS OF THE ACTION OF INTENSE ULTRASOUND ON TISSUE[1]

FLOYD DUNN, PH.D.

It is now well known that intense ultrasound, when properly controlled, can produce unique results in biological systems. An example of this is the high intensity ultrasonic method of producing selective, accurately localized alterations in brain tissue by focusing the acoustic energy in the region to be affected (1, 2, 3). Figure 1 is a tissue section of the brain of a cat which shows a lesion approximately 1.0 mm. in diameter interrupting the mammillothalamic tract. To produce this lesion, converging ultrasonic beams entered the brain from the top and traversed three quarters of the brain thickness before coming to a focus at the site of the lesion. No alteration occurred to intervening tissue. This method now constitutes a unique tool for neurological research and human neurosurgery.

If the full potentialities of such new methods, for fundamental research and for medicine, are to be realized, it is essential that the physical mechanism of the action of intense ultrasound on tissue be understood. In order to elucidate the physical mechanism, a comprehensive investigation involving the determination of the ultrasonic dosage relations required to produce a given functional endpoint in a young mammal has been undertaken at the Biophysical Research Laboratory of the University of Illinois. Dosage relations can be expressed as the time duration of exposure as a function of the acoustic variable, i.e., intensity, pressure, particle velocity, etc., under specified environmental conditions for production of the specific endpoint.

Since mammals are probably the most important animal class to which the new ultrasonic methods will be applied, it is desirable that a representative of this group be used in these investigations. The young mouse, 24 hours after birth, is a convenient preparation for a number of reasons, one of the more important being that it is an essentially poikilothermic animal and as such can be carried through temperature cycles to as low as 0°C. without producing permanent changes. The mice are irradiated at the third lumbar vertebra. This level of the cord is the approximate center of the lumbar enlargement which contains a high density of the motor neurons associated with the femoral, sciatic and obturator nerves. Thus, alteration of the motor neurons of the lumbar enlargement produces motor paralysis of the hind limbs of the animal. Paralysis of the hind limbs then serves as an easily detected and unambiguous functional endpoint for an ultrasonically induced effect in the irradiated animals. Figure 2 shows an animal supported in the mouse-holder in preparation for irradiation.

In these experiments, the variation of the acoustic intensity in the plane traveling wave field over the entire lumbar enlargement is less than 5 per cent. The base temperature of the animal is held to ±0.1°C. These experiments are performed in the absence of cavitation. Results have been obtained at a frequency of 982 kc/s, a hydrostatic pressure of one atmosphere and base temperatures of 2°C., 10°C. and 20°C. The data show that a rather well-defined threshold region exists for each base temperature. The threshold region can be defined as follows: If a large number of animals are irradiated with identical values of the acoustic field variables for various periods of time, and the percentage of animals paralyzed at each duration of exposure is plotted as a function of the reciprocal of the time duration of exposure, a sigmoid curve is obtained. Figure 3 is a typical curve obtained in this fashion. Each plotted point represents approximately 25 animals. The curve through the points is obtained by standard statistical treatment of the data.

---

[1] From the Biophysical Research Laboratory, College of Engineering, University of Illinois, Urbana, Illinois.
This research was supported by Contract Nonr 336(00)NR 119-075 with the Physiology Branch, Office of Naval Research and Contract USPH B-1017 with the Institute of Neurological Diseases and Blindness, National Institutes of Health, U. S. Public Health Service.
Presented as part of the scientific program of the International Conference on Ultrasonics in Medicine, sponsored by the American Institute of Ultrasonics in Medicine, Los Angeles, California, September 6–7, 1957.
Received for publication January 6, 1958.

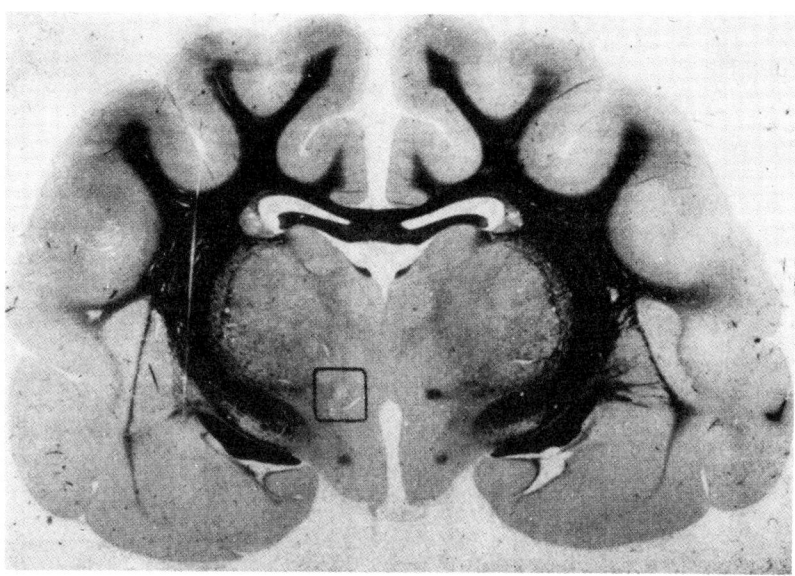

Fig. 1. Interruption of the mammillothalamic tract in the brain of a cat by properly controlled, high intensity, focused ultrasound. Compare the lesion produced by ultrasound on the left side with the untreated right side.

From curves such as this one, a threshold range can be arbitrarily defined as the range of exposure times from 10 per cent of the animals paralyzed to 90 per cent of the animals paralyzed. The collection of these threshold ranges, obtained for various values of a specific acoustic variable, then defines the threshold region. Figure 4 shows the threshold region for the base temperature of 10°C. The ordinate is the reciprocal of the exposure time and the abscissa is the square root of the acoustic intensity. The plotted points (indicated by a +) are obtained from sigmoid curves similar to the one just shown, but obtained at other values for the acoustic intensity. The threshold region is seen to display a "linear portion". For the curve of 50 per cent of the animals paralyzed, the "linear portion" extends from approximately 48 watts/cm^2 (25 seconds time duration of exposure) to at least 160 watts/cm^2 (0.8 second time duration). A statistical analysis indicates that this is a linear relationship to a high degree of accuracy. The width of the threshold region in the linear portion is only 17 per cent. This includes the uncertainties in the physical measurements as well as the biological variation of the animals.

In the course of these studies, measurements were made of the temperature rises in the spinal cords of the mice as a function of ultrasonic dosage (4). This was accomplished by imbedding small thermocouples in the cords. The greatest temperature rises observed are shown in figure 4, plotted at the corresponding dosage coordinates (circles). Figure 5 shows the 50 per cent paralysis curves for base temperatures of the animals of 2°C., 10°C. and 20°C. The maximum observed temperature rises are also shown. The width of the threshold region, for each of the three base temperatures, is less than 18 per cent.

Histological studies have also been carried out on these irradiated animals. These studies show: a) that there is an excellent correspondence between the appearance of a lesion in the lumbar enlargement of the cord and the observed functional change, i.e., all animals which show a loss of motor function of the hind legs also display a lesion in the lumbar enlargement, and b) that in sacrificing animals for various periods of time after irradiation, it is found that the histological lesions appear after approximately 10–15 minutes following irradiation, whereas the functional change is observed instantaneously.

Concerning the present status of our understanding of the physical mechanism of the action of intense ultrasound on tissue, the following statements can be made:

I. Cavitation may be eliminated since these data were obtained in the absence of any phenomena suggesting the presence of cavitation. Also, earlier work (5, 6) performed at this

FIG. 2. An infant mouse supported in the mouse-holder ready for irradiation.

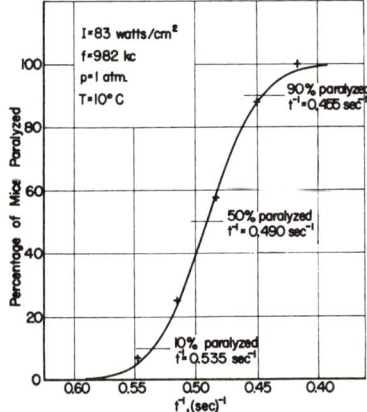

FIG. 3. A typical sigmoidal distribution curve of the percentage of mice paralyzed as a function of the reciprocal of the time duration of exposure.

laboratory, in which animals were irradiated under a hydrostatic pressure sufficiently high to prevent tension forces from occurring in the tissue, demonstrated that this phenomenon does not contribute to the physical mechanism.

II. In the linear portion of the threshold region, the maximum temperatures developed in the cord are considerably less than the normal temperature of the adult animal, viz., approximately 36°C. Hence, a thermal process may be considered unimportant as the primary mechanism in this region.

III. The linear portion of the threshold region displays a relationship showing that the reciprocal of the exposure time is proportional to one of the acoustic field variables, pressure amplitude, particle velocity amplitude, particle acceleration amplitude, etc. However, at the present time there is insufficient information to enable one to make a decision regarding the relative importance of these acoustic variables.

IV. The form of the dosage curve in the non-linear portion suggests that a process different from that obtaining in the linear region may be involved.

V. Since the lesions appear histologically

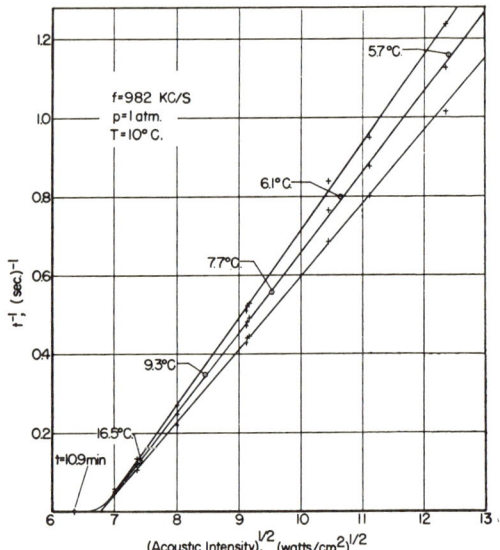

FIG. 4. The threshold region of paralysis of the hind legs of mice under ultrasonic irradiation at the base temperature of 10°C. The indicated temperature rises were measured by thermocouples imbedded in the spinal cords of irradiated mice.

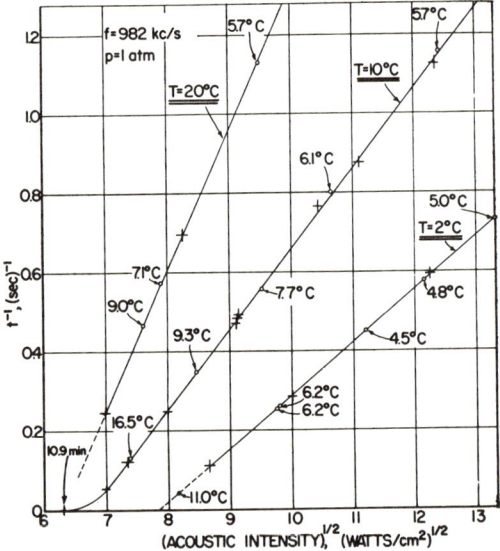

FIG. 5. The 50 per cent paralysis curves for base temperatures of the mice of 2°C., 10°C. and 20°C. The maximum observed temperature rises are also shown.

approximately 10–15 minutes after the loss of function has occurred, it would appear that the site of the physical action is a submicroscopic structure and that the ensuing histologically observed change is associated with secondary processes.

VI. A simple theory (7), which assumes that the observed effects are produced by unidirectional forces of the sound wave, which cause elastic failure in the system when a structural component is displaced from an initial equilibrium position to a second position from which recovery cannot occur, is not supported by the experimental results presented in this paper (8).

REFERENCES

1. BARNARD, J. W., FRY, W. J., FRY, F. J., AND KRUMINS, R. F. Effects of high intensity ultrasound on the central nervous system of the cat. J. Comp. Neurol., 103: 459–484, 1955.
2. BARNARD, J. W., FRY, W. J., FRY, F. J., AND BRENNAN, J. F. Small localized ultrasonic lesions in the white and gray matter of the cat brain. Arch. Neurol. & Psych., 75: 15–35, 1956.
3. FRY, W. J. Ultrasound in neurology. Neurol., 6: 693–704, 1956.
4. FRY, W. J., AND DUNN, F. Ultrasonic irradiation of the central nervous system at high sound levels. J. Acoust. Soc. Am., 28: 129–131, 1956.
5. FRY, W. J., WULFF, V. J., TUCKER, D., AND FRY, F. J. Physical factors involved in ultrasonically induced changes in living systems: I. Identification of non-temperature effects. J. Acoust. Soc. Am., 22: 867–876, 1950.
6. FRY, W. J., TUCKER, D., FRY, F. J., AND WULFF, V. J. Physical factors involved in ultrasonically induced changes in living systems: II. Amplitude duration relations and the effect of hydrostatic pressure for nerve tissue. J. Acoust. Soc. Am., 23: 364–368, 1951.
7. WELKOWITZ, W. Mechanical mechanism of destructive effects of sound on tissue. J. Acoust. Soc. Am., 27: 1142–1144, 1955.
8. DUNN, F. Comments on "Mechanical mechanism of destructive effects of sound on tissue". J. Acoust. Soc. Am., 29: 395–396, 1957.

46

Copyright © 1956 by the Acoustical Society of America

Reprinted from *J. Acoust. Soc. Amer.*, 28(2), 192–201 (1956)

Production of Lesions in the Central Nervous System with Focused Ultrasound: A Study of Dosage Factors*

T. F. HUETER,† H. T. BALLANTINE, Jr., AND W. C. COTTER

Department of Neurosurgery, Massachusetts General Hospital, Boston, Massachusetts

(Received September 5, 1955)

Applications of ultrasonics in neurosurgery involve the production of lesions of a specified size at a specified location, and in a reproducible fashion. Control of the dosage required for a given lesion depends on the geometry of the focal region, the exact determination of focal intensity and the relation between irradiation parameters and biological effect. From an analysis of these factors an optimum frequency for each focal depth is determined and an empirical dosage relationship based on ultrasonically produced paralysis in mice is stated. The mechanism of cell destruction by ultrasound is shown to be a temperature-dependent mechanical effect originating at weak points in the tissue. The difficulties in extrapolating results obtained with experimental animals to applications in human neurosurgery are pointed out.

ULTRASONIC irradiation of live animal tissues at low intensities (1–3 watts/cm^2) and CW, as used in physical therapy, produces effects which are considered to be mainly thermal in origin.[1] At high intensities (50–1500 watts/cm^2) using focused ultrasound, there is some indication[2] that tissue destruction can be produced by a proper choice of the time sequence of irradiation, even if the temperature of the tissues is kept at safe levels. If a different biological response mechanism is operative in each of the two cases the resulting effects will be governed by different relationships between the physical parameters of the ultrasonic irradiation; i.e., the meaning of the term "dosage" will not be the same at high and at low intensities.

Applications of ultrasonics in neurosurgery involve the production of lesions of a specified size at a specified location and in a reliable, reproducible fashion. To do this, one must be able to correlate the size and the location with the measurable physical quantities such as frequency, intensity, duration, and time sequence (for multiple shots). Some amount of reproducibility has been achieved by some investigators for one frequency and for one particular type of equipment.[3] However, no dosage relationships of sufficient general validity have yet been worked out which would enable one to vary the physical quantities such that predictable variations of the degree of destruction can be produced in all types of nervous tissues, and in all parts of the anatomy of the brain.

Dosage control in ultrasonic lesion making depends on the following considerations: (a) Geometry of the focal region. (b) Determination of focal intensity. (c) Relationship between the physical irradiation conditions and the biological effect. (d) Mechanisms involved in tissue destruction by ultrasound.

The value to neurosurgery of the ultrasonic method depends largely on the precision with which tissue destruction can be produced at a predetermined site by focused ultrasonic irradiation of that particular site. It is often difficult to ascertain whether this goal has been achieved by a given irradiation: depending on the kind of nerve tissue irradiated the damage induced by the ultrasound may or may not spread out to areas outside the region of focal irradiation. From our present knowledge it appears that there are two different approaches to the problem of controlling lesion size and shape: (1) One approach would utilize any existing selectivity[4] among the various kinds of nerve tissue, and thus produce damage to structures containing primarily tissue elements that are readily affected by ultrasound. In this case the outline of the lesion is more or less controlled by the anatomical structure of the brain itself [see Fig. 1(a)], provided that the dosage is adjusted to lie within the margins of tissue selectivity. (2) In the other approach one would attempt to control lesion size and shape primarily by the physical properties of the ultrasonic focal spot, irrespective of tissue selectivity [see Fig. 1(b)]. In this case one would be interested in the smallest possible lesion to be placed, if necessary, even in a tissue structure more resistant to ultrasound than its surroundings [see Fig. 1(c)]. This approach is obviously the more difficult one, but it may lead to results of more general validity. Let us, then, adopt the latter view in discussing the aforementioned four groups of problems that arise in connection with dosage.

A. GEOMETRY OF THE FOCAL REGION

For a given intensity and duration of the ultrasonic irradiation both the diameter and the shape of an ultrasonically produced focal spot depend on the beam aperture a/F and the wavelength λ. The lateral diameter (distance between first zeros) of the main lobe of the diffraction pattern which constitutes the focus is

$$d \simeq 1.2 F\lambda/a, \quad (1)$$

---

* Work sponsored by Army Contract DA-49-007-Md-523, and in part by USPHS Contract No. B-816.
† Also at Massachusetts Institute of Technonogy, Acoustics Laboratory.
[1] J. F. Lehmann, J. Acoust. Soc. Am. 25, 17 (1953).
[2] W. J. Fry, J. Acoust. Soc. Am. 25, 1 (1953).
[3] W. J. Fry et al., J. Neurosurg. Psychiat. 11, 471 (1954).
[4] P. D. Wall et al., Science 114, 686 (1951); and W. J. Fry et al., Science 122, 517 (1955).

where F is the focal length and a the transducer radius. The shape of the focal spot is an ellipsoid whose axis ratio $r = l/d$ depends on the solid angle of irradiation. In brain work the maximum angle of convergence β is limited because of surgical considerations‡ as shown in Fig. 2. In a cat brain, for example, the maximum diameter of the skull opening that can be obtained safely is about 3 to 4 cm. For deep sites of irradiation, such as in the thalamus (1.5 to 2.0 cm below the dura) the maximum angle of convergence would then be about $\beta = 90°$ and the focal diameter is then limited to $d \simeq 1.2\lambda$.

As an illustration, Fig. 3 presents some preliminary data obtained by our group using a lens aperture $a/F = 0.33$, and angle $\beta = 37°$ and a frequency of 2.5 mcps. In all cases the region of the ventral anterior nucleus (stereotaxic coordinates $A12$, $L4$, $H+2$) was irradiated,

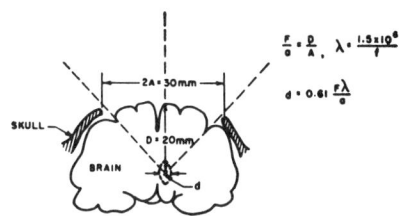

FIG. 2. Dependence of beam aperture on the size of opening in the skull which can be obtained safely, demonstrated for the case of a cat brain.

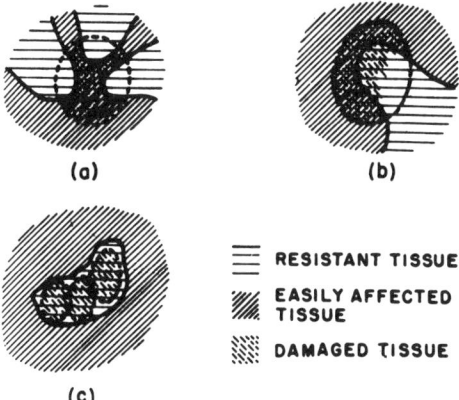

FIG. 1. Influence of tissue selectivity on focusing requirements. (a) lesion size controlled mainly by tissue selectivity; (b) lesion shape determined mainly by focal geometry but modified by tissue selectivity; (c) lesion controlled entirely by focal geometry (triple exposure) against prevailing tissue selectivity.

using different exposure times. The quantities l and d used in this graph express the measured dimensions of the region of total destruction, as evidenced by cystic lesions found seven to nine days after the irradiation. From the upper graph we find for the ratio l/d a value $r \approx 5$, and we note that the lesion becomes more globular with increasing exposure. The lower graph suggests that there may be a minimum dosage, necessary to produce the smallest demonstrable lesion of this kind. It should be noted, however, that in this study no correction was made for variations in the thickness of the intervening brain, which may account for the scatter evidenced by cats 8 and 9.

The deeper the location of the lesion, the larger the

‡ Because of the high absorption of bone at ultrasonic frequencies, portions of the skull have to be removed to provide access to the brain.

minimum size of the lesion becomes for a given focal peak intensity and a given size of the skull opening because of the necessary reduction in beam aperture. If several crossed beams[5] are used instead of a single converging beam of the same over-all aperture, both r and d will be somewhat larger.

According to diffraction theory about 84% of the energy irradiated from a single focusing transducer flows through the ellipsoidal focal region. One way to determine the actual sound intensity to which the tissues are exposed at the focus is by measurements of the wattage W delivered by the transducer and of the focal diameter d. Such measurements are rather difficult to perform for non-plane-wave fields and require the use of calibrated probes which are small enough to resolve the focal field distribution.

Further, the absorption within the medium between the transducer and the focus affects both the amount and geometrical distribution of the ultrasound at the focus. The reduction of focal intensity by the absorption of the intervening tissue can be determined by the nomograph of Fig. 4. For example, for a depth of 2 cm below the surface of the brain the focal intensity is

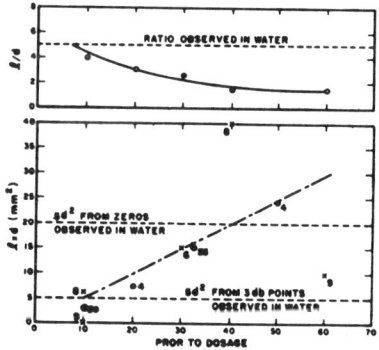

FIG. 3. Some results demonstrating the dependence of lesion size on the time exposure to ultrasound. Pulsed irradiation (0.4 sec "on," 0.6 sec "off") is used at a constant intensity of 950 watts/cm², the abscissa representing the number of pulses.

[5] W. J. Fry and J. W. Barnard, Inst. Radio Engrs., 1954 Convention Record, Part 6, p. 102.

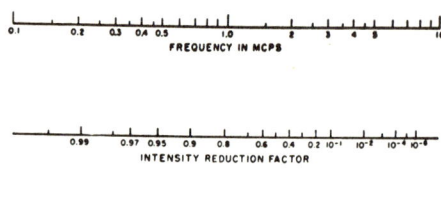

FIG. 4. Nomogram for the determination of the intensity reduction factor from the location of the focus below the brain surface and the frequency.

reduced to 62% at 1 mcps and to 30% at 2.5 mcps from the value which would prevail in the absence of absorption (e.g., in water).

Actually, as the frequency is increased for a given focusing system both the gain of the lens and the absorption of the intervening tissue are increased. The gain Γ may be defined as the ratio $(I_F/I_0)_{av} = 0.84 S_0/S_F$ in which I_F and I_0 are the average intensities at the focus and at the transducer face, respectively, and S_0, S_F are the associated beam cross sections. Using Eq. (1) this becomes

$$\Gamma_{\alpha=0} \simeq 2.4 \frac{a^4}{F^2 \lambda^2}. \quad (2)$$

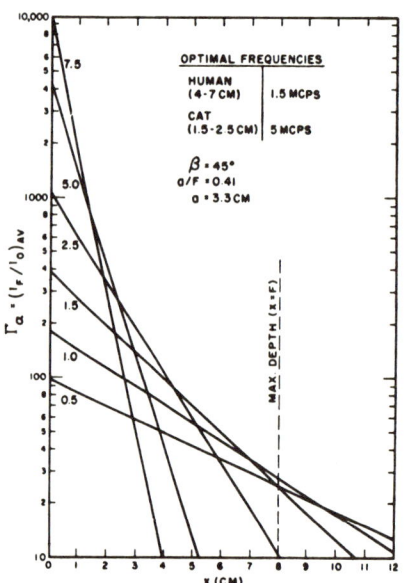

FIG. 5. Focusing gain vs tissue depth of focus for various frequencies. Envelope of straight-line sections determines optimal frequencies.

The effective gain Γ_α will be smaller by a factor $e^{-2\alpha x}$ due to tissue absorption, where α in brain tissue is about 0.12 neper/cm per megacycle,[6] and x is the mean distance between the focus and the brain surface. The quantity Γ_α is plotted *versus* x for a number of frequencies in Fig. 5, assuming a lens system of convergence $\beta = 45°$ and a radius $a = 3.3$ cm. We recognize that an optimal frequency § is associated with a given depth x and that the maximum attainable gain decreases rapidly with depth.

Since the effective gain determines the degree to which tissues outside of the focal region are spared the destructive effects at the focus, the geometrical selectivity of the ultrasonic method [as required, for example, for a lesion of the type depicted in Fig. 1(c)] becomes poor for large depths. This fact results from the geometrical properties of focusing systems, as can be seen from the following argument. Neglecting interference effects in the near field of the transducer and at the focus, we may assume that the average intensity decreases proportionally to the square of the distance from the focus. It is then possible to define the gain in intensity from the brain surface to the focus as $\Gamma_t \simeq \Gamma_\alpha x^2 F^2$ [see Eq. (1)], provided that $x \gg d$. From the optimal gain envelope of Fig. 5 we find that the ratio of the focal depth x to the wavelength λ_{opt} of the optimal frequency is about constant. From Eq. (2) and x/λ_{opt} = const then follows

$$\Gamma_t \simeq 2.4 \frac{a^4}{F^2 \lambda_{opt}^2} \frac{x^2}{F^2} = \text{const} \left(\frac{a}{F}\right)^4. \quad (3)$$

We note that the tissue gain Γ_t is independent of x; for the system analyzed in Fig. 5 we find $\Gamma_t = 28$. Hence, as x becomes larger the slope of the axial intensity distribution toward the focus becomes smaller. The quantity Γ_t/x is thus a measure of the geometrical selectivity, i.e., the sharpness with which a focal lesion can be created.

This finding does not mean that it is impossible to create small lesions at large depths by proper grading of the amount of irradiation, but it indicates that the need for accurate dosage control increases with depth because the intensity drops off more gradually in the directions away from the focus if the corresponding reduction of optimal frequency is made. This must be kept in mind if one is to extend procedures that are found adequate at the smaller depths encountered in cat and monkey brains to the larger depths prevailing in the human brain.

B. DETERMINATION OF FOCAL INTENSITY

The sound pressure at the focus has a lateral distribution of the form $2J_1(x)/x$. The average focal intensity is defined as the power in the main lobe (84% of the

---

[6] R. Esche, Acustica **2**, 71 (1952).
§ Heating effects due to the frequency dependent absorption may be controlled by suitable pulsing.

TABLE I. Comparison between piezoelectric and thermoelectric probes.

	Type of probe	
	Piezoelectric	Sensitized thermoelectric
Smallest possible size for reasonable sensitivity.	BaTiO$_3$ cylinder: 1 mm Crystals with pickup wire: 0.2 mm.	Junction diameter: 0.013 mm.
Scattering artifact at 1 to 3 mcps.	Disturbs focal distribution. Wire probes useful below 2 mc.	Negligible disturbance by junction.
Field disturbance at high amplitudes.	Cavitation at probe interface.	Cavitation at surface of the enclosure containing absorber.
Calibration	*Direct:* reciprocity *Indirect:* comparison with other devices, e.g., radiation pressure.	Indirect: comparison with radiation pressure or using oil of known absorption.
Reading	Instantaneous by VTVM or scope (CW or pulses).	Initial slope of signal *vs* time.
Dependence on base temperature.	None with wire probe, slight with cylinder.	Strongly dependent (temp. control necessary).
Analysis of wave form.	Possible e.g., cavitation noise may be detected.	Not possible
Measurements at site of lesion.	Possible with wire probes mounted in hypodermic needle.	Possible, but calibration depends on knowledge of α at the site of the measurement.

power delivered from the transducers) divided by the area contained within the first zero contour:

$$I_{h\nu} = 0.84W/(d/2)^2\pi. \quad (4)$$

The peak intensity at the center of the focal region then becomes

$$I_p = 4.37 I_{h\nu}. \quad (5)$$

An indirect determination of $I_{h\nu}$ would involve a measurement of the power W which for a given driving RF voltage across the crystal, is delivered by the transducer. Equation (4) implies that the coupling medium between the source and the focus is nonabsorbent and linear, whereas there may be losses, and cavitation-type breakdown may occur at high amplitudes. The latter phenomenon, while well known in liquids, is also observed in tissues.[7] Figure 6 shows some results obtained by us at 2.5 mcps that are indicative of cavitation. A small probe was inserted into the brain at or beyond the focal point and the relationship between the pressure amplitudes generated by the transducer (prop. to plate voltage) and the pressure amplitudes received by the probe was studied. It was found that in each case there was a limiting amplitude beyond which the probe readings became noisy and erratic. Behavior of this kind suggests that the tensile strength of the medium is exceeded beyond a given threshold amplitude.

These findings lead to the need for direct measurements at the focus by suitable probes. Two types of conversion mechanisms may be utilized if an electrical signal proportional to that of the acoustic field pattern is desired: (i) Piezoelectric pickups sensitive to pressure (p).[8,9] (ii) Thermocouples sensitive to absorbed energy (αI).[10] Some characteristics and relative merits of both kinds of probes are demonstrated in Table I. It appears that thermoelectric probes are superior as to resolving power whereas piezoelectric probes give more direct information on the signals received. With both types of probe, difficulties are encountered at high intensities because of imperfect wetting between the probe surface and the surrounding medium, which may lower the cavitation threshold considerably. This threshold in turn, is frequency dependent; it may increase by a factor of 10 as the frequency is varied from 1 to 2.5

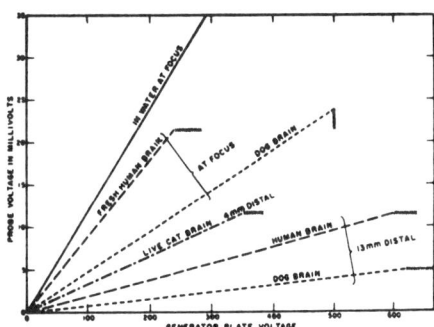

FIG. 6. Probe output voltage detected in tissue at or near focus as a function of generator plate voltage. Shaded regions indicate nonlinear behavior similar to cavitation.

[7] O. Hug and R. Pape, Der Ultraschall l.d. Medizin 7, 42 (1954).
[8] J. Koppelman, Acustica 2, 92 (1952).
[9] E. Ackermann, WADC Tech. Report 53-77 (March, 1953).
[10] W. J. Fry and R. B. Fry, J. Acoust. Soc. Am. 26, 311 (1954).

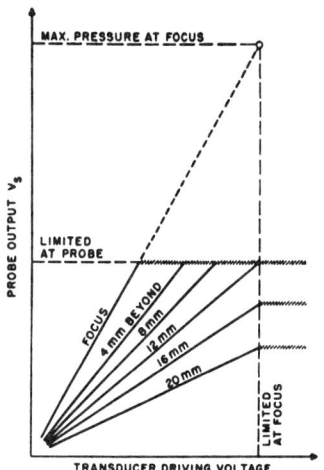

FIG. 7. Extrapolation of focal intensity, required if probe output is limited by cavitation at the probe surface.

mcps. Thus, even with probes sufficiently small to produce negligible field disturbance it is often impossible to measure focal peak intensities exceeding 100 watts/cm^2 directly. Usually the focal intensity must be extrapolated by the procedure shown in Fig. 7.

Again such extrapolations assume linearity over the whole transmission path from crystal to focus. It appears, of course, quite possible to reproduce the irradiation conditions fairly accurately with one given transducer at a given frequency and applied to one type of animal involving a certain range of focal depths. It is rather difficult, however, to correlate the results obtained in one set of such conditions with those of another set, e.g., data obtained on rat cords with those in cat brains, or data obtained at one frequency with those at another.

C. RELATIONSHIP BETWEEN THE PHYSICAL IRRADIATION CONDITIONS AND THE BIOLOGICAL EFFECT

Let us assume that the magnitude and distribution of ultrasonic intensity at the focal point have been determined as accurately as possible, say with an error not exceeding ±10%. There are still many different ways of exposing the tissue to this intensity, and the biological response to a given intensity is found to depend largely on the manner in which the radiation is distributed in time. In this respect the biological response to ultrasound is similar to the response to other physical agents such as ionizing radiation or to some chemicals.

With pulsed ultrasound the temperature rise at the focus can be kept within certain limits[11] by a suitable reduction of the duty factor, i.e., by providing sufficient time between the pulses for the dissipation of heat. Pulsing or similar forms of intermittent irradiation are then a means of exposing the tissue to high mechanical strains without reaching destructive temperatures. To demonstrate this possibility a number of controlled experiments have been devised, of which those of Fry and associates[12] seem to be the most elaborate ones. Whereas heating would destroy all types of tissues

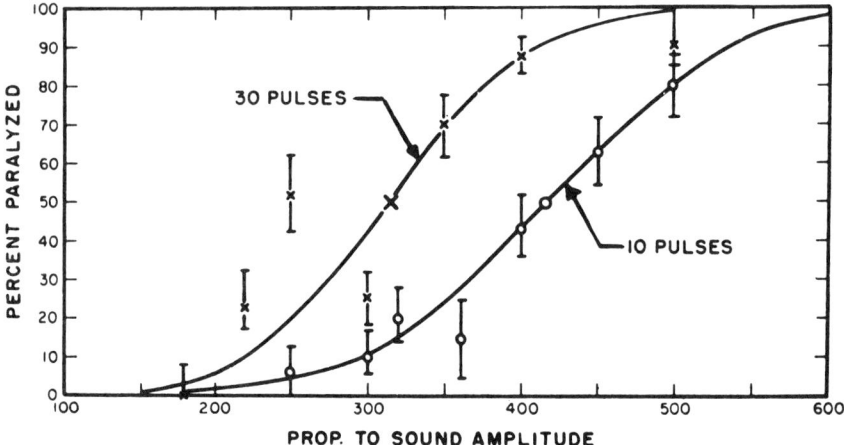

FIG. 8. Probability of paralysis of the hind legs of white mice, after irradiation of the spinal cord, *versus* sound amplitude.

[11] Barth, Paetzold, and Wachsmann, Strahlentherapie 80, 305 (1949).
[12] W. J. Fry et al., WADC Tech. Report 54-152 (September, 1954).

indiscriminately, simply by coagulation, mechanical wear and tear is hoped to affect tissues selectively.

It appears that heating would produce a biological response proportional to $I\alpha(\omega)$ watts/cm², the energy absorbed by the tissue per unit volume and unit time, whereas the effects of mechanical fatigue would be proportional to the strain amplitude $s = 2\pi A/\lambda$ and some function of the time of exposure to the cyclic strains. At a given frequency, dosage could thus be expressed in terms of $I \times t$ in the first case, and as $A \times t$ in the second case. There are, of course, complicating factors such as heat transfer, threshold amplitudes, and recovery phenomena. In any case, some knowledge of how the observed destructive effects are related to these two basic dosage laws is required if lesions are to be produced with a sufficient degree of control.

Such knowledge may be gained with small animals, such as frogs or mice, that are readily available in the large numbers required to obtain statistically meaningful results. We have irradiated the spinal cords of mice with paralysis of the hind legs as a physiological end point. A dual exposure, locating the beam ¾ mm to the right and to the left of the mid-line, was used to ensure that both sides of the cord were equally exposed, the lateral distance between half power points at the focal region being only 0.9 mm at 2.5 mcps. For three different lengths of exposure (10, 30, and 60 pulses, each of 0.4 sec duration, administered at the rate of 1 per sec) the probability of paralysis due to an irradiation defect of the cord was determined as function of strain amplitude (which is proportional to the generator voltage).

Some typical data are shown in Fig. 8, in which each point represents a sample of about 30 mice. The slope of the resulting signoid curves as well as the probable error of each sample illustrate a rather large biological variation of susceptability to the ultrasound. It appears that it would be difficult to derive values for "a maximum amplitude at which no paralysis occurs," or a "minimum amplitude at which all animals are paralyzed," from such data. The best defined point is the median of the distribution curve, as obtained from a reduction of all the data to a best fitting line on probability paper, assuming a normal distribution. Comparing different lengths of exposure we find the relationship between reciprocal exposure time and sound amplitude shown in Fig. 9. The graph reveals that the paralytic effect produced by ultrasound depends more strongly upon amplitude than upon exposure time, suggesting a relationship of the form

$$(A - A_0)^n t = \text{const}, \qquad (6)$$

where $n \geq 2$. This empirical equation obtained with mice at 2.5 mcps, does not agree with the relationship $(A - A_0)t = \text{const}$ postulated by Fry[2] who used cooled frogs at 1 mc. On the other hand, the results obtained by Woeber[13] who irradiated rats with CW ultrasound

[13] K. Woeber, Strahlentherapie 79, 643 (1949).

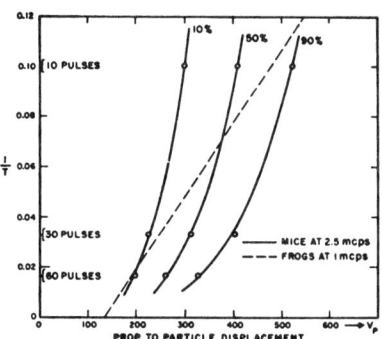

FIG. 9. Plot of reciprocal exposure time versus sound amplitude, derived from reduction on probability paper of data presented in Fig. 8.

at 1 mcps seem to suggest that the product $I \times t$ is approximately constant at low intensity levels ($I \leq 3$ w/cm²) where thermal effects prevail.

Comparing these contradictory results, one may at best conclude that the functional connection of the dosage parameters depends on the temperature level prevailing in the irradiated tissue. For example, Fry states that a single 4.3 sec exposure at an intensity of 35 w/cm² will not paralyze frogs cooled to 2°C, but will paralyze frogs at 25°C, and Lehmann[14] has reported similar results. On the other hand, Fry was able to demonstrate nerve damage at high intensities even if the temperature of the spinal cord was kept well below the levels at which denaturation of proteins occurs.

To gain further insight into the question of whether the primary cause of tissue damage is mechanical strain rather than energy dissipated as heat, we have undertaken a comparison of the intensities required for a given percentage of paralysis at two different frequencies, 1 mcps and 2.5 mcps, for constant exposure time. The result is presented in Fig. 10 from which we note that at 2.5 mcps the median point is located at about 55 w/cm² (average over cord), whereas at 1 mcps the median point of the sigmoid is located at about 7

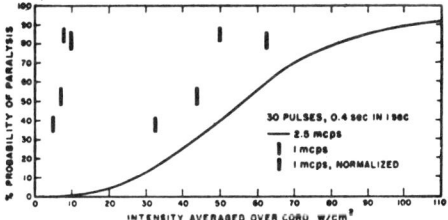

FIG. 10. Probability of paralysis versus average of sound intensity, taken over spinal cord, for 1 mcps and 2.5 mcps.

[14] J. Lehmann, Strahlentherapie 79, 543 (1949).

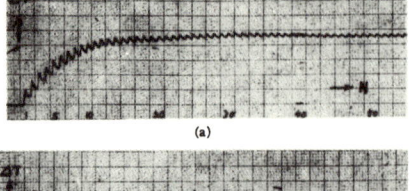

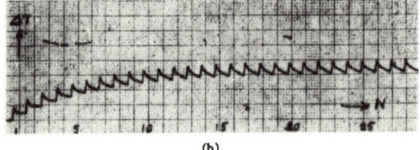

FIG. 11. Recorded temperature response of mouse cord irradiated at 700 w/cm². (a) pulse width 0.1 sec, pulse period 0.5 sec (b) pulse width 0.1 sec pulse period 1 sec.

w/cm² (cord average). The finding that the intensity necessary to produce the paralytic effect becomes higher with increasing frequency seems to rule out heating as the primary cause of cell destruction. The increase of the absorption coefficient α with frequency would call for a decrease in sound intensity to produce the same amount of heat $I\alpha$. Neither can the data be explained simply by associating the cell destruction with a given strain s, or sound pressure p, since

$$s = p/\rho c^2 = (2I/\rho c^3)^{\frac{1}{2}}, \quad (7)$$

which would lead to the requirement that the sound intensity be the same at both frequencies. On the other hand, Fig. 10 shows that normalization of the data with respect to displacement amplitude [by multiplying the abscissa for the 1 mc data by $(2.5)^2$] leads to agreement (within 10% of amplitude) of the results for both frequencies.

It is difficult, however, to visualize a mechanical effect that would depend on displacement amplitude, irrespective of wavelength or frequency. One possible explanation for the observed dependence of the biologically effective intensity on frequency would be in terms of a cavitation-type rupture mechanism.[15] As mentioned above, the sound pressure necessary for rupture in liquids increases rapidly in the ultrasonic megacycle region.[16] This behavior would account for the observed increase in threshold intensity as the frequency is increased at constant displacement amplitude.

The dependence of the biological response on the acoustical field variables, suggested by this experiment need not be in contradiction with the finding that no simple $A \times t$ law seems to hold at all temperature levels. One possible explanation would be that the amount and progression of the mechanical cell damage depends greatly on the local tissue temperature, which in turn is a function of the quantity $I\alpha t$ and the manner of

[15] J. Lehmann and J. F. Herrick, Arch. Phys. Med. Rehab. 86 (February, 1953).
[16] R. Esche, Acustica Akust. Beih. 2AB, 208 (1952).

pulsing. The finally observed degree of tissue damage and the selectivity of the various tissue components with regard to a given amount of ultrasonic energy will then depend on the combined action of mechanical tissue fatigue and tissue heating, and this combination effect need not be governed by the simple types of dosage laws which we have examined so far.

It is then of importance to obtain some knowledge on the magnitude of the temperature rise in the focal region at a given set of irradiation conditions. This may be accomplished by inserting small thermocouples into the irradiated region of the spinal cord of mice. Fry's experiments on cooled frogs at 1 mcps indicate that paralysis does not depend on achieving a critical temperature *level;* in addition, our own results with pulsed ultrasound at 2.5 mcps, administered to mice at normal body temperature, show that for a given product of intensity and total irradiation time the biological effect increases with the excess temperature reached at the end of the irradiation. The data in Table II show that if the sound intensity and the product of pulse width and pulse number are kept constant, the cord damage decreases with diminishing duty factor, which in turn determines the amount of temperature rise in the cord.

Recordings of the temperature rise ΔT in the mouse cord *versus* pulse number N are shown in Fig. 11 for two cases, using pulses of 0.1 sec duration and 0.4 sec pause [Fig. 11(a)], and of 0.1 sec duration and 0.9 sec pause [Fig. 11(b)]. The intensity was 700 w/cm² in both runs and the temperature scale of the graph was 2°C per 5 mm deflection.

We have found that the more gradual temperature rise at the focus which can be achieved by pulsing as compared with CW irradiation brings about greatly improved control over the extent of tissue damage. This is evidenced particularly by the improved repeatability of the size and shape of lesions produced in cat brains. Some data illustrating this point are represented in Table III.

With CW irradiation, small variations in intensity, time, or tissue absorption produce large differences in the temperature rise which have a critical influence on the response of the tissue to the mechanical forces at the focus. With pulsed ultrasound, on the other hand, one may work at equilibrium temperature levels (see Fig. 11) that are both lower and less sensitive to variations of the irradiation parameters. An illustration of the

TABLE II. Effect of duty factor on paralysis of mice.

Run	Peak intensity w/cm²	Number of pulses	Pulse width sec	Pulse period sec	Duty factor %	Animals affected %
1	300	4	1	2.5	40	95
2	300	10	0.4	1.0	40	90
3	300	40	0.1	0.25	40	80
4	300	40	0.1	0.4	25	50
5	300	40	0.1	0.5	20	15

macroscopic appearance of the kind of lesions that can be produced by pulsed ultrasound focused to a point deep within the brain is given in Fig. 12. A detailed discussion of the morphological changes observed by us in the tissues at, and adjacent to the focus will be given in a later publication. The important question of the repeatability with which a given structure within the brain can be hit accurately involves a statistical analysis of the results obtained with a large number of animals irradiated under standard conditions. Such a study is presently in progress.

D. MECHANISMS INVOLVED IN TISSUE DESTRUCTION BY ULTRASOUND

Ultrasonic dosimetry would be much easier if we would have a clearer picture, preferably at a molecular level, of the way in which the tissue and its constituents respond to the applied alternating strains. From the foregoing discussion it seems unrealistic to attempt a clear-cut identification of "nontemperature effects" because most biological reactions are temperature dependent. The complex molecular structure of living tissue may react in a similar way to the disrupting effect of mechanical strain as to the disordering effect of heat. Histological studies of cells affected by either ultrasound or heat do not give many clues on the specific cause of cell death: they reveal the same sequence of progressive decay in both cases.[17]

The main argument for the existence of an interference of the ultrasound with the cell structure that is primarily mechanical rests with the finding[12] that a subliminal dose of ultrasound, which by itself produces neither a substantial temperature rise, nor any histological or physiological effect, has a "priming" effect on the tissue; i.e., the tissue displays a "memory" for such a subliminal dose. It appears that this memory of a subliminal exposure to ultrasound is stored in the form of subtle and reversible structural changes at a submicroscopic level. If the exposure is repeated at suitable intervals the amount of these changes reaches a level where a decay process is initiated. It is reasonable to assume that the changes and their progression are critically dependent on the temperature prevailing at the site of irradiation.

The mechanical hypothesis is corroborated by evidence shown by Peters,[18] that the morphology of brain damage by ultrasound is similar to that following mechanical concussion. Also, the type of destruction in the central nervous system observed by Catchpole and Gersh[19] as a result of decompression sickness is quite similar to the destruction caused by ultrasound. These findings, as well as those by Hug and Pape[7] and by Lindstrom[20] would even suggest that the tissue is affected primarily at weak points within the tissue structure in a similar way as cavitation occurs at weak spots within a liquid. Fry's finding[2] that the application of high hydrostatic pressure reduces the biological effect to a given dose would be consistent with this view.

To obtain a suitable model for a process of this kind we recall our finding that both the frequency dependence and the temperature dependence of the biologically effective sound pressure are similar to that observed in the breakdown of liquids, which is commonly ascribed to cavitation effects. We further note that focal lesions in their earliest stages demonstrate a diffuse lowering of

TABLE III. Effect of focal temperature rise on lesion size in cats.

Cat No.	CW sec	Pulsed (0.4,1) N	I_p	Calculated[a] ΔT (°C)	Lesion size (length ×diameter) mm²
43	2.5		800	26	0
41	3		1100	37	2
49	3		1250	41	3
39	4.5		1150	46	7.5
36	6		1180	52	25
54A		30	910	23	5
55B		30	1070	28	14
56		30	1140	29	17.5

[a] From heat equation, see reference 25.

[17] W. Hoepker, Der Ultraschall i.d. Medizin 5, 178 (1952).

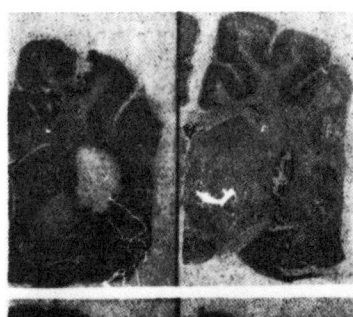
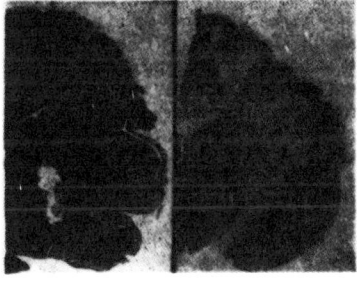

FIG. 12. Stained sections through brains of cats No. 6 and 8. Cats irradiated with pulses of 0.4 sec duration and 1 sec repetition rate. Upper row: Cat No. 8, intensity 950 w/cm², left 40 pulses, right 10 pulses. Lower row: Cat No. 6, 30 pulses, left 950 w/cm², right 710 w/cm².

[18] G. Peters, Strahlentherapie 79, 653 (1949).
[19] H. R. Catchpole and I. Gersh, Physiol. Rev. 27, 360 (1947).
[20] P. A. Lindstrom, Arch. Neurol. Psychiat. 72, 399 (1954).

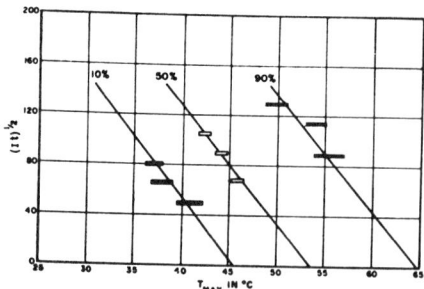

FIG. 13. Plot, according to Eq. (13) of data from Fig. 9 on probability of paralysis in mice.

the so-called "blood-brain barrier"‖ as evidenced by radioactive tracer techniques.[21]

A model based on viscosity, i.e., on the change in configuration of neighboring molecules due to applied stress, thus seems to be promising. Let us conjecture that the processes under consideration are governed by Eyring's theory of viscosity, plasticity, and diffusion.[22] Our approach would be analogous to that of Briggs, Johnson, and Mason[23] in as much as the basic mechanism would consist in the production of "holes" (or breaking of intermolecular bonds) in the medium. The number of holes N_H, assumed to be a fraction of the total number N of molecules involved in the reaction is given by

$$N_H = N e^{-U/kT}, \quad (8)$$

where k is the Boltzmann constant and T the absolute temperature. The activation energy U is defined by

$$U = U_0 + \beta(t) p \Delta v, \quad (9)$$

where U_0 represents the energy associated with intermolecular attraction, p is the external pressure, Δv is the average hole size and $\beta(t)$ is a time dependent coefficient. If p varies with time, as in a sound wave, we assume with Mason that $\beta(t)$ be unity for positive pressure and be much larger than unity, depending on the sound frequency and pulse width, for high negative pressures. An intense sound wave will then increase the number of "holes" N_H until a critical fraction $(N_H/N)_c$ is reached at which the molecular structure of the tissue elements becomes disorganized.[24] To test this hypothesis we may write

$$\ln(N_H/N)_c = -(U_0 + \beta(t) p \Delta v)/kT, \quad (10)$$

‖ This "barrier" is believed to be located at the capillary membranes. Intense ultrasound appears to facilitate leakage of normally retained molecular substances from the capillaries into the nerve tissue.
[21] A more detailed report of this work will appear in a medical journal.
[22] H. Eyring, J. Chem. Phys. 4, 283 (1936).
[23] Briggs, Johnson, and Mason, J. Acoust. Soc. Am. 19, 664 (1947).
[24] K. Altenburg, Naturwiss. 40, 239 (1953).

or, allowing for the rectifying property of $\beta(t)$,

$$-AT = B(t) I^{\frac{1}{2}} - C \quad (11)$$

in which A and C are constants, $B(t)$ is a time-dependent coefficient and I is the sound intensity. Rearrangement of (11) finally yields

$$B(t) I^{\frac{1}{2}} = C - AT = A(T_l - T), \quad (12)$$

where $T_l = C/A$ is that temperature at which tissue structure breaks down without mechanical interference. Equation (12) may be tested by experiments on paralysis of mice by ultrasonic irradiation of the spinal cord.

Our data on 700 mice, presented above in Figs. 8 and 9, may be used for a check on Eq. (12). The temperature T reached at the end of each irradiation was calculated from a solution of the heat equation for the focal point,[25] and confirmed by direct thermocouple measurements. The result is shown in Fig. 13 for the three probabilities of paralysis of Fig. 9. In the absence of any clues with regard to the time function $\beta(t)$ we have tentatively set $B(t) = t^{\frac{1}{2}}$, where t is the total irradiation time. We do not, however, expect this approximation to hold for runs in which the duty cycle D or the pulse width w differ from the values ($D = 0.4$, $w = 0.4$ sec) on which the plots of Figs. 8, 9, and 13 are based.

The fact that our data on paralysis can be presented in the form of Eq. (12) lends some support to the applicability of Eyring's theory to this problem. Moreover, it is gratifying to note that the intersections of the straight lines in Fig. 13 with the abscissa lie in the range of temperatures at which heat alone is found to damage nerve tissues.

CONCLUSION

From all these considerations emerges the picture of a temperature-dependent mechanical effect originating at weak points of the tissue structure and capable of regression or progression depending on the reaction kinetics of the rupture and restitution of molecular bonds. Because of the strong dependence of biological reaction equilibria on temperature, it would be difficult to separate effects that are primarily mechanical and effects that are primarily thermal.

Tissue selectivity may only be expected in a narrow range above dosages that lead to only reversible damage and below dosages at which thermal effects prevail. This range is expected to be a function of such factors as the (frequency dependent) absorption coefficient, the focal geometry, and the location of the site of irradiation. Considering the limited accuracy with which these factors can be evaluated in planning a dosage it appears that lesions based entirely on tissue selectivity cannot be produced with a reasonable degree of control and safety. In applying ultrasonic lesion making to humans, at the present state of our knowledge, it seems preferable to

---

[25] M. D. Rosenberg (private communication).

produce lesions in locations where an extreme amount of control over localization and size is not mandatory.

ACKNOWLEDGMENTS

The authors are grateful to Professor W. J. H. Nauta of the Walter Reed Army Medical Center for histological examination of the irradiated brain material. The help of Mr. M. S. Cohen in design of ultrasonic equipment and of Miss Ann Conant in extensive animal work was invaluable for the success of our investigations We also wish to acknowledge the use of calibration facilities of the M. I. T. Acoustics Laboratory.

Reprinted from *Science*, **127**, 83–84 (Jan. 1958)

Production of Reversible Changes in the Central Nervous System by Ultrasound

For the past several years an intensive research effort has been in progress at the Bioacoustics Laboratory of the University of Illinois on the production of selective lesions in the tissues of the central nervous system by high intensity ultrasound (1). Considerable information has been obtained concerning the dosage conditions required for the production of such lesions, and neuroanatomical studies uti-

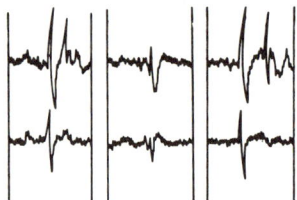

Fig. 1. Cortical potentials evoked by a flash of light (left) before irradiation, (middle) at the termination of irradiation, (right) 30 minutes after irradiation.

lizing this technique are now in progress. Relatively recent electrophysiological investigations indicate that *reversible* suppression of transmission along neural pathways can be accomplished by applying a controlled dosage of ultrasonic radiation at various sites along these pathways (2). By irradiating with ultrasound in the lateral geniculate nucleus it is possible to suppress temporarily the potential usually evoked in the visual cortex in response to a light stimulus. It should be noted that this effect is produced by a dosage of ultrasound which does not cause any histologically observable lesion in the tissue. This ultrasonic technique of producing reversible changes offers unique opportunities for three-dimensional mapping of central nervous system function.

Bipolar recording electrodes are placed in the appropriate cortical areas on both hemispheres to detect the evoked potentials. The focused ultrasonic beam source is used to irradiate the region of one of the lateral geniculate nuclei of the animal (cat) since these nuclei are sites of synaptic stations along the visual pathway. The ultrasonic energy must be transmitted from the irradiator to the brain through degassed Ringer's solution, and the intervening skull bone must be removed.

Stimulation of the eye by light is repeated at fixed time intervals before, during, and after ultrasonic irradiation, and continuous electrical recording is in progress during the course of the experiment. A series of three light flashes, with approximately 3 seconds between flashes, is used to stimulate the eye of the animal. This series of flashes is repeated at variable intervals of time before, during, and after exposure to the ultrasonic radiation. The focus of the sound beam is placed successively in and around the region of the lateral geniculate nucleus. With a suitably chosen sound level and with an exposure time in the range from 20 to 120 seconds, it has been possible to produce reversible suppressions of various components of the elicited electrical response in the visual cortex. The type of result illustrated in Fig. 1 has been obtained in a number of animals. Figure 1 shows the cortical potentials (two electrodes) evoked by a flash of light (i) before ultrasonic irradiation, (ii) at the termination of the ultrasonic exposure period, and (iii) subsequent to irradiation. At the termination of the ultrasonic irradiation period the amplitude of the primary response (upper record) was reduced to less than one-third of its original value. The amplitude of the secondary response (upper record) was reduced to practically zero. Complete recovery of the primary and secondary response was apparent 30 minutes after exposure.

Experiments are in progress to quantify further the conditions for producing controlled reversibility and to determine the site or sites (synapses, axons, cell bodies) of action of the sound (3).

F. J. Fry
Bioacoustics Laboratory,*
University of Illinois, Urbana
H. W. Ades
Division of Neurophysiology and Acoustics, U.S. Naval School of Medicine, Pensacola, Florida
W. J. Fry
Bioacoustics Laboratory,*
University of Illinois

References and Notes

1. W. J. Fry et al., *J. Neurosurg.* 11, 471 (1954); W. J. Fry et al., *Am. J. Phys. Med.* 34, 413 (1955); J. W. Barnard et al., *A.M.A. Arch. Neurol. Psychiat.* 75, 15 (1956); J. W. Barnard et al., *J. Comp. Neurol.* 103, 459 (1955); W. J. Fry, *Neurology* 6, 693 (1956).
2. F. J. Fry, *Abstr. Natl. Biophys. Conf.* (1957), p. 30.
3. This research was supported by contract AF 33(616)-3306 with the Aero Medical Laboratory, Wright-Patterson Air Force Base, Ohio.

* The name of the Bioacoustics Laboratory was recently changed to Biophysical Research Laboratory of the College of Engineering.

14 October 1957

48

Copyright © 1972 by the Acoustical Society of America

Reprinted from *J. Acoust. Soc. Amer.*, **51**(4), Pt. 2, 1333–1351 (1972)

An Analysis of Lesion Development in the Brain and in Plastics by High-Intensity Focused Ultrasound at Low-Megahertz Frequencies

T. C. ROBINSON* AND P. P. LELE

Massachusetts Institute of Technology, Cambridge, Massachusetts 02139

Thermal factors are believed to play a dominant role in the development of the structural and functional effects of irradiation of the nervous system with focused ultrasound at low-megahertz frequencies. Similar mechanisms are postulated to underlie the effects of irradiation in methacrylate, which is frequently used as a test material to evaluate the influence of various factors on the results obtained. This study was undertaken to determine if thermal mechanisms alone can explain the development of trackless focal alterations (lesions) and all of their measurable characteristics in plastic as well as in brain. A purely thermal model is assumed and analytical prediction of lesion development and lesion size and shape for varying values of ultrasonic and thermal constants and controllable variables (frequency, focusing, dosage, target depth, etc.) is attempted. An empirical equation to describe the axial and radial ultrasonic energy distribution at the focus in water is derived. Appropriate heat transfer equations are developed for temperature distributions resulting from ultrasonic irradiation. The computed temperature profiles are plotted against nondimensionalized parameters. Temperatures at the lesion boundary were determined experimentally. Lesion dimensions read off the computed temperature profiles at the measured lesion boundary temperature are compared with experimental data. Agreement of analysis and data shows that, within the range of ultrasonic parameters used in this study, the development of lesions in the brain are explained by thermal mechanisms.

LIST OF SYMBOLS

A	empirical constant (°C)	I_s	intensity in the specimen (W/cm²)
A_B, A_w	output amplitudes of ultrasonic pressure waves in brain and saline, respectively	I_T	intensity at transducer or lens prior to focusing, neglecting attenuation in the lens (W/cm²)
A_{B_0}, A_{w_0}	input amplitudes of ultrasonic pressure waves in brain and saline, respectively	I_w	intensity in water (W/cm²)
a	electrode radius	I_{w_0}, I_{s_0}	maximum intensity within water or the specimen at the focal point (W/cm²)
a_1	radial distance r at $I_w/I_{w_0}=0.5$		
a_2	axial distance z at $I_w/I_{w_0}=0.5$		
$b=k/\rho C$	average thermal diffusivity of the specimen (cm²/sec)	k	average thermal conductivity of the specimen [cal/(sec cm°C)]
C	average specific heat [cal/(g°C)]	l	separation of transducers in brain and saline (cm)
C'	percentage of total power contained in the main lobe of the intensity distribution pattern	P	total acoustic power as measured with a radiation pressure gauge (W)
C_1, C_2, C_3, C_n	empirical constants, nondimensional	Q_m	rate of perfusion through organ divided by organ volume (min⁻¹)
c	acoustic velocity (m/sec)		
f	focal distance, transducer to focal point	q	heat flux density or heat generated in the specimen by ultrasonic energy absorption (W/cm³)
G	ϕ, x the nondimensionalized temperature		
I	intensity of ultrasound (W/cm²)		
I_0	zero-order modified Bessel function of the first kind	q_0	peak heat generation rate at $r=0$ (W/cm³)

r	radial distance from focal point or intensity maximum (cm)	z	axial distance from surface of specimen nearest the lens to a point within the specimen
r_0	effective or maximum radius of heat source	z_f	axial distance within the specimen from surface to focal point (cm)
r_T	outer effective lens radius (cm)		
r_1	radius of curvature of spherical lens	z_i	axial distance from focal point or intensity maximum (cm)
r'	radius of the main lobe of the focal region	α_B	attenuation coefficient for cat brain (Np/cm)
T	temperature axially or laterally at distance r, z_i from the focal point (°C)	α_s	attenuation coefficient of the specimen (Np/cm)
T_I	initial uniform specimen temperature (°C)	α_w	attenuation coefficient in water medium of negligible absorption (Np/cm)
T_L	temperature at lesion edge or boundary (°C)	$\beta_1, \beta_2, \beta_3, \beta_n$	empirical constants, dimensionless
		γ	q/k
t	time from the beginning of the irradiating pulse of ultrasound (sec)	δ	dummy variable
		δ_t	average coefficient of linear thermal expansion, [cm/(cm°C)]
x	$=bt/a_1^2$ or bt/a_2^2 nondimensionalized time parameter	ϕ	q/q_0, a nondimensional relative heat density
x'	dummy variable		
y	$=y_0$ or y_1	μ	intensity transmission ratio or coefficient at specimen–water interface
y_0	$=r_0/a_1$		
y_1	$=r/a_1$	ρ	average density of the specimen
y_2	$=z_i/a_2$	θ	included angle of radiation to the focal point
y'	dummy variable		

INTRODUCTION

In previous papers[1-5] it has been shown that the structural and functional alterations, produced in the nervous system by irradiation with a single beam of focused ultrasound[6] at low-megahertz frequencies, are consistently and reproducibly related to the ultrasonic dosage. Comparable relationships were found to exist in methacrylate[7,8] permitting its use as a test material for rapid evaluation of the influence of physical factors, e.g., lens design, ultrasonic frequency, power, exposure schedules, etc. The experimental data[1,2,4,5,7] suggest that the alterations both in the nervous system and in the plastic, whether structural or functional, permanent or transient, are *probably* thermal in origin. Similar conclusions were reached by others using ultrasound at frequencies of 1 MHz or higher, under different experimental conditions.[9-12] This and other evidence is discussed elsewhere.[13]

The present study was undertaken to determine if thermal mechanisms alone can explain the formation and all of the measurable characteristics of lesions in methacrylate and in the brain. A purely thermal model is assumed and analytical prediction of lesion development and their size and shape for varying values of ultrasonic and thermal constants and controllable variables (acoustic frequency, power, durations of exposure, lens design, target depth) is attempted. These predictions are compared with experimental data. Methacrylate (polymethylmethacrylate, Rohm and Haas Plexiglas G) in ¾-in. square bars and the brain of the cat were used in these experiments. An empirical equation to state the axial and the radial ultrasonic energy distribution of a single focusing transducer or a lens at its focus in water is derived. Appropriate heat transfer equations are developed for temperature distribution, in methacrylate and brain, resulting from ultrasonic irradiation. The computed temperature profiles are plotted nondimensionally for the range of interest of each parameter. Temperatures at the lesion boundary were determined experimentally in methacrylate and brain. Thermal and ultrasonic properties were measured where not available. The expected axial and radial lesion dimensions were read off the curves of temperature distribution at the measured lesion boundary temperature. These predicted lesion dimensions were compared with experimental data to determine if lesion development can be explained by purely thermal considerations.

I. THE ANALYTICAL PREDICTION OF LESION SIZE AND SHAPE

A. Heat Generation in the Specimen

Lesions in methacrylate and in brain are prolate spheroids[5]; thus the lesion maximum radius (or diameter) and the lesion length adequately describe both lesion size and shape (Fig. 1). These two dimensions can be determined from a heat transfer analysis that incorporates the distribution of heat generation. This heat generation is a result of the attenuation of intensity within the specimen. The intensity distribution at the focus in water was determined experimentally.[14] Attenuation of the ultrasonic energy occurring in polyethylene

326

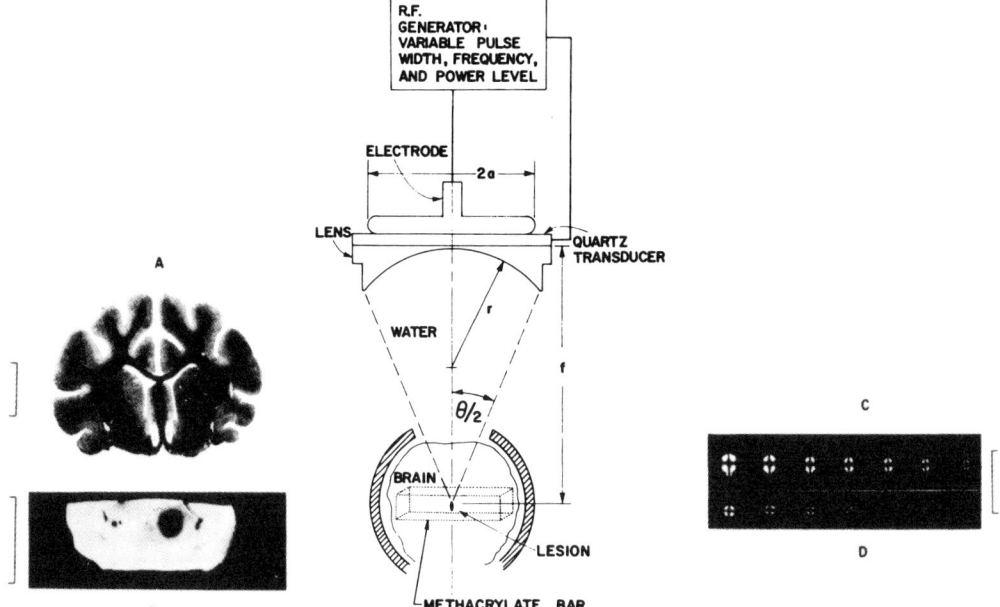

FIG. 1. Equipment used for lesion creation and typical results obtained in the brain and methacrylate. Frequency: 2.7 MHz; lens: polystyrene, spherical plano-concave; $r=2.88$ cm, $f=8.0$ cm, $a=3.0$ cm, $\theta=42°$. Coupling medium: degassed water at 37°C. A and C: side views showing the lesions at right angles to the axis of irradiation; B and D: top views showing the lesion in the axis of irradiation. Cal. marks equal 1 cm.

produces heat which raises the temperature locally. It has been shown that with pulses of ultrasound sufficiently short (0.05 sec) to exclude significant thermal diffusion, the temperature profile accurately reflects that of heat generation and intensity.[15] Typical experimental data have been reported previously.[6] Figure 2 shows the relative intensity distribution in the two axes.

1. The Empirical Equation for Intensity Distribution

An empirical equation has been derived by trial and error to fit the experimental radial and axial intensity distributions for the focusing system used in this work. These are shown in the nondimensionalized plot of Fig. 3. It is seen to agree with the radial intensity distribution rather well; the fit of the symmetrical empirical equation with the axial intensity data is less satisfactory, but acceptable at large distances because of the asymmetry of the axial distributions. The empirical equation is

$$I_w/I_{w_0} = C_1 \exp[-\beta_1 y^2(1-\tfrac{2}{3}\beta_1 y^2)] \\ + C_2 \exp[\beta_2 y^2(1-\tfrac{2}{3}\beta_2 y^2)] \\ + C_3 \exp[\beta_3 y^2(1-\tfrac{2}{3}\beta_3 y^2)], \quad (1)$$

where $y = y_1$ or y_2, $a_1 = 0.034$ cm, $a_2 = 0.25$ cm, $\beta_1 = 0.462$, $\beta_2 = 0.0578$, $\beta_3 = 0.00578$, $C_1 = 0.85$, $C_2 = 0.1415$, and $C_3 = 0.0085$.

The terms of this empirical equation were derived so as to permit solution of an approximate equation for spherical symmetry in Sec. I-B. This equation gives negative values at large distances and so should not be employed to represent the low intensities at distances beyond $y = 5$.

2. The Calculation of Heat Generation

The focused ultrasonic radiation travels primarily in the axial direction toward the focus. The ray from the edge of the lens travels 9% farther in the specimen and is attenuated about 15% more than an axial ray, but the average error will be less. The assumption of purely axial irradiation appears reasonable for the calculation of the attenuation of intensity within the focal region.

The change in intensity with axial distance neglecting ray convergence and divergence is given by

$$I_s = I_{s_0} e^{-2\alpha_s z_i}. \quad (2)$$

The negative intensity gradient or first derivative of intensity with distance is equal to the heat flux density and is due to the attenuation of intensity and consequent energy absorption by the specimen.

$$-\partial I_s/\partial z = 2\alpha I_{s_0} e^{-2\alpha_s z_i} = 2\alpha_s I_s = q. \quad (3)$$

Let us assume that the profile of lateral or radial intensity distribution, which is generally measured in water (Fig. 2), remains unchanged within the specimen. This is true for brain, since little refraction occurs, but less accurate for methacrylate where greater refraction

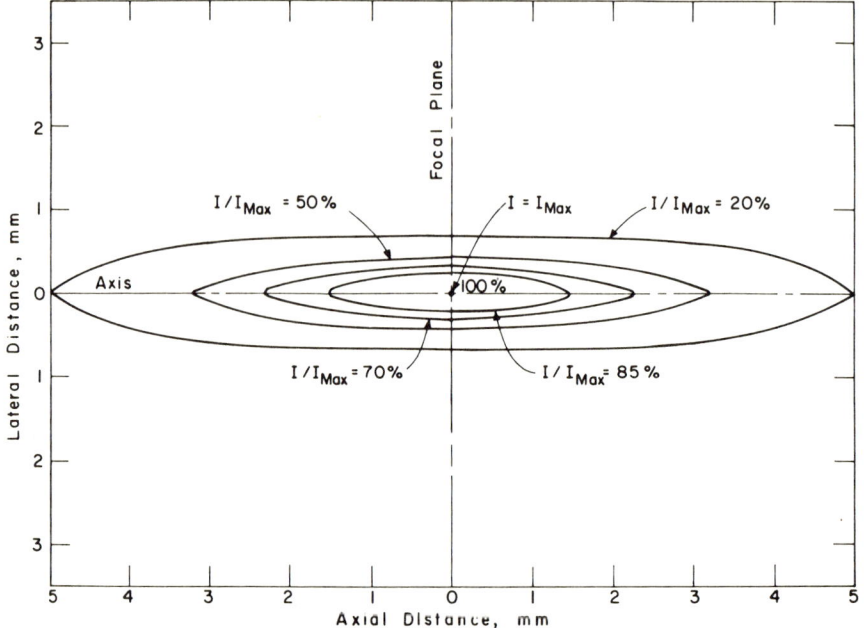

FIG. 2. Map of lines of constant intensity at the focus, showing the relative intensity distribution in the two axes.

occurs. The center of the focus is at some distance z_f in the specimen. Then, at the center of the focus, or focal point,

$$q_0 = 2\alpha_s I_s = 2\alpha_s \mu I_{w_0} e^{-2\alpha_s z_i}. \quad (4)$$

In the absence of any information concerning the attenuation at high intensities in methacrylate and brain, it is assumed that the attenuation at the focus is the same as that measured with plane waves at low intensities.

3. The Relationship between Acoustic Power and Heat Generation

Radiation pressure measurements are made to determine the total acoustic power input.[6] The intensity distribution can be integrated over the focal region to determine the relationship of I_{w_0} and total acoustic power P:

$$C'P = 2\pi I_{w_0} \int_{r=0}^{r'} \frac{I_w}{I_{w_0}} r dr, \quad (5)$$

$$C' \cong 0.84 \text{ (see Ref. 6)}.$$

If we integrate, this becomes

$$C'P = 2\pi I_{w_0} \sum_{n=1}^{3} \frac{C_n a_1^2}{6\beta_n} [(1 + e^{-\beta_n a_1^2 r'^2})(2\beta_n a_1^2 r'^2 - 1)]. \quad (6)$$

For the plano-concave spherical lens used in all of the experiments:

$$r' = 0.18 \text{ cm}.$$

Then, in water:

$$C'P/2\pi = 11.7 \times 10^{-4} I_{w_0}. \quad (7)$$

Thus,

$$I_{w_0} = 114P.$$

Also,

$$P = I_T \pi r_T^2. \quad (8)$$

Combining Eqs. 7 and 8:

$$\frac{I_{w_0}}{I_T} = \frac{114P}{P/12.6}$$

$$= 1440, \text{ the peak intensity gain of the lens.} \quad (9)$$

The peak heat generation rate can be correlated to peak intensity by the relation derived earlier (Eq. 4):

$$q_0 = 2\alpha_s \mu e^{-2\alpha_s z_i} I_{w_0}. \quad (10)$$

Substituting the value for I_{w_0} from Eq. 7,

$$q_0 = 272\alpha_s \mu e^{-2\alpha_s z_i} C'P, \quad (11)$$

or

$$q_0 = 228\alpha_s \mu e^{-2\alpha_s z_i} P2\text{W/cm}^3. \quad (12)$$

This equation relates the peak heat generation rate with the measured total acoustic power.

B. Temperature Distribution in the Specimen

Single pulses of ultrasound are considered; at the beginning of the pulse a uniform known initial tempera-

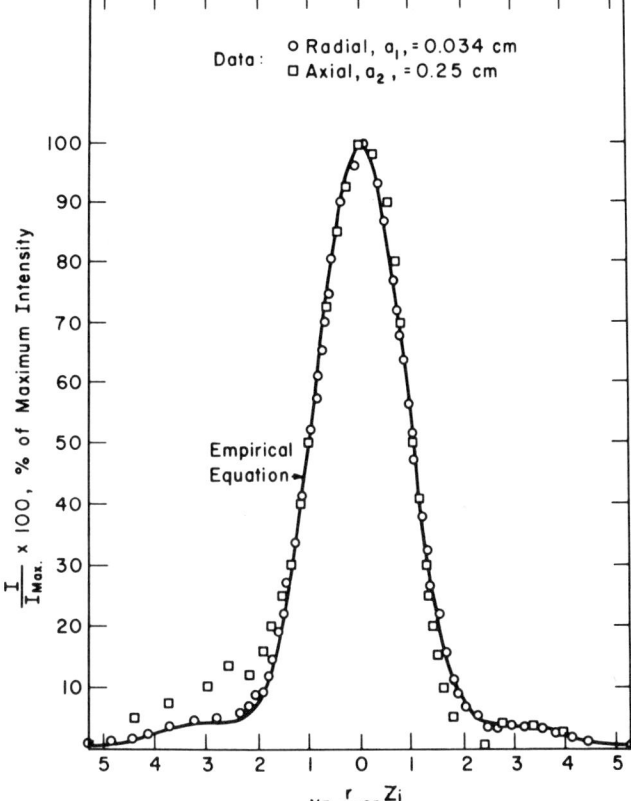

FIG. 3. The empirical relative intensity distribution.

ture exists within the specimen. The nonuniform heat generation within the specimen changes the temperature distribution in it. The temperature increases as the duration of the pulse increases. At small radial distances from the axis of irradiation, a maximum temperature is reached at the moment the pulse is terminated; this is called the primary temperature rise. At larger radii, a maximum temperature is reached some time after the end of the pulse; this, the secondary temperature rise, is because of heat flowing from the center to the surrounding cooler region (see Fig. 18).

A specially useful approach for this heat transfer analysis is to derive nondimensional equations for the temperature distribution. These nondimensional equations and their plotted solutions permit easy determination of the temperature distribution for the thermal properties corresponding to any conditions or materials, for any pulse duration, and for any effective radius of the intensity distribution. In order to make these solutions nondimensional, it is necessary to assume constant average values for the ultrasonic and thermal properties of the material (but see Secs. II and IV and Fig. 14). The axial variation of intensity within the specimen is neglected. The intensity variation from tip to center of lesions in brain is calculated as 30% for the longest lesion (14-mm length) and 12% for the average lesion (6–7-mm length).[1] The percentage deviation in heat generation from the analysis is then an average additional 6% above the focal point and a reduction by 6% below the focal point distal to the lens. It is assumed that the axial and radial intensity distributions are the same in the specimen as the measured distributions in water, but that the peak intensity at the focal point is reduced by reflection at the water–specimen interface and by axial attenuation from this interface to the focal point.

The shape of the intensity profile in Fig. 2 indicates that the radial intensity distribution at and near the focal point is nearly independent of z_i. This cylindrical symmetry at the center of the intensity distribution suggests the use of a heat transfer analysis in cylindrical coordinates, with the axis of the cylinder along the axis of irradiation. This analysis will permit the radial temperature distribution at the focal point or lesion center to be determined, predicting slightly higher temperature than actual because axial heat conduction has been ignored in this calculation.

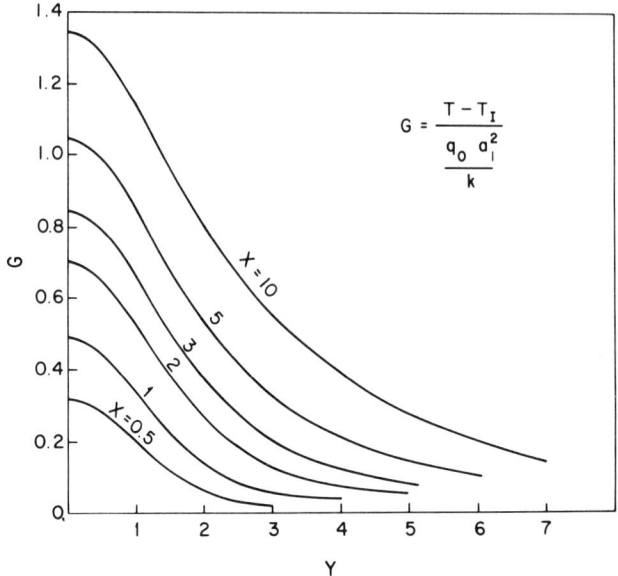

FIG. 4. The nondimensional temperature profile for cylindrical coordinates, $x = 0.5$ to 10.

The axial intensity distribution suggests the use of spherical coordinates for a heat transfer analysis with spherical symmetry, though the temperature thus predicted will be somewhat lower than that actually measured.

The differential equation in cylindrical coordinates for cylindrical symmetry and applicable to determining the radial temperature distribution is

$$\frac{\partial^2 T}{\partial r^2} + \frac{1}{r}\frac{\partial T}{\partial r} + \gamma = \frac{1}{b}\frac{\partial T}{\partial t}. \quad (13)$$

The differential equation in spherical coordinates for spherical symmetry and applicable for determining the axial temperature distribution is

$$\frac{\partial^2 T}{\partial r^2} + \frac{2}{r}\frac{\partial T}{\partial r} + \gamma = \frac{1}{b}\frac{\partial T}{\partial t}. \quad (14)$$

The boundary conditions which must be satisfied are those for an infinite medium at a uniform initial temperature. The assumption of an infinite medium is justified by measurements of temperature distributions in

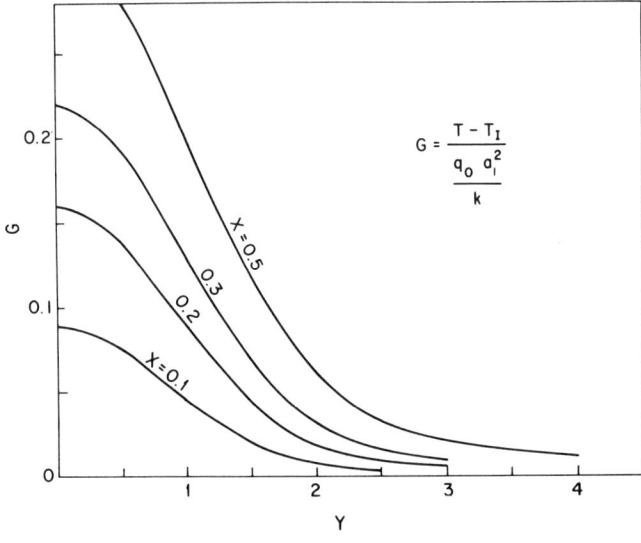

FIG. 5. The nondimensional temperature profile for cylindrical coordinates, $x = 0.1$ to 0.5.

LESION DEVELOPMENT IN BRAIN AND IN PLASTICS

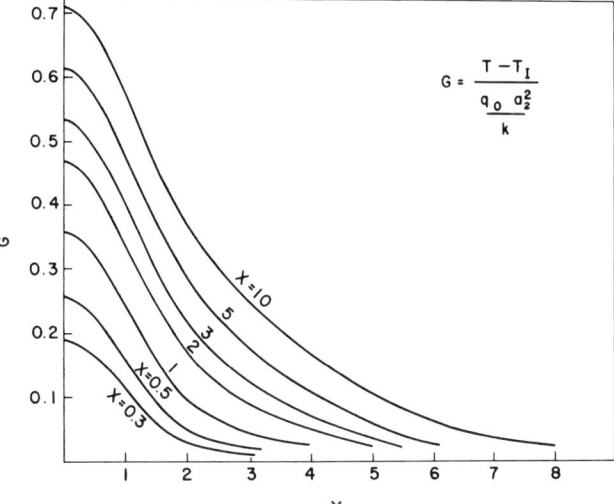

FIG. 6. The nondimensional temperature profile for spherical coordinates, $x=0.3$ to 10.

methacrylate and cat brain which show that the extent of a significant or measurable temperature rise is far less than the dimensions of the specimen; lesions very near a surface of the specimen are not considered. Then the boundary conditions are

T finite for all $t \geqslant 0$, $r \geqslant 0$,
$T \simeq T_I$ for large r when $t > 0$,
T_I = initial uniform temperature (°C).

Analyses are presented below for the solution of the differential equations with these boundary conditions. The differential equations for cylindrical and spherical symmetry with a distributed heat source in an infinite specimen can be solved by integrating the equation for the instantaneous spherical and cylindrical sources[16] with respect to time and the radius over which the heat source acts.

1. An Exact Solution for Cylindrical Symmetry

For conditions during irradiation, the nondimensionalized equation which results is

$$\frac{T-T_1}{(q_0 a_1^2/k)} = \frac{1}{2} \int_{x'=0}^{x} \int_{y'=0}^{y_0} \frac{y' \phi}{(x-x')}$$
$$\times \exp\left[\frac{-(y^2+y'^2)}{4(x-x')}\right] I_0\left[\frac{yy'}{2(x-x')}\right] dy' dx'. \quad (15)$$

It was determined experimentally (Secs. II and IV) that the lesion boundary generally occurs at temperatures which are in that part of the distribution where no secondary temperature rise is found. Then analyses which calculate temperature at the end of the ultrasonic pulse are sufficient to determine the temperature distribution responsible for lesion development.

This numerical analysis was performed over an appropriate range of x and y. The results are plotted in Figs. 4 and 5.

2. An Exact Solution for Spherical Symmetry

The resulting equation can be written nondimensionally[16]:

$$\frac{T-T_I}{(q_0 a_1^2/k)} = \frac{2}{2y(\pi)^{\frac{1}{2}}} \int_{x'=0}^{x} \int_{y'=0}^{y_0} \left\{\exp\left[\frac{-(y-y')^2}{4(x-x')}\right]\right.$$
$$\left. -\exp\left[\frac{-(y+y')^2}{4(x-x')}\right]\right\} \frac{\phi y' dy' dx'}{(x-x')^{\frac{1}{2}}}. \quad (16)$$

This equation with a steady heat source can be integrated with time; the resulting nondimensional equation is

$$\frac{T-T_1}{(q_0 a_1^2/k)} = \frac{1}{2y} \int_{y'=0}^{y_0} \phi y' \left(2\left(\frac{x}{\pi}\right)^{\frac{1}{2}}\right)$$
$$\left\{\exp\left[\frac{-(y-y')^2}{4x}\right] - \exp\left[\frac{-(y+y')^2}{4x}\right]\right\}$$
$$-|y-y'| \operatorname{erfc}\frac{|y-y'|}{2(x)^{\frac{1}{2}}} + (y+y') \operatorname{erfc}\frac{y+y'}{2(x)^{\frac{1}{2}}}\right) dy', \quad (17)$$

where

$$\operatorname{erfc}[\cdots] = 1 - \frac{2}{(\pi)^{\frac{1}{2}}} \int_0^{[\cdots]} e^{-\delta^2} d\delta, \quad (18)$$

the error function. This equation was integrated numerically over an appropriate range of x and y. The results are plotted in Fig. 6.

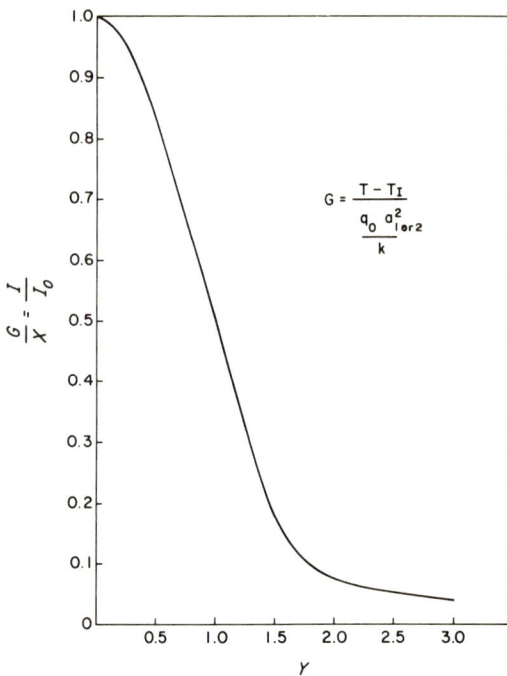

FIG. 7. The nondimensional temperature profile for negligible thermal diffusion, $x<0.01$.

3. The Solution for Negligible Thermal Diffusion

For very short-duration pulses, the cylindrical and spherical cases are reduced to a temperature profile that is the same as the intensity profile. At these short times, all heat goes into raising the local temperature, with insufficient time for significant thermal diffusion to take place. Then

$$qt = \rho C(T-T_1), \quad \text{at any } r;$$

then

$$q_0 \phi(x a_1^2/b) = \rho C(T-T_1) \quad (19)$$

and

$$\frac{T-T_1}{q_0 a_1^2/k} = \phi x = G \quad (20)$$

or

$$G/x = \phi, \quad (21)$$

the intensity distribution.

This is plotted in Fig. 7 and accurately predicts the temperature distribution at the short time intervals, with x less than approximately 0.01.

II. THE PROPERTIES OF METHACRYLATE

It is necessary to determine the attenuation coefficient, ultrasonic velocity and impedance, the thermal conductivity, and the thermal diffusivity of methacrylate which appear in the calculation of heat generation and temperature distribution. The variation of these parameters with temperature must be found in order to ascribe average values over the temperature range encountered in lesion development. Another important parameter is the lesion boundary temperature, to permit the prediction of lesion length and maximum diameter from the analytical temperature distributions

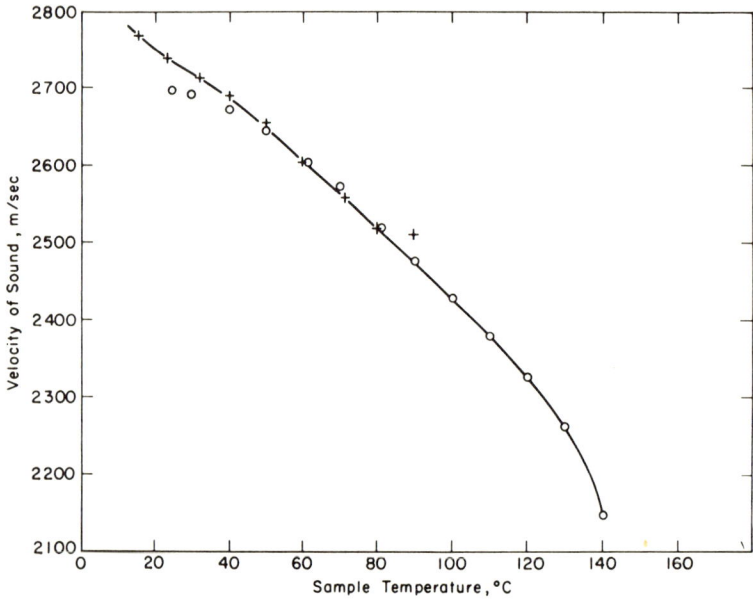

FIG. 8. The velocity of ultrasound (3 MHz) in methacrylate at different temperatures. $+$: in water; $\bigcirc$: in glycerin.

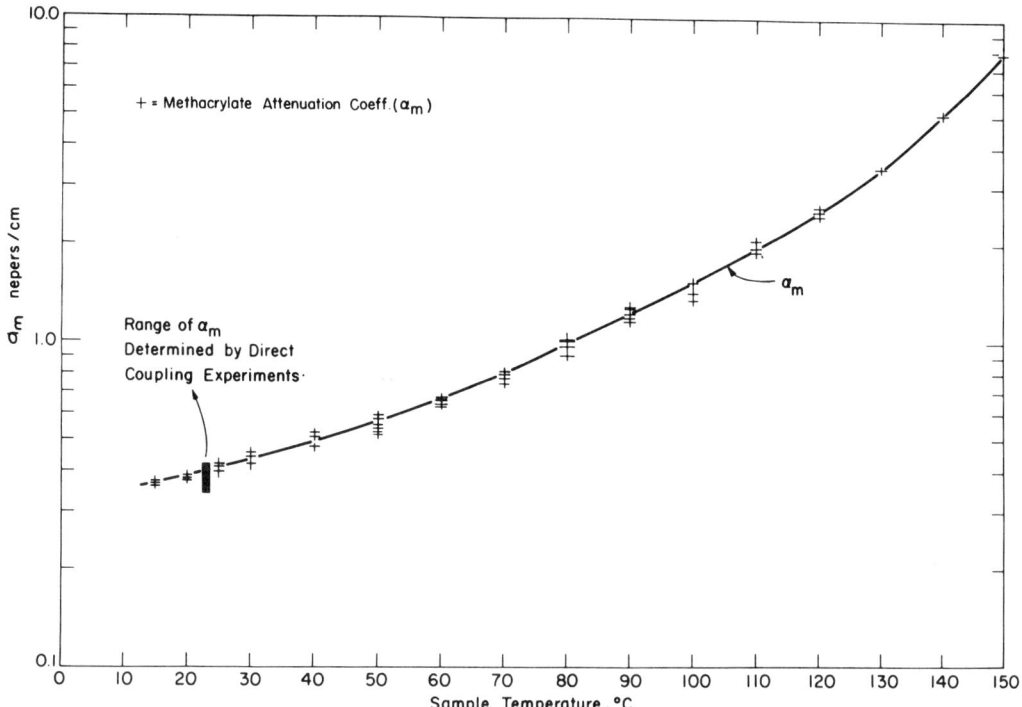

FIG. 9. The attenuation coefficient (3 MHz) in methacrylate at different temperatures. (Abstracted from Ref. 15.)

A. The Thermal Properties

The coefficient of thermal conductivity varies little with temperature. From 0° to 100°C it is 4.4×10^{-4} cal/(sec cm °C).[17]

The density of methacrylate as a function of temperature is

$$\rho = \frac{\rho_0}{[1+\delta_t(T-24°C)]^3} \simeq \rho_0[1-3\delta_t(T-24°C)], \quad (22)$$

where

$\rho_0 = 1.19$ g/cm^3 (at $T_0 = 24$°C),

$\delta_t = 9.7$ cm/(cm °C).

The termal diffusivity (the thermal conductivity divided by the density and by the heat capacity) is 1.03×10^{-3} cm^2/sec at 24°C and 8.82×10^{-4} cm^2/sec at 100°C.

B. The Ultrasonic Properties

The velocity of ultrasound and the attenuation in methacrylate at temperatures from 10° to 160°C at intervals of 10°C were measured. The techniques used are described elsewhere.[15] Figure 8 shows the sonic velocity for methacrylate, the maximum error being less than 1.5% at 40°C and less than 3% at 110°C. Since, with the lens used, the solid angle of radiation is 42°, the acoustic radiation strikes the specimen at angles which increase from zero for axial rays to 21° for rays at the lens periphery. With oblique incidence at the liquid–solid interface, the energy transmitted into the solid is partly longitudinal and partly shear.[18,19] Calculations showed that the difference between the transmission of the axial and peripheral rays was insignificant. Shear-wave transformation of the transmitted wave as well as contributions to the longitudinal wave from secondary reflections were also insignificant.[15]

The attenuation in methacrylate increases considerably with temperature as shown in Fig. 9; from 0.41 Np/cm at 24°C to 7.8 Np/cm at 150°C.

C. The Lesion Boundary Temperature

A lesion in methacrylate, as viewed with crossed polarized light (Fig. 1), has a sharp boundary which appears to be a discontinuity or abrupt change in the index of refraction of the methacrylate and is apparent from the beginning of lesion development.[7,8] Lesion growth is uniform and stops when the ultrasonic irradiation stops, except for large lesions where a small amount of growth occurs for a short period after irradiation followed by a slight shrinkage.

Thermocouple scans of the temperature distribution inside and outside a lesion were performed using a 0.001-in.-diam chromel–constantan thermocouple imbedded 4 mm below the surface of a test block of methacrylate

333

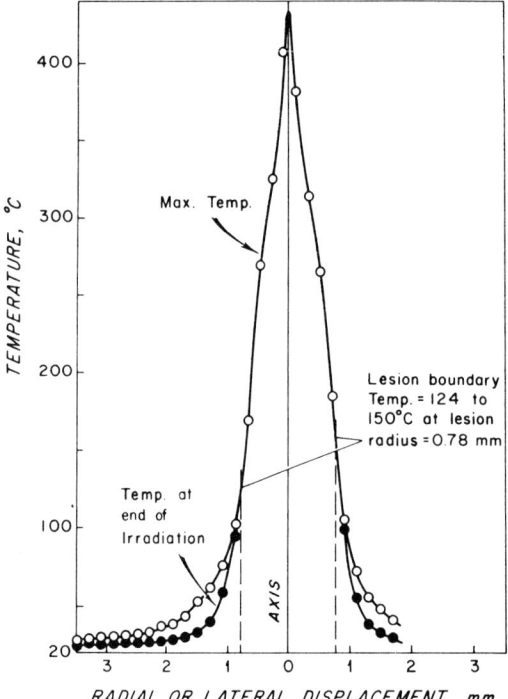

FIG. 10. The temperature profile measured in methacrylate following ultrasonic irradiation. Frequency: 2.7 MHz; pulse duration: 0.7; initial temperature: 24°C; chromel–constantan thermocouple: 0.001-in. diameter. Temperatures at end of irradiation and the maximum temperature were the same at temperature levels above 124°C.

and constant pulse durations (Fig. 10). The temperature at the thermocouple was allowed to reach room temperature between successive irradiations.

It was found that successive irradiations in the same position gave rise to the same ($\pm 2\%$) rise in temperature, and that lesion length and diameter were the same whether made in fresh material or superimposed totally or partially on another lesion.[7] Both these observations indicate that negligible irreversible changes in thermal and ultrasonic properties occur with lesion formation.

From Fig. 10, it is seen that the maximum temperature of 440°C occurs at the lesion center, and very steep temperature gradients occur at the lesion boundary. The high central temperatures will thermally expand the methacrylate, creating high hydrostatic pressures in the central fluid core of the lesion; whereas outside the lesion, the stresses will decrease as the third power of the radius.[20] The lesion boundary temperature, the intercepts of the temperature distribution with the measured lesion radius, is seen to be about 124°C to 150°C.

Figure 11 shows a temperature profile determined with a 0.002-in.-diam iron–constantan thermocouple in methacrylate. In this measurement the distance from the focus to thermocouple bead was fixed, and the pulse duration was increased at each successive pulse to make successively larger lesions. The lesion boundary temperature, determined when the boundary crossed the center of the bead was 107°C.

A third experiment was performed to determine if the lesion boundary temperature is a single fixed point or if it spans a substantial range of temperatures. A kanthal heater wire, 0.020 in. in diameter, was molded into a methacrylate block. The wire was electrically heated to produce cylindrical lesions of increasing diameter, for a wide range of power levels and durations. It was assumed that the heat generated in the wire was uniformly and instantaneously applied at a constant rate to the plastic at the outside surface of the wire as the thermal diffusivity of the wire is about 30 times that of methacrylate. This heat transfer problem was analyzed to determine the temperature distribution by considering the methacrylate to be an infinite medium, and the heat source to be a steady uniform cylindrical surface source.[16] The time, heat input, and lesion radius were measured for the lesions produced experimentally. Then the temperature corresponding to the lesion radius was read from the plot of the analysis. Four cylindrical lesions were created, with power inputs varying by a factor of 3.3, and the durations by a factor of 10.7. The calculated temperature rise varied from 75° to 82.8°C, giving an average lesion boundary temperature of 104.8°C. This agrees fairly well with the observed lesion boundary temperature. The variation of 7.8°C over the wide ranges of time and power inputs appears to confirm that a single lesion boundary temperature can be assumed for all lesions in methacrylate.

Methacrylate has an inflection transition temperature between 109°C and 120°C, where the inflection is the point where the relaxation modulus begins to decrease rapidly with temperature.[21] It is quite close to the glass transition temperature, where the plastic changes from a glassy solid to a rubbery viscoelastic material. The specific volume of the material changes abruptly at this temperature, accompanied by an observable abrupt change in the index of refraction. All the data described here give lesion boundary temperatures in the vicinity of the glass transition temperature, which is apparently the mechanism or discontinuity in material property responsible for the sharp boundary of the lesion.

III. THE COMPARISON OF DATA WITH THE ANALYTICAL PREDICTION OF LESION SIZE AND SHAPE IN METHACRYLATE

The calculation of cylindrical symmetry and the measured lesion boundary temperature are used to predict the lesion diameter. The thermal properties of methacrylate at 24°C (Table I) are used with the nondimensional plots of Figs. 4 and 5 to determine the lesion diameter at different pulse durations. The value of q_0,

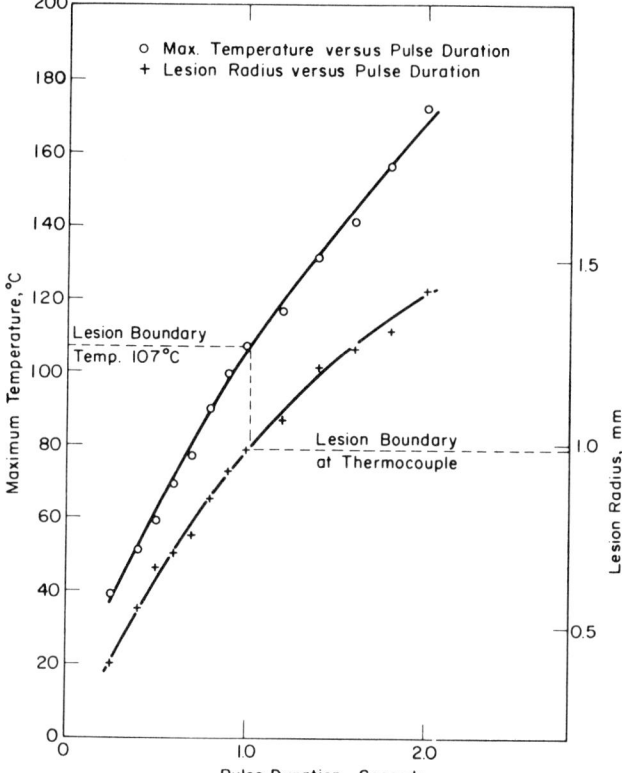

FIG. 11. Temperatures measured in methacrylate for increasing pulse durations. Lesion center to thermocouple: 0.99 mm; iron-constantan thermocouple: 0.002-in. diameter; initial temperature: 24°C.

the peak heat generation, was selected to best fit the data at 20 W of acoustic power. The predicted lesion diameter at other acoustic power levels was calculated with q_0 directly proportional to acoustic power, as determined in Sec. I-A. The resulting fit of analytical prediction to data[7] is shown in Fig. 12. It is obvious that the fit is not perfect with either the cylindrical or spherical calculations. The reason is that attenuation was assumed to be constant with time and temperature in this analysis, though it was subsequently found to vary substantially with temperature. These results suggest that methacrylate should probably be used in these studies at an initial temperature perhaps 20° or 30°(C) below the lesion boundary temperature to obviate the occurrence of large changes in attenuation during lesion development. The temperature distribution in the specimen, resulting from ultrasonic irradiation, among other factors, is also dependent upon its thermal diffusivity, since the pulse durations generally used are long enough for substantial thermal diffusion to take place. The differences between brain and methacrylate in acoustic impedance, lesion boundary temperature, average attenuation over the lesion, and thermal conductivity can be largely compensated by an appropriate level of acoustic power. However, the thermal diffusivity of methacrylate, which is about 36% higher than that for brain, can only be compensated by adjusting the pulse duration (Sec. I-B). No attempt was made to change any of the thermal or geometrical constants in order to better fit the data. Such an empirical fit can probably be performed with some success.

The value of q_0 at 20-W acoustic power is 1690 W/cm³. A value of attenuation of 0.65 Np/cm at 60°C average temperature and 20-W acoustic power, gives 1706 W/cm³ from the relationship between heat generation and acoustic power of Sec. I-A.

IV. THE PROPERTIES OF BRAIN

As in the case of methacrylate, the thermal conductivity, heat capacity, density, acoustic attenuation, and ultrasonic velocity in cat brain, as functions of temperature, and the lesion boundary temperature were determined. The vascularity or length in millimeters of capillaries per cubic millimeter of brain is proportional to and an index of the local blood flow, and varies from 440 mm/mm³ to 1350 mm/mm³ in different regions. In the white matter it varies from 440 to 550 mm/mm³, and the average for the gray matter, which varies widely in the different regions of the brain, is about 1100

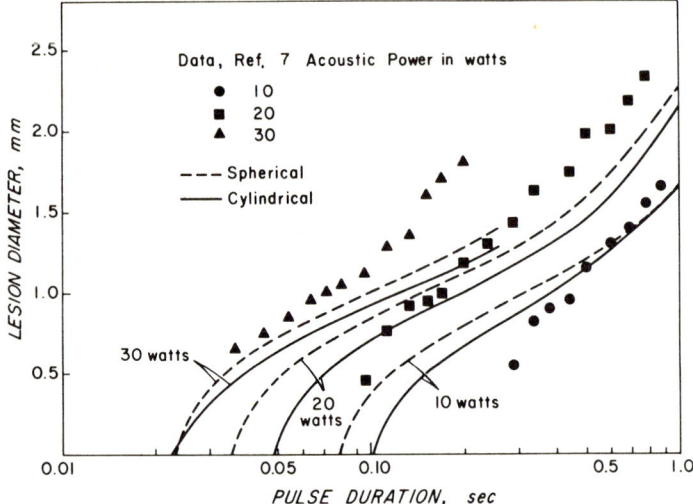

FIG. 12. A comparison of experimental data and analytical results in spherical and cylindrical coordinates for lesion diameter in methacrylate at different pulse durations. $q_0 = 1690$ W/cm³ at 20-W acoustic power, empirical value.

mm/mm³.[22] The mean blood flow for the brain is about 50 cm³/min per 100 g of brain, with a flow velocity in the capillaries of about 0.5 mm/sec. Then for blood flow to have no appreciable effect on lesion size, the pulse duration should be about $\frac{1}{10}$ the lesion radius divided by the flow velocity (0.2 sec times the lesion radius in millimeters). Longer pulse durations relative to the lesion size may result in significant thermal diffusion owing to local blood flow.

A. The Thermal Properties

1. Without Blood Flow

The thermal conductivity of brain has generally been determined by measuring thermal diffusivity and assuming a value for specific heat.[23–25] The values in Table II are for the whole brain, *in situ*, without cranial

TABLE I. A comparison of the thermal and ultrasonic properties of brain and methacrylate.

Property	Brain at 37°C	Methacrylate at 24°C
Thermal conductivity $\left(\frac{\text{cal}}{\text{sec cm °C}}\right)$	1.39×10^{-3}	4.4×10^{-4}
Density (g/cm³)	1.028	1.19
Specific heat $\left(\frac{\text{cal}}{\text{g °C}}\right)$	0.95 (Ref. 29)	0.36
Thermal diffusivity (cm²/sec)	1.40×10^{-3} (Ref. 26)	1.03×10^{-3}
Longitudinal velocity of ultrasound (m/sec)	1.570	2730
Acoustic impedance ρc $\left(\frac{\text{g}}{\text{cm}^2 \text{ sec}}\right)$	1.61×10^5	3.25×10^5
Attenuation coefficient (Np/cm) (2.7 MHz)	0.18–0.32	0.32
Lesion boundary temp (°C)	52–58	108–125

blood flow during the measurements and appear to be quite close to those of water.

The density of brain at different temperatures was determined by measuring its weight in air and weight immersed in water, and calculating the density of brain from the known density of water at the temperature of the measurement. The density of whole fresh cat brain was found to decrease from 1.031 g/cm³ at 26°C to 1.013 at 65°C, a change of less than 1.75%. The value of 1.028 g/cm³ measured at 37°C agrees with that in the literature.[27]

The effect of a temperature increase from 37° to 100°C, the range observed during lesion formation (Sec. IV-C), is seen to increase the thermal diffusivity of water by 11% and the thermal conductivity by 8%. This is approximately the uncertainty of the data for brain; it then appears reasonable to assume that the thermal properties of the brain without blood flow are independent of the temperature within this range.

2. With Blood Flow

The thermal conductivity of various tissues has been observed by several investigators to vary with blood flow or perfusion, though other thermal properties appear to be unaffected. Three investigators[23,28,29] have attempted to measure the changes in conductivity with varying perfusion in experiments with isolated animal organs using the Gibbs heated thermocouple probe[23] calibrated with fluids and gels of known thermal conductivity. The data for sheep spleen perfused with whole blood is representative; an empirical equation representing this data over a physiologic range of perfusion rates is[23]

$$\Delta k / Q_m = 1.39 \times 10^{-3} \text{ (cal min)/(cm sec °C)}.$$

Assuming sheep spleen to be comparable to the cat brain, with about 50-cm³/min average blood flow for

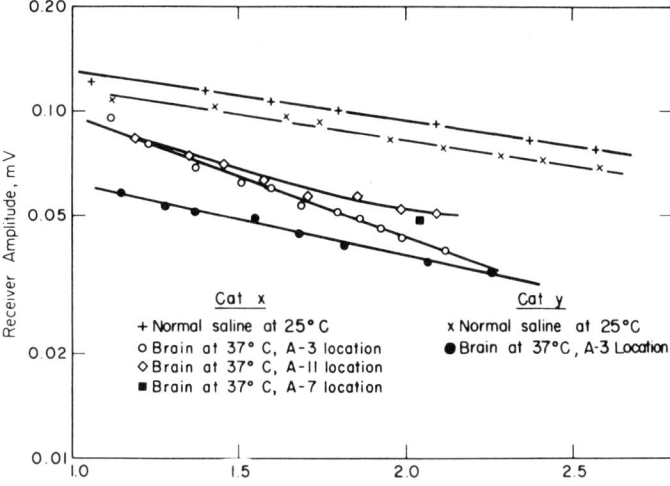

FIG. 13. Relation between received pulse amplitude and piezoelectric probe separation in cat brain and in normal saline. Frequency: 4.2 MHz; input: 420 V peak to peak.

100 cm³ of brain,

$$\Delta k = 6.95 \times 10^{-4} \text{ cal}/(\text{cm sec }°C).$$

This approximate calculation ignores the fact that local blood flow in a tissue increases markedly within a few seconds following an abrupt temperature rise. Also, the spleen is a reservoir of blood; its volume and the blood flow vary over a wide range with varying physiological states, whereas those of the brain as a whole remain relatively constant. Experimentally, the apparent increase in the conductivity is greater when the spleen is perfused with saline instead of whole blood. However, no reliable data on changes in conductivity with blood perfusion are available.

B. The Ultrasonic Properties

The ultrasonic characteristics of cat brain were determined experimentally and compared with the data from the literature. Twenty anaesthetized healthy adult cats were used for these experiments. Lead-zirconate-titanate piezoelectric transducers mounted in the tip of 0.17-mm-diam long bevel hypodermic needles, with transducer faces flat and perpendicular to the needle axis, were used in the measurement of the acoustic properties of brain. Most of the experiments were performed with 0.5-μsec pulses of ultrasound at 4.2 MHz, using the transmission technique. The needle probes were supported in a **U**-shaped bracket guide permitting the placement of the probes at preselected locations in the brain of the cat with transducer separations of up to 5 cm, while maintaining accurate axial alignment. The bracket could be secured to the stereotaxic instrument[1] for *in vivo* experiments. The dura mater at the point of introduction of the probes was slit and the probes advanced slowly to the preselected depth. Thirty minutes were allowed to elapse after the placement of the probes for stabilization before any measurements were made.

Tests made in a saline solution showed that the amplitude and the transit time of the acoustic pulse remained stable for periods of up to 110 min.

1. The Attenuation in Brain in Vivo

Experiments to determine the attenuation of "live" brain (37°C) were performed *in vivo* with ultrasonic probe tips being placed stereotaxically across known structures within the brain. At a constant input amplitude, the receiver output amplitude and delay were measured. These experiments were repeated in normal saline solution as a reference of known attenuation. The resultant data from two representative experiments are shown in Fig. 13.

The majority of the attenuation coefficients calculated from this data fall between 0.28 and 0.50 Np/cm, the likely range in mixed gray and white matter for live brain at 4.2 MHz.

TABLE II. Thermal properties of brain and water at 37° and 100°C.

	Temperature	"Whole" brain without blood flow	Water
Thermal conductivity $\left(\dfrac{\text{cal}}{\text{sec cm }°C}\right)$	37°C	1.39×10^{-3}	1.50×10^{-3}
	100°C	...	1.62×10^{-3}
Thermal diffusivity (cm²/sec)	37°C	1.40×10^{-3} (Ref. 26)	1.50×10^{-3}
	100°C	...	1.67×10^{-3}
Specific heat $\left(\dfrac{\text{cal}}{\text{g }°C}\right)$	37°C	~0.95 (Ref. 24)	1.002
	100°C	...	1.006
Density (g/cm³)	37°C	1.028	0.994
	100°C	~1.00	0.964

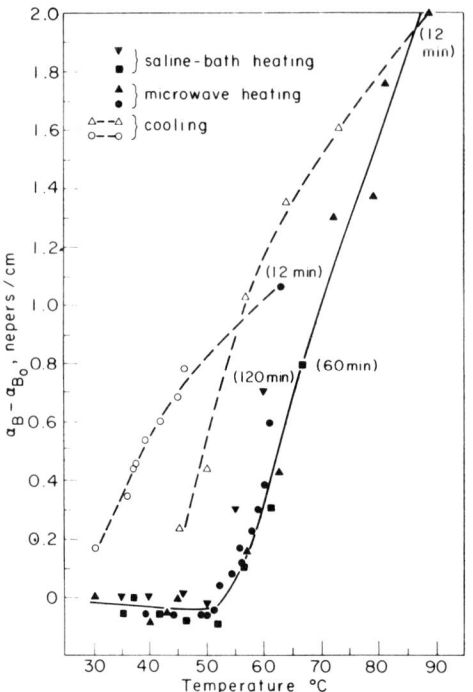

FIG. 14. The changes in the attenuation coefficient in cat brain with temperature. Frequency: 4.2 MHz; α_{B_0}: attenuation at 37°C. The total time to reach the highest temperature in each experiment from the beginning of the heating is shown in parenthesis.

2. The Effect of Death on Ultrasonic Properties

Probes were placed within the brain with transducer separation of 2 cm. Attenuation and velocity were measured after stabilization and were found to remain constant for at least 30 min. The animals were then killed by an overdosage of the intravenous anaesthetic. No significant changes in the attenuation and velocity were found to occur with death and during 4-h post mortem, though the temperatures measured at the probe tip fell from 37° to 24°C over this period. These observations confirming those of others[31] permit determination of acoustic properties of live brain in samples obtained soon after death.

3. The Attenuation and Velocity in Brain at Different Temperatures

Since no significant changes in the attenuation and velocity in the brain were found to occur with death and during 4-h post mortem, the effect of temperature on these properties was measured *in situ* or *in vitro* on specimens obtained within a few minutes after death. In each experiment, probes were placed across known structures in the brain stereotaxically. The temperatures at three adjacent points in the brain were measured continuously. The temperature of the preparation was raised in small increments either by its immersion in a heated stirred saline bath or by microwave irradiation. Figure 14, typical of the results obtained at four different rates of temperature rise, shows that there is little change in the attenuation coefficient to about 50°C, above which it increases markedly. This increase is reversed on cooling, but shows hysteresis. In another series of comparable experiments, changes in attenuation at 55°C with time were determined (Fig. 15). The attenuation was found to increase markedly if the elevation in temperature was sustained over periods of minutes—periods significantly longer than the duration of temperature rise within the lesion during its development (Sec. IV-C). The absence of significant changes in attenuation with temperature rises close to the lesion boundary temperature (*vide infra*) justifies the assumption of constant attenuation in the analysis of lesion formation.

The changes in longitudinal velocity with temperature are shown in Fig. 16, which also shows data from literature[27] for fresh "dead" brain from dog, pig, cow, and horse at frequencies of 1.8–2.5 MHz. The velocity is seen to be slightly higher in brain than in the 0.9% saline solution. The *in vivo* data agrees reasonably well with that *in vitro*.

4. The Effect of Lesion Creation on the Ultrasonic Properties

Changes in the attenuation and velocity were measured during and after lesion formation. Trypan blue was injected intravenously at the end of each experiment to demarcate areas of destruction.[1] Two hours later the animal was killed and the brain fixed by perfusion with 5% formalin. When sufficiently hard, the brain was removed and cut coronally in the axis of irradiation, and the placement and size of lesions and of the probe were determined. With transducer separations of 4–5 mm and lesions of 2-mm diam×9-mm length to 3.5-mm diam×12-mm length, no significant changes in the attenuation or velocity were detected, either during or following lesion creation.

The existence of such changes during lesion formation, however, was demonstrated in the brain and the spinal cord of the cat using the same transducer both for the production and detection of the lesions.[8] In addition to the higher sensitivity of this system due to focusing, the detection occurred along the axis of the lesion over a 4.5 to 5 times longer path length rather than across the diameter as in the present experiments. The changes, however, are so small that constant ultrasonic properties may be assumed in the analysis of lesion formation.

C. The Lesion Boundary Temperature

Thermocouple scans of temperature distribution were performed using a calibrated 0.002-in.-diam chromel–constantan thermocouple threaded through the brain. In each experiment single pulses of 0.2- and 2.0-sec

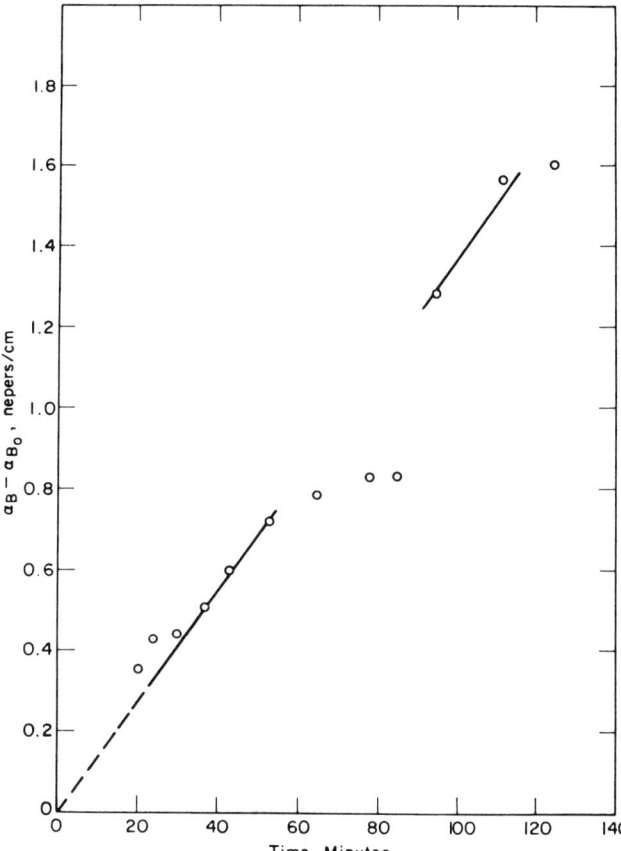

FIG. 15. The change in the attenuation coefficient in cat brain at 55°C with time. Frequency: 4.2 MHz; α_{B_0}: attenuation at 55°C at the beginning of measurements.

duration, respectively, were used at 10-W radiation power level. The 0.2-sec pulse, insufficient to create lesions, was used initially to center the focus on the thermojunction. Then scans were taken across the junction from one side to the other.[6] Typical 0.2-sec scans in white and gray matter are shown in Fig. 17. The initial temperature was 35.0°C in the white matter and 35.6°C in the gray matter.

The temperature rise in white matter is almost double that in gray matter. The ultrasonic intensity in deeper lying gray matter was calculated to be about 29% less than that in the white matter, while the temperature rise is found to be 50% less. The remaining difference (~20%) in temperature rise may be due to the greater blood flow through the gray matter reflected in the slightly greater asymmetry in the scan.

The scans were repeated in each experiment with single pulses of 2.0 sec, sufficient to create lesions in brain. The interval between consecutive irradiations was sufficient for the thermocouple to return to the baseline temperature. The results are shown in Fig. 18. The asymmetry is no greater than that in Fig. 17, indicating that presence of previously created lesions between the thermocouple and the focus only on one side produces no obvious alterations. It was also found that the temperature rise changed only slightly for three irradiations in the same location, with less than a 6% increase for successive irradiations.

Trypan blue was injected following the experiment, and the animals were perfused with formalin 2 h later. Serial sections were cut on a freezing microtome and the lesion diameters were measured. Figure 18 shows the temperature at the lesion boundary. The average is 56.5°C. The error in measurement of lesion diameter may be as much as −5% to +15%, owing to local edema or shrinkage during life or during fixation and freezing. This results in a lesion boundary temperature ranging from 52.0° to 58.0°C.

V. COMPARISON OF DATA WITH THE ANALYTICAL PREDICTION OF LESION SIZE AND SHAPE IN BRAIN

From the parameters and analyses obtained above, the length and maximum diameter of lesions in brain can be calculated and compared with experimental values.[1] The lesion boundary temperature and the cylindrical symmetry analysis is used to calculate the lesion

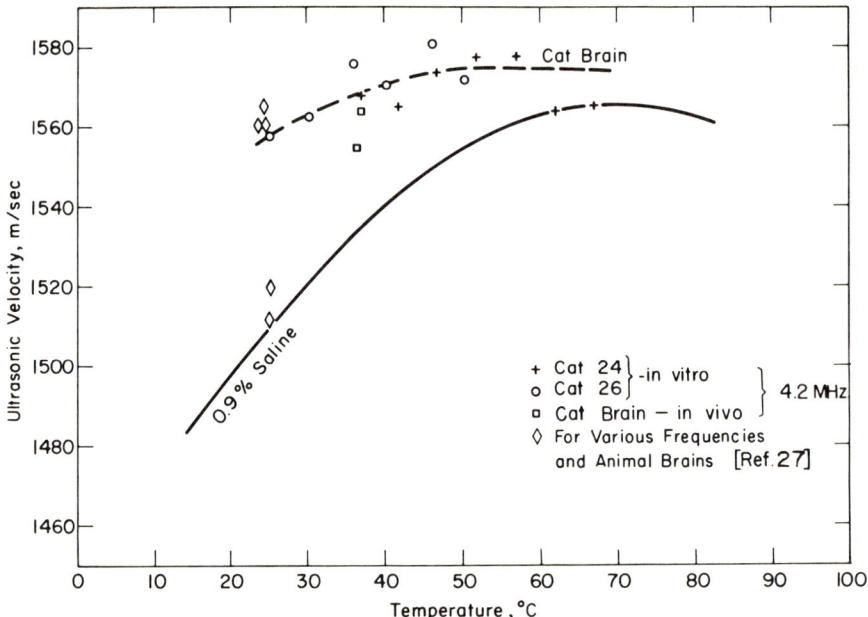

FIG. 16. The change in the velocity of ultrasound in cat brain with temperature. Data for 0.9% saline is shown for comparison.

diameter, and the spherical symmetry analysis is used to determine the lesion length. The spherical case is the closer approximation to the lesion length considering the small radius of curvature at the ends of the intensity distribution (Fig. 2). The fit to the data was performed by using the values of thermal constants (with blood flow) of Sec. IV and selecting a value of q_0 to fit the data at 10-W acoustic power and a 2.0-sec pulse duration. The lesion length was calculated using the same value of q_0 as for lesion diameter. The fit to the data are shown in Figs. 19 and 20, respectively. Calculations predict a substantially greater lesion length than is actually observed. This is expected from the assumption of spherical symmetry which presupposes lower radial heat flow than the sharp-tipped intensity distribution would suggest. The axial intensity changes by 30% from the focus to the tip of a 12-mm-long lesion; this can also lead to substantial error. Another value of q_0 was chosen at 10-W acoustic power and 2-sec pulse duration for better fit. The resulting agreement between calculated and experimental lesion length is better (Fig. 20).

The relative standard deviation in diameter and length for 10 samples[1] is about 5%, with a scatter of 15%. For very small lesions the scatter is greater. Consistent and substantial swelling occurs owing to blood engorgement in the large lesions near the hemorrhage line, which probably explains the poorer agreement between the analysis and data for these lesions.

The values of q_0 determined from the fit to the lesion radius data at 10 W of acoustic power is 338 W/cm³, based on a lesion boundary temperature of 55°C, an initial temperature of 37°C, and the thermal conduc-

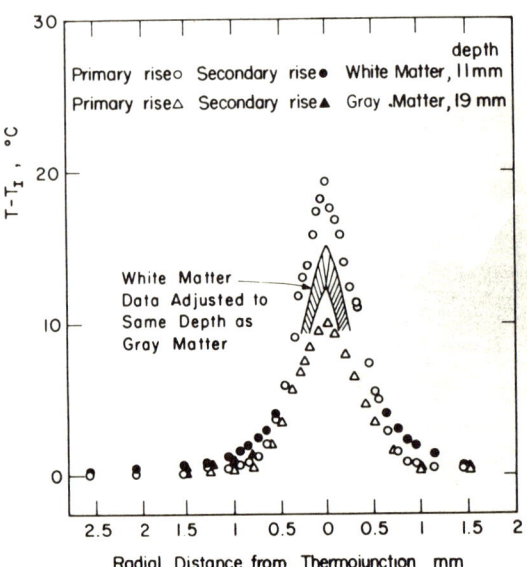

FIG. 17. The temperature profile measured in cat brain *in vivo*. Frequency: 2.7 MHz; pulse duration: 0.2 sec; acoustic power: 10 W; chromel–constantan thermocouple: 0.002-in. diameter; maximum temperature was reached at the end of irradiation (primary rise) with temperature rises of 5°C or more.

FIG. 18. The temperature profile measured in cat brain *in vivo*. Frequency: 2.7 MHz; pulse duration: 2.0 sec; acoustic power: 10 W; chromel–constantan thermocouple: 0.002-in. diameter; Maximum temperature rises of 20°C or more. T_L: lesion boundary temperature.

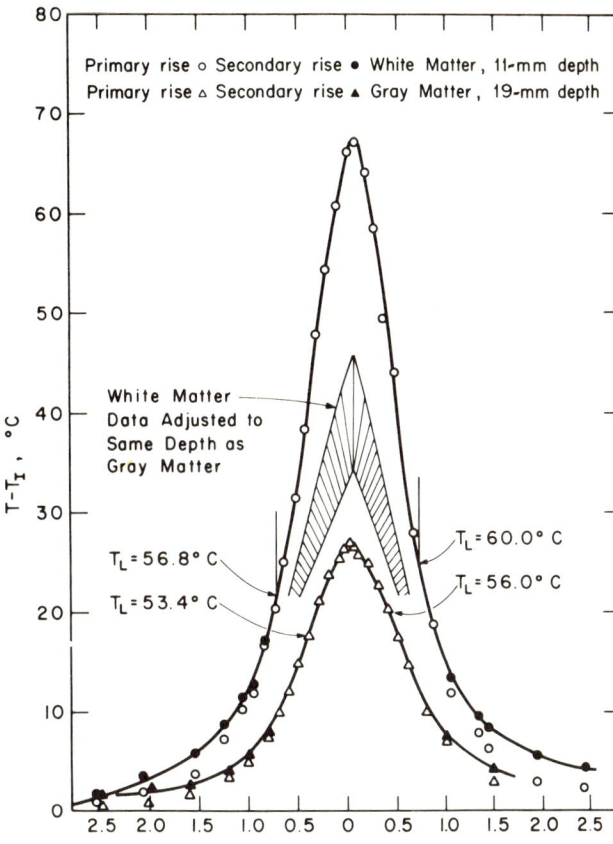

tivity of brain with blood flow. This empirical value of q_0 can be compared to the q_0 which is calculated from the measured acoustic power input and the relationship between this power input and heat generation as determined in Sec. I-A. The calculated value of q_0 is 303 W/cm³, for 10 W of acoustic power; intensity reflection at the water–brain interface is negligible.[15] The empirical value of q_0 is approximately 10% higher than the

FIG. 19. A comparison of experimental data and analytical results in cylindrical coordinates for lesion diameter in cat brain at various pulse durations. $q_0 = 338$ W/cm³ at 10-W acoustic power, empirical value.

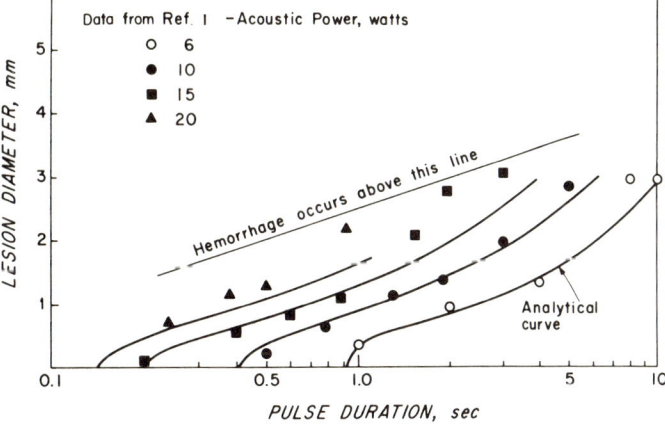

341

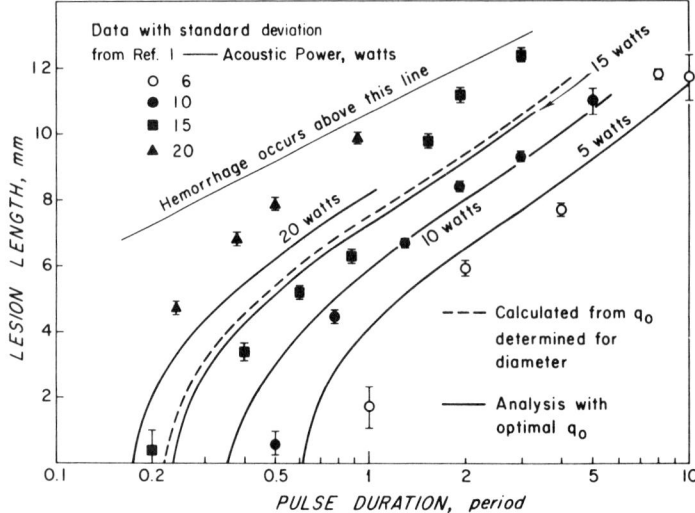

FIG. 20. A comparison of experimental data and analytical results in spherical coordinates for lesion length in cat brain at various pulse durations.

calculated value. The fit of the empirical equation to the measured intensity distribution has been shown in Sec. I-A to be imperfect at large radii. The uncertainties in the values for attenuation and changes in thermal conductivity with blood flow are about $\pm 20\%$ and $\pm 30\%$, respectively. These uncertainties are sufficient to explain the 10% disagreement in the two values for q_0. The attenuation in the dura mater has not been taken into account in this analysis. This attenuation has been observed to be sufficient to cause some nonreproducibility or scatter in lesion size.[1] A correction for this effect would tend to increase slightly the disagreement between the empirical and calculated values of q_0.

The analytical and experimental temperature distributions in nervous tissue are compared in Fig. 21. Here the same q_0 value of 338 W/cm³ was used to fit the data; the result is curve A. Curve B was generated with the same q_0, but with 0.2-sec pulse duration data, although the maximum temperature difference is only 3°C.

Curve C is data taken for the less vascular, white matter corrected in temperature for the difference in depth compared to the data of curve A. Curve C' was generated using the same value of q_0 as for curve A, but using the value of thermal conductivity without blood flow. This lower value of thermal conductivity is seen to shift curve A toward C, but an additional correction in heat generation for C appears necessary.

VI. DISCUSSION AND CONCLUSIONS

The analysis and data for the relation between the lesion diameter and pulse duration in methacrylate do not agree well because of the assumption of constant attenuation in the analysis for the temperature distribution. The analysis will be more accurate for smaller temperature rises (higher initial temperatures) and consequently smaller changes in attenuation in methacrylate.

The agreement between analysis and data for lesion diameter and length in brain is better in spite of the existence of many sources of error previously discussed.

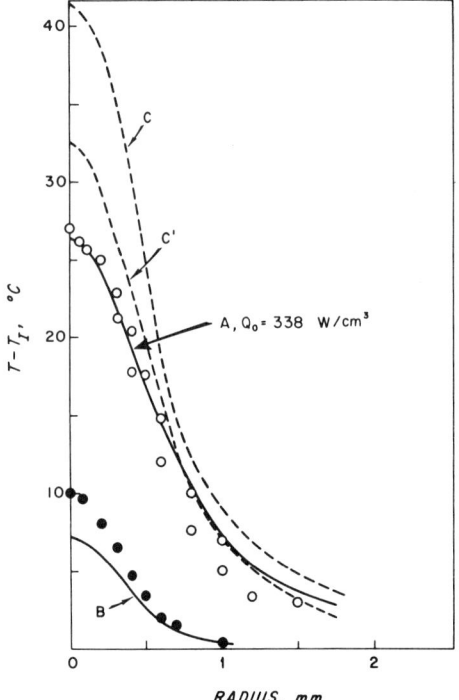

FIG. 21. A comparison of experimental data and analytical results in cylindrical coordinates for radial temperature profile in cat brain. $q_0=338$ W/cm³ at 10-W acoustic power, empirical value. (See text.) ●: *In vivo*, 0.2-sec pulse, 10 W; ○: *in vivo*, 2.0-sec pulse, 10 W.

It is apparent that more accurate data are necessary on the ultrasonic and thermal properties of nervous tissue if greater accuracy is desired in the prediction of lesion size, shape, and location. It is also desirable to determine these properties for white and gray matter separately. The relatively good fit of analytical prediction to data for lesions in brain suggests that the complexity of a more accurate analysis which better describes the three-dimensional shape of the intensity distribution and which takes into account the axial attenuation may be desirable in future investigations. It would also be valuable to calculate the temperature distribution that occurs after the end of the pulse of ultrasound. This would permit the secondary temperature rise to be calculated, which will more accurately determine the lesion boundary for lesions produced at high levels of acoustic power and short pulse durations. This temperature distribution could also be used to calculate a new average initial temperature at the beginning of each pulse for lesions created with more than one pulse of ultrasound. Such an analysis will permit the size and shape of multiple pulse lesions to be calculated.

However, in spite of these deficiencies, the analytical prediction of and model for lesion production and of the size, shape, and location of the resultant lesions in the nervous tissue that has been derived is of sufficient accuracy that the fit of the analytical prediction to experimental data for lesions in the brain is good. This agreement of the analysis and data shows that within the range of ultrasonic parameters used in this study, the development of lesions in the brain and all of their measurable characteristics can be explained by purely thermal mechanisms. This conclusion is also supported by the measured temperature profile and measured lesion boundary temperatures in the brain of the cat.

ACKNOWLEDGMENTS

This work was supported by U. S. Public Health Service Grant NB 08571. The authors are indebted to Professors R. Mann and B. Mikic for their interest and advice. The assistance of Mr. M. Hubelbank in microwave experiments is gratefully acknowledged.

---

* Part of this work was accepted as a PhD thesis at MIT (July 1968).

[1] L. Basauri and P. P. Lele, "A Simple Method for Production of Trackless Focal Lesions with Focused Ultrasound: Statistical Evaluation of the Effects of Irradiation on the Central Nervous System of the Cat," J. Physiol. 160, 513–534 (1962).

[2] P. P. Lele, "Effects of Focused Ultrasonic Radiation on Peripheral Nerve, with Observations on Local Heating," Exp. Neurol. 8, 47–83 (1963).

[3] G. F. Young and P. P. Lele, "Focal Lesions in the Brain of Growing Rabbits Produced by Focused Ultrasound," Exp. Neurol. 9, 502–511 (1964).

[4] P. P. Lele, "Effects of Ultrasonic Periodic Forces (Liquid Coupling)," in *Environmental Biology*, P. L. Altman and D. S. Dittmer, Eds. (Federation of American Societies for Experimental Biology, Bethesda, Maryland, 1966), Chap. III, pp. 205–207.

[5] P. P. Lele, "Production of Deep Focal Lesions by Focused Ultrasound—Current Status," Ultrasonics 5, 105–112 (1967).

[6] P. P. Lele, "A Simple Method for Production of Trackless Focal Lesions with Focused Ultrasound: Physical Factors," J. Physiol. 160, 494–512 (1962).

[7] P. P. Lele, "Irradiation of Plastics with Focused Ultrasound: A Simple Method for Evaluation of Dosage Factors for Neurological Applications," J. Acoust. Soc. Amer. 34, 412–420 (1962).

[8] P. P. Lele, "Concurrent Detection of the Production of Ultrasonic Lesions," Med. Biol. Eng. 4, 451–456 (1966).

[9] T. P. Anderson, K. G. Wakim, J. F. Herrick, W. A. Bennett, and F. H. Krusen, "An Experimental Study of the Effects of Ultrasonic Energy in the Lower Part of the Spinal Cord and Peripheral Nerves," Arch. Phys. Med. Rehab. 32, 71–83 (1951).

[10] J. F. Herrick, "Temperatures Produced in Tissues by Ultrasound: Experimental Study Using Various Techniques," J. Acoust. Soc. Amer. 25, 12–16 (1953).

[11] J. E. Lehmann, "The Biophysical Basis of Biological Ultrasonic Reactions with Special Reference to Ultrasonic Therapy," Arch. Phys. Med. Rehab. 34, 139–152 (1953).

[12] P. Wells, "Some Biological Effects of Ultrasound," PhD thesis, Univ. of Bristol, England (1966).

[13] P. P. Lele and A. D. Pierce, "The Thermal Hypothesis of the Mechanism of Ultrasonic Focal Destruction in Organized Tissues," (to be published).

[14] P. P. Lele, "A Highly Stable, Sensitive Thermocouple Probe for the Determination of Energy Distribution in Ultrasonic Fields," (to be published).

[15] T. C. Robinson, "An Analysis of Lesion Development in Plexiglas and Nervous Tissue Using Focused Ultrasound," PhD thesis, MIT, Cambridge, Mass. (1968).

[16] H. S. Carslaw and J. C. Jaeger, *Conduction of Heat in Solids* (Oxford U. P., England, 1947).

[17] Rohm and Haas, "Plexiglas Design, Fabrication, and Molding Data, Thermal Properties of Plexiglas," Bull. No. PL74d.

[18] L. S. Fountain, "Experimental Evaluation of the Total-Reflection Method of Determining Ultrasonic Velocity," J. Acoust. Soc. Amer. 42, 242–247 (1967).

[19] B. Carlin, *Ultrasonics* (McGraw-Hill, New York, 1960).

[20] R. Roark, *Formulas for Stress and Strain* (McGraw-Hill, New York, 1954).

[21] F. McClintock and A. Argon, *Mechanical Behavior of Materials* (Addison-Wesley, Reading, Mass., 1966).

[22] A. Campbell, "Variation in Vascularity and Oxidase Content in Different Regions of the Brain of the Cat," Arch. Neurol. 41, 223–242 (1939).

[23] J. Grayson, "Internal Calorimetry in the Determination of Thermal Conductivity and Blood Flow," J. Physiol. 118, 54–72 (1952).

[24] M. Lipkin and J. D. Hardy, "Measurement of Some Thermal Properties of Human Tissues," J. Appl. Physiol. 7, 212–217 (1954).

[25] E. Ponder, "The Coefficient of Thermal Conductivity of Blood and of Various Tissues," J. General Physiol. 45, 545–551 (1962).

[26] R. Trezak (1968) (personal communication).

[27] J. Frederick, *Ultrasonic Engineering* (Wiley, New York, 1965).

[28] J. Linzell, "Internal Calorimetry in the Measurement of Blood Flow with Heated Thermocouples," J. Physiol. 121, 390–402 (1959).

[29] W. Perl, "Heat and Matter Distribution in Body Tissues and the Determination of Tissue Blood Flow by Local Clearance Methods," J. Theoret. Biol. 2, 202–235 (1962).

[30] I. Elpiner, *Ultrasound: Physical, Chemical and Biological Effects* (Consultants Bureau, New York, 1964).

[41] T. F. Hueter quoted in F. Dunn, "Ultrasonic Absorption by Biological Materials," in *Ultrasonic Energy*, E. Kelly, Ed. (Univ. of Illinois Press, Urbana, Ill., 1965), Chap. 4.

ERRATA

Page 1333, under "List of Symbols": "a_2" should read "axial distance z_i at $I_w/I_{w_0} = 0.5$."

Page 1334, under "List of Symbols," 12th symbol in column 1 should read: "y, $Y = y_1$ or y_2."

Page 1335, equations (1), (2), and (3) should read:

$$"I_w/I_{w_0} = C_1(1 - \tfrac{2}{3}\beta_1 y^2)e^{-\beta_1 y^2} + C_2(1 - \tfrac{2}{3}\beta_2 y^2)e^{-\beta_2 y^2} + C_3(1 - \tfrac{2}{3}\beta_3 y^2)e^{-\beta_3 y^2}" \quad (1)$$

$$"I_s = \mu I_T e^{-\alpha_s z_t}" \quad (2)$$

$$"-\delta I_s/\delta z = 2\alpha_s \mu I_T e^{-2\alpha_s z} = 2\alpha_s I_s = q" \quad (3)$$

Page 1336, equations (4), (6), (10), (11), and (12) should read:

$$"q_0 = 2\alpha_s I_{s_0} = 2\alpha_s \mu I_{w_0} e^{-2\alpha_s z_t}" \quad (4)$$

$$"C^1 P = 2\pi I_{w_0} \sum_{n=1}^{3} \frac{C_n a_1^2}{6\beta_n}\left[1 + \left\{e^{-\beta_n (r/a_1)^2}\right\}\left\{2\beta_n\left(\frac{r}{a_1}\right)^{'2} - 1\right\}\right]" \quad (6)$$

$$"q_0 = 2\alpha_s \mu I_{w_0} e^{-2\alpha_s z_t}" \quad (10)$$

$$"q_0 = 272\alpha_s \mu C^1 P e^{-2\alpha_s z_t}" \quad (11)$$

$$"q_0 = 288\alpha_s \mu P e^{-2\alpha_s z_t}\ W/cm^3" \quad (12)$$

UHF ACOUSTIC INTERACTION WITH BIOLOGICAL MEDIA

Stephen A. Hawley and Floyd Dunn

UHF Acoustic Interaction with Biological Media*)

It is well known that ultrasound interacts with biological systems both in the presence[1]) and absence[2]) of cavitation phenomena. Until recently, however, the upper frequency range available to investigators was limited to that in the neighborhood of 10 Mc by the desire to operate the electromechanical transducer at the fundamental thickness mode. Recent developments in transducer design and operation, in which the piezoelectric element is excited electrically at the odd harmonics of the fundamental thickness mode, have extended the available frequency range to at least 2000 Mc[3]). The purpose of this note is to describe several initial experiments which illustrate that the range of interaction of ultrasound and biological media extends far beyond that previously considered.

Rotifers[4]), small polynucleated aquatic animals several hundred microns in length, were suspended in physiological saline on the surface of the quartz plate (i.e., the transducer was arranged to radiate vertically upwards). At room temperature (22° C), and with no sound present, the specimens attached their (lower) extremity to the surface of the quartz plate and adopted a nearly vertical posture while executing an undulating-like succession of movements confined to an approximately 60° conical volume. The animals were exposed to single ultrasonic pulses ranging from 0.1 sec to several minutes duration at acoustic intensities of the order of 10^{-3} W/cm^2 in the frequency range from 200 Mc to 600 Mc. Upon irradiation with ultrasound it was observed that only in relatively narrow frequency bands in the neighborhoods of 270 Mc and 510 Mc were these characteristic activities altered. The nature of the change in activity was virtually complete cessation of all movement. The animal, still attached to the quartz plate, assumed a globular configuration and remained dormant for the duration of the acoustic pulse. Upon termination of short acoustic pulses (3 to 30 sec), the specimen recovered the characteristic activity. Numerous rotifers were studied in this manner and any single specimen could be carried through repeated acoustic cycles, throughout the frequency range investigated, without apparent damage. Pulse durations of the order several minutes led to apparent irreversible damage as viability did not return.

Amoeba proteus were also exposed to the radiation at the same frequencies and under similar conditions. Here, however, neither a gross effect on the body of the organism nor a small scale effect, such as perturbation of amoebal streaming, were observed.

Although the nature of the interaction observed with rotifers at 270 Mc and 510 Mc is at present unknown, the following statements can be made. Injury produced in rotifers (and other biological specimens) at lower frequencies (below 1 Mc) has been attributed to cavitation present during ultrasonic exposure[*]). However, the intensities employed in this study are approximately a factor of 10^8 below the threshold of cavitation at these frequencies, viz., 10^5 W/cm², and the acoustic pressure amplitude is approximately 1/100 of the hydrostatic pressure[6]). This fact, together with the findings that suppression of rotifer activity occurs in particular frequency bands and that amoebae are unaffected even in these frequency bands, should eliminate cavitation as the mechanism of interaction. In the absence of more specific information, let it be assumed that the acoustic intensity absorption coefficient per unit path length in the rotifer is the same as the average value observed for the mammalian central nervous system, viz., approximately 0.2 cm^{-1} at 1 Mc, and that it increases linearly with frequency[6]). This leads to an estimate of the time rate of temperature rise in the rotifer of approximately 3×10^{-2} °C/sec. It is seen that for acoustic exposure durations as long as 10^2 sec, the maximum temperature developed in the animal, in the absence of thermal conduction, is but several degrees above room temperature (22° C) and this is not sufficient to be considered seriously for the rotifer, which thrives at temperatures in excess of 35° C [7]). The absorption of sound in the imbedding liquid is sufficiently great, as is the path length in the chamber, such that standing waves of large amplitude are not produced. That this is not important in the alteration of the activity of the rotifer was verified by the observation that changing the acoustic path length had no observable effect upon the experimental results.

As the observed effect appears in the neighborhood of 270 Mc and 510 Mc, two frequency regions nearly integrally related, it is tempting to consider a resonance phenomena as playing a role in the interaction. Further research is in progress.

This research is supported by the Office of Naval Research, U.S. Navy.

Biophysical Research Laboratory, University of Illinois, Urbana, Illinois

Eingegangen am 13. Juni 1964

*) Remarks on some of the initial observations were made at the Symposium on Ultrasound in Biology and Medicine, University of Illinois, June, 1962.

[1]) GRABAR, P., in: Advances in Biological and Medical Physics, ed. by J.H. LAWRENCE and C.A. TOBIAS, vol. III, p. 191. New York: Academic Press 1953. — [2]) DUNN, F.: Am. J. Phys. Med. **37**, 148 (1958). — [3]) DUNN, F., and J.E. BREYER: J. Acoust. Soc. Am. **34**, 775 (1962). — [4]) See for example PENNAK, R.W.: Fresh Water Invertebrates of the United States, chap. 8. New York: The Ronald Press 1953. — [5]) GOLDMAN, D.E., and W.W. LEPESCHKIN: J. Cellular Comp. Physiol. **40**, 225 (1952). — [6]) FRY, W.J., and F. DUNN, in: Physical Techniques in Biological Research, ed. by W.L. NASTUK, vol. IV, chap. 6. New York: Academic Press 1962. — [7]) FINESINGER, J.E.: J. Exptl. Zool. **44**, 63 (1926).

The Role of Heat in the Production of Ultrasonic Focal Lesions

John B. Pond

Guy's Hospital Medical School, London S. E. 1, England

A method was evolved to calculate the temperature cycles at the focus center of an intermittent beam of high-frequency mechanical waves. The computed temperatures were checked in brain tissue, using microthermocouples, for doses just sufficient to damage tissue at the focus center. Similar temperature cycles were produced in brain tissue, in the absence of mechanical wave energy, by applying pulses of electric current to embedded resistance wires in order to find out how far the tissue alteration could be ascribed to heating.

INTRODUCTION

The use of focused ultrasound can provide a method to produce isolated, precise, and repeatable volumes of damage (lesions) in animal tissue.[1,2] An important concomitant of tissue alteration with the ultrasonic-beam characteristics commonly used is the heating of the target following the absorption of wave energy. Thus, Basauri and Lele,[3] working with a focused beam on the brains of cats, found that the lesion dimensions decreased for the same values of acoustic intensity and time duration of exposure when the animals were cooled. Dunn[4] notes a similar dependence on the base temperature of the animal, although the highest temperatures reached were far below that required for thermal damage, viz., the normal temperature of the animal. His experimental endpoint was an impairment of function of mouse spinal cord. Indeed, Lele's[5] repeated pulsing of a focused beam on isolated peripheral nerves immersed in saline or paraffin did not appear to impair function or anatomy by prolonged exposures in the region of 600 W·cm^{-2}, unless the surface temperature of the preparation was allowed to rise. In Ref. 5, the irradiation-induced temperature cycles, associated with nerve function impairment, were considered sufficient to explain the effect.

Such results involve different targets of animal tissue, and different effects of radiation are being observed. However, a dependence upon the irradiation-induced temperature cycles is apparent. In work with beams of high intensity with small focal volume, high rates of heat production per unit volume are associated with rapid heat leakage into the cooler surrounds. Thus, temperature cycles that may have high maxima are critically variable with focus size, intensity, and irradiation timing. In the following, we derive a method of computing the complete temperature cycle at the focus center of an intermittent beam of given intensity distribution, operating into a medium of given absorption. It is further shown that, with beam parameters of particular magnitudes, such information is sufficient to determine the presence of an ultrasonic lesion in rat brain when this effect is considered as alteration, observable under the light microscope in a common method for examining brain tissue.

Consider a solid cylinder in the infinite similar medium that is static and of thermal conductivity K (W·cm^{-1}·°C^{-1}). Suppose this cylinder subject to uniform volume heating so that its temperature would rise, in the absence of heat conduction, at the rate $P/J \cdot \rho C_p$ (°C·sec^{-1}), where P is the absorbed power per unit volume (W·cm^{-3}), and ρC_p is the thermal capacity per unit volume (cal·cm^{-3}·°C^{-1}) and J is the mechanical equivalent of heat.

In such a case, the temperature rise T' on the axis of a cylinder of diameter $2a$ and length $2c$ is given as a function of t, the time in seconds since switch-on, by

$$T' = \frac{P}{J\rho C_p} \int_0^t \left\{ \left[1 - \exp\left(\frac{a^2}{4\kappa u}\right) \right] \times \mathrm{erf}[c \cdot (4\kappa u)^{-\frac{1}{2}}] \right\} \cdot du, \quad (1)$$

where κ is the diffusivity ($\kappa = K/\rho C_p$).

This equation is derived, for example, by Carslaw and Jaeger.[6] In the present context, it may be reduced somewhat. Animal tissues are so largely composed of water that ρC_p may be approximated to unity with an

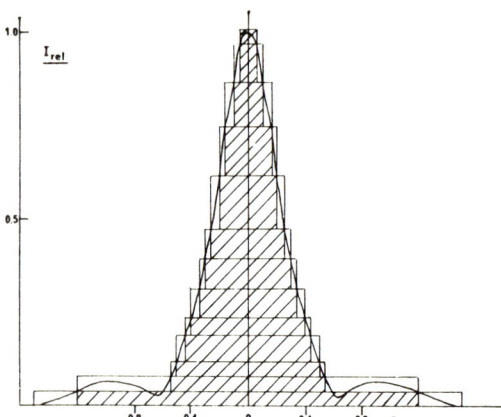

FIG. 1. Transverse central scan of relative intensity in beam focal plane. 1-MHz projector operating at 3 MHz. Cylinder superposition approximation to the intensity distribution is shown with radius, r, in millimeters.

error no greater than approximately 15%. Also, although the thermal conductivity of brain tissue is affected by vascular flows,[7] the effects are less than might be supposed, since, except near large vessels, the flows are fairly isotropic when distances over about 0.2 mm are considered. Generally, for animal soft tissues, thermal conductivities do not vary much from the value for water. For the rat-brain site used as the target in the present work, K has been measured[8] as $K = 6.5 \pm 0.71$ mW·cm^{-1}·°C^{-1} in the range 37°–70 °C. Owing to this low numerical value of K and the fact that the error function is very near unity for arguments exceeding unity, we may simplify Eq. 1 to

$$T' = \frac{P}{J} \int_0^t \left[1 - \exp\left(-\frac{a^2}{4\kappa u}\right) \right] \cdot du, \quad (2)$$

and it accordingly integrates to

$$T' = \frac{tP}{J} \left[1 - \exp\left(-\frac{a^2}{4\kappa t}\right) + \frac{a^2}{4\kappa t} \cdot E_i \frac{a^2}{4\kappa t} \right], \quad (3)$$

where E_i is the exponential integral,

$$-E_i(-x) = \int_0^\infty \frac{e^{-u}}{u} \cdot du.$$

Put

$$T' = (tP/J)\{1 - e^{-x} + xE_i(x)\} = (tP/J) \cdot f(x). \quad (4)$$

Table I gives $f(x)$ in the range $x = 0$ to $x = 4$. Next, put $P = g_h I = g_h I_0 e^{-gd} = g I_0 e^{-gd}$, where g is the wave-energy-absorption coefficient per centimeter, I_0 is the peak unattenuated wave intensity ($W \cdot cm^{-2}$), and d is the field-point depth in the tissue (cm). The equality $g = g_h$ amounts to assuming that all the absorbed wave energy appears as heat in the tissue. Then Eq. 4 becomes

$$T' = I_0 \cdot (g/J) \cdot e^{-gd} \cdot t \cdot f(x). \quad (5)$$

Now, since T' is proportional to I_0, we may consider T, the focus-center temperature rise, as the sum of a number of terms T' arising from cylindrical volumes superimposed coaxially with the beam. The intensity contribution of each such cylinder is constant over its volume, and the radii and associated intensities may be chosen to match the intensity distribution near the focus of a particular beam. We thus have

$$T(t) = \sum_n T' = (I_0 \cdot g/J) e^{-gd} \cdot t \cdot \sum_n b_n f(x)_{a_n, t}, \quad (6)$$

where b_n is the normalized relative intensity and a_n is the radius of cylinder n.

Figure 1 shows the relative-intensity distribution in the focal plane of the beam used in the present work. A cylinder approximation set with normalized intensities is shown. The cylinders are drawn end to end with heights proportional to their intensities. Thus, a correct set with sufficient members would just fill the curve and give an exact approximation. To minimize the computation, one may choose such a set as a_n, b_n (shaded set) and note that $T(t)$ lies between the values obtained from this set and a set a_{n-1}, b_n, where increasing n corresponds to decreasing radius. We therefore define a function $A(t)$ as

$$A(t) = \tfrac{1}{2} \sum_n b_n [f(x)_{a_n, t} + f(x)_{a_{n-1}, t}].$$

TABLE I. Tabulation of the function $f(x) = 1 - e^{-x} + xE_i(x)$.

x	$f(x)$	x	$f(x)$	x	$f(x)$
0.001	0.0073	0.170	0.3870	0.680	0.7573
0.002	0.0133	0.180	0.4004	0.700	0.7651
0.003	0.0187	0.190	0.4132	0.720	0.7723
0.004	0.0238	0.200	0.4256	0.740	0.7795
0.005	0.0286	0.210	0.4381	0.760	0.7862
0.006	0.0332	0.220	0.4493	0.780	0.7929
0.007	0.0377	0.230	0.4610	0.800	0.7992
0.008	0.0428	0.240	0.4715	0.850	0.8141
0.009	0.0461	0.250	0.4822	0.900	0.8277
0.010	0.0503	0.260	0.4926	0.950	0.8299
0.015	0.0692	0.270	0.5066	1.000	0.8516
0.020	0.0867	0.280	0.5122	1.050	0.8621
0.025	0.1022	0.290	0.5220	1.100	0.8717
0.030	0.1069	0.300	0.5310	1.150	0.8807
0.035	0.1210	0.320	0.5484	1.200	0.8889
0.040	0.1357	0.340	0.5673	1.250	0.8973
0.045	0.1566	0.360	0.5810	1.300	0.9037
0.050	0.1719	0.380	0.5961	1.350	0.9100
0.055	0.1809	0.400	0.6107	1.400	0.9161
0.060	0.1961	0.420	0.6244	1.450	0.9217
0.065	0.2029	0.440	0.6376	1.500	0.9269
0.070	0.2177	0.460	0.6507	1.600	0.9319
0.075	0.2290	0.480	0.6619	1.700	0.9442
0.080	0.2388	0.500	0.6735	1.800	0.9512
0.085	0.2488	0.520	0.6843	1.900	0.9572
0.090	0.2586	0.540	0.6948	2.000	0.9630
0.100	0.2773	0.560	0.7048	2.100	0.9670
0.110	0.2950	0.580	0.7145	2.200	0.9710
0.120	0.3127	0.600	0.7238	2.300	0.9740
0.130	0.3286	0.620	0.7327	2.400	0.9776
0.140	0.3437	0.640	0.7415	2.500	0.9801
0.150	0.3590	0.660	0.7496	2.700	0.9846
0.160	0.3729			3.000	0.9913
				4.000	0.9968

$A(t)$ for the present beam is shown as Fig. 2. It may be considered as the ratio of actual temperature rise to that which would have occurred without heat leakage at the focus. Note that $A(t)$ being a function of $a^2/\kappa t$ offers two advantages. *First*, one tabulation may accommodate different diffusivities by the use of a reduced time $t' = t \cdot k/k'$. *Second*, geometrically similar foci of different sizes may be accommodated, since the relation $A_F(t) = A_{nF}(t/n^2)$ holds where F is the fundamental frequency and n is a harmonic number. This is true insofar as the diffraction pattern forming the focus reduces in step with the wavelength and, thus, one may expect to cater for a focusing device operating at different harmonics.

We note that T is linear in I, and it follows that switching off the beam at a time M after switch-on is equivalent to additional heat generation in a negative sense. Thus, for a later time t, we have

$$T(t)_{\text{beam off at }t=M} = T(t) - T(t-M). \quad (7)$$

Figure 3 shows the result up to 7 sec for $M=3$, taking $k=K=6.5\times 10^{-3}$ and $g=0.82$ with the beam of Fig. 1. The value of $g=0.82$ for the present target at $F=3$ MHz was found by experiments[8] involving initial rate of temperature rise where thermal conduction can be neglected.

Next, consider the temperature cycles caused by heated wires. This method of producing field-point temperature cycles offers flexibility:

(1) By consideration of field points farther from the wire and/or arranging the heating-current pulse for a longer time, both shape and length of the cycle are altered.

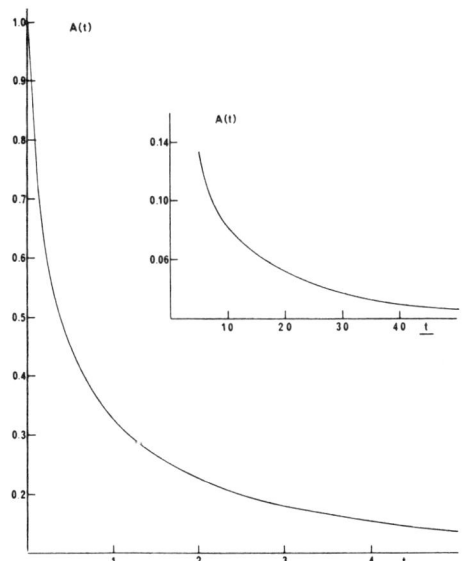

FIG. 2. The function $A(t)$ derived for the beam-energy distribution of the present experiments with time, t, in seconds.

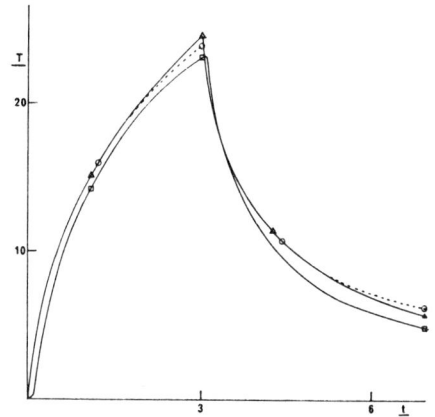

FIG. 3. Temperature rise, T, in degrees Celsius with time, t, in seconds. The curve with triangular points is the mean of 10 experimental determinations of T for the focus center of the beam for a 3-sec pulse at $I_0 = 315$ W·cm^{-2}. The standard deviation for $T_{\max}$ is 4.8°C. The curve with circular points is the corresponding calculated curve from Eqs. 6 and 7. The curve with square points is the temperature cycle calculated for points at radius 0.029 cm from a wire in brain tissue (average diameter of resulting lesion).

(2) Both large and small temperatures near the wire may be arranged to give duplicate temperatures at more remote points, so that the independence of thermal properties from temperature may be checked.

(3) The effects of the different temperature cycles occurring at different radii from the wire may be compared on each perpendicular section of the target and such results averaged by a series of sections down the wire track.

A formula for the temperature rise $T(t)$ at the surface of a relatively "perfect" conductor, statically immersed, which is in the form of a long cylinder, is[9]

$$T(t) = \frac{P'}{4\pi JK}\left\{\frac{2}{x}+\ln(x)-A+\left[\frac{1-2(\rho C_p)_w/\rho C_p}{x/2}\right]\times[\ln(x)-A]\right\}, \quad (8)$$

where $x = 4Kt/a^2\rho C_p$, a is the radius of the cylinder, P' the dissipation rate of the cylinder (W·cm^{-1}), A Euler's constant (0.5772···), and the subscript w distinguishes the cylinder (wire) from the immersing medium. Where r/a is large, a simple formula visualizing the wire as a line source may be used to find $T(r,t)$, where r is the distance of the field point from the wire, thus

$$T(r,t) = P/4\pi JK \cdot E_i(r^2\rho C_p/4Kt). \quad (9)$$

Equation 8 was used to compute and, hence, limit maximum temperatures in the tissue. Equations 9 and 7 were used to compute field-point temperatures. A cycle with relevant parameters is included in Fig. 3, and it shows how a wire cycle $T(r,t,M)$ can be made to simulate a beam cycle at focus center $T(t,M)$.

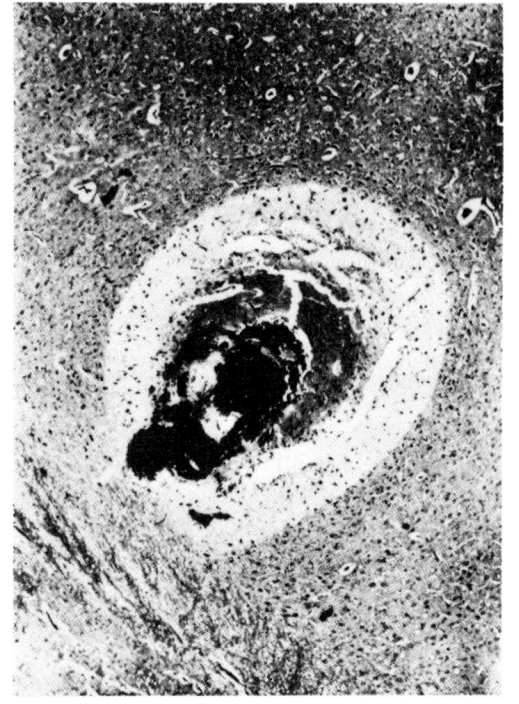

FIG. 4. Section of lesion produced by heated wire. Note *island and moat* structure, as in Fig. 5. The island tissue is better structured than the moat, since the higher temperatures have forestalled decomposition processes. Some blood is shown leaked from the wire track in the center of the island. Average diameter of moat = 0.12 cm. Hematoxylin- and eosin-stained section of rat brain.

The dissipating element used in these experiments was a length of constantan (60% Cu, 40% Ni) wire of 0.125-mm diam. This was butt-welded to bare copper wires of the same diameter. The wire, seen silver against the red copper, was gently pulled through the brain until equal amounts protruded from each side. The protrusions were immersed in saline to prevent excessive temperatures. The ratio of the diameter to the effective length of the element was small enough (1/100) to allow the formulas of interest to be very good approximations. Wires were heated by pulses of current, obtained by way of a relay timer and a rheostat, from a 6-V accumulator. A resulting lesion section is shown as Fig. 4. Note the structure of an island surrounded by a moat. Threshold conditions for visible tissue alteration are at the boundary of the lesion.

We now consider the temperature cycles associated with beam lesions. The focused-beam apparatus has been described elsewhere.[10] The beam angle of convergence was 50°, its focal length was 7.8 cm, and the frequency was 3.02 MHz. The targets for this work were in the brains of anesthetized rats, the beam entering vertically down through a hole abraded in the skull. On either side of the brain they were defined as follows: midway between the Lambda and Bregma skull suture lines, midway between the edge of the brain and the midline, 4 mm down from the brain surface. Resistance or thermocouple wires were threaded through the lines between such points, using holes abraded through the sides of the skull. In the course of inspection for lesions, whether by beam or hot wire, the animal was allowed to live for 1 h after dosage. The brain was then excised, fixed in formol saline, and embedded in paraffin wax. Cut sections were stained in hematoxylin–eosin and examined under the light microscope. The sections were cut perpendicularly to the axis of beam or wire. Temperatures in the brain were measured with thermocouple probes consisting of bare copper and constantan wires butt-welded to form the thermoelectric junction. The wires were of 0.05-mm diam. The output was taken to a differential amplifier and thence to a measuring oscilloscope.

In the present work, a lesion is considered as a tissue alteration visible under the present method of examination. We are particularly concerned with the temperature cycles associated with conditions just severe enough to produce a lesion. Such conditions occur at the *moat* outer edge in Figs. 4 and 5. Note the similarity of the hot-wire lesion (Fig. 4) to the focused-beam lesion (Fig. 5). In the focused-beam case, one may define a threshold dose as one for which conditions are just

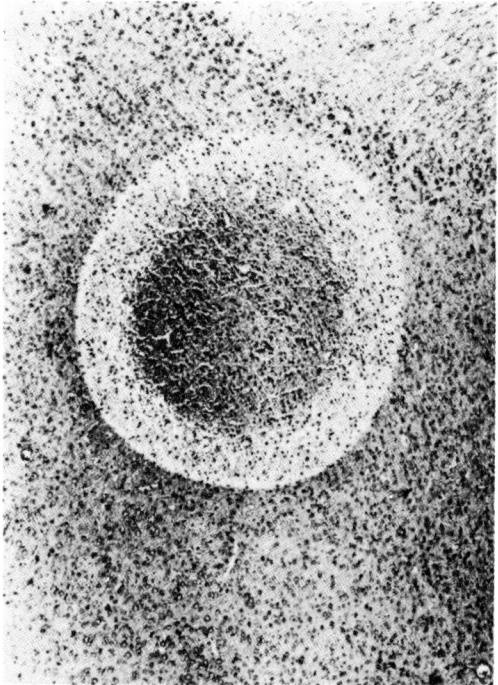

FIG. 5. Section of ultrasonic lesion produced in rat brain by 0.6-sec exposure at 3 MHz, with a focus central intensity of 1500 W·cm^{-2} and a 1-mm-diam lesion. Hematoxylin–eosin stain.

severe enough to cause a lesion at the focus center. In this case, the lesion-threshold-temperature cycle may be predicted from Eqs. 6 and 7. It is of interest to compare it with the measured cycle and also of interest to compare that cycle to a cycle of similar timing that shows lesion threshold without the addition of wave energy.

The intensity required for a threshold-beam dose, corresponding to a single pulse of 3 sec, was determined by making a series of lesions with diminishing intensity, as in Fig. 6. The resulting curve of lesion diameter versus beam intensity reaches 0 diam at the threshold intensity $I_{0\ \text{thresh}}$, being also checked as the 50% point in a curve of percentage lesion incidence versus intensity. A small standard lesion, placed to the side, ensured target location on microscopic examination. Note that I_0 is the unattenuated focus-peak intensity, that is, as measured in water. Thus, the threshold intensity at the target is here $I = I_0 e^{-\nu d}$, which is 226 W·cm^{-2}. Figure 3 shows both measured and calculated temperature cycles. It also shows the threshold cycle (calculated for the edge of the lesion) in the hot-wire case best matching beam-cycle timing.

From Fig. 3 and Table II, we conclude that, with beam parameters in the present regions, (1) the method for calculating focus-center temperature rises gives close agreement with observation; and (2) the brain-tissue damage observed under the present conditions is initiated by heating.

TABLE II. Relation of calculated to measured maximum-beam temperature rises for temperature cycles at the threshold of lesion production $I_0 = 315$, $I = 226$, $M = 3$ sec. Relation of these maxima to those in similar cycles that are threshold for lesion production without wave energy (hot-wire cycles).

	T_{max} (measured)	T_{max} (calculated)
Threshold of lesion production for a beam pulse of 3 sec at $I_0 = 315$ W·cm^{-2}.	24.7 (standard deviation 4.8)	24.1
Edge of wire-lesion at:		
$P = 0.73$ W·cm^{-1} ($r = 0.29$ mm)		23.0
$P = 0.97$ W·cm^{-1} ($r = 0.38$ mm)		24.5
$P = 1.40$ W·cm^{-1} ($r = 0.60$ mm)		21.7

Note that, from similar series of experiments with single pulses of beam energy at 1, 2, 3, 5, and 7 MHz up to $I = 500$, temperature cycles seem adequate to explain lesion onset. However, at 3 MHz for intensities in the region $I = 720$ and beyond, the temperatures are significantly inadequate.[8]

Intensity contours were measured and checked by thermocouples in absorbing targets and by the radiation pressure on small spheres. For this purpose, except for an angular resolution of the momentum, the radiation was considered as plane. Owing to the convergence of the beam, however, there are, as the focus is approached, a phase lag of particle velocity behind acoustic pressure and an increase in the ratio of particle velocity to acoustic pressure and intensity. With the beam convergencies used here and indicated for ultrasonic surgery (some 50° or less included angle), these changes are small.

However, in investigating nonthermal effects in biological tissue, which is, of course, a target with complex possibilities of interaction, it would be interesting to pursue the effect of large separation, both spatial and temporal, or particle velocity and acoustic pressure. This could be done by introducing an artifact into the target to produce a reflected component to the beam.

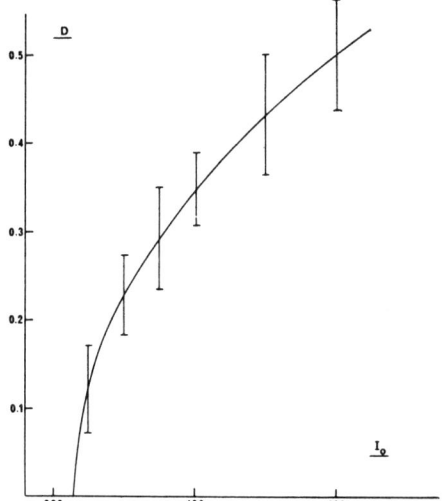

FIG. 6. Lesion diameter (in millimeters) versus peak unattenuated focus intensity (W·cm^{-2}) for 3-sec pulses. Mean plus/minus standard deviation. The threshold intensity conjugate to a 3-sec exposure corresponds to 0 diam.

[1] H. T. Ballantine, Jr., T. F. Hueter, W. H. Nauta, and D. M. Sosa, J. Exp. Med. **104**, 337–360 (1956).
[2] W. J. Fry, in *Advances in Biological and Medical Physics*, J. H. Lawrence and C. A. Tobias, Eds. (Academic, New York, 1958), Chap. 6, pp. 281–348.
[3] L. Basauri and P. P. Lele, J. Physiol. **160**, 513–534 (1962).
[4] F. Dunn, Amer. J. Phys. Med. **37**, 148–151 (1958).
[5] P. P. Lele, Exp. Neurol. **8**, 47–83 (1963).
[6] H. S. Carslaw and J. C. Jaeger, *Conduction of Heat in Solids* (Oxford U. P., London, 1959), p. 265.
[7] W. Perl, J. Theoret. Biol. **2**, 201–235 (1962).
[8] J. Pond, "A Study of the Biological Action of Focused Mechanical Waves," PhD thesis, London University (1968), p. 70.
[9] H. S. Carslaw and J. C. Jaeger, *Conduction of Heat in Solids* (Oxford U. P., London, 1959), p. 265.
[10] R. Warwick and J. Pond, J. Anat. **102**, 387–405 (1968).

Threshold Ultrasonic Dosages for Structural Changes in the Mammalian Brain *

F. J. Fry, G. Kossoff,† R. C. Eggleton,‡ and F. Dunn

Bioacoustics Research Laboratory, University of Illinois, Urbana, Illinois 61801

The relationship between the acoustic intensity and the time duration of exposure, for a single pulse, necessary to produce a threshold lesion in the cat brain was studied. Focused ultrasound of 1, 3, and 4 MHz was employed with intensities ranging from 10^2 to 2×10^4 W/cm² with the corresponding pulse durations from 7 to 2×10^{-4} sec, respectively. Three types of lesions were observed attending three regions. At the lower intensities and long time durations of exposure, the lesion is produced by a thermal mechanism. At the highest intensities and shortest time durations, cavitation is believed to be the mechanism responsible for the sometimes randomly appearing lesions. At intermediate dosages, the lesions are formed by a mechanical mechanism which is thus far not well understood. These results exhibit good agreement with that of other investigators on both the cat and the rat brain.

The use of intense focused ultrasound to produce changes in the mammalian central nervous system has been described in detail by several investigators.[1-5] Characteristically, it has been found that: (1) functional changes occur instantaneously,[6] while histological changes require approximately 10 min after exposure for the first suggestions of lesion formation to appear[7]; (2) heretofore, conditions for cavitation appeared adverse and evidence for its occurrence is sufficiently lacking to discount it as a possible mechanism[6]; (3) thermal mechanisms are important in the lower dosage regions generally employed[6,8,9]; (4) the acoustic properties of white matter are significantly different from those of gray matter and these differences may be associated with vascularity[10]; and (5) direct interaction with molecular processes can be ruled out,[11] but interaction with membraneous structures may well be the site of the acoustic involvement.[12]

Though several investigators have studied ultrasonic dosage (time duration of exposure and acoustic intensity) in the low-megahertz-frequency range, the intensities employed were generally below about 10^3 W/cm², owing to the limitations of the available instrumentation. Thus, it was considered profitable to study the higher dosage range (to 2×10^4 W/cm²) at 1 and 3 MHz and to examine the nature of the lesions so produced.

The irradiation technique and procedure have been described previously.[1] Briefly, the skull of the animal is removed to allow the sound to enter the brain unimpeded, though the *dura mater* is not opened. Degassed Ringer's solution is employed to couple the acoustic energy from the focusing transducer to the brain and the irradiation is performed with the brain temperature stabilized at 37°C. For the portion of the study involving cavitation levels of ultrasound, four irradiation exposures were placed in each animal, i.e., the focal region of the sound beam was, in turn, positioned 15 mm deep, with respect to the dorsal surface of the brain, 5 and 15 mm anterior, and ±5 mm lateral, with respect to ear bar zero. These positions[13] are in regions containing dense white-matter tracts and gray-matter regions in such distribution that the focal volume of the transducer included both white and gray matter. Thus, the threshold for white and gray matter could be examined in a single exposure, though this paper deals only with the threshold doses for white matter. The four single-pulse exposures were generally delivered at the same intensity, but with variable pulse durations. The animals were sacrificed 24 h after the ultrasonic irradiation and the brain was subsequently stained by both Weil and cresylecht violet and examined for histological evidence of damage.

Figure 1, the dosage curve for threshold lesions, at 1, 3, and 4 MHz plus the similar data of Pond on rat brain[8] and Lele on cat brain[3] comprises three ill-defined regions. At the lowest intensities and longest time durations of exposure, the lesions are considered to be

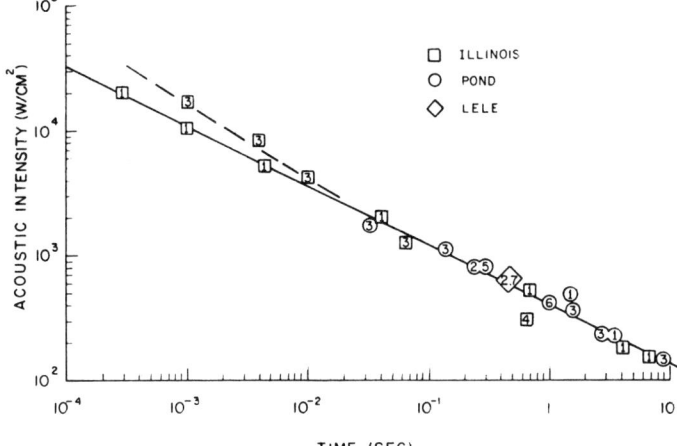

FIG. 1. Acoustic intensity versus single-pulse time duration to produce threshold lesions in white matter of the mammalian brain.

produced by heat, owing to absorption of the ultrasound in the tissue,[6] though there is not universal agreement on the upper intensity boundary.[8] In the intensity range from several hundred to approximately 1500 W/cm², thermal mechanisms do not account for the lesions. Here, the ultrasound is considered to disrupt biological structure subtly by mechanical means.[1,6] For this dosage region, the histological response to the ultrasonic irradiation has been described in great detail.[1] Briefly, white matter exhibits the lowest threshold and the lesion results in demyelination of the axis cylinders. Gray matter is more resistant and the dose must be increased approximately 30% in order to produce lesions of the same volume. Glial structures and the vascular system are more resistant to the action of ultrasound and, for threshold lesions, there is no interruption of the blood supply.

At acoustic intensities above about 2000 W/cm² and time durations less than 40 msec, the threshold lesion is believed to be produced by a cavitation mechanism. Such lesions were produced at 1 and 3 MHz and also by Pond[8] in the low-megahertz-frequency region. The cavitation region for threshold lesions shows (Fig. 1) an increase in cavitation threshold with increasing frequency, a phenomenon expected from all theories and mechanisms of acoustically induced cavitation in liquid media.[16] The character of the cavitation lesions, as discussed below, is considerably different from that due to the noncavitation mechanical and thermal mechanisms at doses near threshold, and the following differences are observed:

(1) For the cavitation mechanism, the lesion results from gross tissue damage, as opposed to the focal lesions which require approximately 10 min following ultrasonic exposure before histological evidence of the lesion emerges (as can be detected by a variety of staining methods and maximum magnification of the light microscope).

(2) Cavitation lesions may not be found at the focus of the transducer, as might be expected on the basis of volume distribution of energy. Instead, the lesions occur at interfaces between neural tissue and fluid-filled spaces, such as ventricles and blood vessels.

(3) Cavitation lesions do not exhibit the tissue selectivity characteristic of the focal lesions. It is generally believed[14] that the presence of cavitation nuclei is essential to produce weak spots in the body of the material exhibiting cavitation. Such nuclei may be very small gas bubbles, which can expand under appropriate stresses to resonant size before sudden collapse occurs. If such an hypothesis is adopted for the results presented here, presumably then there are few if any such nuclei in the body of neural tissue and they are more probably found in the fluid-filled regions such as blood vessels or in the ventricular system. In addition, since the ultrasound in this study was directed to avoid major blood vessels and the ventricles, and the cavitation apparently did not necessarily occur at the focus of the ultrasonic transducer, the intensity at the cavitation sites would be significantly lower than expected at the focal region of the transducer. As there is no information available on the intensity at the cavitation sites, the data plotted in Fig. 1 are those calculated to have been present in the focal region in the absence of cavitation.

Lesions produced with doses of ultrasound sufficient to develop cavitation in the cat brain are shown in Figs. 2, 3, 4, and 5. The adjacent 10-μ-thick brain sections shown are stained with Weil (on the right for white matter) and cresylecht violet (on the left for gray matter). All of these lesions were produced with 1-MHz ultrasound at a peak intensity of 5000 W/cm²

353

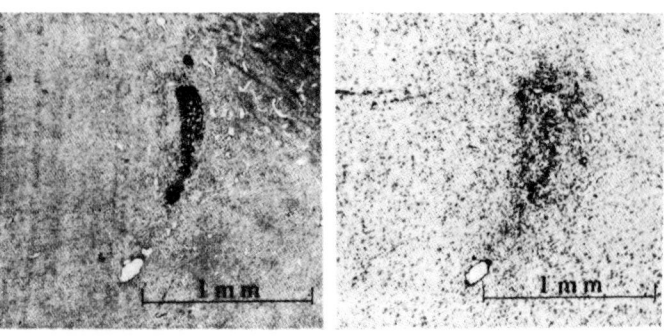

FIG. 2. Ultrasonically induced cavitation lesion in the post subicularis region of the cat brain.

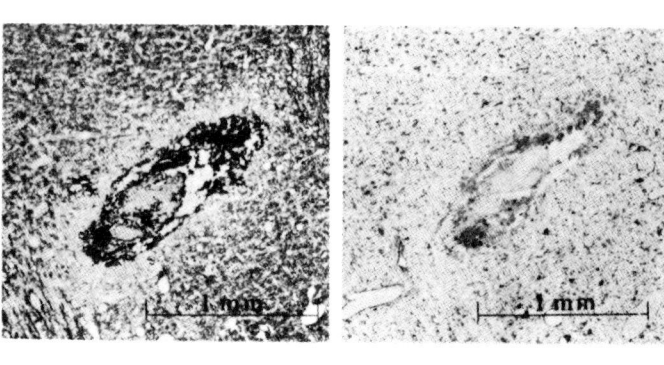

FIG. 3. Ultrasonically induced cavitation lesion in the mesencephalic reticular formation of the cat brain.

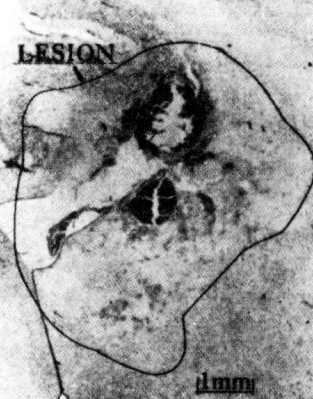

FIG. 4. Ultrasonically induced cavitation lesion at the interfaces of the ventricle, the corpus callosum, and the caudate nucleus of the cat brain.

at time durations of exposure between 25 and 200 msec. The different appearance of the lesions, in terms of a qualitative severity differentiation, is due to different times of exposure of the tissue to the focused ultrasonic beam. As the cavitation mechanism requires nuclei for initiation, a direct correlation between lesion severity and exposure time (for a fixed sound intensity), particularly near threshold, is not necessarily expected. Cavitation-mediated lesions developed at other suprathreshold intensities and exposure times exhibit the same general characteristics.

Figure 2 shows a lesion of 1-mm length in the post subicularis region of the left brain of the cat. Both stains show large erythrocyte concentrations throughout the entire central region of the lesion in which no intact blood vessels appear. Exterior to the central region is an area in which erythrocytes are not present and the blood vessels are intact, though the background matrix is in disarray compared to the adjacent normal tissue. Any identifiable neurons remaining in the lesion region show various degrees of abnormality, one predominant feature being lack of evidence of cytoplasm. The lesion

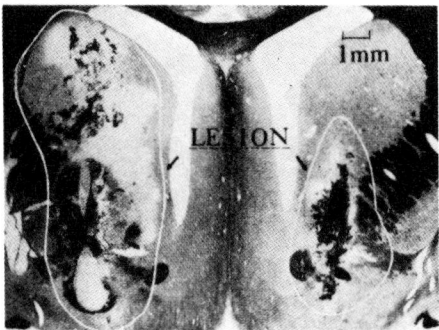 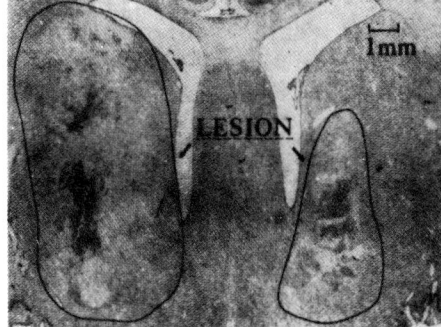

Fig. 5. Ultrasonically induced cavitation lesion in the internal capsule and the caudate nucleus of the cat brain.

exhibits a curved shape, indicating a preferential orientation, due perhaps to small blood vessel placement. This lesion does not show the great disarray of material and the disrupted boundary between normal and altered tissue seen in heavier lesions. In view of the geometrical position of this lesion in the brain, the region of most severe damage appears to be several millimeters posterior to the region of the maximum beam intensity, while in the vertical and lateral directions the lesion appears at the intended maximum beam intensity position.

The lesion shown in Fig. 3 is just lateral to the central griesum in the mesencephalic reticular formation. It is 2 mm long and its axis is at an angle of approximately 45° with respect to the vertical axis of the brain. A preferential position with respect to the blood-vessel orientation is apparent in this case. The lesion has a large central region well populated with erythrocytes, large irregularly shaped holes and tissue spaces in which the background matrix structure is missing. There is a narrow region surrounding the central area in which the matrix is in disarray—an increased number of vacuoles appear—and some small satellite areas appear around blood vessels showing erythrocyte dispersement into the surrounding tissue. These satellite regions are interpreted as possible cavitation sites at the blood-vessel interface. No intact neurons appear in any part of the lesion. Much glial and other cellular debris is seen in the narrow surrounding region, although a few normal-appearing glia are present. The lesion is at the maximum beam intensity position vertically, but is more medial by 3 mm and more posterior by about 2 mm than the maximum intensity region.

A much heavier lesion is shown in Fig. 4. This lesion has no definable boundary and is approximately 6×6 mm in size in the vertical and lateral directions. Regions involved in the lesion are interfaces with the left lateral ventricle, the left corpus callosum, and a considerable portion of the left caudate nucleus. There are many large hemorragic and torn regions intermingled with tissue areas which show the matrix in complete disarray and a large number of vacuoles present. Evidence of blood-vessel disruption in the peripheral lesion region is quite apparent. All neurons are missing from the region in the caudate nucleus except for one interspersed area in which the tissue appears normal. This area is irregular in outline and represents no more than 1% of the entire lesion.

Figure 5 shows two large lesions in the internal capsule with invasion of the caudate nucleus of both sides of the cat brain. Both lesions are centered in the region of highest intensity of the sound beam. Lesion boundaries are nonuniform in shape and large areas of erythrocyte concentration are characteristic. Large clefts and holes also appear and there are no normal cellular elements apparent in the lesion area. Cellular debris is in evidence in the lesion area and blood vessels in the more central regions are disrupted. Many blood vessels in satellite areas show slight (less than 50 erythrocytes) to large numbers of erythrocytes dispersed into the surrounding tissue.

For comparison purposes, the cat-brain section of Fig. 6 is presented to show selective white-matter lesions (noncavitation). The mammillothalamic tract on the right side of the Weil-stained section is completely missing and the fornix tract, on the same side, has been partially interrupted. A detailed study of the tissue section shows that the lesion is restricted to the fiber tracts indicated. This animal was sacrificed much longer after irradiation than those of Figs. 2–5.

The results of the present study, together with the data of Pond[8] and Lele,[3] comprise the most comprehensive set of information available on the histologically observed reaction of mammalian central nervous tissue to intense ultrasound *in vivo*. Thus, it appears profitable to consider these data in terms of the hazard ultrasound presents when employed as a medical diagnostic tool. Hill[15] has determined the output of several ultrasonic instruments manufactured for use as diagnostic tools in medicine and finds that the upper limit of the acoustic output is of the order of 10^2 W/cm^2,

355

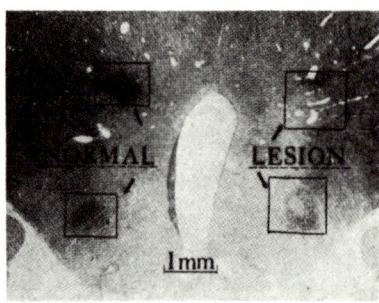

FIG. 6. Ultrasonically induced selective white-matter lesion (noncavitation) in the mammillothalamic tract and the fornix tract of the cat brain.

with pulse lengths of 10^{-6} sec and pulse repetition rates of 10^3 pps. Fry et al.[16] showed that repeated pulses of ultrasound of low duty cycle and short pulse duration do not produce permanent functional changes even after prolonged exposure. Thus, the usual pulse amplitudes and durations employed in diagnostic work appear to be several orders of magnitude below the doses required to produce threshold histologically observed lesions in the mammalian central nervous system and this tissue appears to be among the most sensitive of adult tissues.[9] From these considerations, it would appear that there is little likelihood that ultrasound provides a hazard when employed for medical diagnostic purposes with the regime described above.

---

* Supported in part by a grant from the Institute of Neurological Disease and Blindness, National Institute of Health, and in part by a grant from the Engineering Division, National Science Foundation.
† Present address: Commonwealth Acoustic Laboratories, Sydney, Australia.
†† Present address: Interscience Research Institute, Champaign, Illinois 61820.
[1] W. J. Fry, "Intense Ultrasound in Investigations of the Central Nervous System," in *Advances in Biological and Medical Physics*, J. H. Lawrence and C. A. Tobias, Eds. (Academic, New York, 1958), Vol. 6, pp. 281-348.
[2] W. J. Fry and F. J. Fry, "Fundamental Neurological Research and Human Neurosurgery Using Intense Ultrasound," IRE Trans. Med. Elec. **ME7**, 166-181 (1960).
[3] L. Basauri and P. P. Lele, "A Simple Method for Production of Trackless Focal Lesions with Focused Ultrasound," J. Physiol. **160**, 513-534 (1962).
[4] J. S. Manlapaz, K. E. Astrom, H. T. Ballantine, Jr., and P. P. Lele, "Effects of Ultrasonic Radiation in Experimental Focal Epilepsy in the Cat," Exp. Neurol. **9**, 502-511 (1964).
[5] M. Oka, T. Okumura, H. Yokoi, T. Murao, Y. Miyashita, K. Oka, S. Yoshitatsu, K. Yoshioka, H. Hirano, and Y. Kawashima, "Surgical Application of High Intensity Focused Ultrasound," Med. J. Osaka Univ. **10**, 427-442 (1960).
[6] F. Dunn, "Physical Mechanisms of the Action of Intense Ultrasound on Tissue," Amer. J. Phys. Med. **37**, 152-156 (1958).
[7] J. W. Barnard, W. J. Fry, F. J. Fry, and R. F. Krumins, "Effects of High Intensity Ultrasound on the Central Nervous System of the Cat," J. Comp. Neurol. **103**, 459-484 (1955).
[8] J. Pond, "A Study of the Biological Action of Focused Mechanical Waves (Focused Ultrasound)," PhD thesis, Univ. of London (1968).
[9] T. F. Hueter and W. J. Fry, "Ultrasonics: Central Nervous System Changes Produced by Focused Ultrasound," in *Medical Physics*, O. Glasser, Ed. (Year Book, Chicago, 1960), Vol. III, pp. 671-678.
[10] F. Dunn, P. D. Edmonds, and W. J. Fry, "Absorption and Dispersion of Ultrasound in Biological Media," in *Biological Engineering*, H. P. Schwan, Ed. (McGraw-Hill, New York, 1969), Chap. 3, pp. 205-332.
[11] R. M. Macleod and F. Dunn, "Effects of Intense Noncavitating Ultrasound on Selected Enzymes," J. Acoust. Soc. Amer. **44**, 932-940 (1968).
[12] A. Coble and F. Dunn (to be published).
[13] H. H. Jasper and C. Ajmone-Marson, "A Stereotaxic Atlas of the Diencephalon of the Cat," Nat. Res. Council of Canada (1960).
[14] H. G. Flynn, "Physics of Acoustic Cavitation in Liquids," in *Physical Acoustics*, W. P. Mason, Ed. (Academic, New York, 1964), Vol. I, Pt. B, Chap. 9, pp. 57-172.
[15] C. R. Hill, "Acoustic Intensity on Ultrasonic Diagnostic Devices," Proc. First World Corg. on Ultrasonic Diagnostics in Med. (to be published) (1970).
[16] W. J. Fry, V. J. Wulff, D. Tucker, and F. J. Fry, "Physical Factors Involved in Ultrasonically Induced Changes in Living Systems," J. Acoust. Soc. Amer. **22**, 867-876 (1950).

52

Copyright © 1974 by Pergamon Press Ltd.

Reprinted from *Ultrasound Med. Biol.*, **1**, 133–148 (1974)

THE PRODUCTION OF BLOOD CELL STASIS AND ENDOTHELIAL DAMAGE IN THE BLOOD VESSELS OF CHICK EMBRYOS TREATED WITH ULTRASOUND IN A STATIONARY WAVE FIELD

MARY DYSON, J. B. POND[*], B. WOODWARD[†] and JEANETTE BROADBENT

Anatomy Department, Guy's Hospital Medical School, London, England

(*Received* 14 September 1973; *and in final form* 7 November 1973)

Abstract—Blood cell flow can be arrested in living tissue by exposing it to ultrasound in a stationary wave field. The cells form bands at half wavelength intervals in the blood vessels. The process is generally reversible, and during dissociation the bands assume a parabolic profile in vessels where flow is laminar. Under optimum conditions, in the chick embryo, the minimum intensity required for stasis is less than 0.5 Wcm^{-2} at 3 MHz with continuous irradiation. The threshold intensity varies with the type, size and orientation of the vessel, and with the heart rate. Electron microscopy reveals damage to some of the endothelial cells lining the embryonic vessels in which stasis has occurred. Methods of avoiding stasis are suggested and the mechanisms of stasis are discussed.

Key words: Acoustics, Blood cells, Blood vessels, Blood flow velocity, Chick embryo, Hemostasis, Hemodynamics, Microcirculation, Rheology, Ultrasonics, Ultrasonic therapy.

INTRODUCTION

THE REVERSIBLE arrest of blood cell flow in living embryonic tissues was reported by Dyson *et al.* (1971). The blood cells form bands at half wavelength intervals in a stationary wave field, at frequencies and incident intensities within the therapeutic range. Cell banding has also been obtained by Baker (1972) using whole and diluted human blood *in vitro*. Summer and Patrick (1964) mentioned a similar phenomenon briefly, referring to it as 'parathrombosis'.

The aims of this paper are as follows. (1) To describe the conditions under which blood cell stasis occurs. (2) To demonstrate ultrastructural changes in blood vessels in which stasis has occurred. (3) To suggest how the effect can be avoided during the therapeutic use of ultrasound. (4) To suggest possible applications of the technique in research on the microcirculation. (5) To discuss the mechanisms thought to be responsible for producing blood cell stasis.

MATERIALS AND METHODS

The chick embryo, together with its extra-embryonic membranes, provides a suitable model system for investigating the effects of ultrasound on blood cell flow *in vivo*. Fertile White Leghorn hatching eggs were incubated at 38 ± 1°C for 84 hr. By this time the embryos had generally reached the Hamburger–Hamilton Stage 20 (Hamburger and Hamilton, 1951). Stage 20 embryos can be recognized by the following external features: they have from 40 to 43 somites and an unsegmented tail tip; the cervical flexure is marked; there is a straight mid-line contour and a flexure in the lumbar-sacral region; the eye is faintly pigmented; the allantois is vesicular but of variable size; distinct limb

---

[*] Physics Department, Kingston Polytechnic, Kingston-upon-Thames, England.
[†] Australian Atomic Energy Commission, Sutherland, NSW, Australia.

buds can be seen, the leg buds being clearly larger than the wing buds; and the amnion is completely closed.

The embryo was dissected free from the yolk in such a way as to keep the sinus terminalis intact (Fig. 1) and transferred, together with its extra-embryonal membranes, to an irradiation chamber containing normal saline solution (Fig. 2). The embryo was generally maintained at 38 ± 1°C on a heated stage. Care was taken to avoid damage during transference.

Irradiation procedure

The structure of the irradiation chamber has been described in detail elsewhere (Pond *et al.*, 1971). It consisted of a shallow, rectangular tray of chromium-plated brass with a central viewing area of thin glass over which the embryo was placed. A piezoelectric transducer (Brush–Clevite PZT 4) was fitted to one of the shorter walls of the chamber and held between two phosphor–bronze electrodes. The wall at the opposite end of the chamber acted as a reflector unless masked by an absorber. Movable glass cover slips were placed over the saline to prevent surface rippling. The apparatus used is shown diagrammatically in Fig. 3.

The embryos were irradiated at frequencies of 1, 3 or 5 MHz, with continuous or pulsed ultrasound, at peak intensities in the range of 0·1–12·0 Wcm^{-2}.

Intensity measurement

The ultrasonic field was monitored by means of a probe incorporating an ITT U23 US thermistor, sensitized by a thin layer of Selastoseal rubber compound. The structure of the probe has been described in detail elsewhere (Pond *et al.*, 1971). The position of the probe in the ultrasonic field was controlled by means of a micromanipulator attached to the irradiation chamber. The probe was calibrated by the ball-deflection method (Maidanik, 1957) in free field conditions and field peak intensities were measured in terms of Wcm^{-2}.

When determining threshold intensities for blood cell stasis, the probe was placed immediately over that part of the vessel being viewed microscopically. Incident intensity was increased until the threshold for stasis was reached. Stasis

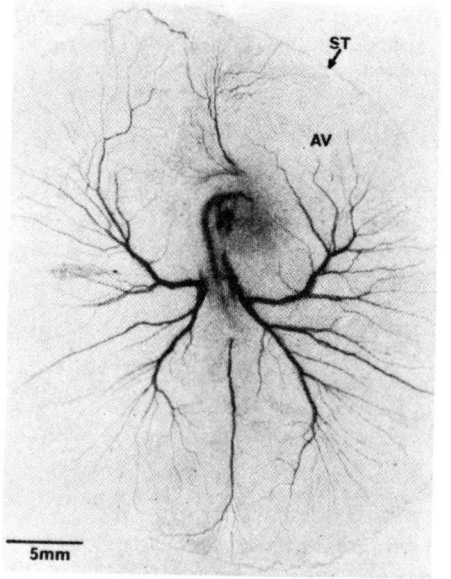

Fig. 1. The chick embryo and extra-embryonal membranes after 84 hr' incubation. AV, area vasculosa; ST, sinus terminalis.

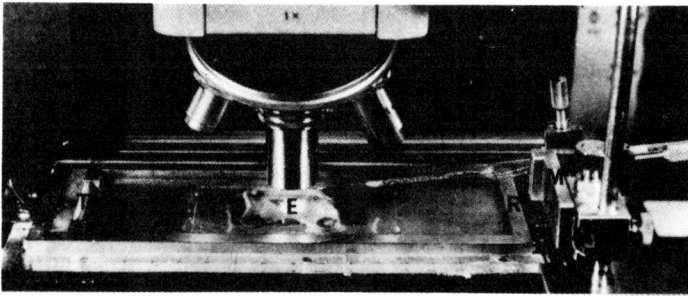

Fig. 2. The irradiation chamber. The embryo (E) lies on a thin glass plate beneath the microscope objective. The micromanipulator (M) holds the thermistor probe. The wall opposite to the transducer (T) acts as a reflector (R) unless masked by an absorber.

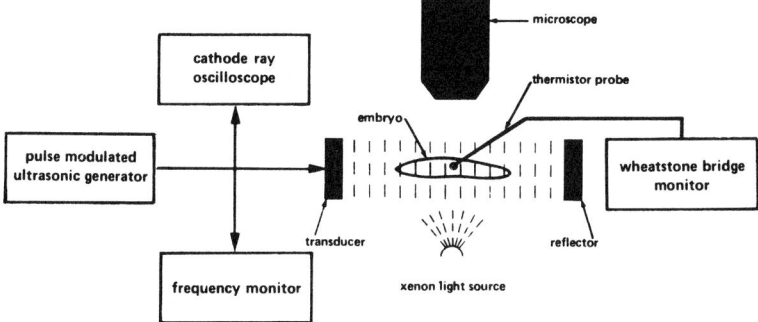

Fig. 3. Diagram of the apparatus used to induce blood cell stasis.

is here defined as the apparent cessation of blood flow when the vessel was examined at ×100 magnification. The minimum intensity required was calculated as the mean of 10 measurements at each point, the coefficient of variation being typically between 5 and 10 per cent.

Observation of blood flow

The effect of ultrasound on blood flow was observed by light microscopy. The process was filmed and subsequently examined in detail with the aid of a motion analysis projector.

Alteration of heart rate

Temporary alteration of the heart rate was attempted in a group of embryos using one of the following methods: (a) incubation in saline of different temperatures, (b) treatment with carbachol applied dropwise over the heart while the embryo was maintained at a constant temperature.

Electron microscopy

Portions of the area vasculosa containing the vessels examined were fixed in glutaraldehyde fixative buffered with 0·1 mole phosphate at pH 7·4 for four hours, followed by post-osmication in Millonig's fixative for one hour after thorough rinsing. The tissue was dehydrated and embedded in either Araldite or TAAB resin. Ultrathin sections were cut with an LKB Ultrotome fitted with a diamond knife, double stained with uranyl acetate and lead citrate, and examined with either an RCA EMU 3E or EMU 4 microscope.

All experimental tissue was irradiated for 15 min before fixation. The ultrasonic output was the minimum required to produce stasis in the vessels selected, all of which were aligned parallel to the direction of propagation of the ultrasound. The tissue was fixed either immediately after irradiation or after an incubation period of 15, 30 or 60 min. During the incubation period the tissue remained in the saline-filled chamber at a temperature of 38 ± 1°C. Control tissue was incubated in the chamber for 15, 30, 45 or 75 min before fixation, i.e. for periods equal to the sum of the irradiation and incubation periods of the experimental tissues.

RESULTS

(a) *The formation and dissociation of blood cell aggregates*

When a chick embryo, together with its extra-embryonic membranes, was irradiated in the manner described earlier, a striking phenomenon occurred: viewed macroscopically, parallel red stripes formed across the embryo and its membranes at right angles to the direction of propagation of the ultrasound. Generally this was a reversible effect, for the stripes disappeared soon after irradiation ceased. Sometimes, however, small vessels seemed permanently blocked.

When the area vasculosa was examined by light microscopy it was seen that the stripes were features of the blood vessels (Fig. 4). Each stripe was a composite structure formed of vessel-bound bands of packed cells. In each vessel the cell bands were separated by columns of virtually cell-free plasma, and the latter formed pale

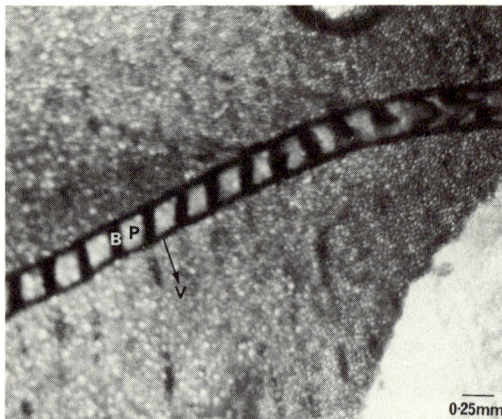

Fig. 4. Surface view of the area vasculosa showing blood cell stasis. The cell bands (B) in the vein (V) are at half wavelength intervals, and are separated by pale areas of plasma (P).

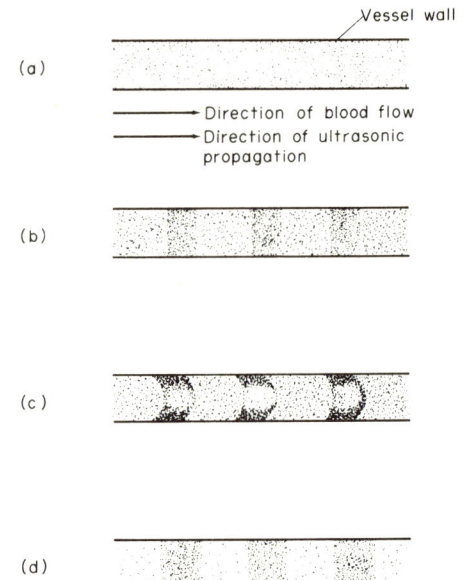

Fig. 5. Stages in the formation of blood cell bands. (a) The blood cells are distributed uniformly. (b) Cell packing occurs at favoured sites. (c) The boundaries of the cell bands arch. (d) Band formation completed.

areas between the cell bands of each vessel. The cell bands of adjacent vessels were in register with one another, and this alignment of alternate red cell bands and pale plasma columns gave rise to the macroscopically visible red stripes.

The distance between the centres of adjacent cell bands in a blood vessel was frequency dependent (Table 1), the centres of adjacent bands being approximately half a wavelength apart.

1. *Motion analysis of cell band formation.* The change from normal cell flow to complete stasis generally occurred in 0·2–0·3 sec, at threshold intensities. Aggregation was most readily observed in veins lying parallel to the direction of ultrasonic propagation, and in which the directions of blood flow and ultrasonic propagation coincided.

The initially homogeneous blood column (Fig. 5(a)) developed an alternate arrangement of light and dark bands, the latter caused by the packing together of blood cells at favoured sites (Fig. 5(b)). At intensities below the threshold level for complete stasis the cells tended to aggregate at preferred positions in the field, forming narrow bands which dissociated and reformed during each heart beat. In conditions above the threshold for complete cell stasis the bands widened to their maximum thickness in less than 0·3 sec.

The shape of the bands of cells changed in the brief interval between normal blood flow and complete stasis. At first the upstream and downstream boundaries of the bands were straight. Then, while the bands were thickening, these boundaries became arched (Fig. 5(c)). The curvature of the arch was presumably related to the differential flow rate of the plasma and represented a velocity profile for the diameter of the vessel. Either just before stasis, or at subthreshold intensities, cells were seen to break free from the downstream boundary of one band only to be trapped at the upstream boundary of the next.

The term 'cell stasis' is used here to describe that condition in which, when a blood vessel is

Table 1. Frequency dependence of cell band separation

Frequency (MHz)	*Wave length in tissue (mm)	Distance between centres of adjacent cell bands ($\bar{x} \pm s_{\bar{x}}$ mm)	N
1	1·50	0·73 ± 0·001	6
3	0·50	0·25 ± 0·01	7
5	0·30	0·12 ± 0·002	6

* Ref: Summer and Patrick (1964).

viewed at a magnification of × 100, in real time, all cell movement appeared to have ceased. In fact, motion analysis showed that some cell movement did still occur. At ventricular systole a few cells migrated from one band to the next as the plasma flowed through the bands, but too few cells were involved for the movement to be obvious under routine low-power microscopy.

The appearance of a blood vessel in which cell stasis, as defined above, had occurred, is shown in Fig. 5(d).

The maximum time that cells have been held in stasis is 30 min, and apparently normal blood flow was resumed after that period, except in a few small vessels, where irreversible occlusion occurred.

2. *Motion analysis of cell band dissociation.* When irradiation ceased, or when the incident intensity was reduced to below the threshold required for stasis in a particular vessel, then the blood cell bands normally dissociated. This was seen most clearly in vessels in which the directions of blood flow and ultrasonic propagation coincided. The time taken for complete dissociation appeared to be directly related to the blood flow rate in the vessel under observation.

The first sign of dissociation was a change in the shape of the upstream and downstream boundaries of each band of cells; these became arched, cell displacement being greater centrally than peripherally (Fig. 6(a)). The central region of each band also became paler, indicating a lower cell concentration than in the rest of the band (Fig. 6(b)). In usually between 0·2 and 0·6 sec, presumably depending on the overall flow rate, cells from the centre of the upstream boundary had passed through the band and had emerged on the downstream side. They then broke free and were carried along in the plasma stream. The arches lengthened, and their curvatures increased until the once-discrete bands fitted into each other like a stack of tumblers (Fig. 6(c)), and, as the blood flow velocity approached normal, the boundaries briefly took on the parabolic profile typical of Poiseuille flow.

Cell release continued in a centrifugal fashion until all the cells had been released, those nearest to the vessel wall being the last to resume normal flow.

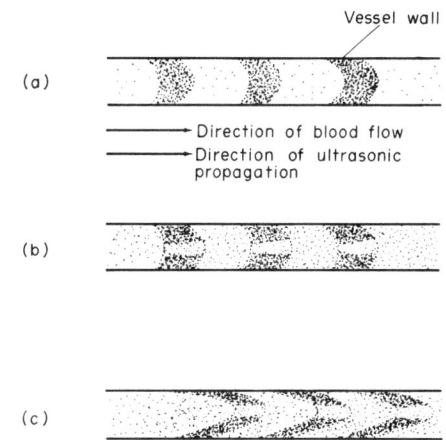

Fig. 6. Stages in the dissociation of blood cell bands. (a) The boundaries of each cell band arch. (b) Cells are forced through the centres of the bands. (c) The bands overlap, as the curvature of the arches increases.

(b) *Factors influencing the threshold intensity for blood cell stasis*

In determining which factors affected intensity thresholds, measurements were restricted to blood vessels lying parallel to the direction of ultrasonic propagation, unless otherwise stated.

1. *Vessel type.* Stasis occurred more readily in veins than in arteries, where the vessels compared were of the same embryo, of similar internal diameter, and were orientated in a similar direction relative to the ultrasonic field vector. Experimental evidence of this is given in Table 2. Each pair of results in the table was

Table 2. Comparison of intensity thresholds in arteries and veins of similar size

Artery		Vein	
Wcm^{-2}	ID (mm)	Wcm^{-2}	ID (mm)
1·58 ± 0·19	0·19	0·79 ± 0·02	0·19
2·45 ± 0·05	0·09	1·00 ± 0·05	0·11
3·33 ± 0·15	0·19	1·79 ± 0·13	0·19
1·04 ± 0·10	0·19	0·89 ± 0·04	0·21
1·87 ± 0·05	0·30	0·82 ± 0·04	0·35
2·53 ± 0·18	0·16	0·91 ± 0·08	0·17
2·13 ± 0·10	0·15	0·37 ± 0·04	0·16
3·41 ± 0·12	0·19	1·08 ± 0·03	0·20
3·22 ± 0·13	0·10	1·89 ± 0·09	0·10

$p < 0.0005$ $d = +1.22$ $t = +5.8568$
Each pair of observations is taken from a different embryo. ID = internal diameter.

obtained from a different embryo, and the significance of the threshold differences between the members of each pair was determined by means of Student's paired '*t*' test. It was found that the threshold for stasis was significantly lower in veins than in arteries ($p < 0.0005$).

2. *Vessel size*. There appeared to be a relationship between vessel size and the threshold intensity for stasis. Within the range of vessels examined (internal diameter *in vivo* = 0·10 to 0·40 mm) it was generally found that the threshold was lower in vessels of larger than of smaller bore (Fig. 7). This relationship was observed most convincingly in veins. Figure 8 shows the threshold intensities for stasis at three points along a single vein. The situation was more complex in arteries, particularly in those close to the heart, for the stability of the blood cell bands changed during each heart beat, and relatively high intensities were required to induce stasis. In some arteries of the area vasculosa a decrease in threshold was found with an increase in size, but in others the reverse occurred (Fig. 9).

3. *Heart rate*. When the threshold intensities for stasis in vessels of similar size and orientation were compared in different embryos, considerable variation was found between them. It was also noted that the threshold at a point over a vessel could vary during the course of an experiment if the temperature of the saline bathing the embryo was allowed to change. Under such conditions the heart rate changes, and this led us to investigate the possibility that there might be some correlation between heart rate and the threshold for stasis.

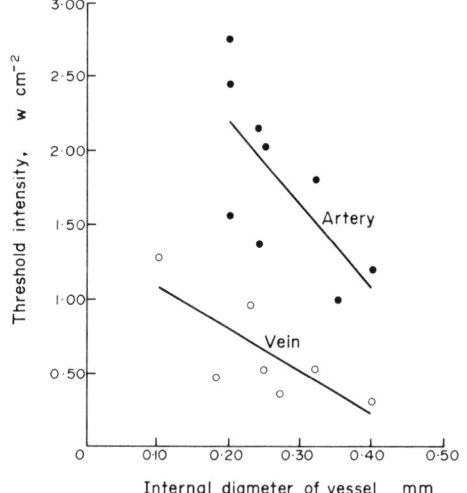

Fig. 7. Relationship between vessel size and threshold intensity for stasis. Each point lies on a different vessel of a single embryo. ○ vein; ● artery.

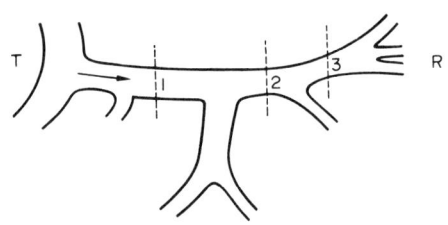

Fig. 8. Relationship between threshold intensity and internal diameter in a vein. T, transducer; R, reflector; → direction of blood flow.

Position	Internal diameter (mm)	Threshold intensity (Wcm^{-2})
1	0·27	2·05 ± 0·06
2	0·22	2·61 ± 0·18
3	0·16	6·33 ± 0·11

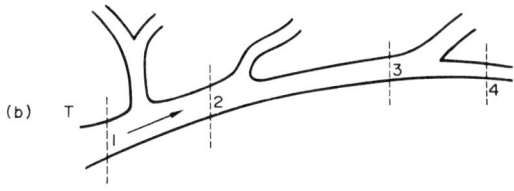

Fig. 9. Relationship between threshold intensity and internal diameter in two arteries. T, transducer; R, reflector; → direction of blood flow.

Position	Internal diameter (mm)	Threshold intensity (Wcm^{-2})
(a) 1	0·21	11·89 ± 0·52
2	0·29	2·77 ± 0·12
(b) 1	0·25	5·04 ± 0·09
2	0·23	4·60 ± 0·52
3	0·20	2·86 ± 0·34
4	0·09	0·88 ± 0·05

Figure 10 summarizes the results obtained when the heart rate was changed by varying the temperature of the saline surrounding the embryo. Within physiological limits, as the heart rate increased the threshold intensity for stasis also increased.

The use of carbachol to depress the heart rate had a similar effect on stasis to that achieved when heart rate depression was obtained by lowering the temperature of the saline. In a typical example, the reduction of heart rate from 63 to 38 beats per min was accompanied by a reduction in the threshold intensity for stasis of 2·43 from 3·62 Wcm^{-2}.

(c) *Intensity variation within the irradiation chamber*

In an attempt to estimate the intensity variation over the viewing area of the irradiation chamber, the field was monitored with a thermistor probe which had been calibrated in free field conditions. The chamber was 160 mm in length and 50 mm wide, while the region within the chamber that was scanned, i.e. the viewing area, was 50 mm in length, 25 mm wide and was covered with saline to a depth of 5 mm.

The scans were made with the probe tip immediately over the glass plate which formed the lower surface of the viewing area. The effect of masking the terminal reflector with an absorber (a roll of cotton gauze) was also determined. Figures 11 and 12 show typical scans, perpendicular and parallel respectively to the direction of ultrasonic propagation.

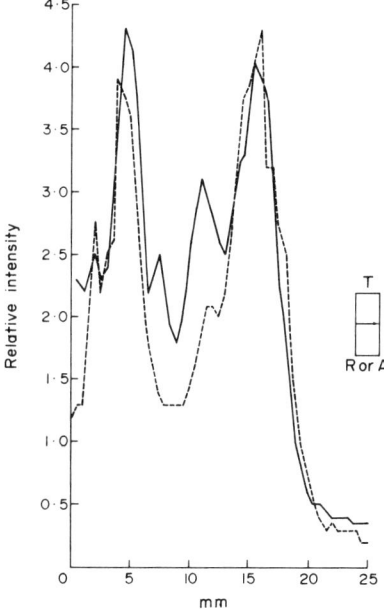

Fig. 11. Intensity profiles perpendicular to the direction of ultrasonic propagation. --- Reflector masked by an absorber. —— Reflector unmasked. T, transducer; R, reflector; A, absorber; → region scanned.

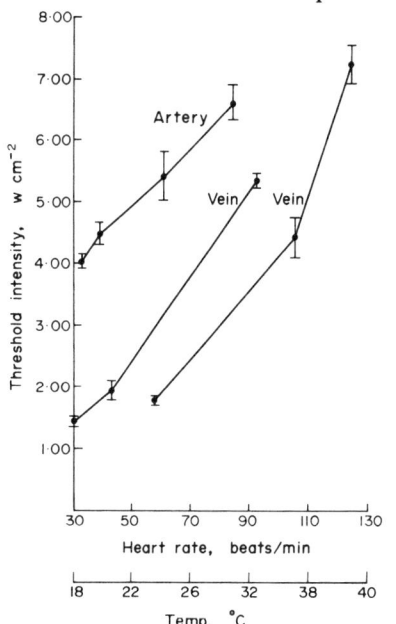

Fig. 10. Effect of altering the heart rate on the threshold intensity required for stasis.

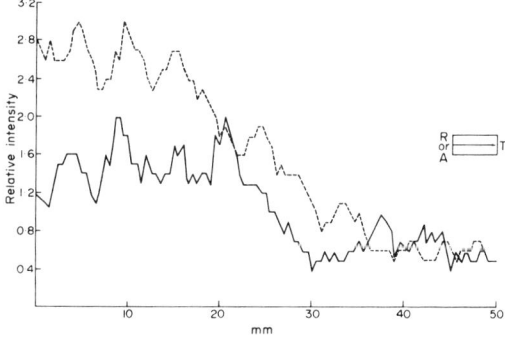

Fig. 12. Intensity profiles parallel to the direction of ultrasonic propagation. --- Reflector masked by an absorber. —— Reflector unmasked. T, transducer; R, reflector; A, absorber; → region scanned.

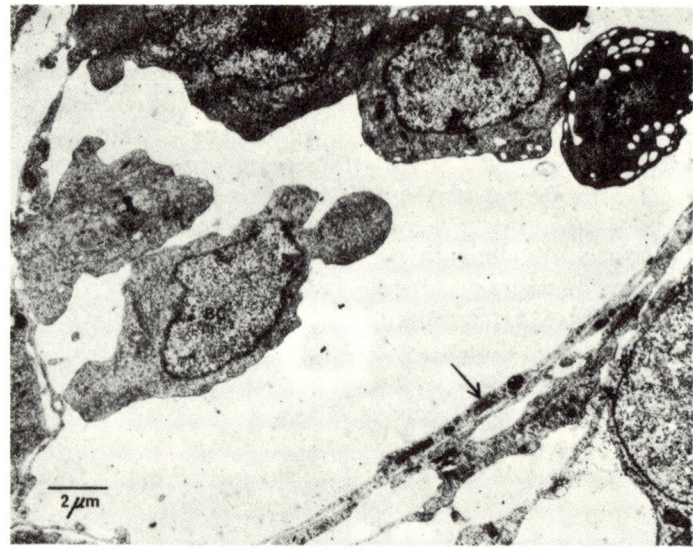

Fig. 13. Control blood vessel. The arrow points to the endothelium and lies in the lumen of the vessel. BC, blood cell.

(d) *Electron microscopy of endothelial cells*

At this stage of development, i.e. after 84 hr incubation, even the largest of the blood vessels of the area vasculosa examined (luminal diameter 50 μm) were essentially endothelial tubes with pericytes scattered irregularly over the external aspect of the endothelium. The vessels thus had the wall structure of capillaries, although those with a luminal diameter of 8 μm or more cannot be classified as such (Rhodin, 1968). An example of one of the larger control vessels examined is illustrated in Fig. 13.

1. *Endothelial cells of control tissues*. In all the samples of control tissue examined, the endothelial cells had similar features. The cells were flattened structures ranging from 3 μm thick in the neighbourhood of the nucleus to 0·1 μm or less toward the periphery of the cell. They were surrounded by a well-defined plasma membrane (Fig. 14). Occasionally, where the membrane was extremely tenuous, 'pores' closed by plasma membrane only were found. The cells rested on an incomplete basement membrane.

The endothelial cytoplasm was rich in ribosomes, giving it a high electron density. There was generally a sparse rough endoplasmic reticulum, consisting of flattened sacs irregularly arranged. Membrane-bound vesicles, fine longitudinal filaments, and microtubules were also

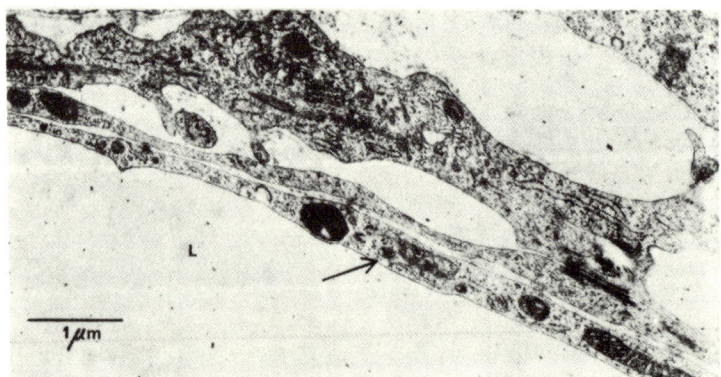

Fig. 14. Portion of control blood vessel. The plasma membrane (arrowed) is clearly defined. L, vessel lumen.

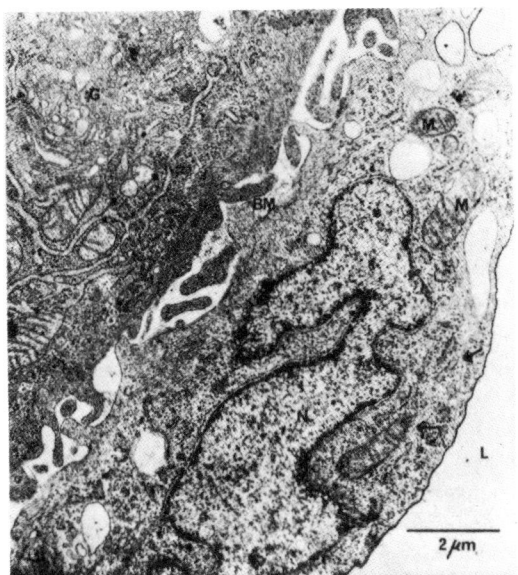

Fig. 15. Control endothelial cell. The cytoplasm contains many ribosomes, and part of the nucleus (N), mitochondria (M) and microtubules (arrowed) can be seen. The vessel lumen (L) and the Golgi apparatus (G) of an adjacent pericyte are indicated. The basement membrane (BM) is incomplete.

present (Fig. 15). Some of the vesicles were of the coated type. Caveolae intracellulares were sometimes found. The mitochondria were generally elongated, and were usually packed with electron-dense granules; in some the cristae exhibited angulations in register. Lysosome-like bodies were present, particularly near the nucleus. Multivesicular bodies were found infrequently, and then usually near the Golgi zone. At junctions between adjacent endothelial cells, flaps of cytoplasm commonly extended from the luminal surface of the cells into the cavity of the vessel.

2. *Endothelial cells of irradiated tissues.* The endothelium of blood vessels from the area vasculosa, fixed immediately after a 15 min irradiation, showed localized damage to the luminal aspect of the plasma membrane (Fig. 16). Sections of the endothelium contained zones where the plasma membrane was grossly disrupted, and cell debris, including fragments of membrane, were found in the vessel cavity. The damaged zones were separated by areas of apparently normal plasma membrane. Where a coated vesicle was partly fused to the plasma membrane (Fig. 17), the invaginated portion of the vesicle was generally undamaged. Erythrocytes were sometimes found adhering to the endothelium.

A further change found in the endothelium of tissue fixed immediately after irradiation was

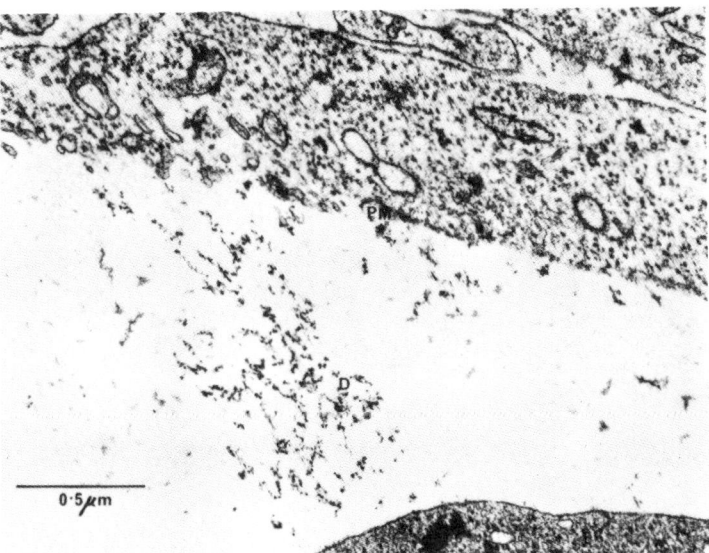

Fig. 16. Endothelium of a blood vessel fixed immediately after 15 min' irradiation. The plasma membrane (PM) has been damaged on the luminal aspect of the cell, and debris (D) lies within the vessel. Part of an erythrocyte (E) can be seen.

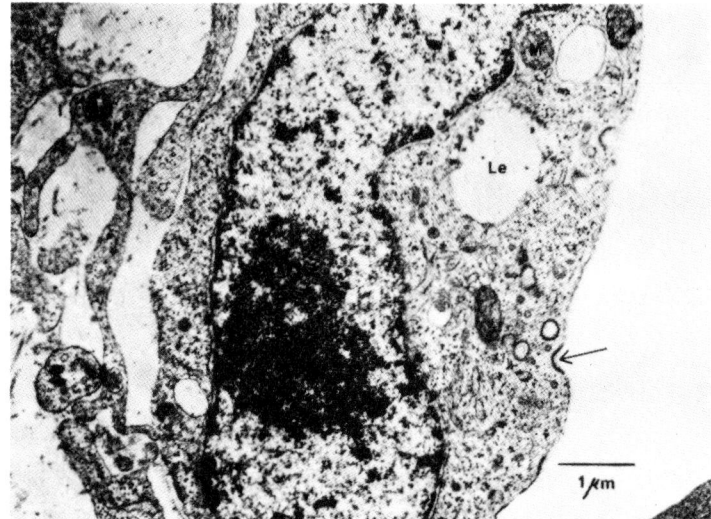

Fig. 17. Undamaged membrane found where a coated vesicle (arrowed) has partly fused with the membrane. An intra-cellular lesion (Le) is also present, and a mitochondrion (M) is indicated.

the presence of ragged cavities, which often had traces of damaged membrane associated with them, in the cytoplasm. In some instances the cavities contained debris (Fig. 18) suggesting that they represented the remains of secondary lysosomes. They differed from normal, undamaged, lysosomes in that the major part of the cavity was not electron dense.

The mitochondria rarely showed signs of damage. Occasionally their limiting membranes were lacking in definition (Fig. 17) but this was probably an artefact caused by the angle of the membrane relative to the electron beam, and similar observations were made in control tissue.

The endothelial cells examined could be divided into two groups: the majority, consisting of cells which appeared completely normal, and the remainder, which had membrane defects and

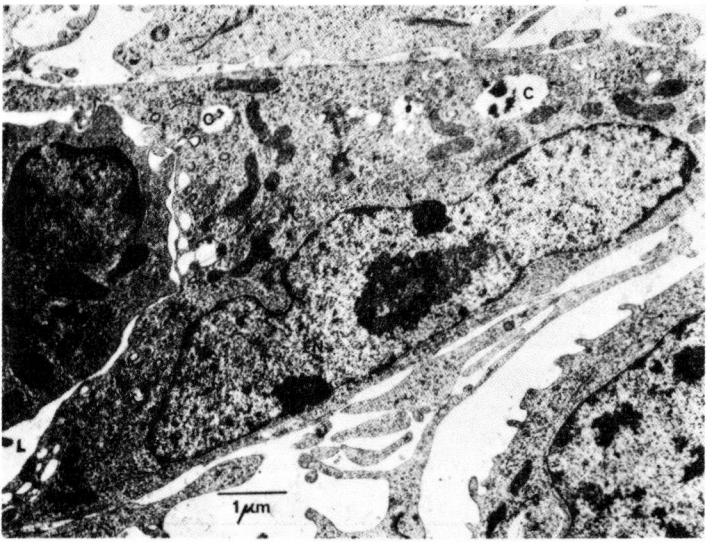

Fig. 18. Cavities (C) containing debris in an endothelial cell fixed immediately after 15 min irradiation. EN, endothelial cell nucleus; L, vessel lumen.

cavities indicative of autolysis. When an incubation period of 15 min separated irradiation and fixation, the degree of damage found in the latter group of cells appeared to have increased. Figure 19 is a representative example of such a cell. The ragged cytoplasmic cavities were generally larger than those found in cells fixed immediately after irradiation. There was, however, a wide variation in cavity size, and this, together with the irregular shape of the cavities and the sampling problems inherent in electron microscopy, made it difficult to assess the situation objectively. Where such a cavity lay next to a mitochondrion that, too, often showed degenerative changes, while other mitochondria appeared to be intact.

In tissue in which the post-irradiation period of incubation had been extended to 30 min, plasma membrane damage was still evident on the luminal aspect of the endothelium in localized sites. Intracellular lesions of the type described above were also found.

After a post-irradiation period of incubation of 60 min, there were indications that cellular repair processes were operating. The majority of the cells were indistinguishable from those of control tissue. In some of the tissue samples examined, fewer examples of damaged cells were found than in tissues incubated for briefer times. The degree of damage was generally slight and restricted to discrete loci within the cells. An example of a damaged cell is shown in Fig. 20.

DISCUSSION

The effects which have been described here, namely blood cell arrest and damage to endothelia, may be potentially hazardous unless adequately controlled. The thresholds below which they do not occur must be established, and to do so in any manner other than empirical, we need to understand the mechanisms producing them.

(a) *Clinical significance of results*

Blood cell stasis is unlikely to occur with the treatment parameters currently used in diagnostic ultrasound, since with both Doppler and Sonar techniques the time-averaged intensities are too low for the effect to occur. However, when ultrasound is used as a therapeutic agent the typical dosage parameters are such that the induction of blood cell arrest is possible unless actively avoided.

Does local, temporary, blood cell arrest constitute a clinical hazard? One result of blood cell arrest is that the supply of oxygen-bearing erythrocytes to the irradiated tissues will be restricted. The effect of this will vary according

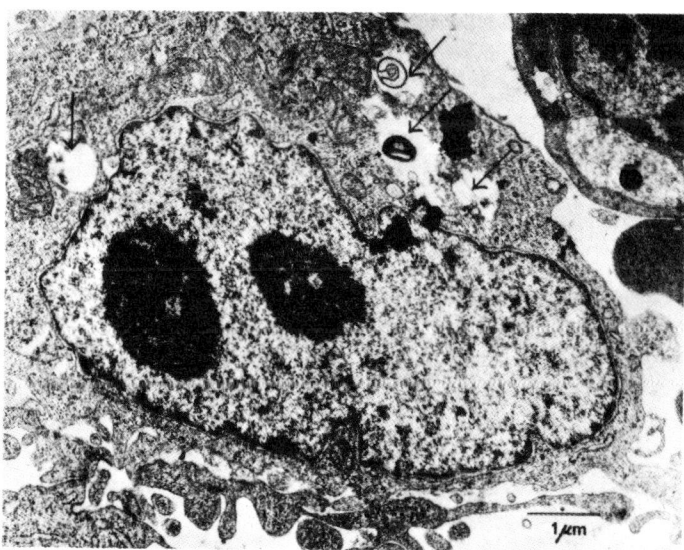

Fig. 19. Part of an endothelial cell incubated for 15 min after irradiation. Ragged cytoplasmic lesions are arrowed.

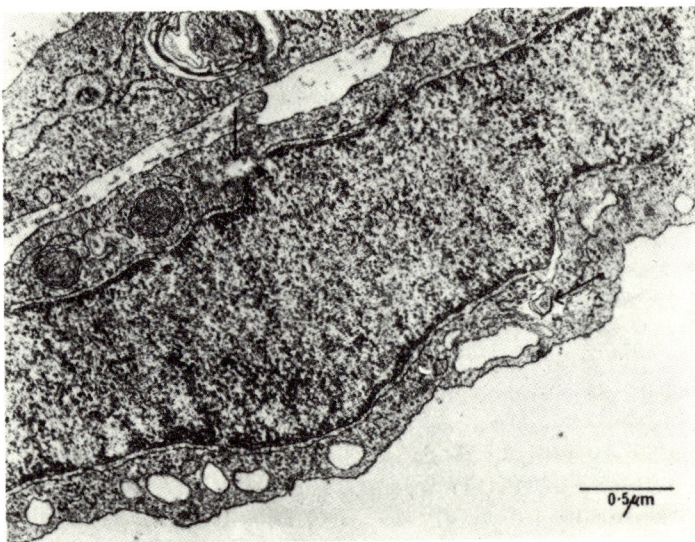

Fig. 20. Part of an endothelial cell incubated for 60 min after irradiation. Two small lesions are arrowed.

to the energy requirements of the tissue concerned. Under conditions of oxygen deficiency, cells can release energy by anaerobic glycolysis, though considerably less than when oxygen is used. Some oxygen is carried in solution in the plasma, and as plasma flow continues during cell stasis, the tissues will not be entirely dependent on anaerobic glycolysis even when red cell arrest occurs. Tissues with a high oxygen consumption, such as kidney, might be expected to be more susceptible to damage than, for example, resting muscle, which has a lower oxygen consumption.

Blood cell arrest can readily be avoided. Complete arrest is favoured in vessels aligned parallel to the direction of ultrasonic propagation. Furthermore, the positions of the cell bands are locked to the position of the transducer. Thus we may avoid vessel blockage by moving the transducer continuously during irradiation, and so maintain a supply of oxygenated blood during treatment. The procedure of moving the transducer is common practice in physiotherapy, for by doing so local tissue heating is reduced and consequently discomfort to the patient is avoided. Even in conditions where the transducer is not deliberately moved, it is usually hand-held and will tend to move slightly during irradiation. We would suggest, however, that to avoid persistent blood cell arrest in tissue, a continuous movement of the transducer should always be employed with the deliberate avoidance of 'dwell time'.

A further observation of possible clinical significance is the damage found in some endothelial cells. In some there was complete disruption of the membrane on the luminal surface. The mechanism causing this is not known, but shear forces would seem to be involved. Red cells have been found adhering to the damaged endothelial membranes, and the danger here is that these might serve as sites of thrombus formation. The irreversible stasis noted in some vessels after prolonged irradiation may be the result of these changes. Similar observations have been made in X-irradiated three-day chick embryos (Stearner and Sanderson, 1969), and have been associated with vessel occlusion, degeneration and circulatory collapse. It should be emphasized, however, that the finding of irreversible stasis in vessels treated with ultrasound was rare, that it affected the smaller vessels only, and that in all the other vessels examined blood cell arrest was reversible.

One feature commonly found in endothelial cells after irradiation at the levels reported here was the development of intracellular lesions. These may represent the sites of damaged lysosomes. An apparent increase in the size of the lesions with time after irradiation supports this hypothesis. Mitochondria were rarely affected unless they lay adjacent to a lesion. In such a situation the mitochondrial damage might result from the action of hydrolytic enzymes released from the lysosomes. Lysosomal damage has also been reported in liver parenchyma after insonation (Taylor and Pond, 1972). Another possibility is that the lesions may represent random incidents of microcavitation and may not be specific for any particular organelle.

It must be emphasized that few endothelial cells showed signs of damage, and that evidence of repair was found in tissues fixed one hour after irradiation. Fewer damaged cells were found in these than in tissue fixed immediately after irradiation, and autophagic vesicles were present in some of the cells, the latter suggesting that normal lysosomal activity was still possible. The results commented on above are of a preliminary nature and a more detailed study of repair phenomena in irradiated endothelial cells is indicated.

It is not yet known how the dosages producing endothelial damage relate to those producing reversible blood cell stasis, and work to determine the threshold for endothelial damage and its dependence on a standing wave field is clearly indicated. The apparent restriction of plasma membrane damage to the luminal aspect of the endothelium is of interest. It would be informative to discover if the loci of damage had a similar periodicity to the cell bands. It may be that constantly varying the transducer relative to the vessels might reduce the incidence of endothelial damage, just as it reduces cell stasis. This is to be investigated experimentally. Much more information is needed concerning the factors producing endothelial damage if it is to be controlled.

(b) *Physical forces involved in the production of blood cell stasis*

In the three-day chick embryo, erythrocytes at various stages of maturation comprise approximately 95 per cent of the circulating cells. In the following discussion only the chick erythrocytes — elliptical, biconvex, nucleated cells of approximately 5 μm radius — are considered.

For the blood cells to aggregate against the drag of the surrounding plasma, they have to overcome a frictional force. Consider banding caused by an intensity little more than the threshold for this effect, where banding is the observable periodic variation in red cell distribution which precedes complete stasis. The time required for band formation will depend on the point in the cardiac cycle at which irradiation commenced. At 3 MHz, and with just suprathreshold intensity, a minimum banding time of approximately 0·05 sec is commonly observed, and it is assumed that this corresponds to a situation in which there is minimum interference from changes in blood velocity and pressure during the cardiac cycle. Since band separation at 3 MHz is approximately 0·03 cm, this suggests an average induced speed of a cell through the plasma of about 0·3 cm sec^{-1}. This is approximately the speed that would have to be induced to overcome the maximum circulation speed in the larger vessels.

The viscous drag on the chick erythrocytes can be assumed to be of the same order as that on spheres of 5 μm radius. For an order of magnitude calculation, the spheres are assumed to be moving through plasma of viscosity 3 c.p. with a relative velocity of 0·3 cm sec^{-1}. Putting this into Stokes drag formula gives a required force of approximately $8·5 \times 10^{-5}$ dyn. We thus have to account for a force of about 10^{-4} dyn, together with the observed band structure, that is, a component of the force must be periodic in half a wavelength of the radiation.

The primary cyclic movements in the present situation are only of the order of 0·01 μm. Moreover, the action is reciprocating, with the result that the disturbance of any quantity is zero in the first approximation. Clearly such motion cannot, by itself, account for the present effect. However, the primary motion alters the mechanical behaviour of the transmitting medium so that one finds non-reciprocating disturbance. This disturbance is complex for a non-homogeneous medium. Here we will briefly consider the leading components relevant to this experimental situation.

1. *Standing wave radiation pressure.* This results in a force on a cell due to its being in a position where there is a time-averaged gradient of pressure or velocity. Nyborg (1967) has given a compact formula applying to rigid spheres and this has been extended by Crum (1971) to cover the case of compressibility near to that of the surrounding medium. These formulae rest on plausible grounds and can be applied to this experimental situation, if the erythrocytes are assumed to be spherical. It should be noted, however, that the treatment takes no account of viscosity. Transposing to our situation, the extended formula becomes, for complete reflection:

$$f = 2 \cdot 8 \times 10^3 \times I \cdot F \cdot \sin(84 F \cdot z) \cdot V \cdot \left\{ \frac{\rho c^2}{(\rho c^2)_s} - \left(\frac{\frac{5\rho_s}{\rho} - 2}{\frac{2\rho_s}{\rho} + 1} \right) \right\}$$

where ρ is density, I is incident intensity (Wcm^{-2}), f is force in dynes, V is volume (cm^3), c is wave velocity, z is distance along the radiation path (cm), F is frequency (MHz), and the suffix 's' relates to the sphere (red cell approximation).

The force is suitably periodic. If the red cell is considered as a sphere of 5 μm radius, the formula gives:

$$f = 1 \cdot 5 \times 10^{-6} \times I \cdot F \cdot \left\{ \frac{\rho c^2}{(\rho c^2)_s} - \left(\frac{\frac{5\rho_s}{\rho} - 2}{\frac{2\rho_s}{\rho} + 1} \right) \right\}$$

for the maximum force in dynes. Banding has been observed when $I = 1$ Wcm^{-2} and $F = 3$ MHz, and the bracketed term is probably 0·3 or less, so that f appears to be about two orders of magnitude too small.

(c) *Average Stokes drag*

This force is also suitably periodic. It arises as follows. In general, the primary cyclic velocity of a particle will not be in phase with the cyclic velocity of its supporting medium. A relatively dense particle, for example, will lag. In a stationary wave, relative velocity will not be in quadrature with the cyclic temperature and pressure variations in the medium. This lack of symmetry will result in a time-average drag force depending on the nature of the variation of the viscosity of the medium with temperature and pressure. This could reach high values in a complex medium, critically temperature dependent.

(d) *Bernoulli attraction*

If the surrounding fluid is flowing past two adjacent particles, then the flow can be faster in a small channel between them. This implies low pressure and consequently inter-particle attraction. This is independent of the sense of the flow and hence leads to a time-averaged force under cyclic conditions. Konig (1891) gave a formula for the attraction between two spheres. This formula, modified for the present situation, becomes:

$$f = 6 \cdot 3 \times 10^2 \times I \cdot r^6 \cdot d^{-4} \text{(dyn)}$$

where the spherical particles have a radius r (cm) and where their centres are separated by a distance d (cm). Where spheres of radius 5 μm are considered, and where a force of 10^{-4} dyn is required, a centre-to-centre distance of 5·6 μm is implied. Thus, when the cells are almost touching, as is the situation in blood, then Bernoulli attraction is of the right magnitude.

The general impression is that the radiation pressure is not sufficient alone to cause the observed banding effect, though it is intended to give further consideration to the effect of viscosity in this connection. Stokes drag forces may play a major role. Bernoulli attraction, which is of the right order of magnitude when the cells are close together, might also be an important factor, with a smaller periodic force giving preferential positions for band formation, in which case blood flow in a vessel would be resisted by the combined attraction of many bands to the walls.

Two approaches suggest themselves to check the importance of Bernoulli attraction. Firstly, is there any hysteresis? That is, having obtained stasis with bands of packed cells, can the incident intensity then be lowered without stasis being lost? Secondly, how does the effect depend on particle separation? We have produced models of blood vessels from fine drawn polythene tubing and intend to use them in investigating the forces involved in the induction and maintenance of blood cell stasis. Since Bernoulli attraction is proportional to intensity, whereas both the periodic possibilities give force values proportional to the product of intensity and frequency, a differentiation may be made by determining the intensity thresholds for stasis at different frequencies. Information on the relative importance of the different forces may also be obtained by comparing the stasis requirements of cells of different sizes in blood vessel models. Human red blood cells are approximately half the size of chick red blood cells; a size difference of this order would be expected to reduce the radiation pressure effect to $\frac{1}{8}$, while average Stokes drag forces would be reduced by a half and Bernoulli effects might be undiminished.

Information is also required on the actual values of the forces to which the blood cells are subjected. Knowledge of the rate of change of the viscosity of plasma with temperature would allow an estimate to be made of Stokes drag force. A measurement of ρc^2 (reciprocal compressibility) could be made from comparison of the wave velocities in the plasma and whole blood and this could be used to put a more accurate value on the last term in the radiation pressure formula.

(d) *Change in stasis threshold with variation in vessel type and heart beat*

It has been found that blood cell stasis can be obtained at lower intensities in veins than in arteries where these are of similar internal diameter and where the heart rate remains constant. Variation in flow velocity, in pressure change during each heart beat, and in wall structure, might account for this.

No information has been found on the relationship of flow velocity and pressure change to the threshold for cell stasis, and this is to be investigated in a model system in which blood flow will be maintained in simulated blood vessels by means of a peristaltic pump.

Although no ultrastructural differences have been detected between the endothelial cells of the extra-embryonic arteries and veins of the chick, there is always the possibility of differences existing at the molecular level which might affect the ability of the blood cells to take up preferred positions at the luminal plasma membrane. Since at the stage of chick development examined the vessels are little more than endothelial tubes distinguishable into arteries and veins on functional rather than structural criteria, the possibility of other wall features affecting the threshold for stasis does not arise. It would be of considerable interest to extend this investigation to include older embryos in which structural differences between arteries and veins have developed.

It has also been shown that varying the heart rate of the chick embryo affects the stasis threshold in the extra-embryonic vessels. As the heart rate decreases, the threshold intensity falls. The problem which arises here is whether the fall in stasis threshold is dependent on the reduction of heart rate directly, or on the means by which this reduction is invoked. It was in an attempt to resolve this difficulty that two alternative methods of altering the heart rate were employed: either the physical method of temperature variation or the pharmacological method of treatment with carbachol. Since the reduction of heart rate by either method was accompanied by a similar fall in the stasis threshold, it would appear that it is the change in heart rate, rather than the method by which this is obtained, which is of significance here.

Temperature variation affected other aspects of the experimental situation as well as the heart rate of the chick. Changes in blood viscosity would be expected, though the extent of these in the blood of the chick embryo at this stage of development is not known. Such changes might also have affected the stasis threshold. Further investigations are clearly indicated.

(e) *Applications of ultrasonically induced red cell aggregation and arrest*

The technique of inducing blood cell arrest could provide a useful tool in biorheology. For example, in vessels where flow is laminar, the rate of change of the parabolic front of the blood cell bands during their dissociation could be used to estimate and compare flow rates across individual vessels of the microcirculation. The absence of a parabolic profile implies that flow is non-laminar, and those portions of vessels in which flow is of this type could thus be identified.

If the phenomenon of ultrasonically induced blood cell arrest is to be applied to biorheology, the physical forces involved in producing it must be investigated and, where possible, defined. The results of experiments planned toward this end will be the subject of further papers.

Acknowledgements—We wish to thank Professor R. Warwick for his advice and encouragement during this investigation, Professor J. Joseph for his cooperation, and the Medical Research Council and Wellcome Trust for financial assistance. We are also grateful to Mr. K. Fitzpatrick, and the Department of Medical Illustration, for their assistance in preparing the illustrations.

REFERENCES

Baker, N. V. (1972) Segregation and sedimentation of red blood cells in ultrasonic standing waves. *Nature, Lond.* **239,** 398–399.

Crum, L. A. (1971) Acoustic force on liquid droplet in an acoustic stationary wave. *J. acoust. Soc. Am.* **50,** 157–163.

Dyson, M., Woodward, B. and Pond, J. B. (1971) Flow of red blood cells stopped by ultrasound. *Nature, Lond.* **232,** 572–573.

Hamburger, V. and Hamilton, H. L. (1951) A series of normal stages in the development of the chick embryo. *J. Morph.* **88,** 49–92.

Konig, W. (1891) Hydrodynamisch-akustische untersuchungen. *Ann. d. Phys.* **42,** 549–563.

Maidanik, G. (1957) Acoustical radiation pressure due to incident plane progressive waves on spherical objects. *J. acoust. Soc. Am.* **29,** 738–742.

Nyborg, W. (1967) Radiation pressure on a small rigid sphere. *J. acoust. Soc. Am.* **42,** 947–952.

Pond, J. B., Woodward, B. and Dyson, M. (1971) A microscope viewing irradiation chamber. *Phys. Med. Biol.* **16,** 521–524.

Rhodin, A. G. J. (1968) Ultrastructure of mammalian venous capillaries, venules and small collecting veins. *J. ultrastruct. Res.* **25,** 425–500.

Stearner, S. P. and Sanderson, M. H. (1969) Early vascular injury in the X-irradiated chick embryo: an electron-microscope study. *J. Path.* **99,** 213–218.

Summer, W. and Patrick, M. K. (1964) *Ultrasonic Therapy—A Textbook for Physiotherapists*. Elsevier, London.

Taylor, K. J. W. and Pond, J. B. (1972) In: *Interaction of Ultrasound and Biological Tissues. Workshop Proceedings* (Edited by J. M. Reid and M. R. Sikov), p. 87. DHEW Publication (FDA) 73-8008.

53

Copyright © 1972 by the British Journal of Radiology

Reprinted from Brit. J. Radiol., 45, 343–353 (May 1972)

A study of the production of haemorrhagic injury and paraplegia in rat spinal cord by pulsed ultrasound of low megaHertz frequencies in the context of the safety for clinical usage

By K. J. W. Taylor, B.Sc., M.B., B.S.

Department of Anatomy, Guy's Hospital Medical School, London, S.E.1

and J. B. Pond, M.Sc., Ph.D.

Department of Physics, Kingston Polytechnic, Kingston, Surrey

(*Received July, 1970 and in revised form October, 1971*)

Abstract

The spinal cords of adult rats were irradiated with ultrasound using peak intensities of 25 or 50 W cm^{-2} at frequencies of 0·5 to 6 MHz. Delivery of energy was pulsed to avoid thermal effects. In most experiments, 10 ms pulses were separated by intervals of 100 ms. Such treatment resulted in paraplegia and/or gross haemorrhage into the cord. The appearance of haemorrhage was found to be a more consistent occurrence and this was used to compare the effects of ultrasound of varying parameters. Damaging ability was maximal at the lowest frequency employed (0·5 MHz)—it decreased with increasing frequency to 5 MHz, at which frequency neither paraplegia nor haemorrhage could be produced.

The same method was used to investigate the effects of hypoxia when it was found that an arterial partial pressure of oxygen of 50 mm rendered the tissue more vulnerable to ultrasonic damage by a factor of 40 per cent.

The effects of changing the duty cycle were similarly investigated. It was found that haemorrhage occurred whenever an accumulated dose-time had been received which time was characteristic of each frequency and independent of the changed time-averaged intensity resulting from the changed duty cycle.

The results are discussed in terms of the known effects of ultrasound, indicating possible action mechanisms. Also considered are the possible relevance of the results to present and future applications of ultrasound in medicine.

Ultrasonic irradiation can act upon living tissues in many ways (Brown and Gordon, 1967). Such energy is normally given to the human for two purposes—therapy and diagnosis. For the latter, ultrasound is employed in three ways—continuous wave Doppler, pulsed sonar and Doppler-sonar. All these applications employ frequencies mainly in the low megaHertz region (1Hz=1 cycle s^{-1}). A patient may be given many such doses over a period of time on the present assumption that effects do not summate to produce damage. This failure of dose summation is thus crucial for the safety of applications of ultrasound in medicine. In the work reported here, the dose is greatly increased by extending the length of the pulses when accumulation of effects is clearly shown. There should be concern that this does not also pertain to present, or more likely, future regimes used in clinical practice. Further, diagnostic techniques used on pregnant women form an important and expanding area of clinical application where targets would appear to be especially delicate and subject to delayed effects, perhaps in terms of many generations. An analogy exists between the histories of ionizing radiation and of many therapeutic medications. Such agents had been extensively used long before their side effects became apparent. Likewise, all the effects of ultrasound are not yet known—but this should not prevent its use when current evidence indicates that it is safer and more useful than present alternative techniques.

Heat and cavitation are the most widely accepted effects of ultrasound and have been responsible for most of the biological phenomena described in previous reports. The effects on peripheral nerve were found to be identical to those of heat alone (Herrick, 1953; Shealey and Henneman, 1962; Lele, 1963). However, there has for 20 years been evidence of another mechanism of action. Fry (1952), reviewing the effects of ultrasound on spinal cord of frogs, noted that this produced an immediate and permanent paraplegia. A non-thermal, non-cavitational mechanism was indicated, since both factors were suppressed by pre-cooling and increased pressure during irradiation (Fry et al., 1950; 1951). Histological examination of the spinal cords of frogs so irradiated showed that the large motor neurones were most easily damaged, but that higher intensities injured all neurones. Still higher intensities were required to damage neuroglia, but even the highest left blood vessels intact.

Several groups have also used ultrasound to produce focal lesions in brain, both for ablative treatment in patients and as a method of neuro-anatomical investigation in animals (Fry et al., 1964). In these focal lesions, it was found that white matter was more susceptible than grey. Again, Fry (1952) reported extreme resistance of blood vessels to ultrasonic injury. This selective sensitivity could devolve both

373

from the higher coefficient of absorption of white matter and from the richer perfusion of the grey; the latter was supported by Basauri and Lele (1962) when they reported that the size of focal lesions could be increased by obstruction of blood supply. These authors reported haemorrhages in their cerebral lesions, but these only occurred at the highest intensities (above 420 Wcm^{-2}). Pond (1968; 1970) also produced evidence that focal lesions in rat brain could be attributed to heat alone by intensities between 75 and 600 Wcm^{-2} at a frequency of 3 MHz, since identical results could be obtained through the same temperature cycle without ultrasonic irradiation. Results of prolonged investigation by the Illinois group (Dunn and Fry, 1971) show that a linear relation exists between log. intensity and log. pulse length for production of focal lesions by a single pulse. They propose that three mechanisms are involved: long single pulses at relatively low intensities produce a lesion by heat alone; higher intensities but short pulses do so by a mechanical factor; the highest intensities (10^4 Wcm^{-2}) and shortest pulses (10^{-4}s) produce a disruptive lesion due to cavitation.

Since sonar diagnostic machines employ peak intensities of only 1–96 Wcm^{-2} (Hill, 1971), with pulses of microseconds duration, the authors quoted above (Dunn and Fry, 1971) suggest that these machines are probably safe. By extrapolation from their results to single pulses of 1 μs duration, an intensity of 10^6 W cm^{-2} would be required to produce a lesion. A safety margin of at least 100 is proposed. However, this is an enormous extrapolation and is particularly uncertain when biological targets are involved. Even if true for single pulses this may well not be true of the diagnostic regime when rapid, repetitive pulsing is used. Until recently there have been no reports of positive effects due to pulses of microseconds duration. Recent work by Taylor and Newman (1971) has demonstrated reduction in electrophoretic mobility of malignant cells using pulses of only 20 μs duration, indicating modification of the cell surface.

It has been found in the experiments reported here that under conditions of rapid pulse repetition, mechanical strains may accumulate so that intensity thresholds may be very much lower than those indicated by the single-pulse regime. Threshold intensities for tissue damage should be determined for any pulse regime used. This work shows that both paraplegia and vascular damage in the spinal cord of rats occurs at much lower intensities than those previously established. The advent of such damage has been used to compare the effects of various parameters of ultrasound, and the consequence of tissue hypoxia.

MATERIAL AND METHODS

A. *Production of paraplegia*

Adult rats of the Wistar strain were anaesthetized by ether and a mid-line dorsal incision made over the thoracolumbar region. The spinous processes of four adjacent vertebrae were removed and the extensor muscles dissected off the laminae. Two adjacent laminae were removed thereby exposing about 10 mm of spinal cord. Laminectomies were performed to prevent excessive absorption of radiation by bone. The complete exposure is shown in Fig. 1. The skin was clipped over the wound and the animal allowed to recover from the anaesthetic. Tone and voluntary movements of the hind legs and tail were observed to be normal before the animal was re-anaesthetized and radiation treatment commenced. Operative damage to the cord was thus precluded. After irradiation, the wound was repaired in two layers with 4/0 Mersilk and the animals observed for paraplegia for 14 days postoperatively. Animals were killed if they were paralysed; otherwise they would have succumbed to urinary tract infection. At killing, they were anaesthetized and perfused, *via* an aortic cannula, with 10 per cent formalin. Those not paralysed were similarly treated on the 14th day. The irradiated segment of the cord was removed and processed for paraffin sections. These were stained with haematoxylin and eosin. The gross appearance of the cord after clearing in cedar wood oil was noted. Twelve control animals were treated in the same way except for irradiation.

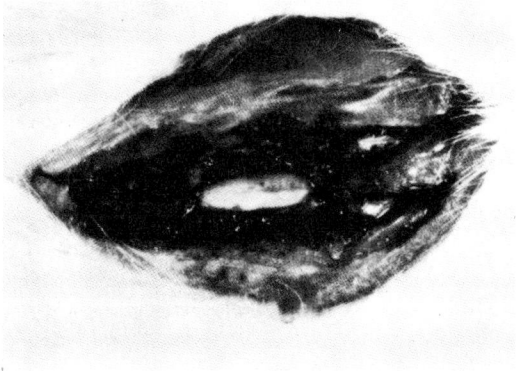

FIG. 1.
The completed exposure of the spinal cord before irradiation achieved by reflection of musculature and a double laminectomy.

B. Production of gross haemorrhage into the cord

Since haemorrhage was noted to be a common feature of the cleared cords obtained from many of the paralysed animals, and since this was an immediate and consistent phenomenon whenever a characteristic pulse time sum had been received, it was used as the endpoint in further experiments. In this series of experiments, operative procedure was as before. Half the animals were injected with 1 ml. 1 per cent trypan blue to visualize any abnormal vascular permeability before gross defects allowed extravasation of red cells. After irradiation, the animals were kept anaesthetized for five minutes and then perfused. Thus the cord was fixed *in situ*, the blood vessels washed out, and the cord removed without the possibility of post-irradiation haemorrhage due to trauma. The cord was sectioned longitudinally and inspected for blue staining and gross haemorrhage. Irradiation times were gradually increased until haemorrhage was produced. Thus the irradiation times required for gross haemorrhage were established for each frequency.

C. Effects of hypoxia

The animals were rendered hypoxic by supplying them with a gas mixture containing 8 per cent oxygen, 7 per cent carbon dioxide and 85 per cent nitrogen. This was the requisite amount of carbon dioxide to prevent hypocapnia due to hypoxia stimulation ventilation. Aortic blood gas analyses after breathing this mixture for five minutes resulted in an arterial partial pressure of oxygen of 47 to 54 mm Hg, together with a normal partial pressure of carbon dioxide. Under these conditions of hypoxia, total dose times required for appearance of haemorrhage were determined as in the previous experiments.

D. Irradiation procedure

Details of the ultrasonic generator have already been published (Pond, 1968; Warwick and Pond, 1968). Peak intensity was measured by observing the deflection of a lead ball 2 mm in diameter at the end of the transducer tube. Calibration was carried out in a water tank in which the walls were lined with an acoustic absorber to prevent standing wave formation. The variations in the field were further defined using miniature thermistors sensitized with silicone rubber, which were standardized for each frequency—comparing the deflection of suspended spheres by radiation pressure in regions of even field. A large number of line scans were required to describe the field adequately, the most important ones being those passing through maximum intensity regions in a plane normal to the beam path. Two such scans are shown in Figs. 2 and 3 for 0·5 and 6 MHz, respectively. They demonstrate the large variations in intensity across the beam, and were plotted at 4·6 cm from the transducer face. Since these variations attend all ultrasonic beams in the near field, a radiation balance giving simply the total power in the beam is quite insufficient to describe the field.

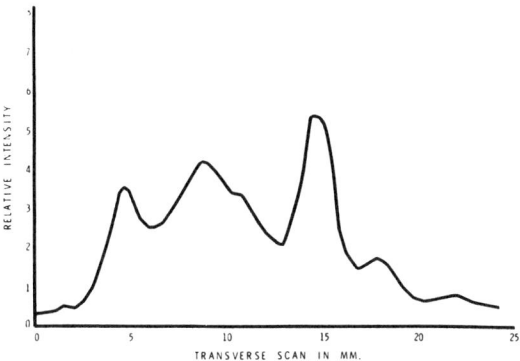

FIG. 2.
Transverse scan across the end of the transducer tube showing the intensity profile at a frequency of 0·5 MHz.

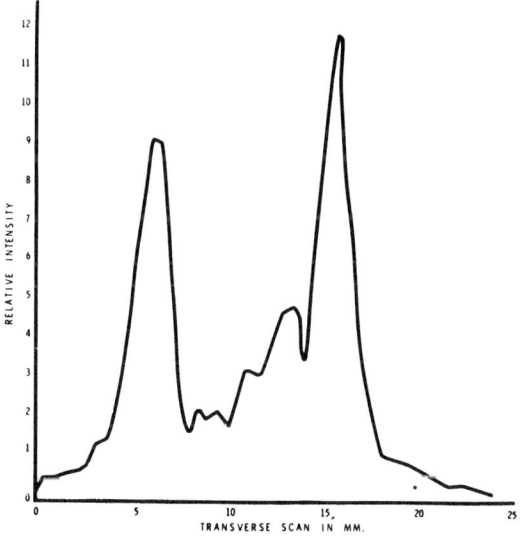

FIG. 3.
Transverse scan across the end of the transducer tube showing the intensity profile at a frequency of 6 MHz.

A frequency range of 0·5 to 6 MHz and intensities of 25 to 50 Wcm^{-2} (intensities at peak position in the field, I_p) were used in these experiments. A plane beam was used, but the radiation was pulsed so that pulses (duration of pulse, mark = M ms) were separated by intervals (space duration, S ms) both of which could be varied. E denotes the total exposure duration—that is, the sum of all the marks and spaces. The time-averaged intensity, in peak field position, is denoted by I_a and is given by:

$$I_a = I_p . M/M + S$$

Hence, this value is approximately the peak intensity multiplied by the mark/space ratios if the latter is small.

A final parameter of dosage is the effective dose time, which is the sum of all the pulses received during the total exposure duration. This is denoted by $\int M$, and may be expressed as:

$$\int M = E.M/M + S$$

A Perspex tube was fitted over the transducer and filled with degassed water and sealed with an acoustically transparent membrane (Saran wrap). Final coupling to the cord was achieved using degassed, sterile, normal saline.

During irradiation temperature changes within the cord were measured using a copper/constantan microthermocouple, threaded through the cord using a fine needle. Three animals were used for this purpose at each frequency and these animals were, of course, killed after this procedure. The temperature change observed over five minutes total exposure duration was denoted by $\triangle T_5$ °C.

Results

A. Production of paraplegia

These results are given in Table I. The symbol PP denotes the animal was paraplegic immediately after irradiation. P denotes that there was delayed onset of paresis occurring between one and seven days after irradiation. In this series of experiments it was noted that:

(1) There was a slight indication of a decreasing efficacy of irradiation as the frequency increased, other parameters being maintained constant. This was followed by a large decrease between 3·5 and 6 MHz, at which frequency irradiation was totally ineffective in producing paresis.

(2) Decreasing peak intensity from 25 to 50 W cm^{-2} (with time-averaged intensity reduced *pro rata*), did not prevent paresis when tried at 0·5 and

TABLE I
Conditions for paresis

F (MHz)	I_p (W cm^{-2})	$M:S$	I_a (W cm^{-2})	E (min)	Number of animals		
					O	P	PP
0·5	50	10:100	5·0	5	–	3	3
0·5	50	10:100	5·0	3	–	1	–
0·5	50	10:100	5·0	2	–	1	1
0·5	50	10:100	5·0	1	–	1	–
0·5	25	10:100	2·5	3	–	2	–
0·5	25	10:100	2·5	2	–	2	–
0·5	25	10:100	2·5	1	–	2	–
1	50	10:100	5·0	5	–	2	2
1	50	10:100	5·0	4	–	4	–
1	50	10:100	5·0	3	2	1	–
1	50	10:100	5·0	2·5	1	3	–
2	50	10:100	5·0	5·0	–	–	6
2	50	10:100	5·0	4	–	2	–
2	50	10:200	2·5	5	2	–	–
2	50	10:200	2·5	10	–	2	–
3·5	50	10:100	5·0	5	–	5	–
3·5	50	10-100	5·0	4	1	5	–
3·5	25	10:100	2·5	5	–	3	–
3·5	25	10:100	2·5	4	–	2	–
3·5	50	10:200	2·5	10	–	2	–
3·5	50	10:200	2·5	8	1	1	–
3·5	50	10:200	2·5	6	1	1	–
6	50	10:100	5·0	5	7	–	–
6	50	10:100	5·0	15	3	–	–
0	0	Controls			12	–	–

$M:S$ is the duration of a single pulse and the interval between pulses in ms.
O—No paresis. P—Subsequent paresis. PP—Immediate paresis.

3·5 MHz, nor did it affect the total exposure duration required for occurrence of paresis. It follows that the threshold peak intensity for this phenomenon is below 25 Wcm^{-2}, and, in view of the very large variations in field intensity shown by the field plots, it seems likely that the threshold is very much lower than this.

(3) When the time-averaged intensity was reduced by increasing the intervals between the adjacent pulses, the same $\int M$ was needed to produce equal effect—even though (by doubling S) the total exposure duration had been doubled. None of the twelve control animals subjected to the same operation showed signs of paraplegia, and were perfused 14 days post-operatively. Despite perfusion, it was noted that many of the cords from the paralysed animals were partially necrotic and softened. After clearing in cedar-wood oil, it was noted that the cords of most of the paralysed rats showed a number of haemorrhages as seen in Fig. 4. A typical section of cord is seen in Fig. 5. The cord was partly necrotic and the commonest site of haemorrhage was noted to be the junction between grey matter and posterior columns. The columns themselves were necrotic, and many of the cell bodies in the adjacent grey matter were in a state of advanced chromatolysis. Small clusters of red cells were frequently observed throughout the cord, occurring mainly in the grey matter. Since vascular damage appeared prominent, the condition of the blood vessels was carefully observed. The largest vessels appeared intact, but in animals killed only six hours after irradiation a large number of capillary endothelial cells were found, either singly or in small groups as shown in Fig. 6—suggesting that the capillary endothelial cells are an important site for ultrasonic damage.

In some animals having delayed paresis, there was no haemorrhage. This suggested that although the main cause of paresis was due to interference with blood supply, other structures (so far not identified) might also be injured.

B. *Observation of gross haemorrhage into the cord*

The first series of experiments which used paraplegia as a criterion of damage sustained, gave only indications as to both the effects of differing frequencies, and the significance of peak- or time-averaged intensities. More precise comparison was

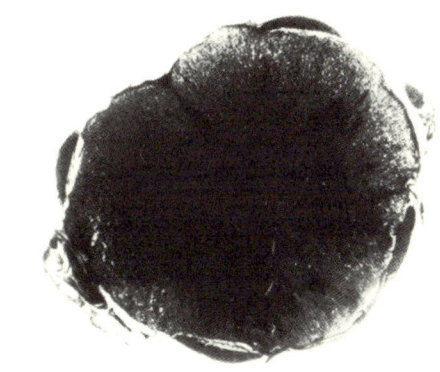

FIG. 5.
Transverse section of spinal cord from paraplegic rat four days after irradiation at a frequency of 1 MHz for five minutes with delivery of energy pulsed 10:100 ms (25).

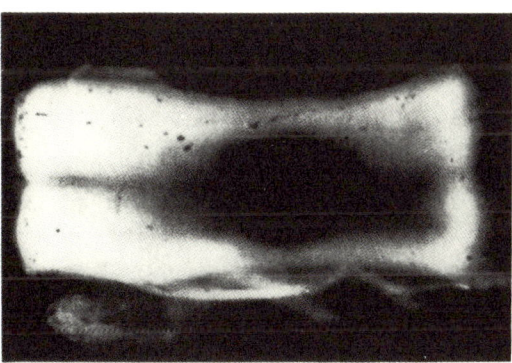

FIG. 4.
Perfused cord from paraplegic rat after clearing in cedar wood oil showing extensive haemorrhage into the cord (×5).

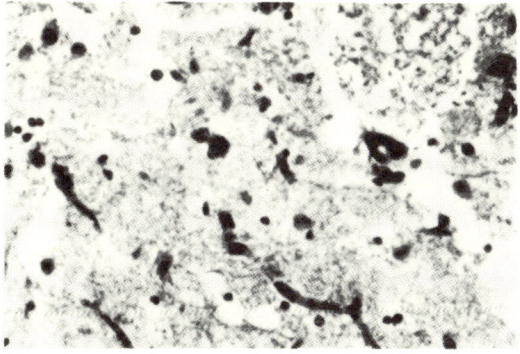

FIG. 6.
Capillary endothelial disruption stained with haematoxylin and eosin six hours after irradiation at a frequency of 1 MHz for five minutes, pulsed 10:100 ms (×250).

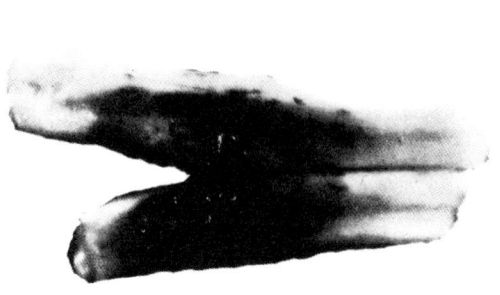

FIG. 7.
Longitudinal section of spinal cord of rat perfused immediately after irradiation at a frequency of 3·5 MHz for five minutes showing gross haemorrhage into the grey matter (×4).

prevented, since paresis was either frequently delayed in onset or when individual variation was apparent near the threshold for the total exposure duration required for damage. However, it was found here that if the cords were observed immediately after irradiation for signs of gross haemorrhage as in Fig. 7, there was a completely uniform integrated pulse time ($\int M$) required, and this was characteristic of each frequency.

The results of this series of experiments are grouped together in Table II. At a frequency of 0·5 MHz, a sudden and consistent occurrence of haemorrhage can be seen to occur as $\int M$ increased to a value of 18 seconds. This value, $\int M$, was established for each of the frequencies investigated from 0·5 to 4·9 MHz. Clearly, the damaging ability of any frequency is inversely related to the time required for injury. Hence, a "damaging ability index" can be ascribed to each frequency, expressed as $100/\int M$. These results are abstracted in Table III. For damage to occur, there appears to be a prerequisite threshold of between 4·2 and 4·9 MHz. This is shown in Fig. 8 which is a graph of damaging ability index against frequency. It shows a large increase in index between 0·5 and 1 MHz, a relative plateau between 1 and 3·5 MHz, followed by a sharp decrease approaching zero at the frequency threshold. Also shown, on the upper line in this figure, are the temperature changes induced by a five-minute irradiation, as given in Table IV. Thus, under these conditions, damaging ability decreases with increasing frequency, whereas thermal stresses rise.

A further series of experiments was undertaken to assess the relative importance of peak intensity and time-averaged intensity within this dose regime. Time-averaged intensity was varied in two ways: firstly peak intensity was reduced while maintaining the same mark/space ratio, and secondly the same

TABLE II
CONDITIONS FOR HAEMORRHAGE

F (MHz)	I_p (W cm^{-2})	M:S	I_a (W cm^{-2})	E (min)	Number of animals			$\int M$ (sec)
					O	OH	HH	
0·5	50	10:100	5·0	2	1	–	–	12
0·5	50	10:100	5·0	2·5	–	1	–	15
0·5	50	10:100	5·0	3	–	–	4	18
1	50	10:100	5·0	3·5	1	–	–	21
1	50	10:100	5·0	4	–	–	5	24
2	50	10:100	5·0	4	1	–	–	24
2	50	10:100	5·0	4·25	1	–	–	25·5
2	50	10:100	5·0	4·5	–	–	3	27
3·5	50	10:100	5·0	3	1	–	–	18
3·5	50	10:100	5·0	4	-	1	..	24
3·5	50	10:100	5·0	4·5	-	1	..	27
3·5	50	10:100	5·0	5	–	–	3	30
4·2	50	10:100	5·0	5	1	–	–	30
4·2	50	10:100	5·0	6	1	–	–	36
4·2	50	10:100	5·0	7	2	–	–	42
4·2	50	10:100	5·0	7·25	–	1	–	44
4·2	50	10:100	5·0	7·5	–	–	3	45
4·9	50	10:100	5·0	10	–	1	–	60
4·9	50	10:100	5·0	12	–	1	–	72
4·9	50	10:100	5·0	15	2	–	–	90
4·9	50	10:100	5·0	20	2	2	–	120

M:S is the duration of a single pulse to the interval between pulses in ms.
O—No haemorrhage. OH—Minute haemorrhage. IIII—Gross haemorrhage.

TABLE III
DAMAGING ABILITY AT VARYING FREQUENCIES

Frequency (MHz)	$\int M$	$100/\int M$
0·5	18	5·6
1	24	4·2
2	27	3·7
3·5	30	3·3
4·2	45	2·2
4·9	>120	>0

TABLE IV
TEMPERATURE CHANGE IN THE CORD ($\triangle T_5°C$) INDUCED BY IRRADIATION AT EACH FREQUENCY WITH A PEAK INTENSITY OF 50 W CM^{-2}, PULSED 10:100 FOR FIVE MINUTES

Frequency (MHz)	$\triangle T_5°C$
0·5	5·5
1·0	6·6
2·0	9·0
3·5	10·0
6·0	11·0

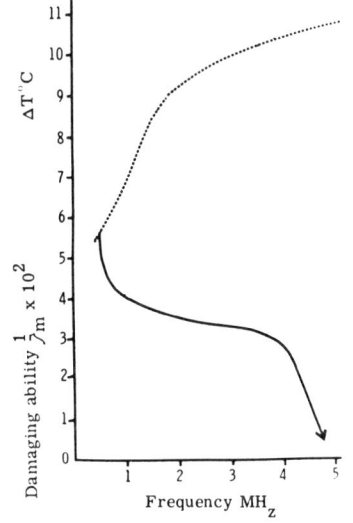

FIG. 8.

The upper dotted line is the change in temperature induced in the cord at each frequency investigated using a peak intensity of 50 W cm^{-2}, with delivery pulsed 10:100 ms. The lower solid line is a graph of damaging ability against frequency showing that the latter decreases as thermal stresses increase.

peak intensity was maintained while increasing the intervals between adjacent pulses. The results are collated in Table V, and were obtained at a frequency of 3·5 MHz. It was found that:

(1) When I_a was halved by reducing I_p from 50 to 25 W cm^{-2}, the integrated pulse lengths required for damage remained unaltered.

(2) When I_a was progressively reduced by increasing the space between pulses, $\int M$ was still unaltered.

These results confirmed more precisely the trend indicated by the first series of experiments: that 25 W cm^{-2} was above the threshold for damage and that mechanical strains induced by irradiation showed short-term accumulation. Hence, occurrence of damage was independent of time-averaged intensity but required a suprathreshold peak intensity for an integrated effective dose time ($\int M$), the latter being characteristic for each frequency.

Ten rats were treated with shorter pulses of 1 ms or 200 μs duration, maintaining the same mark/space ratios. Six of the animals so treated showed haemorrhages which were smaller than those produced by the longer pulses and required a longer irradiation time to be effective. Thus, if there is a protective factor operating in the direction of shorter pulses, it is not decisive in this regime. This aspect requires further experimental study, since the pulses used in the sonar regime are, of course, very much shorter still.

All animals injected with trypan blue were inspected for signs of the vascular leakage to be expected before gross haemorrhage. In no instance did this occur and it was apparent that blood vessels showed "all or nothing" reactions: either they were even impermeable to such small molecules as trypan blue, or they suddenly developed gross defects which allowed extravasation of red cells.

C. Effects of hypoxia

The results of these experiments are shown in Table VI. Spinal cords were irradiated under the same conditions as in the previous experiments but the animals had been rendered hypoxic. A peak intensity of 25 W cm^{-2} was used, which was pulsed 10:100 ms. The damaging ability index was assigned as before to the frequencies in the range 0·5 to 4·9 MHz.

At a frequency of 0·5 MHz the index was increased from 5·6 to 10 (using occurrence of haemorrhage as the end point) and again at the other frequencies, except at 4·9 MHz where it approached zero as before. The results are shown graphically in Fig. 9 together with the analogue for tissues under normal oxygen pressure. The minute haemorrhages noted at 4·9 MHz were probably due to the effects of prolonged hypoxia itself.

D. Temperature changes during irradiation

Temperature changes during five minutes'

TABLE V

EFFECT OF INCREASING INTERVALS BETWEEN ADJACENT PULSES

F (MHz)	Ip (W cm^{-2})	M:S	I_a (W cm^{-2})	E (min)	Number of animals			∫M (sec)
					O	OH	HH	
3·5	50	10:100	50	3	1	–	–	18
3·5	50	10:100	50	4·5	–	1	–	24
3·5	50	10:100	50	4·5	–	1	–	27
3·5	50	10:100	50	5	–	–	3	30
3·5	25	10:100	2·5	4	–	2	–	24
3·5	25	10:100	2·5	5	–	–	4	30
3·5	25	10:200	1·25	8	2	–	–	24
3·5	25	10:200	1·25	10	–	–	2	30
3·5	25	10:300	0·83	12	2	–	–	24
3·5	25	10:300	0·83	15	–	–	2	30
3·5	25	10:400	0·63	16	2	–	–	24
3·5	25	10:400	0·63	20	–	–	2	30

M:S is the duration of a single pulse to the interval between pulses in ms.
O—No haemorrhage. OH—Minute haemorrhage. HH—Gross haemorrhage.

TABLE VI

EFFECTS OF HYPOXIA

F (MHz)	Ip (W cm^{-2})	M:S	I_a (W cm^{-2})	E (min)	Number of animals			∫M (sec)
					O	OH	HH	
0·5	25	10:100	2·5	1·5	2	–	–	9
0·5	25	10:100	2·5	1·67	–	–	5	10
1	25	10:100	2·5	2·0	1	–	–	12
1	25	10:100	2·5	2·25	–	1	–	13·5
1	25	10:100	2·5	2·5	–	–	4	15
2	25	10:100	2·5	2·5	2	–	–	15
2	25	10:100	2·5	2·67	–	–	4	16·5
3·5	25	10:100	2·5	2·5	2	–	–	15
3·5	25	10:100	2·5	3	–	–	4	18
4·2	25	10:100	2·5	4·5	–	1	–	27
4·2	25	10:100	2·5	5	–	1	–	30
4·2	25	10:100	2·5	5·25	–	–	3	31·5
4·9	25	10:100	2·5	20	–	4	–	120

M:S is the duration of a single pulse to the interval between pulses in ms.
O—No haemorrhage. OH—Minute haemorrhage. HH—Gross haemorrhage.

irradiation varied considerably from animal to animal, the three readings at each frequency being averaged in Table IV. The symbol $\triangle T_5 °C$ denotes this change. The results show the expected trend whereby higher frequencies produce greater heat. After operative exposure and cooling with saline used as the coupling medium, the initial temperature in the cord was between 28° and 30°C. This initial cooling prevented the maximum temperature from exceeding 40°C, with the exception of one result at 6 MHz. There were many sources of variation involved in this temperature change, apart from the initial temperature, including: a variable heat "sink" due to the differing quantities of saline required for coupling; variation in cooling by blood flow; and variation due to the site of the thermocouple within the ultrasonic intensity profile.

The results are summarized as follows:

(1) Using a peak intensity of 50 W cm^{-2} and a mark/space ratio of 10:100 ms, shorter dose durations were needed for damage at the lower frequencies in the range 0·5 to 4·9 MHz despite the fact that lower frequencies involve much less thermal stress.

(2) Variation in interpulse spacing between 100 and 400 ms, other parameters being held constant, involved a *pro rata* increase in total exposure time to be effective. This meant that the integral of pulse lengths or effective dose duration required for damage to occur was constant for any given frequency.

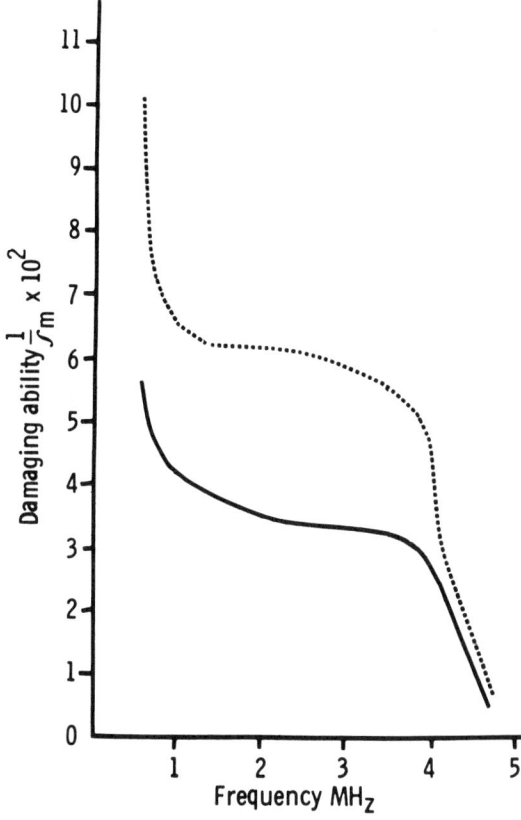

FIG. 9.
This shows a graph of damaging ability against frequency. The upper dotted line was obtained from hypoxic animals whilst the lower line shows the analogue for animals under normal oxygen pressure.

(3) Using the appearance of gross haemorrhage as an end point, a "damaging ability index" could be assigned to each frequency, calculated as $100/fM$. This index decreased with increasing frequency approaching zero at 4·9 MHz.

(4) The index was not altered merely by reducing peak intensity from 50 to 25 W cm^{-2}.

(5) There was only a slight change in index when the pulse lengths were reduced to 0·2 ms.

(6) Damage to blood vessels of the spinal cord appeared to be an all-or-none phenomenon.

(7) Histological evidence suggested that a primary site of ultrasonic damage was the capillary, and this was supported by the observation that tissue hypoxia increased the susceptibility to vascular damage.

DISCUSSION

For measurement of blood velocity, the continuous Doppler dose form is used with intensities in the region of of mW cm^{-2}. In the therapeutic dose form higher intensities up to some 5 W cm^{-2} are used, but at the higher intensities delivery is pulsed to reduce time-averaged intensities and hence painful heating effects (Sumner and Patrick, 1964). There is no effort to make these pulses short except compared to bulk thermal time constants.

In the sonar dose form, much higher intensities: 50 and up to 96 W cm^{-2} may be used (Hill, 1971). However, the pulses are necessarily short—of the order of 10^{-6} second with pulse intervals in the region of 10^{-3} second. Time-averaged intensities are therefore low and effective exposure times short. In a related dose form—which may be called sonar-Doppler, and is otherwise similar—pulse intervals may be lower by an order of magnitude to increase resolution (Baker, 1970).

Therapeutically, ultrasound has been widely used for the past 30 years. At one time this dose form was employed in the treatment of superficial malignancies, but its use was abandoned when it was noted that some tumours became widely disseminated after such treatment (Horvath, 1944). It appears likely that vascular defects could have occurred, as in the experiments reported here, allowing tumour emboli to enter the circulation.

Other effects of the therapeutic dose form described from this laboratory have been marked stimulation of tissue regeneration (Dyson et al., 1970) and modification of blood flow during irradiation (Dyson, Woodward and Pond, 1971). However, relatively long clinical experience with this dose form indicates that it should be safe when used with care. Reservations remain about its use, especially if prolonged or repeated, where gonads or developing embryos are involved. Concerning the former, a recessive mutation may not make itself apparent for several generations, while with the latter, streaming movements could produce a redistribution of primary organizers resulting in abnormal development. However, such targets are not within the usual scope of therapeutic ultrasound.

The continuous Doppler dose form may be safe because of the very low intensity levels employed. A recent report on chromosomal aberrations following *in vitro* irradiation with this dose form (Macintosh and Davey, 1970) has not been substantiated by later workers (Coakley et al., 1971; Boyd et al., 1971). With such low intensities the chief danger might appear to be due to the effects of streaming on developing embryos (*vide supra*). Indeed, vigorous streaming movements have been observed in cells using intensities of the order of 0·2 W cm^{-2} (Selman and Counce, 1953).

The experiments reported here, on the effects of ultrasound on intact tissue when delivery of energy is repetitively pulsed, have most relevance to the sonar dose form and this is now discussed in more detail:

Sonar is extensively used in obstetrics on the most delicate targets and instantaneous peak intensities are high. It is commonly argued that this is safe because the individual pulses are short and are separated by relatively long intervals, reducing time-averaged intensities to very low levels. Indeed, time-averaged intensities—not peak ones—are quoted by the manufacturers. However, observations on patients following irradiation *in utero*, animal studies using commercial equipment (Smyth, 1967), and with more powerful laboratory apparatus (Woodward, Pond and Warwick, 1970), have not disclosed injurious effects. Nevertheless, some targets may be rendered more vulnerable by, for example, incidental foetal hypoxia. Adventitious focusing effects can occur—notably when the beam is reflected from bony concavities. The possibility of recessive mutations and their delayed manifestations particularly concerns pregnant patients when embryonic ovaries are irradiated. Thus, to establish completely regions of safety for clinical use, the neighbouring regions of dose parameters should be examined to establish the thresholds for effects. Any effects so found should be explained as fully as possible.

The present experiments show damage consequential to the accumulation of effective radiation time of the order of 20 seconds. If an extrapolation is valid to a typical sonar dose form, in which the pulses are of the order of 1 μs and the mark/space ratio approximately 10^{-3}, the total exposure duration would have to be between three and six hours (depending on the frequency and physiological state of the tissue) before subliminal effects could summate to produce observable damage. Although such effects can be shown to accumulate over a period of hours, it is technically difficult to demonstrate whether such doses, repeated over a period of weeks, could similarly summate. Equally, summation has not been shown for the effects of such short pulses used in the sonar regime, and no time scale elucidated for any reversion of changes.

Mechanisms for the present results are now considered:

The effects of ultrasonic radiation on any target can be considered, quite generally, to be the result of the following factors.

(1) Heating due to the absorption of wave energy.
(2) Oscillatory disturbances. These are cyclic variations of pressure, temperature and forces on inhomogeneities which oscillate about their ambient values but the time-averaged displacements are zero.
(3) Time-averaged disturbances. These are due to non-linearities in the response of the target such that the oscillatory disturbances (which are the prime result of irradiation) give rise to displacement with a time-averaged value or which increase with time. These are streaming movements and associated forces and also displacements of chemical equilibrium points.
(4) Cavitation disturbances. These result from the enhanced movement around any cavities which are caused to vibrate strongly or to collapse.

The present results are now considered with respect to these categories:

Heat may be excluded as a primary damage factor. Although a positive temperature-time coefficient for damage may be expected, another mechanism is overriding since damage occurs more rapidly when parameters are changed to give less thermal stress. When considering oscillatory disturbances, it must be remembered that the target is not homogeneous but contains, of course, many complex structures. Despite this complexity, a way to proceed which gives a good approximation, is to consider the general oscillatory motion in the target to be that of a homogeneous one with the same bulk mechanical responses. Special disturbances to structures then, are due to such action occurring around them. For many purposes, a homogeneous target equivalent to soft tissue is water. In that case, calculating from the acoustic impedance and taking 50 W cm^{-2} at 0·5 MHz, the cyclic pressure variations have a maximum (amplitude) of 12 bar, the cyclic point displacements that of 0·26 μm, cyclic velocity 82 cm s^{-1} and cyclic acceleration $2·6 \times 10^5$ g. At 5 MHz and the same intensity, the respective figures are 12 bar, 0·026 μm, 82 cm s^{-1} and $2·6 \times 10^6$ g. Thus accelerations rise directly with frequency and point displacements fall. Cyclic temperature variations are a small fraction of 1°C.

To explain the observed results, it is necessary to search for a damage factor which increases in effect as the frequency falls from 5 to 0·5 MHz. But, over this range, cyclic temperature, pressure and velocity magnitudes do not, to a first approximation, alter. Indeed, accelerations decrease. Thus, if cyclic disturbances are the direct cause of damage, mechanisms must be considered in which point displacement is paramount and accelerations adequate. This suggests a line of experiment in which frequency and intensity are altered in concert so that damaging ability may be related to point displacement in isolation. The greatest forces and movements occur at certain points in the larger structures and this should direct histological examination.

Time averaged disturbances are now considered. The observed frequency dependence of damaging ability argues against a mechanism moderated by shifts in chemical equilibria, since, to a first approximation, the magnitudes of cyclic temperature and pressure variations are not altered. Thus reactions with time constants similar to the cycle time of the radiation would have to be involved.

For radiation of a single pure frequency, streaming motions and forces increase with frequency, since the scattering and absorption (from which they must be derived) increase with the first and higher powers of the frequency. Further, boundary layers are thinner at higher frequencies and hence the shear in them is increased. Thus, an effect which increases with reduced frequency is unlikely to be due to such forces. However, one possibility remains under this heading: distortion of the radiated waves due to a component of an even harmonic as well as the fundamental frequency. A very small component can give rise to relatively large time-averaged forces (Oseen forces). Different projectors and transducer plates were used for the different frequencies allowing, perhaps, significant though small differences in wave distortion. Further experiments are in preparation in which the target is at the intersection of two beams, thus allowing production of large Oseen forces which can be carefully controlled. These forces can be made spatially periodic thus allowing easy identification of target damage. Such experiments should be informative about time-average forces in general.

Finally, cavitation disturbances are considered. Especially in less viscous media, these can produce marked biochemical changes with very low intensities (El'Piner, 1967). They form a damage mechanism which does, generally, become more effective with decreasing frequency. Also there would be a tendency for damage to occur at discrete weak points in the target suggesting a linearly cumulative mechanism which again is in accord with the present experiments.

To obtain sufficient energy from the radiation field, a cavity needs to be near resonant in size. For bubbles an approximate formula is:

$$R = \frac{1}{2\pi F}\left(\frac{3\gamma(p_0 + 2\sigma/R) - 2\sigma/R}{\rho}\right)^{\frac{1}{2}}$$

where γ is the principal specific heat ratio of the cavity fluid
σ is the interfacial tension (N.M^{-1})
ρ is the surround density (Kg.M^{-3})

R is the bubble radius (M)
p_0 is the ambient pressure (N.M^{-2})
F is the lowest resonant frequency (MHz).

Thus, in the MHz region, the radius turns out to be only μm or so. It is more difficult for small bubbles to exist, due to surface tension which tends to force the enclosed fluid into the surround. However, once such bubbles are formed and capable of extracting energy from the radiation field, only small intensities are needed to enable them to be the sites of high energy density (Skinner, 1968). Cavities, not necessarily spherical, may exist more easily in the presence of hydrophobic structures.

The action of cavities in the target may occur in two different ways. In stable cavitation, the cavity pulsates, driven by the pressure variations in the field. Energy is then concentrated near the surface of the cavity where large shear gradients may arise. In collapse cavitation, a cavity is caused to collapse by the positive part of the pressure cycle. However, the cavity has absorbed energy from the field and, as the cavity gets smaller, the energy density gets larger. With rather incompressible surrounds, enormous values of local energy density result from complete collapse. The resulting secondary radiation may open up other cavities and this may lead to a chain reaction, marked disruption of the target and crackling sounds. This chain reaction may be described as interaction cavitation as opposed to discrete cavitation. Lack of histological evidence of gross disruption in tissue excludes the occurrence of interaction cavitation but not necessarily discrete cavitation.

Not only the absence of large scale target disruptions, but also the delay in the onset of functional damage discriminates against the occurrence of interaction cavitation, but discrete cavitation, whether stable or collapse, remains a possibility. Such cavitation is difficult to detect, since, due to the smallness of the cavities, very high local values of energy density are rapidly reduced by inverse square law propagation before a detector is reached. Further experiments are required in which the target and radiator are placed in a vessel allowing alteration of ambient pressure to which cavitation is very sensitive.

In conclusion, it appears to be necessary to increase greatly both pulse length and total exposure time over those currently used in diagnosis, before damage is produced. This supports the lack of clinical evidence of ill-effects. Nevertheless, it remains important to identify thresholds for damage since greatly increased dosage could occur by:

(1) technical malfunction;
(2) increased intensity and p.r.f. to improve resolution, *e.g.* as in the sonar-Doppler regime;
(3) adventitious focusing effects from bony concavities or at tissue-air interfaces;
(4) Increased pulse lengths for holographic techniques.

In view of these possibilities in present and future applications in ultrasonography, it appears to be important to continue basic experimental studies to define safety limits accompanied by epidemiological surveys of infants subsequent to irradiation, noting especially the period of gestation, the output parameters of the machine employed, and the number and length of scanning examinations performed.

ACKNOWLEDGMENTS

We wish to thank Professor Warwick for constant encouragement, Raymond Wright for expert technical assistance, and Kevin Fitzpatrick for photographic help.

REFERENCES

BAKER, D. W., 1970. Pulsed ultrasonic doppler blood-flow sensing. *Transactions of the IEEE on Sonics and Ultrasonics SU-17*, 3, 170–185.

BASAURI, L., and LELE, P. P., 1962. A simple method for the production of trackless focal lesions with focused ultrasound. *Journal of Physiology*, 160, 513–534.

BOYD, E., ABDULLA, U., DONALD, I., FLEMING, J. G. E., HALL, A. J., and FERGUSON-SMITH, M. A., 1971. Chromosome breakage and ultrasound. *British Medical Journal*, 2, 501–502.

BROWN, B., and GORDON, D., 1967. In *Ultrasonic Techniques in Biology and Medicine* (Iliffe, London).

COAKLEY, W. T., HUGHES, D. E., SLADE, J. S., and LAURENCE, K. M., 1971. Chromosome aberrations after exposure to ultrasound. *British Medical Journal*, 1, 109–110.

DUNN, F., and FRY, F. J., 1971. *Transactions of the IEEE on Bio-Medical Engineering*, 18, 2, 53–65.

DYSON, M., POND, J. B., JOSEPH, J., and WARWICK, R., 1970. *Transactions of the IEEE on Sonics and Ultrasonics SU-17*, 3, 133–140.

DYSON, M., WOODWARD, B., and POND, J. B., 1971. Flow of red blood cells stopped by ultrasound. *Nature*, 232, 572–573.

EL'PINER, I. E., 1967. Ultrasonic waves in molecular biophysics. *Ultrasonics*, July, 181–185.

FRY, W. J., 1952. *Journal of the Acoustical Society of America*, 25, 1.

FRY, W. J., FRY, F. J., MALEK, R., and PANKAU, J. W., 1964. *Journal of the Acoustical Society of America*, 36, 1795.

FRY, W. J., WULFF, V. J., TUCKER, D., and FRY, F. J., 1950. *Journal of the Acoustical Society of America*, 22, 867.

HERRICK, J. F., 1953. *Journal of the Acoustical Society of America*, 25, 16.

HILL, C. R., 1971. Proceedings of International Congress on Ultrasonic Diagnostics (Vienna, 1969).

HORVATH, J., 1944. Ultraschallwirkung beim menschlichen Sarkom. *Strahlentherapie*, 75, 19.

LELE, P. P., 1963. Effects of focused ultrasonic radiation on peripheral nerve with observations on local heating. *Experimental Neurology*, 8, 47–83.

MACKINTOSH, I. J. C., and DAVEY, D. A., 1970. Chromosome aberrations produced by an ultrasonic fetal pulse detector. *British Medical Journal*, 4, 92.

POND, J. B., 1968. A study of the biological action of focused mechanical waves. Ph.D. Thesis, University of London; 1970. *Journal of the Acoustical Society of America*, 47, 1607–1611.

SELMAN, G. G., and COUNCE, S. J., 1953. Abnormal embryonic development in drosophila induced by ultrasonic treatment. *Nature*, 172, 503–504.

SHEALEY, C. N., and HENNEMAN, E., 1962. Quantitative neuro-anatomic studies implemented by ultrasonic lesions. *Archives of Neurology*, 6, 374.

SKINNER, L. A., 1970. *Journal of the Acoustical Society of America*, 47, 327–331.

SMYTH, M. G., 1966. In *Diagnostic Ultrasound* (Plenum Press, London).

SUMNER, W., and PATRICK, M. K., 1964. In *Ultrasonic Therapy* (Elsevier, London).

TAYLOR, K. J. W., and NEWMAN, D. L., 1972. Effects on electrophoretic mobility of Ehrlich cells exposed to ultrasound of variable parameters. *Physics in Medicine and Biology*, 17, March.

WARWICK, R., and POND, J. B., 1968. Trackless lesions in nervous tissues produced by high intensity focused ultrasound. *Journal of Anatomy*, 102, 387–405.

WOODWARD, B., POND, J. B., WARWICK, R., and CONNOLLY, C. C., 1970. How safe is diagnostic sonar? *British Journal of Radiology*, 43, 719–725.

AUTHOR CITATION INDEX

Abdulla, U., 383
Ackerman, E., 126, 180, 208, 317
Adams, C. E., 180
Ajmone-Marson, C., 356
Alexander, P., 224
Alfrey, C. P. Jr., 198, 208
Allis, J. W., 96
Altenburg, K., 322
Amberson, W. R., 282
Ambrose, J. S., 306
Anderson, T. P., 208, 343
Andreae, J. H., 85, 96
Angerer, 41
Aoki, K., 97
Applegate, K. R., 13, 84, 86, 100, 103, 116
Applequist, J., 89
Argon, A., 343
Astrom, K. E., 356
Atkinson, G., 86
Atter, M. F., 181

Baker, D. W., 383
Baker, N. V., 372
Ballantine, H. T., Jr., 30, 31, 181, 309, 351, 356
Bang, F. B., 155
Baranowski, T., 210
Barksdale, A., 89
Barlow, A. J., 53
Barnard, J. W., 209, 284, 309, 313, 315, 324, 356
Barnes, D. K., 174, 181, 232, 251
Barth, G., 41, 281, 318
Basauri, L., 343, 351, 356, 383
Bass, R., 85

Basurmanova, O. K., 166
Bauld, T. J., 13, 61, 99
Baumgartner, E., 86
Bebchok, 126
Becker, G., 281
Behnke, A. R., 27
Bejdl, W., 166
Bell, E., 166, 181
Benjamin, T. B., 129
Bennett, W. A., 208, 343
Beranak, L. L., 79
Bergmann, L., 25
Best, C. H., 79
Beuthe, H., 228
Biancani, E., 14, 166
Biancani, H., 166
Bier, M., 214
Bílková, B., 228
Blackshear, P. L., Jr., 198
Blair, H. A., 273
Blake, F. G., Jr., 131, 180, 251
Blandamer, M., 88
Blitz, J., 216
Block, R. J., 196
Bloomfield, V., 97
Blum, H. F., 79
Bock, J., 69
Bolling, D., 196
Bolt, R. H., 30, 33, 126, 212, 241
Bondy, C., 252
Born, H., 22, 34, 281
Borovyagin, V. L., 166
Borysko, E., 155
Bostelmann, W., 217

Bowers, B., 240
Boyd, E., 383
Boyer, P. D., 224
Bradfield, G., 181
Bragg, J., 89
Braginskaya, F. I., 89, 224
Brandt, 41
Brennan, J. F., 309, 313
Bresjanac, M., 209
Brett, H. W. W., 224
Breyer, J. E., 346
Briggs, H. B., 125, 322
Brohult, S., 252
Brown, B., 383
Brown, D. E. S., 250
Brown, R. L., 166, 173
Bryant, J. C., 241
Bullen, M. A., 181
Burgi, E., 195
Burke, J. J., 13, 80, 88, 100, 116
Burns, V. W., 195
Busnel, R. G., 273, 306
Buzzel, J. G., 92, 98, 104

Callis, P. R., 224
Campbell, A., 343
Canadau, S., 88, 92
Carlin, B., 343
Carslaw, H. S., 208, 306, 343, 351
Carstensen, E. L., 13, 31, 33, 34, 38, 43, 46, 48, 53, 61, 91, 93, 99, 100, 101, 116, 210
Castellan, G., 116
Castle, W. W., 208
Catchpole, H. R., 321
Cerf, R., 13, 88, 89, 92
Ceriotti, G., 196
Chalmers, R., 166
Chambers, L. A., 166, 195
Champion, J. V., 208, 241
Chauchard, P., 273, 306
Chaudhuri, S., 84
Chou, P., 84
Choungran, C., 198
Christman, C. L., 13
Claeys, 17
Clark, L. C., 224
Clarke, P. R., 224, 229, 230, 241
Coakley, W. T., 208, 224, 240, 241, 383
Coble, A., 356
Cohen, M. S., 33, 309
Cohn, E. J., 45, 92, 116
Collingham, R. E., 198
Colombati, S., 33
Colvin, J. E., 103
Conn, H. J., 63
Connolly, C. C., 196, 208, 383

Constantin, T., 228
Cooper, N. W., 231
Cooper, K. W., 174, 232, 251
Corelli, J. C., 145, 150, 170, 201
Cota-Robles, E. H., 181
Counce, S. J., 383
Courtenay, V. D., 230
Crook, E. M., 210
Crothers, D. M., 196
Crowe, M. R., 224, 229, 241
Crum, L. A., 372
Cullis, A. F., 101, 102
Cunningham, L. W., Jr., 214
Curry, B., 306
Curtis, J. C., 166, 167

Dalton, G. A., 181
Damodaran, M., 210
Darrow, M. A., 63
Davey, D. A., 383
Davidson, J. N., 224
Davidson, N., 224
Davis, C. M., 83, 96
Davison, P. F., 172, 195, 196
Debye, P., 116
De Forest, R. E., 282
De Lateur, B. J., 69
de Maeyer, L., 88, 91, 116
Devin, C., 79, 196, 241
Dittmar, C., 281
Djukanovic, A., 209
Dognon, A., 14, 166
Doi, Y., 306
Donald, I., 383
Dorman, F. D., 198
Doty, P., 181, 195, 212, 224
Dounce, A., 196
Drášil, V., 228
Dreyer, L. L., 62
Drummond, D. S., 196
Dunn, F., 13, 61, 62, 69, 79, 80, 89, 91, 96, 99, 100, 102, 116, 123, 167, 173, 196, 208, 209, 211, 217, 224, 240, 241, 248, 313, 343, 346, 351, 356, 383
Dussik, K. T., 30, 33
Dyer, H. J., 139, 149, 150, 151, 166, 167, 173, 181
Dyro, J. F., 53, 61, 99
Dyson, M., 240, 372, 383

Eckart, G., 250
Edmonds, P. D., 53, 61, 69, 80, 86, 91, 96, 99, 103, 116, 210, 241, 356
Edsall, J. T., 45, 116
Eggleton, R. C., 166, 208, 241
Eigen, M., 53, 88, 91, 116

Eigner, J., 224
Eisenberg, H., 86
Elder, S. A., 126, 127, 145, 172, 180, 196, 198, 208, 232
Elis, A., 116
El'piner, I. E., 89, 166, 209, 224, 240, 343, 383
Emara, M., 86
Embleton, T. F. W., 157, 169
Epstein, P. S., 41
Errera, 17
Esche, R., 33, 34, 125, 134, 306, 316, 320
Exner, M., 130
Eyring, E., 116
Eyring, H., 48, 95, 322

Faikin, I. M., 166
Fedorova, N. M., 195
Feissel, H. J., 281
Ferguson-Smith, M. A., 383
Ferry, J. D., 96
Fiedler, 33
Field, E. C., 102
Finesinger, J. E., 346
Firestone, F. A., 23
Fischer, G. A., 230
Fitzgerald, M. E., 232
Fleming, J. G. E., 383
Flemming, E., 208
Flosdorf, E. A., 195
Flynn, H. G., 118, 174, 180, 229, 356
Fontaine, M., 273
Foster, J. F., 92, 95, 96, 97, 101
Foster, M., 88
Fountain, L. S., 343
Fox, F. E., 129, 198, 208, 251, 254
Fox, M., 224
Francis, W. L., 282
Franke, E. K., 79
Fred, R. K., 6
Frederick, J., 343
Freifelder, D., 196
French, C. S., 181
Frenkel, J., 195
Freund, 41
Freundlich, H., 208, 252
Freundlich, M. F., 181
Frey, R. R., 196
Frings, 126
Frucht, A. H., 31
Fry, F. J., 62, 166, 167, 208, 209, 240, 241, 253, 259, 260, 265, 306, 309, 313, 324, 356, 383
Fry, R. B., 33, 62, 69, 79, 306, 309, 317
Fry, W. J., 33, 41, 61, 62, 69, 79, 80, 91, 99, 166, 167, 173, 180, 209, 211, 240, 241, 259, 265, 284, 306, 309, 313, 314, 315, 317, 318, 324, 346, 351, 356, 383

Gaines, N., 135
Galf, J. K., 24, 93, 100
Garay, K., 166, 306
Gerendas, M., 166, 306
Gersh, I., 27, 321
Gersten, J. W., 306, 307
Gessler, U., 166
Giacomini, A., 306
Gibbs, J., 89
Giese, A. C., 91
Gill, S. J., 224
Gillings, D. W., 208, 250
Glasstone, S., 48
Gligorijevic, J., 273, 306
Gloggengiesser, W., 166
Gohr, H., 228
Goldman, D. E., 31, 61, 62, 79, 139, 208, 346
Gomori, G., 210
Gooberman, G., 195, 224
Gordon, D., 383
Gordon, H. T., 254
Gor'kov, L. P., 169
Gould, R. K., 180, 208
Gourke, M., 89
Grabar, P., 180, 224, 346
Gramberg, H., 45, 53, 61
Granath, K., 82
Graves, 45
Grayson, J., 343
Green, N. M., 214
Greenspan, M., 93
Gregg, E. E., Jr., 249, 264
Griffing, V., 126, 232, 254
Grossman, M. G., 102
Grundfest, H., 250
Guttner, W., 33, 41, 63, 69

Haefely, W., 166
Hagnauer, G., 89
Hall, A. J., 383
Hall, C. E., 224
Ham, T., 208
Hamburger, V., 372
Hamilton, H. L., 372
Hamman, S., 116
Hammes, G. G., 13, 53, 80, 83, 84, 88, 100, 116
Hammick, J. W., 224, 229, 241
Hampe, W., 130
Hampton, D., 208
Harding, C. V., 155
Hardy, J. D., 343
Harrington, R. E., 196
Harrington, W. F., 104
Harvey, E. B., 139, 145, 150, 153
Harvey, E. N., 123, 139, 145, 150, 166, 167, 174, 181, 232, 243, 249, 251, 264, 307

Author Citation Index

Hasserodt, U., 103
Haurowitz, F., 43
Hawkinson, 27
Hawley, S. A., 80, 91, 93, 96, 100, 116, 123, 173, 196, 209, 210, 224
Head, L. H., 166
Heasell, E., 85
Heilbrunn, L. V., 153, 156
Hellums, J. D., 198, 208
Henneman, E., 383
Hermans, J. J., 210
Herrick, J. F., 208, 281, 282, 306, 320, 343, 383
Hershey, A. D., 195
Herzfeld, K. F., 91, 129, 198, 208
Hidden, N., 88
Hiedemann, E. A., 41
Hill, C. R., 224, 229, 230, 233, 241, 356, 383
Hipp, 45
Hirano, H., 356
Höber, R., 282
Hoepker, W., 321
Hogeboom, G. H., 61
Hohlfeld, R., 281
Hollander, A., 251
Holtzmark, J., 170, 201
Hopkins, J. C., 181
Horatz, K., 166
Horton, J. P., 208
Horvath, J., 22, 282, 383
Hovorka, F., 126, 224
Howkins, S., 196
Howry, D. H., 79
Hrazdira, I., 228, 230, 232
Hsi, E., 306
Hsieh, D. Y., 232
Hueter, T. F., 22, 25, 30, 33, 61, 62, 64, 69, 79, 126, 212, 241, 282, 309, 343, 351, 356
Huf, E., 282
Hug, O., 317
Hughes, D. E., 166, 170, 172, 181, 196, 198, 208, 228, 229, 232, 383
Hughes, W. L., 92
Hugo, W. B., 180
Hummel, B. C. W., 210
Hussey, M., 61, 86, 99, 103, 116
Hüter, T., 30, 34, 250, 282

Iernetti, G., 241
Ingard, K. U., 201
Inoue, H., 86, 89

Jackson, F. J., 145, 150, 155, 170, 180, 196
Jaeger, J. C., 306, 343, 351
Jaenicke, W., 281
Jagannathan, V., 210

James, A. R., 181
Janes, J. M., 282
Jarman, P., 180
Jasínska, J., 228
Jasper, H. H., 356
Jellinek, H. H. G., 224
Jennings, B. R., 97
Jirgensons, B., 96, 214
Johnsen, J., 170
Johnson, C. H., 139, 145, 150
Johnson, I. B., 125, 201
Johnson, J. B., 322
Johnson, P., 104
Johnson, W. R., 195
Joseph, J., 383

Katchalsky, A., 86
Kawahara, K., 102
Kawashima, Y., 356
Kay, E., 196
Keech, D. B., 181
Kelly, E., 166
Kessler, L. W., 13, 61, 89, 91, 93, 96, 99, 100, 101, 116
Kielley, W. W., 224
Kikuchi, Y., 79
King, L. V., 168, 251
Kinsler, L. E., 196
Kinsloe, 126
Kirkwood, J. G., 116
Kirschner, A. G., 102
Kittel, C., 250
Klair, J., 307
Klein, H., 282
Knapp, R. T., 251
Koester, E., 281
Kolb, J., 126, 127, 180
Koller, S., 282
Konig, W., 372
Koppelman, J., 317
Korn, E. D., 240
Kornberg, A., 210
Kornfeld, M., 126, 180
Kossoff, G., 208, 241
Kostic, I., 209
Kovrov, B. G., 228
Kremkau, K. N., 13
Krumins, R. F., 209, 309, 313, 356
Krusen, F. H., 208, 282, 343
Kuffler, S. W., 307
Kunitz, M., 212

Laidler, K. J., 48
Lamb, H., 41, 150, 208, 251
Lamb, J., 53, 85, 91

Lambert, E. H., 282
Lane, C., 133
Lang, J., 13, 84, 90, 116
Larson, E., 156
Laurence, K. M., 383
Lawrence, J. H., 346
Lawrence, N. S., 181
Layne, E., 224
Leach, W. M., 6
Lee, W. A., 195
Lehmann, J. F., 69, 208, 281, 282, 306, 314, 319, 320, 343
Lehninger, A. L., 217
Lele, P. P., 69, 167, 208, 343, 351, 356, 383
Lepeschkin, W. W., 139, 208, 346
Lester, W. W., 64
Leverett, L. B., 208
Levinson, M. S., 228
Levinthal, C., 172, 195
Lewis, T. B., 13, 80, 83, 88, 92, 100, 116
Li, K., 31, 33, 38, 43, 61, 99
Liebermann, L. N., 250
Lim, S., 116
Lindstrom, O., 180, 231
Lindstrom, P. A., 321
Linzell, J., 343
Lipkin, M., 343
Litovitz, T. A., 64, 83, 91, 96
Litt, M., 224
Loomis, A. L., 123, 139, 145, 150, 167
Love, A. E. H., 150
Lowry, C., 214
Ludwig, G. D., 23, 30, 31, 34
Luft, J. H., 166
Lund, E. J., 282
Lynch, E. C., 198, 208
Lynn, J. S., 249, 264
Lyon, T., 64

McClintock, F., 343
McElroy, W. D., 174, 181, 232, 251
McGill, B. B., 181, 195
McKellar, J. F., 96
Mackintosh, I. J. C., 383
Macleod, R. M., 123, 173, 196, 209, 210, 217, 224, 356
McMeekin, 45
McSkimin, H. J., 250
Maidanik, G., 372
Makarov, 126
Malek, R., 209, 383
Mallams, J. T., 79
Manlapaz, J. S., 181, 356
Marinesco, N., 252
Mark, H., 252

Marr, A. G., 181
Marshland, D. A., 250
Mason, S. G., 172
Mason, W. P., 30, 46, 125, 151, 250, 322
Mathias, A. P., 210
Matthes, K., 281, 307
Matthews, M. M., 181
Mayback, E. J., 198
Mayer, A., 99, 116
Mazoué, H., 273, 306
Melville, H. W., 209, 224
Metcalf, A., 30
Meyer, E., 126
Meyers, R., 62, 209
Michaelis, E., 282
Michalowski, A., 228
Michels, B., 88, 89, 92
Mikhailov, I. G., 195
Millar, D. B. S., 214
Miller, J. E., 79
Miller, W., 89
Milner, H. W., 181
Minnaert, M., 136, 196
Miyashita, Y., 356
Mode, E. B., 224
Morse, P. M., 201, 250
Mosberg, W. H., 309
Mosher, W. A., 307
Mostafa, M. A. K., 224
Mowry, S. C., 24
Muirhead, H., 101, 102
Müller-Landau, F., 99, 116
Murao, T., 356
Murray, A. J. R., 209, 224
Musa, R., 85

Naimark, G. M., 307
Nauta, W. H., 351
Neiderland, T., 210
Neilands, J. B., 210
Nelson, P. A., 208
Neppiras, E. A., 180, 181, 196, 232
Neurath, H., 214
Nevaril, C. G., 198
Newman, D. L., 383
Nitsch, W., 281
Noguchi, H., 88, 89
Nolle, A. W., 24
Noltingk, B. E., 180, 196
Nomoto, O., 232
Nord, F. F., 214
North, A. C. T., 101, 102
North, P. F., 208, 241
Northrup, J. H., 212

Nossal, P. M., 181
Noyes, R. F., 62
Nozaki, Y., 103
Nyborg, W. L., 118, 126, 127, 139, 145, 149, 150, 151, 155, 157, 166, 167, 169, 170, 174, 180, 181, 196, 198, 201, 208, 228, 229, 232, 372
Nylund, R., 89

O'Brien, J. R. P., 102
O'Brien, W. D., Jr., 6, 13, 61, 83, 99, 102, 116
Oestreicher, H. L., 79
Oka, K., 356
Oka, M., 356
Okni, S., 232
Okumura, T., 356
Okuyama, M., 196
Olson, A. R., 139, 145, 150
Oncley, J. L., 48
Orttung, W. H., 116
Ossoinig, K., 69
Ostreicher, H. L., 30
Ottewill, R. H., 104
Otto, J., 281
Overend, W. G., 195

Pankau, J. W., 209, 383
Pape, R., 317
Paret, G., 224
Parker, R. C., 13, 84, 86, 100, 103, 116
Parow-Souchon, E., 14, 249, 264
Parrack, H. O., 79
Patrick, M. K., 372, 383
Pattison, S., 86, 103, 116
Pätzold, J., 22, 33, 318
Pauly, H., 61, 100
Peacocke, A. R., 172, 195, 196, 208
Pease, D. C., 174, 181, 232, 251
Pekeris, C. L., 129
Pellam, J. R., 24, 93, 100
Pennak, R. W., 346
Pennell, R. B., 35
Perl, W., 343, 351
Perutz, M. F., 45, 101, 102
Peters, G., 321
Petralia, S., 33
Pfander, F., 30, 31, 33, 282
Pfleiderer, H., 224
Philippoff, W., 82
Phillips, R. A., 25
Pierce, A. C., 343
Pincus, G., 309
Pinkerton, J. M. M., 93
Plesset, M. S., 232
Pohlman, R. R., 14, 22, 24, 25, 30, 33, 34, 249, 264

Polet, H., 103
Polotsky, I. G., 228
Pond, J. B., 240, 351, 356, 372, 383
Ponder, E., 42, 343
Ponzio, M., 166
Pospíšilová, J., 228
Preston, B. N., 196
Price, C. C., 195
Pritchard, N. J., 172, 196, 208
Prosser, C. L., 257, 262, 268
Prudhomme, R. O., 224, 228
Ptitsyn, O., 89
Pumphrey, R. J., 282
Putnam, T. J., 249, 264

Rahbun, 27
Rand, R. P., 208
Randerath, K., 215
Rands, D. G., 92, 104
Raney, W. P., 145, 150, 170, 201
Rebello, M. A., 166
Rech, W., 281, 307
Reichmann, M. E., 103
Reid, J. M., 79, 126
Rhodin, A. G. J., 372
Rice, S. A., 181, 195
Richards, J. R., 31
Richards, O. C., 224
Richards, W. T., 252
Richter, R., 14, 249, 264
Riddiford, C. L., 97
Rifkind, J., 89
Roark, R., 343
Roberts, P., 84, 116
Roberts, W., 224
Robin, B. R., 210
Robinson, R., 87
Robinson, T. C., 69, 343
Rodgers, A., 181, 198, 228, 229
Rohm and Haas, 343
Rooney, J. A., 169, 198, 201, 208
Rosenberg, M., 126, 128, 180, 322
Rosenberger, H., 282
Rossman, M. G., 101
Rouse, P., 82
Rumscheidt, F. D., 172
Rust, K. H., 233
Ryabchenko, N. I., 224

Sack, 17
Saksena, T. K., 88, 92
Samosudova, N. V., 166
Sanderson, M. H., 372
Saroff, H. A., 98
Sasabe, H., 13, 100
Schachman, H. K., 210, 214

Schellman, J. A., 212
Scheraga, H. A., 84
Schimmel, P. R., 83, 116
Schlichting, H., 201, 208
Schmid, G., 224
Schmitt, F. O., 123, 139, 145, 150, 154, 166, 167, 252
Schmitz, W., 166
Schneider, F., 99, 116
Schneider, W. C., 61
Schnitzler, R. M., 166, 170, 173
Schultes, H., 228
Schultz, D. F., 62
Schwan, H. P., 13, 31, 33, 34, 38, 43, 48, 53, 61, 91, 99, 100, 101, 116, 210
Schwarz, G., 13, 85, 92, 100
Schwert, G. W., 210, 214
Selman, G. G., 383
Sewell, C. J. T., 41
Shan, 208
Shavit, N., 86
Shealey, C. N., 383
Sheraga, H. A., 210
Shore, M. L., 6
Sichel, F. J. M., 166, 170, 173
Sikkeland, T., 170, 201
Silverman, D. R., 69
Simmons, N., 196
Simpson-Gildemeister, V. F. W., 196
Singh, A., 198
Singh, K., 210
Sistrom, W. R., 181
Skavlem, S., 170, 201
Skinner, L. A., 383
Skvorstov, A., 89
Slade, J. S., 383
Slutsky, L. J., 13, 84, 86, 100, 103, 116
Smith, A., 61
Smith, F. D., 252
Smith, R., 89
Smith, R. E. P., 180
Smith, W. C., 35
Smoluchowski, M. V., 116
Smyth, M. G., 383
Snipp, R., 89
Snyder, S., 231
Sollner, D., 252
Sollner, K., 252
Sosa, D. M., 351
Soška, J., 228
Spector, W. S., 61
Statnikov, Y. G., 174
Stearner, S. P., 372
Stefanovic, V., 209
Steinback, J. H., 198
Steinhardt, J., 103

Stellwagen, E., 214
Stier, T. J. B., 309
Stokes, R., 87
Strasberg, M., 129
Struthers, F. W., 23
Stuehr, J., 89
Sturm, J., 84, 90
Sullivan, J., 232
Summer, W., 372, 383
Sumwalt, M., 282
Suvorov, L., 126, 180
Swanson, S. A., 92, 104
Symons, M., 88
Szent-Gyorgi, A. G., 195

Takenaka, Y., 210
Tanaka, K., 79
Tanford, C., 89, 91, 92, 98, 102, 103, 104, 116
Tanford, D., 102, 103
Taylor, A. B., 282
Taylor, G. I., 172
Taylor, J. F., 214
Taylor, K. J. W., 240, 372, 383
Taylor, N. B., 79
Theismann, H., 30, 31, 33, 282
Thompson, D. S., 224
Thrasher, R., 133
Tietze, F., 214
Tobias, C. A., 346
Tomono, M., 13, 100
Tondre, C., 84, 90, 116
Treanor, W. J., 282
Trezak, R., 343
Trinstram, C., 89
Tschiegg, C. E., 93
Tseitlin, P. I., 224
Tucker, D., 167, 240, 241, 259, 265, 306, 309, 313, 356, 383
Tyler, A., 152

Uchida, R., 79
Uhlemeyer, B., 252
Urick, R. J., 25, 31, 41, 97
Urnes, P., 212
Ussing, H. H., 282

Van Slyke, D. D., 25
Velasevic, K., 209
Vierordt, H., 27
Vinograd, J., 103
von Grierke, H. E., 79
von Wittern, W. W., 79
Vorschütz, R., 282

Wachsmann, F., 281, 318
Wada, Y., 13, 100

Author Citation Index

Wagai, T., 79
Wakim, K. G., 208, 343
Walker, S. M., 307
Wall, P. D., 284, 314
Warwick, R., 351, 383
Weare, J. H., 92
Weber, G., 97, 98, 104
Webster, E., 196, 231
Weissler, A., 198, 224, 231
Welkowitz, W., 313
Wells, P., 343
Welsh, J. H., 254
Westervelt, P. J., 133, 145, 150, 170, 201
White, F. H., Jr., 215
White, R. D., 86, 103, 116
Whiteley, A. H., 174, 181, 232, 251
Wick, R. F., 151
Wiercinski, F. J., 166, 173
Wiercinski, W. L., 170
Wilcox, F. W., 306
Wild, J. J., 79, 306
Will, G., 101
Willard, G. W., 126, 180, 198, 241

Williams, A. R., 123, 198, 201, 208, 224, 241
Williams, E. J., 97
Wilson, W. L., 156, 166, 170, 173
Woeber, K., 319
Woodward, B., 372, 383
Wulff, V. J., 167, 240, 241, 259, 265, 306, 309, 313, 356, 383
Wyt, L., 30

Yang, J. T., 84, 89, 95, 96
Yeager, E., 126, 224
Yokoi, H., 356
Yoshioka, K., 356
Yoshitatsu, S., 356
Young, G. F., 343
Young, L. B., 97

Zana, R., 84, 88, 90, 92, 116
Zelníček, E., 228
Zimm, B. H., 89, 196
Zimny, M. L., 166
Zivanovic, D., 209
Zorina, O., 89

SUBJECT INDEX

Absorption, 8–9, 249, 250, 258 (see also Attenuation)
alanine, 103
albumin, 34–37, 59, 80, 91–100, 102
amino acid, 12, 103–104
O-amino benzoic acid, 110–111, 116
amplitude dependence
   spinal cord, 62–64
   tissue, 8, 11
animal tissue (see Tissue)
anisotropy, 9, 10, 250
arginine, 103, 104
aspartic acid, 103, 104
blood, 10, 34–37, 38–42, 99
bone, 8, 10, 28–30, 32–33, 66
brain, 32–33, 66, 316, 328, 336–338, 373–374
carbon disulfide, 16–17
cell size, 40
chemical factor, 108, 114
classical, 19
concentration dependence
   albumin, 93–98
   blood, 34, 39–42
   dextran, 81–83
   hemoglobin, 43, 44, 101–104
   liver, 56
   poly-l-lysine, 85
cysteine, 103–104
deoxyribose nucleic acid, 102
dextran, 9, 12, 80–83, 91, 96, 100
enthalpy dependence, 108, 112
fat, 16, 32–33
frequency dependence
   albumin, 80, 91–98, 102

biopolymer, 57–58
blood, 34–37, 38–40
bone, 8, 28–30, 66
brain, 316
deoxyribose nucleic acid, 102
dextran, 80–83
gelatin, 100
hemoglobin, 43–49, 50–53, 80, 99–104, 105–116
liver, 54–58
lung, 66, 69
ovalbumin, 102
plasma, 35, 38
polyethylene glycol, 82–83
poly-l-glutamic acid, 80
poly-dl-lysine, 84–90
poly-l-lysine, 84–90
poly-l-ornithine, 84–90
spinal cord, 65–69
tissue, 9–11, 14–18, 31–33, 62, 66, 250
gelatin, 54, 57–59, 61, 100
glutamic acid, 103–104
glycine, 103
heart, 19–22, 32–33, 250
heat denaturation, 57
heating, 234
hemoglobin, 8–9, 11–12, 34–37, 40, 43–49, 50–53, 59, 80, 91, 96, 99–104, 105–116
   dependence, 99–104
   solubility, 45–46
histidine, 103–104
human tissue, 14–18 (see also Tissue)
hydration, 57
intact cells, 38–42

393

Subject Index

interface, 263
kidney, 19–22, 32–33
liver, 8, 19–22, 32, 33, 54–61
lung, 9, 11, 66, 69–79
lysine, 103–104
measurement (see Measurement)
medulla oblongata, 32–33
molecular conformation dependence, 12
molecular contribution in tissue, 11
molecular structure, 57
molecular-weight dependence, 80–83
muscle, 14–18, 25, 32–33, 125, 250
nerve, 32–33
nucleic acid, 13 (see *also* Deoxyribose nucleic acid)
ovalbumin, 102
pH dependence
  albumin, 91–98
  amino acid, 103–104
  biopolymer, 57
  gelatin, 57–59, 61, 100
  hemoglobin, 43, 45, 99–104, 105–116
  liver, 54, 57–59
  poly-*dl*-lysine, 84–90
  poly-*l*-lysine, 84–90
  poly-*l*-ornithine, 84–90
  protein, 57
plasma, 32–33, 34–37, 38, 59, 99
polycrystalline metal, 29
polyethylene glycol, 12, 82–83
poly-*l*-glutamic acid, 12, 80, 100
poly-*dl*-lysine, 84–90
poly-*l*-lysine, 12, 84–90, 100
poly-*l*-ornithine, 84–90
polysaccharide, 12, 80–83, 91, 96, 100
protein, 8, 10–12, 34–38, 43–49, 50–53, 54–61, 80, 99–104, 210
  contribution, 54–61
  dependence, 99–104
radiation force, 169
salt dependence, 43, 45–46
scattering, 96
serine, 103
shear viscosity, 48
spinal cord, 32–33, 62–64, 65–69
structural contribution, 11, 54–61
tabular summary, 31–33
temperature, 216, 249, 257, 258, 262–263, 264
temperature dependence, 34–37, 95
  albumin, 95–96
  brain, 66, 338
  hemoglobin, 43–44, 101–102
  poly-*l*-lysine, 85
  spinal cord, 62–64, 65–69
  tissue, 8, 11, 249

  water, 249
temperature gradient, 280
threonine, 103
tissue, 8–11, 14–18, 19–22, 31–33, 54–61, 62–64, 65–69, 240, 249–250, 263, 308, 320
  antisotropy, 19–22
  dependence, 9
  summary, 31–33
tongue, 19–22, 250
tyrosine, 103–104
volume dependence, 108, 112
water, 83, 93, 249
Absorption mechanism
  bulk viscosity, 95, 96, 250
  cavitation, 15
  charge transfer, 116
  chemical equilibrium, 48, 250
  chemical relaxation, 105, 111–115
  conformational change, 100, 105
  enzyme, 217
  Epstein's scattering theory, 38, 41–42
  helix-coil transition, 12, 84–90, 92, 100
    vs. helix-coil, 84–90
  hemoglobin, 99
  hydration layer perturbation, 102, 104
  hysteresis, 46, 250
  intecullular cavitation, 18
  intermolecular interaction, 13, 94, 101
  intermolecular transfer, 111–115
  ionization, 105
  mode conversion, 96
  molecule impurity, 97
  plastic slip, 250
  vs. proton-transfer, 84–90
  proton-transfer reaction, 9, 57–59, 84–90, 99–100, 102–104, 105–116
  relative motion, 41–42, 96
  relaxation process, 41–42, 44, 46–48, 51–53, 54–61, 102
  scattering, 96
  shear and sliding effects, 21
  shear viscosity, 64, 95, 96, 250
  single relaxation, 58–59
  sliding, 21
  solvation, 92, 105
  solvent–solute interaction, 12, 91, 99–100, 102
  streaming, 15, 238
  structural relaxation, 48, 96, 105
  viscous damping, 250
  viscous interaction, 10–11, 34–42
  viscous losses, 82
  viscous process, 48
  viscous slip, 250
ac boundary layer, 148
ac flow, 147–149

Subject Index

ac stress, interaction mechanism, 122, 202–208
Acanthamoeba (see Amoeba)
Acceleration force, interaction mechanism, 382
Acoustic boundary layer, 129–130, 133–134
Acoustic streaming (see Streaming)
Action potential
  bioeffect, 245–255
  measurement, 287–289
  muscle, 246, 285–306
  pressure dependence, 292–296
  temperature dependence, 257–258, 285, 296–302
Activation energy
  frequency dependence, 96, 101–102
  pH dependence, 96
  relaxation theory, 48, 54, 322
  water, 96, 101–102
Active site, 210
Active transport, 275
Aggregation
  blood cell, 357, 359–363, 369–371
  intracellular motion, 146
Alanine, 103
Albumin
  absorption, 34–37, 59, 99, 100
    concentration dependence, 93–98
    frequency dependence, 80, 91–98, 102
    pH dependence, 91–98
    temperature dependence, 95–96
  activation energy, 96, 101–102
  bulk modulus, 91, 97–98
  conformation, 91–98, 104
  conformational dependence, 9, 91–98
  intrinsic viscosity, 96–98, 104
  N–F ' transition, 91, 104
  proton-transfer reaction, 90
  rotational relaxation time, 98
  velocity, 34–37, 99
    frequency dependence, 95–97
    pH dependence, 91–98
Aldolase, 209–217
Allium cepa, 146–149
Ambient pressure dependence, 260–264
Amino acid (see *specific amino acid*)
O-amino benzoic acid, 110–112, 116
Amoeba
  bioeffect, 120, 234–240, 345–346
  cavitation, 120, 234
  destruction
    cavitation dependence, 236–240
    intensity dependence, 235
    temporal dependence, 235–236
  growth, 234–240
  streaming, 122

  survival, 122
  vibrating wire, 122
Amoeba proteus (see Amoeba)
Amplitude dependence
  absorption, 62–64
  bioeffect, 159–166
  boundary layer, 171
  bubble action, 175–180
  streaming, 144–145, 153–156
Anisotropy, 19–22, 25, 250
Antinodes, 130–131
Application (see Use)
Arbacia egg (see Sea urchin egg)
Arbacia punctulata (see Sea urchin egg)
Arginine, 103–104
Aspartic acid, 103–104
Asterias egg (see Starfish)
Attenuation (see *also* Absorption)
  brain, 328, 336–337
  dura mater, 342
  heat generation, 326–328
  measurement (see Measurement)
  methacrylate, 328, 332–333, 335–336
  muscle, 25
  polyethylene, 326–327
  temperature dependence, 338
Attenuation coefficient (see Absorption)

Backbone scission, 122, 219
Beam
  directivity, 24
  pattern
    focus, 314–316, 348
    measurement, 253–254, 286–287, 298, 309
  radius dependence, 238
  width, 240, 260
Beef, 27
Bernoulli's principle, 147–149, 168, 370–371
Bioeffect
  action potential, 254–255
  amoeba, 120, 234–240, 345–346
  amplitude dependence, 159–166
  arbacia egg, 173
  bacteria, 2
  blood vessel, 283
  bone marrow, 120, 225–228
  brain, 244–247, 283–284, 314–323, 325–343, 347–351
  cavitation, 122, 202, 229, 251–253
    tissue, 124–125
  cell, 2, 119, 120–121, 170, 175–180, 184, 202, 229, 374
  chemical, 1, 318
  chemiprotective, 229
  chromosome aberration, 381

Subject Index

cumulative, 257, 261, 263, 373–374, 378–383
degradation, 172–173
deoxyribose nucleic acid, 119–122, 172–173, 179, 182–195, 209, 218–224
direct, 120, 122, 225–228, 229–233, 239
direct vs. indirect, 122, 225–228, 229–233
duty-cycle dependence, 373, 376–377, 378–381
endothelial cell, 357, 364–367, 377
enzyme, 120, 122, 209–217
erythrocyte, 2, 176–177, 189–190
Escherichia coli, 176–177, 189–190
evoked cortical potential, 247, 324
fish, 2–3
frequency dependence, 247–248, 345–346, 376, 381
frog, 2–3
functional, 216
hemorrhage, 245, 373–383
hydrodynamic shear, 172
hypoxia, 379–381
indirect, 120, 122, 225–228, 230–233
intensity dependence, 270–273, 373, 376–380
intracellular motion, 146, 149, 167
ionizing, 318, 373
lesion, 245, 247, 283–284, 310, 314–323, 325–343, 347–351, 352–356, 373–374
limb paralysis, 209
macromolecule, 121
mechanical, 1–2, 229
membrane, 246, 274–281
memory, 321
microorganism, 2, 175–180, 209, 234
mouse, 2, 308–309, 310–313, 318–320
muscle, 159–166, 244, 246, 285–306
mutation, 381
nerve, 249–258, 259–263, 264–269, 270–273
oxygen dependence, 379–381
paralysis, 173, 209, 240, 245, 256–258, 259–263, 266–269, 308, 309, 319, 373, 376–378
paramecium, 2
pressure dependence, 250–251, 260–263
reversible, 245, 247, 324
shear, 20, 120, 176, 202–208
sonochemical, 229
spinal cord, 245, 373–383
spirogyra, 2
spontaneous activity, 254–258, 259, 265–266
stasis, 247, 357–372
structural, 216
summation, 257, 261, 263
synergism, 248–373

temperature dependence, 249–258, 262–263, 267–269, 378–380
temporal dependence, 159–166, 270–273, 367–381
Tetrahymena pyriformis, 176–177
threshold, 246, 260, 261–263, 270–271
tissue, 124–125, 249
tissue regeneration, 381
Biopolymers, 57–58
Biopotential, 275–281
Biorheology, 371–372
Blood
  absorption, 8, 32–33, 34–37, 38–42, 99
  absorption mechanism, 38–42
  cavitation threshold, 120, 125
  nuclei, 124
  velocity, 8, 31, 34–37, 99
Blood-brain barrier, 322
Blood cell (see Erythrocyte)
Blood flow, lesion size dependence, 336
Boltzmann constant, 322
Bone
  absorption, 8, 10, 28–30, 32–33, 66
  temperature dependence, 250, 257
  velocity, 31
Boundary flow, 144–145, 148–149
Boundary layer
  amplitude dependence, 171
  bubble, 187, 197–198, 204, 206
  density dependence, 171
  distortion, 171–172
  frequency dependence, 171
  hemolysis, 171
  radiation force, 169
  streaming, 171
  velocity gradient, 171–174
  vibrating wire, 201, 204
  viscosity dependence, 171
Bovine hemoglobin (see Hemoglobin)
Bovine serum albumin (see Albumin)
Brain
  absorption, 32–33, 66, 316, 328, 336–338, 373–374
  bioeffect, 244–247, 283–284, 314–323, 325–343, 347–351
  blood flow, 336
  density, 27, 336–337
  gray matter, 246, 283–284
  impedance, 27, 336
  lesion, 180, 245, 247, 283–284, 310, 314–323, 325–343, 347–351, 352–356, 373–374
  lesion boundary temperature, 336
  measurement (see Measurement)
  specific heat, 336–337
  temperature distribution, 326, 328–331

thermal properties, 336–337, 342, 347–348
velocity, 26–27, 31, 336
  temperature dependence, 338, 340
white matter, 246, 283–284
Breaking strength, 172
Brillouin scattering, 5
BSA (see Albumin)
Bubble (see also Cavitation)
  ac stress, 122
  acoustic boundary layer, 129, 130, 133, 134
  amplitude dependence, 175–180
  boundary layer, 187, 197–198, 204, 206
  cavitation, 125
  dc stress, 122
  displacement amplitude, 203
  eddying motion, 176, 184
  erythrocyte, 197–198
  gas, 126
  growth, 251
  hemolysis, 119, 122, 197–198, 199, 202
  interaction mechanism, 175, 232, 238
  microstreaming, 176
  nuclei, 353
  oscillatory stress, 122
  pressure amplitude, 189
  pressure dependence, 251
  regime I (Elder), 132
  regime I (PHP), 184
  regime II (Elder), 131–132
  regime III (Elder), 132
  regime IV (Elder), 132
  resonance, 126–136, 145, 174, 176, 178, 187, 232, 353
  scatter, 185–187
  shearing force, 182
  source, 137, 167–174, 178, 179, 202–203
  stable, 197–198
  streaming, 172, 197, 199
  streaming field, 130–135
  surface displacement, 128
  surface energy, 126
  surface velocity, 128, 130–132
  surface viscosity, 129–130
  surface wave, 197, 203
  temperature dependence, 203, 207–208, 251–252
  vapor, 126
  velocity amplitude, 185–187
  velocity gradient, 121, 184, 187, 189, 191–195, 198, 199
  viscosity dependence, 127
  vortex motion, 197
Bulk modulus, 91, 97–98
Bulk viscosity, absorption mechanism, 250

Calf thymus deoxyribose nucleic acid
  (see Deoxyribose nucleic acid)
Calorimeter, 270, 275
Carcinoma, 31
Cat, 283, 308, 315, 318, 326, 336, 347, 352
Cavitation (see also Bubble)
  absorption mechanism, 15
  amoeba, 120, 234
  bioeffect, 124–125, 202, 251–253
  blood, 120, 125
  cell damage, 122, 229
  cell survival, 122, 234–240
  chemical effect, 218
  chemical reaction, 252
  chemoluminescence, 175, 177
  cleaning, 127, 135
  collapse, 117, 353
  degradation, 218
  deoxyribose nucleic acid, 119, 121–122, 182–195, 217, 218–224, 229
  detection, 219–224, 229, 230, 235, 236–238, 252–253, 304
  effective volume, 238
  electrical discharge, 175
  emulsification, 252
  enzyme, 120, 122, 209–217
  erythrocyte, 189–190
  free radical, 117, 175, 178–179, 202, 218, 229–230, 232
  frequency dependence, 124–125, 134, 183, 317
  gas dependence, 124–125, 183, 252–253
  gaseous, 175, 180
  heating, 234
  hemocyanin, 252
  hot spot, 202, 232
  intensity dependence, 252–253
  interaction mechanism, 117, 121–122, 159, 165–166, 167–174, 175, 183, 209–217, 218–224, 229–233, 234–240, 246–247, 249, 251–253, 258, 259–263, 268–269, 285–286, 298, 303–304, 308, 310–311, 317, 320–321, 345–346, 352–355, 373–374, 382–383
  lesion, 352–356
  liquefaction, 252
  liquid, 119, 124–125
  measurement, 219–220
  microstreaming, 119, 126–136
  muscle, 125, 285–286
  nuclei, 174, 229, 238, 251
  nuclei dependence, 125, 183
  paramecia, 126
  pressure, 175
  pressure dependence, 126

397

Subject Index

rate process, 126
rectified diffusion, 131, 176, 229, 232
sonoluminescence, 175, 177, 178
source, 184–186
stable, 117, 122
streaming, 229, 232
structural viscosity, 252
subharmonics, 222–223, 229, 237
suppress, 218, 219
temperature, 175
temperature dependence, 126
threshold, 120, 124–125, 134, 176, 183–184, 212, 236, 317, 346
tissue, 120, 124–125
transient, 117
viscosity dependence, 125, 212
water, 120, 124–125, 252–253
Cavitation dependence, 218–224, 236–240
Cavitation nuclei, 174, 229, 238, 251
Cavitation streamer, 177–178, 180
Cell
  bioeffect, 120, 121, 170, 175–180, 202, 229, 374
  culture, 229
  death, 122, 229
  disruption, 119, 175–180, 184
  intracellular motion, 119, 137–139, 140–145, 146–149, 151–158
  lysis, 225
  model, 119, 137–139, 144–145, 148–149, 168–169
  survival
    cavitation, 122, 234–240
    intensity dependence, 230–233
    pH dependence, 227
    pulse width dependence, 230–233
  suspension, 167, 171
Charge transfer, absorption mechanism, 116
Chemical
  bioeffect, 1–2, 318
  interaction mechanism, 120, 225, 228, 229–233
Chemical effect, 218
Chemical equilibrium, absorption mechanism, 48, 250
Chemical factor, 108, 114
Chemical reaction, 252
Chemical relaxation, absorption mechanism, 105, 111–115
Chemical relaxation mechanism, 103
Chemiprotective effect, 229
Chemoluminescence, 175, 177
Chick, 357–372
Chromosome aberration, 381
Chronaximetry, 270–273
α-Chymotrypsin, 209–217

Circulatory movement, interaction mechanism, 167–174
Clam egg, 151–158
Cleaning, 127, 135
Collapse cavitation (see Cavitation)
Compressibility, 81
Concentration dependence
  absorption
    albumin, 93–98
    blood, 34, 39–43
    dextran, 81–83
    hemoglobin, 40, 43–44, 101–104
    liver, 56
    poly-l-lysine, 85
  denaturation, 213, 215
  velocity, hemoglobin, 43–44
Conformational dependence
  absorption mechanism, 100, 105
  albumin, 91–98, 104
  hemoglobin, 92, 102–103
  polyglutamic acid, 92
  poly-dl-lysine, 84, 86
  poly-l-lysine, 84, 86
  protein, 9, 210
Connective tissue, 225
Covalent bond, 183
Crayfish, 254, 264
Critical stress
  temperature dependence, 206–207
  temporal dependence, 204, 205
  viscosity dependence, 204
Cumulative effect, 373–374, 378–383
Curarization, 287–288
Current dependence, 349–351
Cylinder streaming, 201
Cysteine, 103, 104

Damage (see Bioeffect)
Damage ability index, 373, 378–381
Damping constant, 129–130, 134, 136
Damping ratio, 129
dc flow, 147–149
dc stress, interaction mechanism, 122, 202–208
Death, 122, 229
Debye–Eigen equation, 88
Debye–Sears effect, 19, 216
Debye–Smoluchowsky theory, 105, 110
Decompression, 321
Degas, 212
Degradation (see also Bioeffect)
  backbone scission, 122
  cavitation dependence, 218–224, 229
  deoxyribose nucleic acid, 119–122, 172–173, 182–195, 209, 218–224
  frequency dependence, 213, 216–218

hydrodynamic shear, 183
intensity dependence, 218–224
interaction mechanism, 183
molecular-weight dependence, 219–224
velocity gradient, 121
viscosity dependence, 218–219, 222
Dehydrogenase, 209–217
Denaturation
  aldolase, 209–217
  α-chymotrypsin, 209–217
  concentration dependence, 213, 215
  dehydrogenase, 209–217
  enzyme, 122, 209–217
  frequency dependence, 213, 216–217
  intensity dependence, 209, 213, 216–217
  lactate, 209–217
  pH dependence, 209–210, 213, 215
  ribonuclease, 209–217
  temperature dependence, 209–210, 214–216
  trypsin, 209–217
Density
  beef, 27
  brain, 27, 336–337
  human tissue, 27
  kidney, 27
  liver, 27
  lung, 72–73
  methacylate, 333, 336
  spleen, 27
  tissue, 25, 27
  water, 27, 337
Density dependence
  boundary layer, 171
  bubble, 128
  intracellular motion, 157–158
  radiation force, 169
Deoxyribose nucleic acid
  absorption, 102
  backbone scission, 122, 219
  bioeffect, 119–122, 172–173, 179, 182–195, 209, 218–224
  breaking strength, 172
  cavitation, 122, 217, 218–224, 229
  degradation, 119–122, 172–173, 182–195, 209, 217, 218–224
  hydrodynamic shear, 172, 182
  hyperchromicity, 179
  intrinsic viscosity, 182, 189, 191, 195, 219
  molecular weight, 219–224
  relaxation time, 223–224
  stable cavitation, 119, 121, 182–195
  streaming, 223–224
  time-averaged force, 223–224
  viscosity, 179, 218–219, 222
Detection (see Measurement)

Dextran
  absorption, 9, 12, 80–83, 91, 96, 100
  intrinsic viscosity, 82
Diagnostic equipment (see also Use)
  desirable frequency range, 24
  intensity, 374, 381
Dielectric dispersion, 43, 48
Diffraction, 81, 93, 314–316 (see Measurement)
Diffusion, 275–281
Dilational viscosity (see Bulk viscosity)
Direct effect
  bioeffect, 120, 122, 225–228, 229–233, 239
  interaction mechanism, 239
Dispersion (see Frequency dependence; Velocity)
Displacement amplitude, bubble, 203
  interaction mechanism, 167–174
  vibrating wire, 199–200, 203
  viscosity dependence, 203
Disruption (see also Bioeffect)
  cavitation dependence, 236–240
  intensity dependence, 235–236
  temporal dependence, 235–236
Doppler (see Use)
Dura mater, 342, 352
Duty cycle, 308
Duty-cycle dependence, 309, 320, 373, 376–377, 378–381

Eddying motion, bubble, 176, 184
  hydrodynamic stress, 197–198
  interaction mechanism, 121, 175–180
  intercellular, 121, 137, 142–145, 146–149, 167, 169–171, 173
  shear, 176
  vibrating wire, 165, 200
Effective volume, 238
Elastic hysteresis, absorption mechanism, 250
Electrical discharge, 175
Electrophoretic mobility, 374
Elodea leaf cells, 137–139, 140–145, 146–149
Embleton's theory, 121, 157–158, 169
Emulsification, 252
Endothelial cell, 357, 364–367, 377
Enthalpy dependence, 108, 112
Enzyme
  absorption mechanism, 217
  active site, 210
  cavitation, 120, 122, 209–217
  conformation, 211
  denaturation, 122, 209–217
    concentration dependence, 213, 215
    frequency dependence, 213, 216–217
    intensity dependence, 209, 213, 216–217

pH dependence, 209, 210, 213, 215
  temperature dependence, 209, 210, 213, 215–216
Epidemiology, 383
Epstein's scattering theory, absorption mechanism, 38, 41–42
Equilibrium kinetics, 80
Erythrocyte
  aggregation, 357, 359–363, 369–371
  bioeffect, 176–177, 189–190
  bubble, 197–198
  cavitation, 125, 189–190
  shear stress, 237
  streaming, 176–177
  vibrating wire, 165, 199–201
Escherichia coli, 176–178
Evoked cortical potential, 247, 324
Extracellular motion, 165–166
Eyring's theory, 322

Fat
  absorption, 16, 32–33
  velocity, 31
Field distribution (see Measurement)
Finite wave effect, 64
Fluid motion (see also Measurement)
  laminar, 357
Focus, beam pattern, 314–316, 348
Free-draining coil, 82–83
Free radical
  cavitation, 117, 175, 178–179, 202, 218, 225, 229–230, 232
  interaction mechanism, 183
Frequency dependence
  absorption
    albumin, 80, 91–98, 102
      biopolymer, 57–58
      blood, 34–37, 38–42
      bone, 8, 28–30, 66
      brain, 316
      deoxyribose nucleic acid, 102
      dextran, 80–83
      gelatin, 100
      hemoglobin, 43–49, 50–53, 80, 99–104, 105–116
      liver, 54–58
      lung, 66, 69
      ovalbumin, 102
      plasma, 35, 38
      polyethylene glycol, 82–83
      poly-*l*-glutamic acid, 80
      poly-*dl*-lysine, 84–90
      poly-*l*-lysine, 84–90
      poly-*l*-ornithine, 84–90
      spinal cord, 65–69
      tissue, 9–11, 14–18, 31–33, 62, 66, 250

activation energy, 96, 101–102
bioeffect, 247–248
  hemorrhage, 373, 375, 378–381
  paralysis, 376–377
  rotifers, 345–346
boundary layer, 171
cavitation, 124–125, 134, 183, 317
damping constant, 129–130, 134, 136
degradation, 213, 216–218
lens gain, 316
lesion threshold, 247, 352, 356
relaxation process, 57–59
velocity
  albumin, 95–97
  blood, 34–37
  hemoglobin, 43–49, 105
Frog, 173, 260–264, 274, 308, 319, 373
Functional change, 216, 246, 308–309, 310–313, 326, 352

Gas dependence, cavitation, 124–125, 183, 252–253
Gelatin, 54, 57–59, 61, 100
Glutamic acid, 103–104
Gluteal musculature, 14–18
Glycine, 103
Gray matter, 246, 283–284
Guanidine hydrochloride, hemoglobin denaturation, 12

Half-value layer, 16, 19–21, 32–33
Hartmannella castellanii (see Amoeba)
Hazard, 355–356
Heart, 19–22, 32–33, 250
Heart-rate dependence, stasis, 357, 360, 362–363, 371
Heat
  interface, 274, 279–281
  lesion, 247, 347–351
  synergism, 122
Heat capacity, 63
Heat-conductivity coefficient, 262–263
Heat denaturation, absorption, 57
Heat generation, 326–328, 332
Heat flow, general equation, 299
Heat transfer, 240, 319, 326
Heating, 234
  interaction mechanism, 121, 159, 183, 192, 234, 240
  interfacial, 259, 262–263, 274, 279–281
Helix-coil reaction
  absorption mechanism, 9, 84–90, 92, 100
  nucleation parameters, 88–89
  rate constant, 89
  relaxation time, 89, 92
Hemocyanin, 252

Hemoglobin
  absorption, 8–9, 11–12, 34–37, 43–49, 50–53, 59, 80, 91, 96, 99–104, 105–116
    concentration dependence, 40, 43–44, 101–104
    frequency dependence, 43–49, 50–53, 80, 99–104, 105–116
    pH dependence, 43, 45, 99–104, 105–116
    temperature dependence, 43–44, 101–102
  activation energy, 101–102
  conformation, 92, 102–103
  dielectric measurement, 43
  intrinsic viscosity, 102–103
  velocity, 34–37, 43–49, 99, 105
    frequency dependence, 45–46
    temperature dependence, 36–37, 43–45, 101–102
Hemoglobin dependence, absorption, 43–44, 99–104
Hemoglobin solubility, absorption, 45–46
Hemolysis
  boundary layer, 171
  bubble, 119, 122, 197–198, 199, 202
  shear, 121–122, 197–198, 199–201
  shear stress, 202
  temperature dependence, 206–207
  vibrating needle, 202
  vibrating wire, 119, 198, 199–201
Hemorrhage, 245, 373–381
  duty-cycle dependence, 373, 378–381
  frequency dependence, 373, 375, 378–381
  intensity dependence, 373, 378–381
  oxygen dependence, 379–381
  temperature dependence, 378–380
  temporal dependence, 373, 375, 377–381
High-intensity ultrasound, 62
Histidine, 103–104
History, 1–6
Hot spots, 202, 232, 246, 259, 263
Human tissue, 8, 14–18, 27–30
Hydration, 57, 98, 102
Hydration-layer perturbation, absorption mechanism, 102, 104
Hydrodynamic flow, interaction mechanism, 302
Hydrodynamic force, 183
Hydrodynamic shear, interaction mechanism, 172, 182, 183, 191–195
Hydrodynamic stress, 197–198
Hyperchromicity, 172, 192
Hypoxia, 248, 373–375, 379–381
Hysteresis, 370–371
  absorption mechanism, 46, 250

Impedance
  beef, 27
  brain, 27, 336
  kidney, 27
  liver, 27
  lung, 72–73
  methacrylate, 332, 336
  muscle, 23–27
  spleen, 27
  tissue, 8, 10, 23–27
  water, 27
Indirect effect, 120, 122, 225–228, 230–233
Intensity
  diagnostic device, 374, 381
  distribution, 326–327
  measurement, 67, 226, 230, 236, 253–254, 259, 270, 275, 286–287, 309, 316–318, 358–359, 363, 375–376
  therapeutic device, 381
  threshold, 230–233, 374, 377
Intensity dependence
  Bernoulli attraction, 371
  cavitation, 252–253
  cell survival, 230–233
  degradation, 218–224
  denaturation, 209, 213, 216–217
  destruction, 235
  hemorrhage, 373, 378–381
  lesion, 352–356
  paralysis, 308–309, 319, 375–377
  stasis, 357, 360–363, 369–370
  streaming, 238
  temperature, 347–351
Interaction mechanism
  ac stress, 122, 202–208
  acceleration force, 382
  Bernoulli attraction, 370–371
  biophysical, 117
  bubble, 175, 232, 238
  cavitation, 117, 121, 122, 159, 165–166, 167–174, 183, 197–198, 199–201, 202–208, 209–217, 218–224, 229–233, 234–240, 246, 247, 249, 251–253, 258, 259–263, 268–269, 285–286, 298, 303–304, 308, 310–311, 317, 320–321, 345–346,, 352–355, 373–374, 382–383
  chemical, 120, 225, 228, 229–233
  circulatory movement, 167–174
  dc stress, 120, 202–208
  degradation, 183
  direct, 239
  direct vs. indirect, 122, 225–228, 229–233
  displacement, 167–174
  eddying motion, 121, 175–180
  free radicals, 183
  gaseous cavitation, 175

Subject Index

heating, 183, 192, 234, 240
hydrodynamic flow, 302
hydrodynamic shear, 172, 182, 183, 191–195
interface heating, 121, 159
mechanical, 1–2, 124, 218, 229, 246, 247, 274, 302, 303–305, 314–323, 352–356, 374, 382
membrane, 244, 246, 274–281
microcavitation, 369
molecular, 352, 371
nonthermal, 119, 167–174
oscillatory force, 303
oscillatory stress, 122, 202–208
Oseen force, 303, 382
particle acceleration, 268
physical, 120, 229
pressure, 249, 250–251, 264, 268
radiation force, 165, 169
radiation pressure, 167, 168–169, 171–174, 249. 251. 264. 268. 302, 370
rate process, 259, 263
relative motion, 122, 303
shear, 120, 121, 175–180, 183, 202–208, 368
solute–solvent, 195
sonochemical, 229
stable cavitation, 122, 229, 231
steady stress, 167–174
stirring, 246, 274–281
Stokes' drag, 369, 370–371
streaming, 121, 122, 137–139, 140–145, 146–149, 151–158, 167, 170–174, 184, 191–195, 197–198, 199–201, 238, 264, 381–382
structural viscosity, 250–251
summation, 257, 261, 263, 268
synergism, 121, 202, 207, 373, 378–381
temperature dependence, mechanical, 246–247, 314–323
thermal, 1, 7, 121, 124, 159, 165–166, 167, 209, 231, 243, 244, 245–246, 247, 249–258, 259, 262–263, 264–269, 272, 274–281, 285, 303–304, 308–309, 312, 314–323, 325–343, 346, 347–351, 352–353, 373–374, 382
time-averaged force, 223–224
transient cavitation, 229, 231
unidirectional forces, 209, 251, 303, 313
velocity gradient, 191–195, 198, 199, 202–207, 232
viscosity, 249
viscosity stress, 198, 202–207
volume heating, 121, 159
Interfacial heating, 259, 262–263, 274, 279–281
interaction mechanism, 121, 159
Interfacial tension, 172
Intermolecular interaction, absorption mechanism, 94, 101, 111–115
Intracellular motion
 ac boundary layer, 148
 aggregration, 146
 Allium cepa cell, 146–149
 Arbacia punctulata egg, 151–158
 Asterias egg, 151–158
 Bernoulli's principle, 147
 bioeffect, 146, 149, 167
 boundary flow, 144–145
 cell model, 119, 137–139, 144–145, 148–149
 clam egg, 151–158
 density dependence, 157–158
 eddying, 121, 137, 142–145, 146–149, 167, 169–171, 173
 elodea leaf cells, 137–139, 140–145, 146–149
 Embleton's theory, 121, 157–158
 general theory, 146–147
 membrane-tension dependence, 148
 muscle, 165–166
 radiation pressure, 121, 149, 165
 sea urchin egg, 151–158
 Spirogyra cells, 146–149
 Spisula soidissima egg, 151–158
 starfish egg, 151–158
 steady stress, 147
 streaming, 120–121, 137–139, 140–145, 146–149, 165–166
 tissue, 149, 173
 viscoelastic dependence, 146–147
 viscosity dependence, 138–139, 153–156
 whirling, 142
Intracellular stress, 173
Intrinsic viscosity
 albumin, 96–98, 104
 deoxyribose nucleic acid, 182, 189, 191, 195, 219
 dextran, 82
 hemoglobin, 102–103
Ionization, absorption mechanism, 105
Ionizing radiation, 318, 373

Jet layer, 134, 135

Kidney, 19–22, 26–27, 31–33
Kinematic viscosity (see Viscosity)
King's theory, 251
Kirkwood–Tranford theory, 106

Lactate, 209–217
Laminar flow, 357

402

Lens, 315
Lens gain, 316, 328
Lesion
  brain, 180, 245, 247, 283–284, 310, 314–323, 325–343, 347–351, 352–356, 373–374
  cavitation, 352–356
  current, 247, 349–351
  heat, 247, 347–351
  histology, 311
  mechanical, 352–356
  neurosurgery, 314
  plastic, 245, 325–326
  thermal, 325–343, 347–351, 352–356
  threshold
    frequency dependence, 247, 352–356
    intensity dependence, 352–356
    temporal dependence, 352–356
Lesion boundary temperature, 326, 331, 334, 336, 338–339 (see also Measurement)
Lesion size, 325–343, 347–351
Limiting velocity, 200
Liquefaction, 252
Liquids, 119, 124–125
Liver
  absorption, 8, 19–22, 32, 33 54–61
    concentration dependence, 56
    frequency dependence, 54–58
    pH dependence, 54, 57–59
  density, 27
  impedance, 27
  velocity, 26–27, 31
Luminescence, 229
Lung
  absorption, 9, 66, 69–79
  density, 72–73
  impedance, 72–73
  measurement, 11, 70–73
  reflection, 9, 70–79
Lymphoma cell, 230
Lysine, 103–104
Lysis, 225

Marrow, 120, 225–228
Mason horn (see Vibrating needle)
Measurement
  absorption, 14–16, 19–21, 34–35
    albumin, 92–93
    bone, 28
    brain, 237
    dextran, 80–81
    hemoglobin, 100–101
    interferometer, 85
    lung, 11, 70–73
    optical method, 25

    pulse technique, 85, 100–101
    poly-*dl*-lysine, 85
    poly-*l*-lysine, 85
    poly-*l*-ornithine, 85
    thermocouple probe, 62–63, 66–69, 71–72
    tissue, 250, 258
  acoustic pressure, 286–287
  action potential, 287–289
  beam pattern, 253–254, 286–287, 298, 309
  calorimetry, 270, 275
  cavitation, 219–224, 229–230, 231, 235, 236–238, 252–253, 304
  cell survival, 226
  density, 72
  diffraction, 81, 93, 314–316
  dilatometric, 89
  field distribution, 363, 375
  fluid motion, 127, 138
  intensity, 67, 226, 230, 236, 253–254, 259, 270, 275, 286–287, 309, 316–318, 358–359, 363, 375–376
  intensity distribution, 326–327
  lesion boundary temperature, 338–339
  particle displacement, 127–128
  particle velocity, 128
  potential, 275
  power, 275, 328
  pressure, 127–128
  pressure balance, 275
  pulse technique, 23
  radiation pressure, 67, 309, 328
  reflection, 70–73
  sonoluminescence, 231
  standing wave, 72
  streaming velocity, 131
  temperature, 250, 254, 262, 265, 276, 296–298, 309, 311, 350–351, 376
  thermocouple probe, 71–72
  velocity, 34–35
    albumin, 92–93
    dextran, 81
    hemoglobin, 100–101
    lung, 72–73
    tissue, 24–25
Mechanical
  bioeffect, 1–2, 229
  interaction mechanism, 124, 218, 229, 246, 247, 274, 302, 303–305, 314–323, 352–356, 374, 382
    temperature dependence, 246–247, 314–323
  lesion, 352–356
  stirring, 246, 274
Medical use (see Use)

Subject Index

Medium dependence, 148
Medulla oblongata, 32–33 (see also Brain)
Membrane, 167–170, 173, 219, 352
  bioeffect, 246, 274–281
  charge, 246, 278–281
  connective tissue, 225
  interaction mechanism, 244, 246, 274–281
  potential, 244, 246, 274–281
    temperature dependence, 276–277, 279
  strength, 197
  temperature, 244, 274
  temperature gradient, 244, 274–281
  vibrating, 169, 173
Membrane-tension dependence, 148
Membrane thickness, 138, 145
Memory, 321
Meningioma, 31
Metachronosis, 270–271
Methacrylate, 326, 328–336
Methanolic solution, 111
Microbubbles, 173–174, 197 (see also Cavitation; Streaming)
Microcavitation (see also Cavitation)
  interaction mechanism, 369
Microorganism, 175–180, 209, 234
Microstreaming, 119, 126–136 (see also Streaming)
Migration, 251
Mode conversion, absorption mechanism, 96
Molecular, interaction mechanism, 352, 371
Molecular structure, absorption, 57
Molecular-weight dependence
  absorption, 80–83
  degradation, 219–224
Monkey, 283, 308
Morphology, arbacia egg, 173
  muscle, 119, 159–166
  nerve, 256
Mouse, 308–309, 310–313, 318–320 (see also Paralysis)
Multiple relaxation process, 46–48
Muscle
  absorption, 14–18, 25, 32–33, 125, 250
  action potential, 246, 285–306
  bioeffect, 159–166, 244, 246, 285–306
  cavitation, 285–286
  cavitation threshold, 125
  extracellular motion, 165–166
  intracellular motion, 165–166
  morphology, 119, 159–166
  permeability, 173
  relaxation frequency, 64
  temperature dependence, 250, 257
  thermal-conductivity coefficient, 299
  velocity, 23–27, 31
Mutation, 381

N–F′ transition, 91, 97, 104
Nerve
  absorption, 32–33
  bioeffect, 249–258, 259–263, 264–269, 270–273
  morphology, 256
  temperature dependence, 250, 255–257
Neuroanatomy (see Use)
Neurology (see Use)
Neurosurgery (see Use)
Newtonian fluid, 137, 138, 171, 173
Nodes, 131
Nondestructive testing (see Use)
Nonthermal, interaction mechanism, 119, 167–174
Normal mode, 82–83
Nucleation parameter, 88–89
Nuclei
  blood, 124
  bubble, 353
  cavitation, 174, 229, 238, 251
Nuclei dependence, 125, 183
Nucleic acid, 13 (see also Deoxyribose nucleic acid)

Optical rotation, 85
Oscillatory flow, 147–149
Oscillatory force, interaction mechanism, 303
Oscillatory stress, interaction mechanism, 122, 202–208
Oseen force, interaction mechanism, 303, 382
Ossification, 308
Ovalbumin, 102
Oxygen dependence, 379–381

Paralysis, 173, 209, 240, 245, 256–258, 259–263, 266–269, 373, 376–378
  ambient-pressure dependence, 260–264
  duty-cycle dependence, 309, 320, 376–377
  frequency dependence, 376–377
  intensity dependence, 308–309, 319, 375–377
  particle-acceleration dependence, 312
  particle-displacement dependence, 319
  particle-velocity dependence, 308, 312
  pressure dependence, 260–263, 308, 312
  pulse-width dependence, 308–309, 320
  temperature dependence, 257, 267–269, 308–309, 310–313, 320
  temporal dependence, 256, 308–309, 318–319, 320, 376–378
  threshold, 261–263, 305, 309, 310–313
Paramecium, 2, 126
Parathrombosis, 357

Subject Index

Particle acceleration, interaction mechanism, 268
Particle-acceleration dependence, 312
Particle-displacement dependence, 319
Particle-velocity dependence, 130–132, 308, 312
Permeability, 173, 225, 274, 279–280
pH dependence
  absorption
    albumin, 91–98
    amino acid, 103–104
    biopolymer, 57
    gelatin, 57–59, 61, 100
    hemoglobin, 43, 45, 99–104, 105–116
    liver, 54, 57–59
    poly-*dl*-lysine, 84–90
    poly-*l*-lysine, 84–90
    poly-*l*-ornithine, 84–90
    protein, 57
  activation energy, 96
  cell survival, 227
  conformation, 91–98, 102–104
  denaturation, 209, 210, 213, 215
  intrinsic viscosity, 96–98, 102–104
  optical rotation, 85
  potential, 274, 278
  velocity
    albumin, 91–98
    hemoglobin, 43
Piezobase, 271
Piezoelectric probe, 317
Plasma
  absorption, 32–33, 34–37, 38, 59, 99
  velocity, 31, 34–37, 99
Plastic, lesion, 245, 325, 326
Plastic slip, absorption mechanism, 250
Poikilothermic, 62, 67, 308, 310
Poiseuille flow, 361
Polycrystalline metal, 29
Polyelectrolyte, viscoelastic interaction, 80
Polyethylene, 326–327
Polyethylene glycol, 12, 82–83
Polyglutamic acid, 92
Poly-*d*-glutamic acid, 89
Poly-*l*-glutamic acid, 12, 80, 88–89, 100
Polylysine, 9
Poly-*dl*-lysine, 84–90
Poly-*l*-lysine, 12, 84–90, 100
Polymer, 218
Polyornithine, 9
Poly-*l*-ornithine, 84–90
Polypeptides, 9
Polysaccharide, 12, 80–83, 91, 96, 100
Potential
  diffusion, 275–281
  measurement, 275
  membrane, 244, 246, 274–281

permeability, 280
pH dependence, 274, 278
stirring, 277–281
Poulsen arc oscillator, 1–3
Power
  heat generation, 328
  measurement, 275, 328
  radiation pressure, 238
Pressure
  bubble, 189
  cavitation, 126, 175
  interaction mechanism, 249, 250–251, 264, 268
  measurement, 286–287
  streaming, 120
Pressure-amplification factor, 186
Pressure balance, 275
Pressure chamber, 259–260
Pressure dependence
  action potential, 292–296
  bioeffect, 250–251, 260–263, 308, 312
  bubble, 251
  paralysis, 260–264, 308, 312
Protein (*see* Absorption)
Protein dependence, 34–37, 99–104
  absorption mechanism, 100
Proton-transfer reaction
  absorption mechanism, 57–59, 84–90, 99–100, 102–104
  vs. helix-coil, 9, 105–116
  chemical relaxation mechanism, 103
  rate constant, 84, 87–88
  relaxation time, 88
  volume change, 84, 87–88, 106
Pulse-echo technique, 367, 373, 382 (*see* Use)
Pulse-width dependence, 230–233, 308–309, 320

Quartz wind, 202

Radiation force (*see also* Radiation pressure)
  absorption, 169
  density dependence, 169
  interaction mechanism, 165, 169
Radiation pressure, 15, 125, 127, 149 (*see also* Radiation force)
  extracellular motion, 165
  interaction mechanism, 167, 168–169, 171–174, 249, 251, 264, 268, 302, 370
  intracellular motion, 121, 149, 165
  measurement, 309, 328
  power, 328
    unidirectional force, 251, 303
Radicals, 229
Rana esculenta, 275
Rana pipiens, 159–160, 254, 264, 287
Rat, 245, 270, 318, 347, 352, 373–374

Rate constant, 91
  helix-coil, 89
  proton-transfer, 84, 87–88
Rate process
  cavitation effect, 126
  interaction mechanism, 259, 263
Recovery, 319
Rectified diffusion, 131, 176, 229, 232
Red cell (see Blood; Erythrocyte)
Reflection, 9, 70–79
Regime I (Elder), 132–133
Regime I (PHP), 184
Regime II (Elder), 131–135
Regime III (Elder), 132, 135
Regime IV (Elder), 132, 135
Regime threshold, 132
Relative motion
  absorption mechanism, 41–42, 96
  interaction mechanism, 122, 303
  oscillatory force, 303
  rotational, 48
  translational, 48
Relaxation distribution, 46–47
Relaxation frequency, 44, 46–48, 58–59, 83 (see also Relaxation time)
  O-amino benzoic acid, 110–112
  helix-coil, 89
  methanolic solutions, 111
  muscle, 64
  spinal cord, 69
Relaxation process, 44, 46–48, 80, 83, 91
  absorption mechanism, 41–42, 51–53, 54–61, 102
  activation energy, 54
  frequency dependence, 57–59
Relaxation spectra (see Frequency dependence)
Relaxation theory, 11, 46–48
Relaxation time, 47, 80, 91, 95, 96, 100, 105 (see also Relaxation frequency)
  deoxyribose nucleic acid, 223–224
  helix-coil, 92
  hemoglobin, 53
  N–F' transition, 97
  proton-transfer, 88
  rotational, 98
Resolution, 24
Resonance, 127, 130, 133, 134, 145, 174, 176, 178, 187, 232, 353
Reversible effect, 245, 247, 324
Reynold's number, 135
Ribonuclease, 209–217
Rotifers, 345–346

Salt dependence, 43, 45–46
Scattering, 96, 125, 185–187

Sea urchin egg, 123, 151–158, 173
Serine, 103
Shear
  bioeffect, 120, 176, 202–208
  eddying motion, 176
  hemolysis, 121–122, 197–198, 199–201
  interaction mechanism, 120, 121, 175–180, 183, 202–208, 368
Shear dependence, 187
Shear force, 182
Shear gradient, 199–201
Shear stress, 202, 237
Shear viscosity, 48
  absorption mechanism, 64, 95, 96, 250
Sheep, 336–337
Single relaxation, 46, 83, 95
Soft-tissue visualization (see Use)
Solvation, absorption mechanism, 92, 105
Solvent–solute equilibrium
  absorption mechanism, 91, 99, 100, 102
  interaction mechanism, 195
Sonochemical, interaction mechanism, 229
Sonoluminescence, 175, 177–178, 231
Source
  bubble, 137, 167–174, 178, 197, 202–203
  cavitation, 184–186
  pulsating, 167–168
  translating, 167–168
  ultrasound, 211, 219, 226, 230, 235, 253–254, 270, 275, 286–287, 309, 345, 358–359, 375–376
  vibrating membrane, 168
  vibrating needle, 120–121, 137, 140–142, 146, 151–153, 159–160, 176–178
  vibrating wire, 200, 202–203
Specific absorption (see Absorption)
Specific gravity, 55 (see also Density)
Specific heat, 336–337
Spinal cord
  absorption, 32–33, 62–64, 65–69
  bioeffect, 245, 373–383
  relaxation frequency, 69
Spirogyra, 2, 146–149
Spisula solidissima egg, 151–158
Spleen, 26–27, 31, 336–337
Spontaneous activity, 254–258, 259, 265–266
Stable cavitation (see Cavitation)
Standing wave ratio, 72, 251
Starfish, 151–158
Stasis
  Bernoulli attraction, 371
  blood, 247, 357–372
  definition, 358–359, 360–361
  significance, 367–369
  threshold

heart-rate dependence, 357, 360, 362–363, 371
intensity dependence, 357, 360–363, 369–370, 371
temperature dependence, 371
temporal dependence, 360, 369
vessel-orientation dependence, 357, 360
vessel-size dependence, 357, 362
vessel-type dependence, 357, 360–362, 371
Steady stress
  interaction mechanism, 167–174
  intracellular motion, 147
  streaming, 229
Stirring, 246, 274–281
Stokes' drag, interaction mechanism, 264, 369, 370–371, 381–382
Streaming
  absorption dependence, 15, 238
  ac flow, 147–149
  amoeba, 122
  amplitude dependence, 144–145, 153–156
  beam-radius dependence, 238
  boundary flow, 144–145, 148–149
  boundary layer, 171
  bubble, 172, 176, 197, 199
  cavitation, 229, 232
  dc flow, 147–149
  deoxyribose nucleic acid, 223–224
  erythrocytes, 176–177
  Escherichia coli, 176–177
  heat transfer, 240
  intensity dependence, 238
  interaction mechanism, 121–122, 137–139, 140–145, 146–149, 151–158, 167, 170–174, 184, 191–195, 197–198, 199–201, 238
  intracellular motion, 120–121, 137–139, 140–145, 146–149, 165–166
  medium dependence, 148
  oscillatory flow, 147–149
  particle-velocity dependence, 130–132
  pressure dependence, 120
  quartz wind, 202
  regime I (Elder), 132
  regime II (Elder), 131, 132
  regime III (Elder), 132
  regime IV (Elder), 132
  shear stress, 229
  surface dependence, 148
  Tetrahymena pyriformis, 176–177
  velocity, 174
  velocity gradient, 182, 202
  viscosity dependence, 120, 130–132, 134
  vortex ring, 131–132
Streaming field, 130–135

Streaming mechanism, 133–135
Streaming velocity, 131, 134, 135, 223
Stress
  intracellular, 173
  synergism, 122
Stromata, 43
Structural change, 216, 245, 326, 352–356
Structural equilibrium, 48
  absorption mechanism, 105
Structural relaxation, absorption mechanism, 96
Structural viscosity, 252
  interaction mechanism, 250–251
Subharmonic signal, cavitation, 222–223, 229, 237
Summation
  bioeffect, 257, 261, 263
  interaction mechanism, 257, 261, 263, 268
Supramacromolecular disorganization, 182
Surface dependence, 148
Surface energy, 126
Surface tension, 172
Surface velocity, 128–132
Surface wave, 197, 203
Surgery (*see* Use)
Survival, 122, 227, 230–233
Synergism, 248, 373
  interaction mechanism, 122, 202, 207, 373, 378–381
  stress vs. heat, 122

Temperature
  absorption, 216, 249, 257, 258, 262–263, 264
  cavitation, 175
  current dependence, 349–351
  intensity dependence, 347–351
  lesion boundary, 326, 331–332, 333–334
  measurement, 250, 254, 262, 265, 276, 296–298, 309, 311, 350–351, 376
  membrane, 244, 274
  threshold, 277, 280
Temperature dependence
  absorption, 34–37, 95
    albumin, 95–96
    brain, 66, 338
    hemoglobin, 43–44, 101–102
    poly-*l*-lysine, 85
    spinal cord, 62–64, 65–69
    tissue, 8, 11, 249
    water, 249
  action potential, 257–258, 285, 296–302
  bioeffect
    hemorrhage, 378–380
    tissue, 249–258, 262–263, 267–269
  bone, 250, 257
  bubble, 203, 207–208, 251–252

407

Subject Index

cavitation, 126
critical stress, 206–207
denaturation, 209, 210, 213–216
duty cycle, 308
hemolysis, 206–207
interaction mechanism, mechanical, 246–247, 314–323
lesion size, 325–343, 347–351
muscle, 250, 257
nerve, 250, 255–257
paralysis, 257, 267–269, 308–309, 310–313, 320
permeability, 274, 279
potential, 276–277, 279
stasis, 371
threshold, 310, 371
velocity
  albumin, 36–37
  blood, 36–37
  brain, 338, 340
  hemoglobin, 36–37, 43–45, 101–102
  plasma, 36–37
Temperature distribution, 326, 328–331, 332
Temperature gradient, 244, 274–281
Temporal dependence
  bioeffect
    muscle, 159–166
    nerve, 270–273
  critical stress, 204–205
  destruction, 235–236
  hemorrhage, 373, 375, 377–381
  lesion, 352–356
  paralysis, 256, 308–309, 318–319, 320, 376–378
  stasis, 360, 369
  threshold, 352–356, 360, 369
Tensile force, 182, 195
Tension, 172, 285, 312
Tetrahymena pyriformis, 176–178
Therapeutic device (*see also* Use)
  intensity, 381
Therapeutic ultrasound, bioeffect, 124
Therapy (*see* Use)
Thermal
  interaction mechanism, 1, 121, 124, 159, 165–166, 167, 209, 231, 243, 244, 245–246, 247, 249–258, 259, 262–263, 264–269, 272, 274–281, 285, 303–304, 308–309, 312, 314–323, 325–343, 346, 347–351, 352–355, 373–374, 382
  lesion, 325–343, 347–351, 352–356
Thermal capacity, 68, 347–348
Thermal conductivity, 299, 300, 332–333, 336–337, 342, 347–348
Thermal diffusivity, 327, 332–333, 335, 336–337, 347–349

Thermistor, 358, 363
Thermocouple, 250, 254, 265, 286, 296–298, 309, 311, 317, 322, 333–334, 338–339, 347, 350, 351, 376
Theronine, 103
Threshold
  bioeffect, 246, 260, 261–263, 270–271
  cavitation, 176, 183–184, 212, 236, 346
    blood, 120, 125
    frequency dependence, 134, 317
    muscle, 125
    tissue, 120, 125
    water, 120, 124–125
  defined, 309, 311
  frequency dependence, 134, 247, 317, 352–356
  heart-rate dependence, 357, 360, 362–363, 371
  intensity, 374, 377
  intensity dependence, 352–356, 357, 360–363, 369, 370
  lesion, 247, 352–356
  paralysis, 261–263, 305, 309, 310–313
  stasis, 357, 360–363, 369–371
  surface velocity, 131
  temperature, 277, 280
  temperature dependence, 310, 371
  temporal dependence, 352–356, 360, 369
  vessel-orientation dependence, 357, 360
  vessel-size dependence, 357, 362
  vessel-type dependence, 357, 360–362, 371
Thromus, 368
Time-averaged force, interaction mechanism, 223–224
Tissue (*see also specific tissue*)
  absorption, 8–11, 14–18, 19–21, 54–61, 240, 263, 308, 320
    frequency dependence, 31–33, 62, 66, 250
    measurement, 250, 258
    tabular summary, 31–33
    temperature dependence, 249
  bioeffect
    cavitation, 24, 249
    temperature dependence, 249–258, 262–263, 267–269
  cavitation threshold, 120, 125
  heat conductivity coefficient, 262–263
  intracellular motion, 149
  specific gravity, 25
  velocity, 8, 10, 23–27, 31–33
Tissue gain, 316
Tissue regeneration, 381
Tongue, 19–22, 32–33, 250
Torque, 169
Transient cavitation (*see* Cavitation)

Trypsin, 209–217
Tyrosine, 103–104

Ultrasonic absorption (see Absorption)
Ultrasonic spectroscopy, 5
Ultrasound (see Source)
Unidirectional force
  interaction mechanism, 209, 251, 303, 313
  Oseen force, 303, 382
  radiation pressure, 251, 303
Use, 70, 243, 285
  biorheology, 371–372
  diagnostic, 240
  Doppler, 367, 373
  neuroanatomy, 209
  neurology, 283, 308, 310, 314
  neurosurgery, 314
  nondestructive testing, 10
  pulse-echo technique, 23, 367, 373, 382
  surgery, 180, 308, 310, 314
  therapy, 274, 314, 357, 367, 373, 381
  visualization, 70

Velocity
  albumin, 34–37, 91–99
  blood, 8, 31, 34–37, 99
  bone, 31
  brain, 26–27, 31, 336, 338, 340
  carcinome, 31
  concentration dependence, 43, 44
  fat, 31
  frequency dependence, 45–46, 95–97, 105
  hemoglobin, 34–37, 43–49, 99, 101–102, 105
  kidney, 26–27, 31
  light, 216
  liver, 26–27, 31
  lung, 72–73
  measurement, 24–25, 34–35, 72–73, 81, 92–93, 100–101
  meningioma, 31
  methacrylate, 332–333, 336
  migration, 251
  muscle, 23–27, 31
  pH dependence
    albumin, 91–98
    hemoglobin, 43
  plasma, 31, 34–37, 99
  salt dependence, 43
  spleen, 26–27, 31
  streaming, 174
  temperature dependence
    ablumin, 36–37
    blood, 36–37
    brain, 338, 340
    hemoglobin, 36–37, 43–45, 101–102
    plasma, 36–37
  tissue, 8, 10, 23–27, 31–33
  tissue anisotropy, 23–27
  tissue dependence, 31–33
  water, 25, 37
Velocity amplitude, 185–187
Velocity gradient
  boundary layer, 171–174
  bubble, 121, 184, 187, 189, 191–195, 198, 199
  interaction mechanism, 191–195, 198, 199, 202–207, 232
  streaming, 182, 202
  vibrating wire, 199–201
Vessel-orientation dependence, 357, 360
Vessel-size dependence, 357, 362
Vessel-type dependence, 357, 360–362, 371
Vibrating needle, 120, 137, 140–142, 146, 149, 151–153
  hemolysis, 202
Vibrating wire
  boundary layer, 201, 204
  displacement amplitude, 199–200, 203
  eddying motion, 200
  hemolysis, 119, 198, 199–201
  shear gradient, 199–201
  source, 200, 202–203
  velocity gradient, 199–201
Viscoelastic behavior, 80
Viscoelastic dependence, 146–147
Viscosity
  absorption mechanism, 82–83
  cavitation, 252
  deoxyribose nucleic acid, 179
  interaction mechanism, 249
  shear dependence, 187
Viscosity dependence
  acoustic boundary layer, 129
  boundary layer 171
  cavitation, 125, 212
  cell model, 138
  critical stress, 204
  damping ratio, 129
  degradation, 218–219, 222
  displacement amplitude, 203
  intracellular motion, 138–139, 153–156
  Newtonian fluid, 138
  resonant bubble, 127
  streaming, 120, 130–132, 134
  surface velocity, 130–132
Viscous damping, absorption mechanism, 250
Viscous interaction, absorption mechanism, 38–42
Viscous process, absorption mechanism, 48
Viscous slip, absorption mechanism, 250
Viscous stress, interaction mechanism, 198, 202–297

Subject Index

Volume change, 84, 87–88, 106, 108, 112
Volume heating, interaction mechanism, 121, 159
Vortex motion, 197
Vortex rings, 130–132

Water
　absorption, 83, 93, 249
　activation energy, 96, 101–102
　cavitation, 120, 124–127, 252–253
　density, 27, 337
　impedance, 27
　specific heat, 337
　thermal properties, 263, 300, 337, 347–348
　velocity, 25, 37
Weiss's law, 271
Whirling motion, 142
White matter, 246, 283–284
Willard phenomenon, 178

Young's modulus, 200

About the Editors

FLOYD DUNN was born in Kansas City, Missouri, on April 14, 1924. He took his B.S., M.S., and Ph.D. degrees in 1949, 1951, and 1956, respectively, all in electrical engineering from the University of Illinois, Urbana–Champaign, where he specialized in bioacoustics.

Dr. Dunn has been a member of the faculty of the University of Illinois since 1956 and became Professor in 1965. He holds joint appointments as Professor of Electrical Engineering in the Department of Electrical Engineering and as Professor of Bioengineering in the Faculty of Bioengineering, both in the College of Engineering, and as Professor of Biophysics in the Department of Physiology and Biophysics, College of Liberal Arts and Sciences.

Professor Dunn's main research interests deal with the interaction of ultrasound and biological media, about which he has published more than sixty journal articles and book chapters. He was a visiting professor in the Department of Microbiology at University College, Cardiff, during the academic year 1968–1969, after having been awarded an NIH Special Research Fellowship. During the academic year 1975–1976 he was a visiting senior scientist at the Institute of Cancer Research, Sutton, Surrey, England, after having been awarded an American Cancer Society–Eleanor Roosevelt–International Cancer Grant. Professor Dunn is an Associate Editor of the *Journal of the Acoustical Society of America*, having responsibility for the field of bioacoustics. He is also on the editorial boards of *Ultrasound in Medicine and Biology* and of *Radiation and Environmental Biophysics*.

WILLIAM D. O'BRIEN, JR. is Assistant Professor of Electrical Engineering and of Bioengineering at the University of Illinois, Urbana–Champaign. From 1971 to 1975 he was a research scientist with the Bureau of Radiological Health, Food and Drug Administration. Professor O'Brien received his B.S., M.S., and Ph.D. degrees in electrical engineering from the University of Illinois in 1966, 1968, and 1970, respectively. His research deals with examining the interaction processes between ultrasonic energy and biological material.

QP
82.2
U37
U47

JAN 6 1977